季生平

Ji Shengping

著

数据质量

汉英双语版
Chinese-English
bilingual edition

DATA
QUALITY

中国出版集团
中译出版社

图书在版编目（CIP）数据

数据质量：汉英对照 / 季生平著. —北京：中译
出版社，2025.2. —ISBN 978-7-5001-8176-7

Ⅰ. TP274

中国国家版本馆 CIP 数据核字第 2025YM8663 号

数据质量（汉英双语版）
SHUJU ZHILIANG（HAN-YING SHUANGYU BAN）

出版发行/中译出版社
地　　址/北京市西城区新街口外大街 28 号普天德胜主楼 4 层
电　　话/（010）68359827，68359303（发行部）；68359725（编辑部）
邮　　编/100088
传　　真/（010）68357870
电子邮箱/book@ ctph. com. cn
网　　址/http://www. ctph. com. cn

出　版　人/刘永淳
出版统筹/杨光捷
总　策　划/范　伟
策划编辑/刘瑞莲　钱屹芝
责任编辑/钱屹芝

排　　版/凯高教育
封面设计/郭盟盟
印　　刷/河南龙华印务有限公司
经　　销/新华书店
规　　格/880 毫米×1230 毫米　1/16
印　　张/33
字　　数/839 千字
版　　次/2025 年 2 月第 1 版
印　　次/2025 年 2 月第 1 次

ISBN 978-7-5001-8176-7　定价：198.00 元
中 译 出 版 社

前言

这个时代是数据的时代，把握好对于数据的采集、分析、运用等过程，可以更好地为各项管理工作服务。当数据铺天盖地到来时，如何识别和提高数据的质量，需要掌握一定的方法，这也是写作本书的初衷。

数据有很多特性，有些数据具有连续性、波动性，可以用直方图表示；有些数据具有趋势性，可以用趋势图表示；有些数据之间具有相关性、回归性，可以对其进行相关分析、回归分析；有的数据具有隐藏性，需要分层分类、不断挖掘，并通过假设检验来得到真相；等等。

数据质量体现在很多地方，它可以真实反映客观对象的特征，可以通过统计分析辅助正确推论、决策；利用数据可以进行建模，并提高模拟的质量。但数据也会出现很多异常，如由于观察、测量、试验不正确等原因导致数据失真；由于条件限制、错用等缘由而造成数据缺失；由于样本容量过小、过程能力不足等形成数据风险等。客观世界中的大数据、多元数据是所有与数据打交道的人必须面对的事实，它们对于数据质量提出了更高的要求。

本书共分 6 章。第 1 章数据的含义，介绍了数据的概念、数据的分类、数据的分布、数据的用途等内容。第 2 章数据的来源，包含了从观察法、调查法、测量法、抽样法、试验法、模拟法、计算法、统计法等过程中获得数据的途径。第 3 章数据的质量体现与保证，介绍了高质量的数据可以真实反映客观对象，可以由抽样反映全貌、可用于预测趋势；描述了如何保证数据的质量，包括统计及计算过程准确、推论过程正确、正确观察、正确的逻辑关系、测量系统保证、试验及试验系统保证、可以高度模拟、形成大数据等。第 4 章数据的异常概括了导致数据发生异常的情形，包括观察与记录异常、测量异常、试验异常、抽样异常、模拟异常、统计分析及计算过程异常、统计推断异常等。第 5 章介绍了数据缺失的情形，包括不能完整记录数据、数据采集的过程存在缺陷、删失数据、数据资料遗失、数据错用、缺失数据的弥补方法等。第 6 章数据的风险，阐述了产生数据风险的几种情景，包括抽样的风险、样本容量过小的风险、应用统计方法不适宜、过程能力不足、假设检验的风险、数据人员职业技能的影响、多元数据的存在等内容。

在写作本书过程中参阅的大量文献已列入参考文献中，在此向所有的中外作者、译者等表示衷心感谢。同时感谢家人自始至终的支持，是家人给了我持续写作的动力。由于本人能力和专业知识有限，书中一定存在很多不足，请读者指正。

本书适合与数据工作有关的人员阅读与使用，包括质量管理、统计管理、计划管理、生产运营管理、市场营销、材料管理、综合管理、运维管理等相关人员，其他专业的人员亦可以作为参考。

2024 年 5 月 2 日

This is the era of data. A good grasp of the process of data collection, analysis and application can serve for various managing work better. When the data is overwhelming, we need to master certain methods to distinguish and improve data quality. This is also the original intention of writing this book.

Data has many characteristics. Some data can be represented by histogram because of their continuity and volatility; some data have tendency and can be represented by run chart; some data are correlated and regressive, so correlation analysis and regression analysis can be conducted; some data are hidden and need to be analyzed by layers and classification continuously, and the truth can be obtained through hypothesis test, etc.

Data quality appears in many places. It can truly reflect the characteristics of the objects, and can help proper inference and decision-making through statistical analysis. We can use data to establish models and to improve the quality of simulation. But there are many abnormalities in data, such as data distortion due to incorrect observation, measurement, test and other reasons; data missing due to limited conditions, misuse and other reasons; data risk due to the too small sample size, insufficient process capability, etc. The big data and multivariate data in the objective world are facts that all people dealing with data must face. They have put forward higher requirements for data quality.

This book is divided into six chapters. Chapter 1 Meaning of Data introduces the contents of data concept, data classification, data distribution and use of data, etc. Chapter 2 Sources of Data includes ways to obtain data from observation, survey, measurement, sampling, test, simulation, calculation, statistics, etc. Chapter 3 Quality Representation and Assurance of Data introduces that high-quality data can truly reflect the objectives, the overall perspectives by sampling, and can be used to predict the trend. This chapter describes how to ensure the quality of data, including accurate process of statistics and calculation, correct inference process, correct observation, correct logical relationship, measurement system guarantee, test and test system guarantee, high simulation and forming big data, etc. Chapter 4 Data Abnormalities summarizes the situations that cause data abnormalities, including observe and recording abnormalities, abnormal measurement, abnormal test, sampling anomaly, abnormal simulation, abnormal statistical analysis and calculation process, statistical inference abnormalities, etc. Chapter 5 Data Missing includes incomplete data recording, defects in the process of data collection, censored data, data loss, data misuse, and remedy methods for missing data, etc. Chapter 6 Data Risk elaborates some situations that create data risk, including sampling risk, risk of too small sample size, inappropirate application of statistical method, insufficient process capability, risk of hypothesis test, influence of professional skills of data personnel and multivariate data existence, etc.

A large number of references have been listed in "Bibliography" in the process of writing this book. Here I'd like to express my gratitude to all the Chinese and foreign authors, translators, etc. In the mean time, I would like to thank my family for their support from beginning to end. It is my family that gives me the motivation to continue writing. Due to my limited ability and professional knowledge, there must be much insufficiencies in the book, and readers' advice is appreciated.

This book is suitable for reading and using by personnel related to data work, including quality management, statistical management, plan management, operation management, marketing, material management, comprehensive management, operation and maintenance management and other releverent personnel. Other professionals can refer to it.

Ji Shengping

May 2, 2024

目 录 / CONTENTS

Chapter 1
Meaning of Data

第 1 章
数据的含义

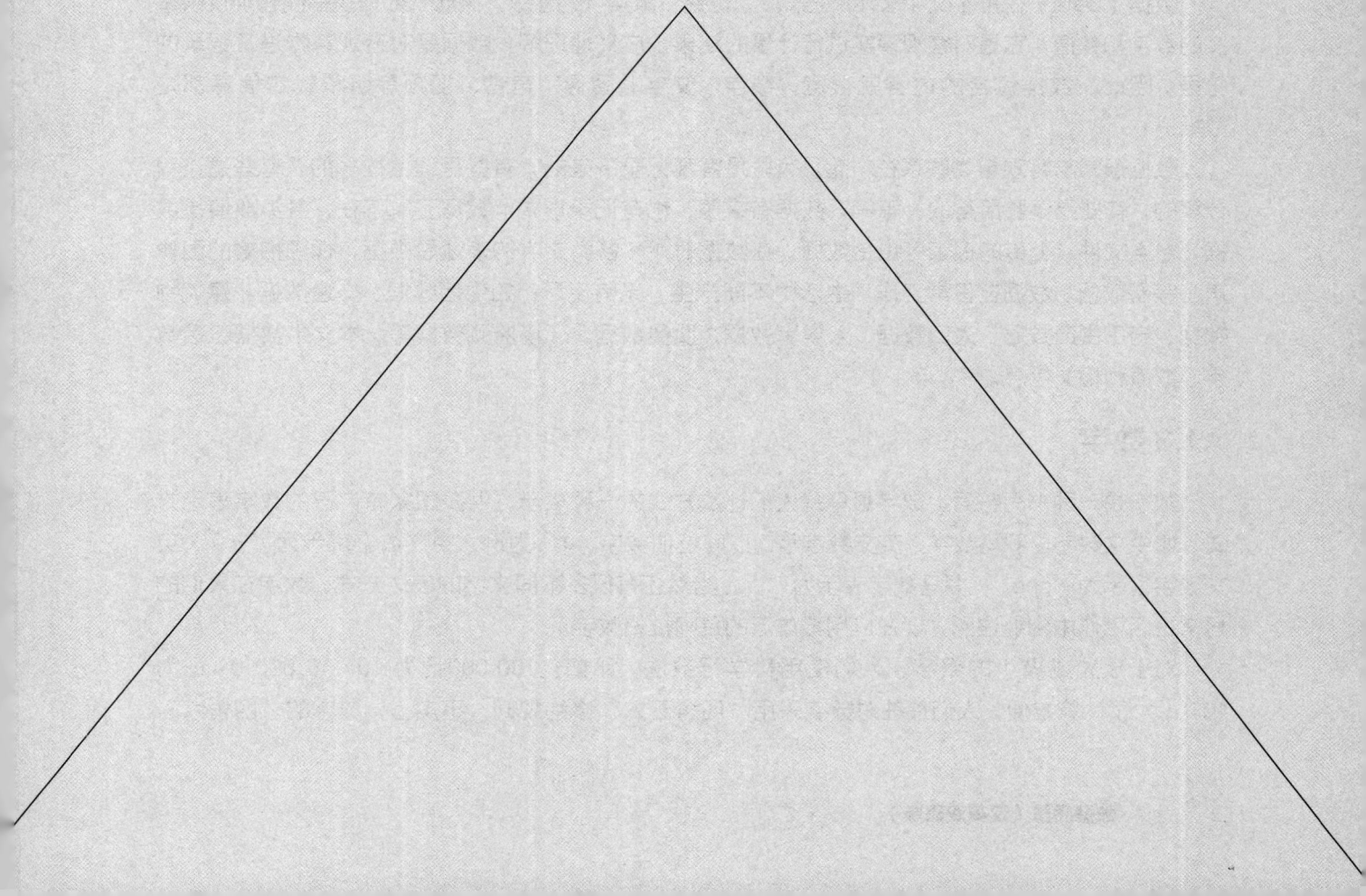

1.1 数据的概念

1.1.1 数

公元前 6 世纪，毕达哥拉斯学派有个说法：数是人类思想的向导和主人。后来，人们为了研究数，于是出现了符号。数字、数值、数据、信息这几个词都可以归为符号，它们的含义各不相同，却又有着紧密的联系。毕达哥拉斯认为，音乐能净化人们的心灵，随后他发现二弦琴两条琴弦的音程之比越简单，和音就越和谐。而和谐概念的提出，引发了毕达哥拉斯学派的核心理念，即万物皆数，指世界的本源是数。当然，$\sqrt{2}$ 的出现对"万物皆数"提出了挑战。

数是一个抽象概念，它被用于计数、标记、量度等。数有很多种分类方法，比如：实数与虚数、整数小数与分数、循环小数与不循环小数、奇数与偶数、完全数（指一个数等于它的真因子之和，至 2018 年完全数只发现了 51 个）、有理数与无理数、自然数（零与正整数）与负整数等。

1.1.2 数据

数据（Data）在拉丁文中表示"已知"，也是"事实"的意思。因此，数据是指能客观反映事实的数字和数值，它是对客观事实进行计量的结果。广义地理解，数据是对客观事物进行记录的符号。因此，数据包含的内容就很宽，数字、文字、图像、声音、视频等所承载的信息都是数据。

数据通常与客观事物联系在一起，有些是客观记录下来的，有些是理论推导的，有些是过程计算的，有些是试验所得的，等等。数据有来源，也有记录形式。数据无所不在，各类数值是数据，各类文件、上传的报表等也是数据。在数据时代，各类文字的表述是数据，视频拍摄的图像声音等是数据。数据在日常工作、生活中不断产生，无穷无尽，如生活数据、交通数据、建筑物数据、施工生产数据、人口数据、入学率数据、质量数据、环境监测数据等。本文中数据包括数字、数值和相关信息。

1.1.3 数字

数字是一种书写符号。数字最早是人们计数方式的一种表示，很多国家有自己的数字表示方式，如中文数字、罗马数字、英文数字等。以阿拉伯数字"1"为例，中文数字表示为"一"，英文数字表示为"one"，罗马数字表示为"I"，当然还有很多不同文字的表达方式。数字在人们的日常生活工作中到处遇到，现在应用最广泛的是阿拉伯数字。

对于极大或极小的数字，人们使用科学记数法，比如：100 000 记为 10^5，0.000 01 记为 10^{-5}。为了计算方便，人们往往对数字采用"四舍五入"修约规则，事实上，数字的"四舍六入五单双"修约规则更符合实际需要。"四舍六入"好理解，比如：1.14 修约为 1.1，1.16 修约为 1.2。"五单双"是指小数点后面的 5 处于奇数位上时"进一"，处于偶数位上时"舍去"，比如：1.145 修约为 1.15，1.1425 修约为 1.142。

1.1.4 数值

数值是在数字后面加上单位名称，是为了表明某个物体、平面、空间等所具备的具体性状，比如表示物体重量，使用"公斤""kg"等单位，其表现出的数值有"1 公斤"等，准确地说，"1"由"数字"的身份转换成了"数值"。所以，数值是指用数目表示的一个量的多少，简单地说，数字与计量单位结合后，就具备了数值的属性。

1.1.5 数据的精确水平

为了获得数据，就需要对客观事物进行计量或测度，针对不同的事物进行计量或测度的程度是不同的，按照对客观事物测度的精确水平，可以把计量尺度从粗略到精确分为四个层次：

一是定类尺度，它的计量尺度比较粗略，是按照某种属性对客观事物进行平行分类，这样得到的数据称为名义数据，名义数据的观察值是事物的属性而不是一个具体的数。比如人的性别为男或女、未婚与已婚、各种颜色、形状等。此时可以用某一个数字来代表某种属性，假如"1"代表红色，"2"代表蓝色，"3"代表白色等等，但这里的"1""2""3"之间不能比较大小，不存在任何数学运算。

二是定序尺度，它是对事物之间等级或顺序差别的一种测度，这样得到的数据称为有序数据，有序数据的观察值代表了事物的属性有一个顺序的关系，如文化程度从小学、中学、大学等，分别用"1""2""3"来表示，但这里不存在"2-1=1"之类的数学运算。

三是定距尺度，它不仅能定类、定序，还能准确地计量出客观事物之间的差距，其计量尺度通常为物理单位，比如当天的气温等。

四是定比尺度，它不仅能定类、定序、定距，还可以计算测量值之间的比值，比如不同货车载重量的比较等。定比尺度中必须有绝对固定的、非任意确定的零值。通常把定类和定序尺度形成的数据称为定性数据，也称属性数据；把定距和定比尺度形成的数据称为定量数据。

1.1.6 信息

信息其实也是客观反映事实的数字、数值、资料等。数据与信息的概念交织在一起，有联系又有区别。严格来说，数据是信息的表现形式和载体，数据是符号，是物理性的。信息则是数据内涵的解释。信息会对数据进行加工，是逻辑性和观念性的。在数字化时代到来之前，信息与数据的区别比较明显，信息的含义更广，它包含了各种数据类的资料，也包含了各种非数据类的资料，信息是传递一定的含义的，人们说的话、写的字、图形、色彩、几何形状、尺寸等等，都是信息。随着数字化时代到来，数据和信息有融合的趋势，基本是等同的概念了，人们传输数据，基本就是传输信息；人们传输信息，也可理解为传输数据。IT 之"I"，是 information，指信息；而 IT 又意味着数据时代。

1.1.7 大数据

大数据是指需要处理的信息量极大，超过了一般电脑在处理数据时所使用的内存量。有时特指人们可以获取或使用的全面数据、完整数据、系统数据。用来表达大数据存储量的单位已经从拍字节（1 PetaByte $=2^{50}$ 字节）到艾字节（1ExaByte $=2^{60}$ 字节），从泽字节（1ZetaByte $=2^{70}$ 字节）到优字节（1YottaByte $=2^{80}$ 字节），再到波字节（1BrontoByte $=2^{90}$ 字节）乃至更巨大。当今时代已经处于大数据时代，"机器学习"就是把数学算法、统计分析用到了大数据的分析上。

国际商用机器公司 IBM 提出了大数据的 5 个 V 特点：海量规模（Volume）、高速度流转（Velocity）、多样的类型（Variety）、低价值密度（Value）、数据的真实性（Veracity）。据 IBM 研究称，人类文明所获得的全部数据中，90%是过去两年内所产生。据 2022 年 8 月 22 日长江商报报道，从 2017 年到 2021 年，中国的大数据产业规模从 4700 亿元增加到 1.3 万亿元；数据产量从 2.3 ZB 增加到 6.6 ZB，位居世界第二。

大数据正日益产业化，大数据的产业基础设施建设不断巩固，大数据的产业链基本形成，大数据的生态体系不断优化，区域集聚显著，对于数据的处理加工能力需求日盛。在这样的背景下，云计算应运而生。云计算指通过"网络云"（"云"是对网络形象的称呼）把海量数据的计算处理程序分解成无数个小程序，然后通过多部服务器组成的系统进行处理和分析，计算结果进行合并后返回用户。

如今，对于数据的管理和使用正成为企业新的核心竞争力。与此同时，数据质量也引起了高度关注。在生产经营过程中，高质量的数据将极大减少原始数据与客观事实之间的差距，极大减少数据分析与原始数据之间的差距。

1.1.8 数据化与数字化

很多人往往把数据化与数字化的概念混淆在一起，以为数据化就是数字化，其实不然。数据化与数字化有着紧密的联系，又有着明显的区别。全球著名的咨询公司麦肯锡（Mckinsey）最早提出了大数据时代的概念，主要意思指数据是一个重要的生产要素，数据已经渗透到了当今时代的每一个行业和领域。在大数据时代，海量的数据资源推动着所有的行业和领域开启量化的进程。面对当今量化进程中的数据，人们只能借助于计算机来处理信息，这就需要把数据予以数字化。

显然，数据化是指把客观事实转变为用数据来制表分析的量化过程。数字化是指把模拟数据转换为计算机可读的二进制码（用 0 和 1 表示）的过程。我们正处于数据化时代，说明这个时代的数据已经无处不在；我们正处于数字化时代，说明这个时代对于数据的处理能力在极速提升。

1.1 Data Concept

—

1.1.1 Number

In the 6th century BC, the Pythagorean School said: number is the guide and master of human thought. Later, in order to study numbers, symbols appeared. Numbers, numerical values, data and information can all be classified as symbols. Their meanings are different, but they are closely related. Pythagoras believed that music could purify people's minds. Later, he found that the simpler the interval ratio between the two strings of a two-stringed lute, the more harmonious the consonance is. The proposition of the concept of harmony initiated the core idea of Pythagoras School, that is, everything is numbered, meaning that the origin of the world is number. Of course, the appearance of $\sqrt{2}$ challenged the idea that everything is numbered.

Number is an abstract concept, which is used for counting, marking, measuring, etc. There are many ways to classify numbers. For example, there are real and imaginary numbers, integers, decimals and fractions, cyclic and non cyclic decimals, odd and even numbers, perfect numbers (it means that a number is equal to the sum of its true factors and by 2018, only 51 perfect numbers have been found), rational numbers and irrational numbers, natural numbers (zero and positive integer) and negative integers, etc.

1.1.2 Data

Data means "known" in Latin, and also means "fact". Therefore, data refers to the numbers and numerical values that can objectively reflect the facts. It is the result of measuring the objective facts. In a broad sense, data is a symbol for recording objective things. Data contains a wide range of contents and all the information carried by numbers, words, images, sounds, videos, etc. are data.

Data is usually associated with objective things. Some are recorded objectively, some are derived theoretically, some are calculated in process, and some are obtained from experiments. Data has sources. Data has forms of record. Data is ubiquitous. All kinds of numerical values are data. All kinds of files and uploaded reports are data. In the era of data, the descriptions of various words are data, and the images and sounds captured in video are data. Data is generated in daily work and life, which is endless, such as life data, traffic data, building data, construction data, population data, enrollment rate data, quality data, environmental monitoring data, etc. Data in this book includes numerals, numerical value and relevant information.

1.1.3 Numerals

Numerals are graphic symbols. Numerals were originally a representation of people's counting methods. Many countries have their own numeral representations, such as Chinese numerals, Roman numerals, English numerals. Take the Arabic numeral "1" as an example, the Chinese numeral is expressed as "一" while the English numeral is "one" and the Roman numeral is "I". Of course, there are many different ways of expression. Numerals are encountered everywhere in people's daily life and work, and Arabic numerals are now the most widely used.

For extremely large or extremely small numbers, people use scientific notation, such as 10^5 for 100000 and 10^{-5} for 0.00001. For the convenience of calculation, people often adopt the "half adjust" rounding rule for figures. In fact, the rounding rule of "Banker's rounding" is more in line with actual needs. If the number is 4 or less than 4, round off; if the number is 6 or greater than 6, go into one digit. For example, 1.14 is rounded off to 1.1, and 1.16 is rounded off to 1.2. If the number is 5, we need to observe that whether the "5" is on odd digit or on even digit. When 5 is on the odd bit, go into one digit, and when 5 is on the even bit, round off. For example, 1.145 is rounded to 1.15 and 1.1425 is rounded to 1.142.

1.1.4 Numerical value

The numerical value is the number added with the unit name, which indicates the specific properties of an object, plane, space, etc. For example, the unit "kg" is used to express the weight of the object, and the displayed value includes "1kg" and so on. To be exact, "1" is converted from the identity of "number" to "value". Therefore, numerical value refers to the number of a quantity. It can be said that after the combination of number and UOM (unit of measure), the number has the attribute of value.

1.1.5 Accuracy of data

In order to obtain data, it is necessary to meter or measure objective things. The degree of metering or measurement for different things is different. According to the accurate level of measurement of objective things, the measurement scale can be divided into four levels from rough to accurate.

The first is the nominal scale. Its measurement scale is relatively rough, which classifies objective things in parallel according to a certain attribute. The data thus obtained is called nominal data. The observed value of nominal data is an attribute of a thing rather than a specific number. For example, there are male and female, unmarried and married person, various colors, different shapes, etc. At this time, a number can be used to represent a certain attribute. "1" represents red, "2" represents blue, "3" represents white, etc., but the sizes of "1" "2" and "3" here cannot be compared, and there is no mathematical operation.

The second is the ordinal scale, which is a measure of the level or order difference between things. The data obtained in this way is called ordinal data. The observed values of ordinal data represent the attributes of things and have a sequential relationship. For example, the educational level from primary school, middle school and university is represented by "1" "2" and "3" respectively, but there is no mathematical operation such as "2−1 = 1".

The third is the interval scale, which can not only classify and order, but also measure the gap between objective things accurately. Its measurement scale is usually a physical unit, the temperature of the day is an example.

The fourth is the ratio scale, which can not only determine the category, sequence and distance, but also calculate the ratio between measured values, such as the comparison of different truck weights. There must be an absolutely fixed and non-arbitrarily determined zero value in the fixed scale. Generally, the data formed by nominal scale and ordinal scale is called categorical data, also known as attribute data, and the data formed by interval scale and ratio scale is called quantitative data.

1.1.6 Information

In fact, information is numbers, numerical values and data that reflect the facts objectively. The concepts of data and information are intertwined, related and different. Strictly speaking, data is the form and carrier of information, while data is a symbol and physical. Information is the explanation of the data connotation. It will process the data and it is logical and conceptual. Before the digital age, information and data are different obviously. The connotation of information is broader. It includes all kinds of data and all kinds of non-data materials. Information conveys a certain meaning. Saying writing, graphics, colors, geometric shapes, dimensions, etc. are all information. But with the advent of the digital age, data and information tend to converge and they are basically the same concept. When people transmit data, they actually transmit information; and the information transmitted can also be understood as data. "I" of IT refers to information, and IT also means data era.

1.1.7 Big data

Big data refers to the amount of information which exceeds the amount of memory a typical computer uses to process data. Sometimes, it especially refers to the comprehensive data, complete data, systematic data that people can access or use. The units used to express big data storage have ranged from petabytes ($1PB = 2^{50}$ bytes) to exabytes ($1EB = 2^{60}$ bytes), from zettabytes ($1ZB = 2^{70}$ bytes) to yottabytes ($1YB = 2^{80}$ bytes), and from brontobyte ($1BB = 2^{90}$ bytes) to enormous. Nowadays, we are in the age of big data. "Machine learning" refers to the application of mathematical algorithms and statistical analysis to the analysis of big data.

International Business Machines (IBM) put forward five "V" characteristics of big data, which are volume, velocity, variety, value and veracity. According to IBM research, 90% of all data obtained by human civilization were generated in the past two years. It is reported *by Changjiang Times* on 22nd August 2022 that the scale of big data industry of China had increased from 470 billion yuan to 1.3 trillion yuan from 2017 to 2021. The data output had increased from 2.3 ZB to 6.6 ZB which ranked second in the world.

Big data is becoming increasingly industrialized, and the industrial infrastructure construction of big data is constantly consolidated. The industrial chain of big data has basically formed and its ecosystem has been continuously optimized. The regional agglomeration is remarkable. The demand for data processing capacity is growing. Against this background, cloud computing came into being. Cloud computing refers to discomposing the calculation process programme of mass data into countless small programmes through Network Cloud ("cloud" is a term for network image), then the data are processed and analyzed by a system composed of multiple servers. The calculation results would be combined and returned to the user.

Nowadays, the management and use of data have become the new core competitiveness of enterprises. At the same time, data quality has also been concerned greatly. In the process of production and operation, high-quality data will greatly reduce the gap between original data and objective facts, and greatly reduce the gap between data analysis and original data.

1.1.8 Datamation and digitization

Many people confuse the concepts of datamation and digitization, and think that datamation is the same as digitization, but it is not. Datamation has close connection with digitization and has obvious difference as well. McKinsey, a world-famous consulting company, was the first to put forward the concept of the big data era, which mainly means that data is an important production factor, and data has penetrated into every industry and field in today's era. In big data era, the massive data resources have given impetus to all industries and fields to start the process of quantification. Facing the data in the process of quantification, people can only use computers to process information, which requires digitizing the data.

Obviously, datamation refers to a quantitative process that turns objective facts into tabulated data. Digitization refers to a process that converts analog data into computer-readable binary code (represented by 0 and 1). We are in the era of data, which shows that data in this era has been everywhere, and the processing capacity of data in this era is increasing rapidly.

1.2　数据的特点与分类

——

1.2.1 数据的特点

数据是客观存在的，是对于客观事物特征的描述。大千世界存在着无数的数据，不同的数据有着不一样的特征。把各种数据进行归纳，通常具有以下一些特点：

（1）数据的产生、获取、分析等需要相应的投入。这里的投入包括人工、机器与设备、培训、时间等方面。

（2）数据可复制、可共享。在一个地方所获得的数据，可以被复制到另一个地方使用、存储，可以被其他人出于不同的目的而共享。

（3）数据可以被解读。在通常的情况下，人们所获得的数据可以通过直观的方式、工具、仪器、计算机等予以解读，成为解释客观对象的重要内容。

（4）数据具有时效性。很多情况下，数据只能反映一定时间范围内的客观情况，过了某个时点，就是新的数据。

（5）数据具有分散性。数据的记录是分散的，分别记录不同的客观事物的相应状态。

（6）数据具有多样性。数据记录的形式呈现多样性，记录下来的形式有文本、图形、图像、视频、音频等。

（7）数据具有可感知性。有些数据可以看得到，有些数据可以听得到，有些数据可以被其他方式所感知。

（8）规模数据的规律性。单个数据是独特的，有时候仅仅观察单个数据所得到的信息量是相当有限的。当数据量达到一定的规模，那么可以从中找到相应的规律。大数规律就是如此得来。

（9）数据具有一定的概率。在很多情况下，某个数据的出现具有其自身的概率。

（10）数据分析的预见性。对于系列数据、规模数据等进行分析，可以进行统计推论，在一定的概率下得出相应结果。

（11）数据可以成为资产。在数据时代到来之后，数据已经成为重要的资产。数据资产不会折旧，越使用就有越多的数据产生。数据还会不断地增值。从数据质量角度看，需要去除重复数据、伪数据，提高数据资产的质量。

（12）数据的记录方式多样化。模拟数据的记录方式有：表格、影像、磁带、地图等。数字数据的记录方式主要有两种，一种是以文件格式记录在文件中，这种数据可以长期保存、随时查阅、修改、增减。另一种是以数据库的形式记录，这种数据共享程度高、冗余度低，容易扩充数据，数据的独立性高，容易增加、删减、修改、查询，方便使用，但数据的存储量大。

1.2.2 数据质量的特点

数据质量指数据内容的质量。数据的选择、类别、数量、采集方法、详细程度等，取决于数据应用的目的。不论出于什么目的,对于数据的质量都要求比较高,也就是使用者所使用的数据能完全、真实地反映产生数据的对象。目前对于数据的主要使用方式有:获取、存贮、搜索、共享、分析、可视化等,这些步骤中的任何一个环节的质量,构成了数据的质量。而数字时代对于数据处理过程的误操作、病毒植入、软件硬件的故障、系统崩溃、断电、黑客侵入等,对数据的质量提出了很大的挑战。

数据质量的特点主要指数据的内容质量所具备的一些特点,主要有:

1. 数据本身的相关性、准确性、及时性

这里的相关性指数据满足使用者需求的程度,它与使用者最关心的主题密切相关。这个相关性会随着使用目的的改变而改变。

准确性指观测值或估计值与未知的真实值之间的接近程度,通常可以用统计误差来衡量。准确性是数据质量的核心内容。

及时性指数据在使用者做出决策之前已经传递到使用者。及时性是数据能否满足使用者需求的重要特征。

2. 数据传递内涵的可比性、可衔接性、可理解性

可比性指同一类型的数据在时间上、空间上的可比程度,必须概念相同、取得方法稳定、不同地区有相同的统计规制和分类标准,在计算方法上保持一致,数据所使用的单位一致。

可衔接性指不同的统计调查对象之间、不同的统计方、以及统计方与国内外组织之间的衔接程度,要有统一的统计框架体系、分类标准、编制数据的方法,数据的加工程序统一,采用国际统计标准等。

可理解性指统计数据便于使用者正确理解和使用的程度。统计方在提供数据时,对于相关数据的分类、数据收集和加工过程中使用的方法等予以明示,便于使用者了解数据的性质,能够正确使用数据,解读出正确的数据信息。

3. 数据受到规制约束的可取得性、有效性

可取得性指使用者取得统计数据的便利程度。获得数据的途径、获得哪些数据、获得数据需要承担的成本、对于数据的统计方式等,组成了获得数据的便利程度。

这里的有效性指利用统计数据所产生的收益大于取得该数据的成本。在保持数据基本质量不受影响的前提下,要尽可能提高数据利用的有效性。

1.2.3 数据的分类

数据按照不同的区分方式可以有多种分类。对数据进行分类,目的是更好地对数据加以利用,提高数据使用效率。

1. 按照显示特性分类

数据可以根据其显示特性分为模拟数据和数字数据。

模拟数据是指在取值范围中呈现连续性质的数值。模拟数据中包含了符号数据(如:字母、中文)、图形数据(如:点、线与线段、面)、文字数据(如:用中文记载的事实、经过,用阿拉伯数字记录的温度变化)、图像数据(如:录像带录制的图像)等。模拟数据的保存载体是有形的,如果数量多,往往会一大堆。模拟数据如果数量多,则不易携带和传递。模拟数据的保存需要专门库房,如图书馆、档案室、文件柜等,保存条件比较高。

数字数据是指在取值范围中呈现离散的数值。目前保存在计算机中的数字数据是以二进制的形式保存。数字数据中，符号数据、图形数据、文字数据、图像数据等只是它们的表现形式，数字数据几乎囊括了各类类型的数据，根据需要呈现不同的形式。数字数据的保存载体有计算机硬件、软件、内存、U 盘、网络云等。数字数据容易复制，容易携带和传输。数字数据也容易被破坏，如病毒入侵、数据损毁等。数字数据的隐蔽性比较高。数字数据保存所占用的空间小。

2. 按照时间和空间维度分类

数据根据其取得的时间和空间分为横截面数据、时间序列数据、合并数据、纵向数据。

横截面数据，又被形象地称为时点数据。是指在某一个时点上所收集到的数据，它是在这个时点上所有对象的数据集合。时点数据中个体的差异性比较大，其表现往往是无规律的，类似于在某个时点上拍照定格的数据。

时间序列数据，是指针对同一个对象在不同的时间内连续观察所取得的数据。它主要是研究对象在时间顺序上的变化，可以发现其中的规律。类似于某一个人每年拍一张证件照片。

合并数据，是指把横截面数据与时间序列数据联合在一起，它既有不同的横截面数据，又有不同的时间序列数据。它是数据的大集合。

纵向数据，是指同一个横截面单位在不同时间点上的数据集合。纵向数据是合并数据中的特殊形式。它类似于拍摄的胶片电影，由若干定格的胶片所组成，而每张横截面的胶片组成内容不断在发生变化。

3. 按照数据表达和计算机使用的方式分类

现在很多的数据是由计算机来处理，因此按照数据的表达和计算机使用的方式，数据可以分为结构化数据、非结构化数据和半结构化数据。这样的分类主要是用于信息的管理。结构化数据是指由二维表结构来逻辑表达和实现的数据，有预先定义的数据模型，也称为行数据，亦称数据库，比如一个财务管理系统中的数据库就是结构化数据。非结构化数据是指数据结构不规则或者不完整，没有预先定义的数据模型，比如各类报表、图像、音频与视频、办公文档等是非结构化数据。半结构化数据，是具有一定的结构性，比如一个管理信息系统中的数据存储，保存在一个指定的关系数据库中，这样的数据就是半结构化数据。

通过结构化数据标记，可以让网站在搜索中有比较好的内容展示，各类搜索引擎都支持标准的结构化数据标记。相应的标记方式有 HTML、微数据标记。

4. 计量数据与计数数据

数据根据取值情况的不同，可以分为计量数据和计数数据两种类型：

计量数据，也称为计量值，它是可以连续取值的数据，也是可以用测量工具测出小数点以下数值的数据，比如重量、速度、长度、容积、电压、电流、强度、化学成分等。计量数据的一个关键词是"连续性"，比如长度为 1.0 米、1.1 米、1.2 米等，也可以为 1.00 米、1.01 米、1.02 米等，还可以继续细分下去而保持连续性的特征。在质量管理活动中，计量数据一直在产生，如分析牵引力用到了牵引力量数据、分析效率用到了产出和时间数据等。

计数数据，也称为计数值，它是人们用计数装置或人工计数等方法而得到的非连续性的数据，或者说使用测量工具不能得到小数点以下数值的数据。计数数据通常表现为正整数，比如人数、交通事故次数、航班次数、合格品数、不合格品数、起降次数、投诉次数、发货批数等。在质量管理小组活动中常用到的检验批数、质量问题的不合格数、小组活动次数等都是计数数据。

计数数据还可以分为计件数据（计件值）和计点数据（计点值），计件数据是按件进行计数的数据，如产品数、不合格品数等；比如在一个材料检验批次中有 3 件不合格，这个"3 件"就是计件值。计点数据是按不合格项计数的数据，它可以表现为某一个产品、某一个过程、某一个单位产品等上面发生的、具有某个质量特性的数据，如气泡数、砂眼数等；比如在机电安装施工中出现 6 处不合格点，这个"6 处"就是计点数据。

在数据的统计应用实践中，一般也不用刻意留心所采集的数据为计量值还是计数值。但是，在某些应用场合就需要引起注意，比如在使用控制图时，计量值的控制图与计数值的控制图是有区别的。

1.2　Data Characteristics and Data Classification

1.2.1 Data characteristics

Data is objective, which is a description of the characteristics of objective things. There are countless data in the world, and different data have different characteristics. After summarizing various data, it is found that they usually have the following traits:

(1)The production, acquisition and analysis of data need corresponding inputs. These inputs include labor, machinery & equipment, training, time, etc.

(2)Data can be copied and shared. The data acquired in one place can be used and saved in another place by being duplicated, can be shared by another person for different purposes.

(3)Data can be interpreted. In the common circumstances, the data obtained can be interpreted through intuitive ways, tools, instruments, computers, etc., and become the important contents to explain the objective things.

(4)Data has timeliness. In many cases, data can reflect the objective things within a certain time range only. After a certain time point, it will be new data.

(5)Data has dispersibility. The records of data are dispersive. They record the corresponding status of different objective things respectively.

(6)Data has diversity. The forms of data records are diverse. There are many kinds of recorded forms, such as texts, graphics, pictures, videos and audios, etc.

(7)Data has perceptibility. Some data can be seen, some can be heard, some can be perceived by other methods.

(8)A certain scale of data has regularity. A single datum is unique. Sometimes we can get quite limited information by observing single datum only. When the data quantity achieves a certain scale, we can find correspondent regularity among them. The law of large numbers comes from this phenomenon.

(9)Data has a certain probability. The appearance of a certain data has its own probability in many cases.

(10)Data analysis has predictability. After analyzing a series of data or a scale of data, the statistical inference can be made, and the corresponding results can be concluded under a certain probability.

(11)Data can become assets. After the arrival of the data era, data has become an important asset. Data assets will not be depreciated. The more data are used, the more data are generated. Data will be added value continually. From the view of data quality, duplicated data and false data should be deleted to improve the quality of data assets.

(12)The recording methods of data are diversified. The recording methods of analog data include table, image, tape, map, etc. There are two main ways to record digital data. One is to record in the file format, which can be saved for a long time, consulted at any time, modified, increased or decreased. The other is to record in the form of a database, which is characterized by high degree of data sharing, low redundancy, easy data expansion, high data independence, easy addition or deletion or modification, convenient query and use, but has a large amount of data storage.

1.2.2 Characteristics of data quality

Data quality refers to the quality of data contents. The selection, category, quantity, collection method, detail level of data depend on the purpose of the data application. For whatever purpose, the requirement of data quality is high, which means the data used by users can reflect the object generating data fully and truly. Up to now, the main ways to use data include acquisition, storage, search, sharing, analysis, visualization, etc. The quality of any one of these steps constitute the data quality. In the digital era, the misoperation of data processing, virus implantation, software and hardware failures, system crashes, power outages, and hacker intrusion pose great challenges to the data quality. The characteristics of data quality mainly refer to some features of data content quality, which mainly include the following traits:

1. The correlation, accuracy and timeliness of data

Data correlation refers to the extent to which the data meets the needs of users, which is closely related to the topic most concerned by users. The correlation will change with the purpose of use.

Data accuracy refers to the closeness between an observed or estimated value and an unknown true value, which can usually be measured by statistical error. Data accuracy is the core content of data quality.

Data timeliness means that the data has been transferred to the user before the user makes a decision. Timeliness is an important characteristic of whether the data can meet the needs of users.

2. Comparability, connectability and understandability of the connotation transmitted by data

Comparability refers to the comparable degree of the same type of data in time and space. All kinds of data must have the same concept, stable acquisition method, the same statistical regulation and classification standard in different regions, the same calculation method, and the same data units.

Connectability refers to the cohesion degree between different statistical survey objects, between different statistical parties, between different statisticians and domestic and foreign organizations. There should be a unified statistical framework system, classification standards, data compilation methods, unified data processing procedures and adoption of international statistical standards.

Understandability refers to the degree to which statistical data are easy for users to understand and use correctly. The statistician should express the classification of relevant data, the methods used in data collection and processing when they provide data, so that users can understand the nature of the data conveniently, use the data correctly, and interpret the correct data information.

3. Availability and validity of data subject to regulatory constraints

Availability refers to the convenience of users to obtain statistical data. The way to obtain data, which data to obtain, the cost of obtaining data, and the statistical method of data constitute the convenience of obtaining data.

The effectiveness here refers to that the benefit generated by using statistical data is greater than the cost of obtaining the data. On the premise that the basic quality of data is not affected, the effectiveness of data utilization should be improved as much as possible.

1.2.3 Data classification

There are various classifications of data according to different separate modes. The purpose of data classification is to better use the data and improve the efficiency of data use.

1. Classification by display properties

Data can be divided into analog data and digital data according to its display properties.

Analog data refers to the numerical value that is continuous in the value range. Analog data includes symbolic data (such as letters and Chinese), graphic data (such as point, line, segment, and plane), text data(such as facts and process written by Chinese, temperature variations recorded by Arabic numerals), picture data(such as recorded video), etc. The storage carrier of analog data is tangible. If the number is large, there will be a lot, and they are not easy to carry and transfer. The storage of analog data requires special conservatory, such as libraries, archives, filing cabinets, etc., and the storage conditions are relatively high.

Digital data refers to the numerical value that is discrete in the value range. At present, digital data in computers are stored in binary form. Symbolic data, graphic data, text data and picture data are the expressive forms of digital data. Digital data almost includes all kinds of data, and presents different forms as required. The storage carrier of digital data includes computer hardware, software, computer memory, U disk and network cloud, etc. Digital data can be duplicated, carried and transfered easily. Digital data can also be destroyed easily, such as virus intrusion, data corruption, etc. Digital data has high invisibility and small saving space.

2. Classification by dimensions of time and space

Data can be divided into cross section data, time series data, combined data, vertical data according to the time and space when it is acquired.

Cross section data, also known as time point data, refers to the data collected at a certain time point, which is the data collection of all objects at this time point. The individual differences in time point data are relatively large, and their performances are often irregular. They are similar to taking pictures at a certain time point, which means the data obtained by continuously observing the same object at different times. It mainly studies the changes in the chronological order, and can find the rules therein. It is similar to a person taking an ID photo every year.

Time series data refers to the data obtained by observing the same object continuously at different times. It mainly studies the changes in the chronological order of the research object and can discover the rhythm within it. It's similar to someone taking an annual ID photo.

Combined data combines cross section data with time series data. It has different cross section data and different time series data. It is a large collection of data.

Vertical data refers to the data collection of the same cross section unit at different time points. Vertical data is a special form of combined data. It is similar to a photographic film, which is composed of several fixed film, and the film composition of each cross section is constantly changing.

3. Classification by data representation and style of computer usage

Nowadays, a lot of data is processed by computers, so according to the way of data expression and computer use, data can be divided into structured data, unstructured data and semi-structured data. This classification is mainly used for information management. Structured data refers to the data that logical expressed and implemented by two-dimensional structure, it has pre-definited data model, also called row data or database. For example, the database in a financial management system is structured data. Unstructured data refers to the undefined data model which data structure is irregular or incomplete. For example, all kinds of report forms, images, audio and video, office documents, etc. are unstructured data. Semi-structured data has a certain property of structure. For example, the data storage in a management information system stored in a designated relational database is semi-structured data.

Through structured data tags, websites can have better content display in search. All search engines support standard structured data tags. The corresponding marking methods are HTML and micro-data symbol.

4. Metering data and counting data

Data can be divided into metering data and counting data according to different values.

Metering data, also known as metering value, is data that can be taken continuously. It is the data that can be measured below the decimal point with measurement tools as well, such as weight, speed, length, volume, voltage, current, strength and chemical composition. A key word of metering value data is "continuity". For example, the length can be expressed as 1.0 m, 1.1 m, 1.2 m, which can also be expressed as 1.00 m, 1.01 m, 1.02 m, etc. The length can be further subdivided and its feature of continuity can be maintained. In the activities of the quality management, the metering value data has generated all the time. For example, traction force data is used to analyze the tractive force, the output and time data are used to analyze the efficiency.

Counting data, also known as counting value, is discontinuous data obtained by counting device or manual counting, or data with values below the decimal point can not be obtained by measuring tools. The counting data is usually expressed as positive integers, such as the number of people, the number of traffic accidents, the number of flights, the number of qualified products, the number of unqualified products, the number of take-off and landing, the number of complaints, the number of delivery batches. The number of inspection batches, the number of unqualified quality problems and the number of quality control circle activities commonly used in quality management activities are counting data.

Counting data can be divided into piece data (piece value) and point counting data (point counting value). Piece data is the data counted by piece, such as the number of products and unqualified products. For example, if 3 pieces in a material inspection lot are unqualified, the "3 pieces" is the piece value. Point counting data is the data counted by unqualified items. It can be expressed as the data with certain quality characteristics on a product, a process, a unit product, etc., such as bubble number and sand hole number. For example, there are 6 unqualified points in the mechanical and electrical installation construction, and this "6 points" is the point counting data.

In the application of data statistics, it is not necessary to deliberately pay attention to whether the collected data is metering values or counting values. However, in certain application scenarios, we should pay attention to the value type of data. For example, there is a difference between the control chart for metering values and the control chart for counting values when we apply a control chart.

1.3　数据的分布

————

单个数据只能表达该数据本身的涵义，而多个数据就能传递出更多的信息，有的数据会呈现集中趋势，有的呈现离散趋势。因此，对于数据的分布有着不同的描述，而众多的数据还会呈现出不同的分布形式。

1.3.1　数据分布的描述

关于数据分布的描述，主要分为数据的集中趋势、离散程度和分布形状。

1. 数据集中趋势的描述

（1）平均数。又称均值，在统计中大量应用。平均数可以消除观测值的随机波动，但容易受到极端值的影响，会出现"被平均"的现象。如果数据基本呈现对称分布，则平均数的代表性就比较好。

（2）中位数。一组数据通过由大到小排序后，取最中间的值即为中位数。如果一组数据为偶数，则取最中间两个数据的平均数为中位数。中位数不会受到极端值的影响。当数据分布的偏斜程度较大时，中位数的代表性就比较好。

（3）四分位数。一组数据由大到小排序后，位于 25％和 75％位置上的值即为四分位数。四分位数不受极端值的影响。

（4）众数。一组数据中出现次数最多的值即为众数。众数不受极端值的影响，适合于数据量比较多的时候使用。在一组数据中，可能会有若干个众数，也可能没有众数。当数据分布的偏斜程度比较大的时候，而且数据中有明显峰值的时候，众数的代表性比较好。

2. 数据离散程度的描述

（1）极差。极差是指一组数据中的最大值与最小值之差。它是表述数据离散程度最简单的一种数值，很容易受到极端值的影响。极差没有考虑到数据的分布情况。

（2）四分位差。又称四分间距、内距，它是一组数据中上四分位数与下四分位数之差。四分位差反映了数组中间的 50％数据的离散程度，用来衡量中位数的代表性。四分位差不会受到极端值的影响。

（3）方差与标准差。方差与标准差反映了各数据与平均值之间的平均差异，是反映数据离散程度最常用的测度值。根据总体数据计算得到的平均差异称为总体方差或总体标准差；根据样本数据计算得到的平均差异称为样本方差或样本标准差。

（4）离差。离差是指每个数据与它们平均值的差异。

（5）标准分数。标准分数是对某一个值在一组数据中相对位置的度量，可以用于对变量的标准化处理，用来判断一组数据是否有离群点。标准分数的计算方式是（原始数据−平均值）/标准差。标准分数的平均值为 0，方差为 1。标准分数相当于把原始数据进行了线性变换，它并没有改变某个数据在该组数据中的位置，也没有改变该组数据的分布形状。

（6）离散系数。离散系数是一组数据的标准差与数据平均值的比值。离散系数消除了数据水平高低和计量单位的影响，可以用于对不同组别数据离散程度的比较。离散系数不存在置信区间。但数据的平均值趋向于 0 的时候，具体的解释会有很大的敏感性。

（7）异众比率。异众比率是指非众数组的频数占总频数的比率。它是衡量众数对一组数据的代表程度，如果异众比率高，则说明非众数占总频数的比重大，那么众数的代表性就差。

3. 数据分布形状的描述

（1）偏态。偏态是对一组数据偏斜形状的描述，反映这一组数据是否对称、偏斜的程度。偏态可以用偏态系数表达。偏态系数等于 0 时，一组数据呈对称分布；偏态系数大于 0 时，呈现右偏分布；偏态系数小于 0 时，呈现左偏分布。偏态系数大于 1 或小于 −1 时，呈现高度的偏态分布；偏态系数在 0.5~1 之间或在 −1 ~ −0.5 之间时，为中等偏态分布；偏态系数接近 0 时，偏斜程度就很小。

（2）峰态。峰态是对一组数据高低形状的描述，反映这一组数据扁平的程度。峰态可以用峰态系数表达。峰态系数等于 0 时，一组数据的扁平峰度适中；峰态系数小于 0 时，呈现扁平分布；峰态系数大于 0 时，呈现尖峰分布。

1.3.2 数据的分布

具有稳定输出的一组数据，呈现出不同的数据分布，如正态分布、二项分布等。鉴于数据主要分为连续型数据和离散型数据，故数据的分布主要分为连续型数据的分布、离散型数据的分布两大类。

1. 连续型数据的分布

（1）均匀分布。又称矩形分布。均匀分布是对称的概率分布，它在相同长度间隔的分布概率是等可能的。均匀分布由数据中的最小值 a、最大值 b 来定义。均匀分布的密度函数为

$$f(x) = \frac{1}{b-a}, a < x < b$$

$$f(x) = 0, x < a \text{ 或 } x > b$$

（2）正态分布。又称高斯分布、常态分布。正态分布的概率密度函数为

$$f(x) = \frac{1}{\sigma \cdot \sqrt{2\pi}} \cdot \exp\{-(x-\mu)^2 / 2\sigma^2\}, -\infty < x < \infty$$

正态分布可以用来估计总体方差已知的总体的均值。

（3）t 分布。t 分布可以用来根据小样本估计总体的均值，而该总体呈正态分布且方差未知。

（4）伽马分布。伽马分布是一种连续概率函数。指数分布和 χ^2 分布是伽马分布中的特例。伽马分布的密度函数为

$$f(x, \beta, \alpha) = \frac{\beta^\alpha}{\Gamma(\alpha)} x^{\alpha-1} e^{-\beta x}, x > 0$$

（5）指数分布。指数分布是描述泊松过程中事件之间的时间的概率分布，表示事件以恒定的平均速率连续且独立发生的过程。它是伽马分布的一个特例。指数分布的密度函数为

$$f(x) = \lambda e^{-\lambda x}, x > 0$$

$$f(x) = 0, x \leq 0$$

（6）χ^2 分布。χ^2 分布的概率密度函数为

$$f(k,\ x) = \frac{(1/2)^{k/2}}{\Gamma\ (k/2)}\ x^{k/2-1}\ e^{-x/2}, x > 0$$

$$f(k,\ x) = 0, x \leqslant 0$$

2. 离散型数据的分布

（1）二项分布。二项分布指重复 n 次独立的伯努利试验，发生的结果只有两个。伯努利试验是指只有两种结果的单次随机试验。

二项分布的概率密度函数为

$$P\ (X = i\ ) = \binom{n}{i} P^i\ (1\ -p)^{n-i},\ \ i = 0, 1, \cdots, n。$$

（2）伯努利分布。又称两点分布。伯努利分布是二项分布在 $n = 1$ 时的特例。

（3）泊松分布。泊松分布是指在连续时间或空间单位上发生随机事件次数的概率分布。它的意思是基于过去某个随机事件在某个时间段或某个空间内发生的平均次数，从而预测该随机事件在未来同样长时间段或同样大空间内发生 n 次的概率。指数分布与泊松分布互补。泊松分布的概率密度函数为

$$P\ (X = i\ ) = e^{-\lambda}\ \lambda^i/i!, i = 0, 1, 2, \cdots,\ \lambda > 1。$$

1.3 Data Distribution

One single datum can express the connotation of itself only, while multiple data can transmit more information. Some data will show a centralized trend which some will show discrete trend. Therefore, there are different descriptions for data distribution, and many data will show different distribution forms.

1.3.1 Description of data distribution

The description of data distribution mainly includes centralized trend, discrete degree and distribution forms.

1. Description of centralized trend of data

(1) Average, also called mean value. It is widely used in statistics. Average can eliminate the random fluctuation of observed value, but can be influenced by extreme value easily and the phenomenon of "passive average" will appear. If the data basically presents symmetrical distribution, the representation of average is good.

(2) Median. A set of data is sorted from large to small. And the middle value is the median. If a set of data is even, the average of the middle two data is the median. Median will not be influenced by extreme values. When the skew degree of data distribution is large, the representation of median is good.

(3) Quartile. A set of data is sorted from large to small. The values at 25% and 75% are quartiles. Quartile will not be influenced by extreme values.

(4) Mode. The value with the most occurrences in a set of data is the mode. Mode will not be influenced by extreme values. It is suitable for use when there is a large amount of data. There may be several modes or no modes in a set of data. When the skew degree of data distribution is large and there are obvious peaks in the data, the representation of modes is quite good.

2. Description of discrete degree of data

(1) Range. Range refers to the difference between the maximum value and the minimum value in a group of data. It is the simplest numerical value to express the degree of data dispersion, and is very vulnerable to extreme values. Range does not take into account of data distribution.

(2) Interquantile range, also called quarter spacing or inner spacing. It is the difference between upper and lower quartiles of a set of data. The interquartile range reflects the dispersion degree of 50% data in the middle of the array, and is used to measure the median. Interquantile will not be affected by extreme value.

(3) Variance and standard deviation. Variance and standard deviation reflect the average

difference between all the data and their average value. It is the most commonly used measure-value to reflect the degree of data dispersion. The average difference calculated according to the population data is called the overall variance or overall standard deviation; and the difference calculated according to the sample data is called sample variance or sample standard deviation.

(4) Deviation. Deviation is the difference between each data and their average.

(5) Standard score. Standard score is a measure of the relative position of a value in a group of data. It can be used to standardize variables and judge whether there are outliers in a group of data. The calculation method of standard score is (original data-average)/standard deviation. The average of standard score is 0 and the variance is 1. The standard score is equal to the linear transformation of the original data. It doesn't change the position of a certain data in this group of data, nor does it change the distribution shape of this group of data.

(6) Discrete coefficient. Discrete coefficient is the ratio between the standard deviation of a group of data and their average. Discrete coefficient eliminates the impacts of data level and metering unit, and can be used to compare the degree of data dispersion of different groups. There are no confidential intervals in discrete coefficients. When the data average tends to 0, the specific explanation will have great sensitiveness.

(7) Hetero mode ratio. Hetero mode ratio refers to the ratio of the frequency of non mode sets to the total frequency. It is a representative degree of measuring the mode versus a set of data. If the hetero mode ratio is high, it means that the proportion of non-mode in the total frequency is large and the representativeness of the mode is poor.

3. Description of data distribution shape

(1) Skewness. Skewness is the description of skewed shape of a group of data, and reflects the degree of symmetry and deviation of this group of data. The skewness can be expressed by skewness coefficient. When the skewness coefficient is 0, this group of data presents symmetrical distribution; If the skewness coefficient is greater than 0, it presents right-skewed distribution; If the skewness coefficient is less than 0, it presents left-skewed distribution. When the skewness coefficient is greater than 1 or is less than -1, it presents highly skewed distribution; When the skewness coefficient is between 0.5 and 1 or between -1 and -0.5, it is medium skewed distribution; If the skewness coefficient is close to 0, the skewed degree is very small.

(2) Kurtosis. Kurtosis is a description of the high and low shapes of a group of data, and reflects the degree of flatness of this group of data. The kurtosis can be expressed by kurtosis coefficient. When the kurtosis coefficient is equal to 0, the flat kurtosis of a group of data is moderate; When the kurtosis coefficient is less than 0, it presents flat distribution; If the kurtosis coefficient is greater than 0, it presents peak distribution.

1.3.2 Data distribution

A group of data with stable output appears different distribution, such as normal distribution and binomial distribution. Because data is divided into continuous data and discrete data, the data distribution can be divided into continuous data distribution and discrete data distribution.

1. Distribution of continuous data

(1) Uniform distribution, also called rectangular distribution. Uniform distribution is a symmetrical probability distribution. Its distribution probability at the same length interval is equally possible. Uniform distribution is defined by the minimum value a and the maximum value b in data. The density function of the uniform distribution is

$$f(x) = \frac{1}{b-a}, a < x < b$$

$$f(x) = 0, x < a \text{ or } x > b$$

(2) Normal distribution, also called Gaussian distribution. The probability density function of normal distribution is

$$f(x) = \frac{1}{\sigma \cdot \sqrt{2\pi}} \cdot \exp\{-(x-\mu)^2/2\sigma^2\}, -\infty < x < \infty$$

Normal distribution can be used to estimate the mean value of the population with known population variance.

(3) t distribution. t distribution can be used to estimate the average of population according to the small sample size, while the population appears normal distribution with unknown variance.

(4) Gamma distribution. Gamma distribution is a continuous probability function. Exponential distribution and x^2 distribution are special cases in Gamma distribution. The density function of Gamma distribution is

$$f(x, \beta, \alpha) = \frac{\beta^\alpha}{\Gamma(\alpha)} x^{\alpha-1} e^{-\beta x}, x > 0$$

(5) Exponential distribution. Exponential distribution is a probability distribution that describes the time between events in Poisson process, which indicates the process of event that occurs continuously and independently at a constant average rate. It is a special case in Gamma distribution. The density function of exponential distribution is

$$f(x) = \lambda e^{-\lambda x}, x > 0$$

$$f(x) = 0, x \leq 0$$

(6) x^2 distribution. The probability density function of x^2 distribution is

$$f(k, x) = \frac{(1/2)^{k/2}}{\Gamma(k/2)} x^{k/2-1} e^{-x/2}, x > 0$$

$$f(k, x) = 0, x \leq 0$$

2. Distribution of discrete data

(1) Binomial distribution. Binomial distribution refers to independent Bernoulli test repeated n times, and there are only two results. Bernoulli test refers to single random test with only two results.

The probability density function of binomial distribution is

$$P(X = i) = \binom{n}{i} P^i (1-p)^{n-i}, i = 0, 1, \cdots, n$$

(2) Bernoulli distribution, also called two point distribution. Bernoulli distribution is a special case of binomial distribution when $n = 1$.

(3) Poisson distribution. Poisson distribution refers to the probability distribution of event

numbers which occurred randomly at the continuous time or space unit. It means the average frequency based on a past certain random event that occurred on a certain time period or a certain space, and thus to forecast n times of probability of this random event in the same period of time or in the same space in the future. Exponential distribution and Poisson distribution complement each other. The probability density function of Poisson distribution is

$$P(X = i) = e^{-\lambda} \cdot \lambda^i / i!, i = 0, 1, 2, \ldots, \quad \lambda > 1$$

1.4　数据的作用

———

众所周知，生产力由生产要素所构成。在传统的经济中，生产要素主要指土地、资金、劳动力、技术等。在当今时代，数据成为新的生产要素。曾经的数据毫不起眼，收集到的数据更多的是为了单一目的而体现价值。如今的数据不仅可以进行个性化的分析，还可以通过大数据来进行各种用途的分析与使用。

1.4.1 数据的用途

数据的用途非常广泛，从管理的角度看，数据与统计时时刻刻相生共存，在这基础上，数据的用途不断地在拓展、延伸。数据的用途主要有：

（1）记录瞬间发生的事情。很多客观事情的发生都在转瞬即逝的瞬间，通过数据的记录，可以对该瞬间的事情予以证实。

（2）还原客观事物的原貌。在没有原物或原管理状态下，通过数据可以准确还原曾经发生的管理过程。尤其是在人们难以触及的时间和空间上，由数据而窥知真相。

（3）为统计分析提供了第一手资料。各种数据为统计方法而产生，统计方法为数据而存在。现在的统计分析方法已经有了极大的发展，从图示技术描述统计、语言文字描述统计，到各种试验设计、可靠性分析等，无不深刻地丰富着统计的世界，而统计世界中又蕴含和运作着无穷无尽的数据资源。

（4）对管理的全过程进行控制。数据在管理中起到了特别重要的控制作用：把各项管理分解为各子目标和相应指标后，通过全面监测运营指标，实施过程管控。当管理问题发生后，及时依靠数据支持找到运营中的症结，减少对人为经验判断的依赖，减少个人倾向化的决策。在后方总部管理与前方一线运营生产之间形成有效管理闭环，及时反馈上传下达的各类信息，提高管理运行效率。管理的每一个操作环节都有数据作为依据。

（5）对未知的管理进行假设和推测。窥一斑而见全豹，这句话用在数据上有着贴切的比喻。当管理者手中有着数据，哪怕是少量的数据，经由统计分析，可以对原有的管理进行假设，可以对未来的发展趋势进行推测。比如删失数据的存在，管理者可以通过可靠性分析及相应的数据分布曲线来知晓产品的寿命。尤其是借助于计算机软件等，各种建模、仿真等为管理决策的科学性做出巨大的推动。

（6）衡量管理效率。管理好投入产出比是管理者的重要工作。当数据从管理的产业链、价值链中源源不断生成时，管理效率即刻形成，与同行的比较数据也同时出台，基于各种管理目的对于管理的调整可以进入管理的全过程。数据驱动型的管理模式为企业提供高效协同基础，管理效率由此提升。

（7）为提高产品和服务质量提供依据。在服务经济不断深化的今天，提供个性化的产品和服务是管理者在市场中胜出的绝招，也是管理者的不懈追求。经由各种数据的收集和分析，可以及时改进服务的过程，提高服务品质。由数据而产生的实时可视化生产过程、服务过程，为提高产品和服务质量再增加新动力。

（8）推动管理创新。随着数据价值的发现、数据使用中数据价值的叠加，推动管理者在技术、经营、生产、服务等各流程中进行创新，为客户提供更精准的高质量服务，再进一步推动创新能力的发展和数据价值的提升。人工智能、机器学习等的深度应用，使各类管理创新有了无限可能。

（9）在应用中提高数据质量。数据价值的开发在于应用，而数据质量决定着其应用价值、推广可能。数据的质量受到了各种挑战，但管理者可以在数据应用的过程中发现存在的问题，客观对待，解决数据质量的痛点，从而不断提升数据质量。

1.4.2 数据使用的过程

数据在使用的过程中体现其真正的价值。数据使用的过程主要包括数据体系的构建、数据的收集、存储、清洗、分析、应用、可视化等。数据使用过程的质量是数据质量的核心内容，面对着海量的数据、快速变化的数据，使用者需要应对的方法，排除疑难，提高数据的使用价值。

（1）数据体系的构建。数据管理者根据管理的目的，构建相应的数据体系。数据体系是一个完整的数据管理框架，它包括了收集数据对象的明确、数据存储的方式、数据清洗的要求、数据分析的开展、数据应用等一系列需要明确的数据管理相关的内容。

（2）数据的收集是大数据的源头，数据采集的质量直接决定了后续所有关于数据应用的工作质量。通过多维数据的不断积累，数据资产不断增加。数据资产的体量越大，其价值同步提升。数据收集包括内部收集和外部收集两种。内部收集是在数据系统中进行各类数据的采集。外部收集是向外界收集数据，包括购买数据、交换数据等行为。

（3）数据的存储。这里的数据存储的方式可以是纸质文本上的记录，现在更主要的指存储在计算机上的方式，包括存储的顺序、链表、固定长度输出（散列）、索引等。数据存储的对象包括数据流在加工过程中产生的各类文件和信息。存储在计算机上的数据需要记录在存储介质上，数据的存储单位是比特。

（4）数据的清洗。数据中会存在残缺数据、错误数据、异常数据、重复数据等。数据清洗是对数据进行重新审查和校验的过程，以删除重复的信息，纠正存在的错误，保证所提供数据的一致性。在计算机应用中，指发现并纠正数据文件中可识别错误的最后一道程序，包括检查数据的一致性，处理无效值和缺失值等。一致性主要指根据每个变量的合理取值范围和相互关系，检查数据是否符合要求，对超出正常取值范围的数据、逻辑不合理的数据等予以核对、纠正、处置。对无效值、缺失值的处理包括估算、删除等。

估算中比较简单的方法是用样本均值、中位数、众数来代替无效值、缺失值，这种方法误差较大。也可以根据调查对象对其他问题的答案，通过变量之间的相关分析、逻辑推理来估计。删除中包括：整例删除，就是把含有缺失值的样本删除，它会导致有效样本量减少；变量删除，适用于某一变量的无效值、缺失值比较多，且该变量不是很重要，则删除该变量，这个方法不会减少样本量；成对删除，指用一个特殊码来代表无效值、缺失值，保留数据集中的全部变量和样本，比如用 99 来代表。

（5）数据分析是数据利用的重要环节，也是数据价值化的重要过程。以企业来说，在发展初期的数据量相对较少，仅有的数据中难以发现问题或规律性事物。而随着企业的发展和管理的深入，管理数据不断产生，数据分析的价值显现。数据分析的方式主要有两种，一是从统计学的角度去分析，另一个是机器分析，也是机器学习。通过数据分析，可以让管理者看清当前的管理现状，预测将来的发展趋势，达到精细化管理的程度。

（6）数据应用是数据价值的直接体现，数据应用的对象可以是人，也可以是一种智能物体。数据应用的内容包括数据的标准化、数据报表、数据应用系统、专项数据分析、数据的自动化等。数据标准化是对数据进行统一口径的管理，是在发生数据变化、人员变动时的主动、有效管理。数据应用的标准化包括数据指标的标准化，如统一的指标名称、规范的数据解释、标准的数据逻辑计算。数据应用的标准化还包括数据分析的标准化，通过标准的报表、应用系统等实现数据的模块化、自动化，确保数据应用流程在专业分工下开展，不受人员变动的干扰。

数据报表是应用者所开发及使用的各类报表，以保存数据、分析数据、维护数据。

数据应用系统是面向内部使用人员的数据产品，使用者可以根据使用目的进行数据的组合与分析。管理者根据需求分析，设计相关的数据产品方案，界定相关的指标，然后开发数据应用系统，在过程中予以优化、维护。数据产品开发过程中会产生大量文件，包括开发数据产品的目的、数据产品的系统构成、各类指标解释、使用手册等。数据产品开发的过程文件要及时整理、归档，以便查阅。

数据的自动化应用是面向未来的数据应用。机器学习、深度学习等通过对模式的识别、模块的自动化，从而实现数据自动化的应用，可以解决人工无法穷尽的模式，挖掘人工难以获得的有效价值。

（7）数据的可视化。可视化是对数据使用的重要表现形式，它借助于图形化的手段，清晰、有效地传达信息。数据的可视化与信息图形、信息可视化、统计图形等密切相关，以直观的形式传递数据的关键信息和特征。

1.4　Data Role

———

As we all know, productivity is composed of production factors. In the traditional economy, production factors refer to the land, capital, labor, technology, etc. In today's era, data has been a new production factor. Data used to be unconspicuous. The data collected is more for a single purpose to reflect the value. Nowadays, data can not only be analyzed individually, but also be analyzed and utilized for various uses through big data.

1.4.1 Use of data

The use of data is extensive. From the perspective of management, data and statistics live together all the time. The use of data expands and extends continuously on this base. The main uses of data are as follows:

（1）Record what happened instantly. Many objective things happen in an instant. The instantaneous event can be confirmed by the data record.

（2）Restore the original appearance of the objective things. In the absence status of original objects or original management, the ever management process could be restored accurately through data. Especially in the time and space that people can hardly reach, the truth can be got from the data.

（3）The firsthand material will be supplied for the statistics. All kinds of data are generated for the statistical method, and the statistical method exists for data. Up to now, the statistical analysis methods have been greatly developed, from graphic technology description statistics, language description statistics, to various experimental designs, reliability analysis, etc. All of them have profoundly enriched the world of statistics. Meanwhile, the statistical world contains and operates endless data resources.

（4）Control the whole process of management. Data has played a very important role of controlling in the management. It disassembles the management into each sub objective and corresponding indicator, implement process control through comprehensive monitoring of operational indicators. When management problems occur, we can rely on data support timely to find the crux of the operation, reduce dependence on human experience judgment, and reduce personal biased decision-making. We can form an effective management loop between the headquarters management and the front line operation and production, timely feed back all kinds of information uploaded and issued, and improve the management and operation efficiency. Each operation sector of management is based on data.

（5）Assume and speculate on unknown management. The sentence of "look at one spot to see the whole picture" is a good metaphor for data. When the managers have data, even a small amount of data, they can make assumptions on the original management and speculate on the development trend through statistical analysis. For example, because of the existence of censored data, the managers can know the life of product by reliability analysis and corresponding data distribution curves. Especially, with the help of computer software, various modeling and simulation have made a huge push for the scientific management decisions.

（6）Judge the management efficiency. It is the manager's important work to manage the input-output ratio. When data is continuously generated from the managed industrial chain and value chain, the management efficiency is formed immediately. At the same time, comparative data with peers are also released. The management adjustment based on various management purposes can enter the whole process. The data-driven management model provides enterprises with a foundation for effective collaboration. Therefore, the management efficiency has been improved.

（7）Provide basis for improving product quality and service quality. With the deepening of the service economy, providing personalized products and services is a unique way for managers to win in the market, and it is also the unremitting pursuit of managers. Through the collection and analysis of various data, the service process can be improved in time to improve the service quality. Real time visual production process and service process generated by data add new impetus to improve the quality of products and services.

（8）Promote management innovation. With the discovery of data value and super imposition of data value in use, data promotes the managers to innovate in the processes of technology, business, operation and service, etc., provides customers with more accurate and high-quality services, and further promote the development of innovation ability and the improvement of data value. Deep application of artificial intelligence, machine learning, etc. has made all kinds of management innovation possibile.

（9）Improve the data quality in application. The development of data value lies in application, and the data quality determines its application value and promotion possibility. The data quality has been challenged, but the managers can find the problems in the process of data application, objectively treat the problems and solve the pain points of data quality, so as to continuously improve the data quality.

1.4.2 Process of data use

Data reflects its real value in the process of use. The process of data use mainly includes construction of data system, data collection, storage, cleaning, analysis, application and visualization. The process quality of data use is the core content of data quality. For massive data and rapidly changing data, users need methods to cope with difficulties and improve the use value of data.

（1）Construction of data system. The data managers construct the corresponding data system according to management purpose. Data system is a complete data management framework which includes a series of data management related contents that need to be clarified, such as the definition of data collection object, the method of data storage, the requirements for data cleaning, the development of data analysis, and data application, etc.

（2）Data collection is the source of big data. The quality of data collection directly determines the work quality of all subsequent data applications. The data assets have increased constantly through the unceasing accumulation of multi-dimension data. The larger the volume of data assets, the greater their value. There are two kinds of data collective methods. They are internal collection and external collection. Internal collection is the method to collect various data in the data system. External collection is the way to collect data from outside, including data purchase and data exchange.

(3)Data storage. The data storage method here refers to the record on the paper text and now it mainly refers to the way of storage on the computer, including storage order, linked list, fixed length output (hash), index, etc. The objects of data storage include all kinds of files and information generated during the process of data flow. The data stored in the computer need to be recorded on storage medium. The storage unit of data is bit.

(4)Data cleaning. There will be incomplete data, wrong data, abnormal data, duplicate data, etc. Data cleaning is the process of re-reviewing and verifying data to delete the duplicated data and correct the existing errors, so as to ensure the consistency of the data provided. In the computer application, data cleaning refers to the last procedure which can find and correct identifiable errors in the data files, including checking the data consistency, handling invalid and missing values, etc. Data consistency mainly refers to checking whether the data meets the requirements according to the reasonable value range and mutual relationship of each variable. The data beyond the normal value range and the data with unreasonable logic shall be checked, corrected and disposed. The handling of invalid and missing values includes estimation, deletion, etc.

The relatively simple method in estimation is to replace invalid values and missing values with mean, median and mode sample. This method has a large error. It can also be estimated through correlation analysis and logical reasoning among variables according to the answers of respondents to other questions. The following three types are included in the deletion: The first is whole case deletion. It is to delete samples with missing values, which will reduce the effective sample size. The second is variable deletion. If there are many invalid and missing values applicable to a variable and the variable is not very important, delete the variable. It is applicable when there are many invalid and missing values of a variable and the variable is not very important. Then delete the variable. This method will not reduce the sample size. The third is paired deletion. It refers to the use of a special code to represent the invalid and missing values, and the retention of all variables and samples in the dataset. For example, the symbol 99 is used to represent the missing value.

(5)Data analysis is an important link of data utilization and an important process of data value. For enterprises, the amount of data at the initial stage of development is relatively small, and it is difficult to find problems or laws in the small amount of data. With the development of enterprises and the deepening of management, management data is constantly generated, and the value of data analysis appears. There are two main ways of data analysis: one is to analyze from a statistical perspective, the other is machine analysis, which is also called machine learning. Through data analysis, managers can see the current management status, predict the future development trend and achieve the degree of refined management.

(6)Data application is the direct embodiment of data value. The object of data application can be a person or an intelligent object. The contents of data application include data standardization, data reports, data application system, special data analysis, data automation, etc. Data standardization is a unified management of data. It is an active and effective management of data changes and personnel changes. The standardization of data application includes the standardization of data indicators, such as unified indicator names, standardized data interpretation, and standard data logical calculation. It also includes the standardization of data analysis. The modularization and automation of data are realized through standard reports, application systems, etc., to ensure that the data application process is carried out under professional division of labor and is not disturbed by personnel changes.

Data reports are all kinds of reports developed and used by users to save, analyze and maintain data. The data application system is a data product for internal users. Users can combine and analyze data according to the purpose of use. The managers can design relevant schemes of data products and define relevant indexes according to the requirement analysis, then develop the data application system, optimize and maintain the system in the process. A large number of documents will be generated during the development of data products, including the purpose of developing data products, the system composition of data products, various indicator explanations, user manuals, etc. The process documents of data product development shall be timely sorted and archived for reference.

The automatic application of data is a future oriented data application. Machine learning, deep learning, etc. realize the application of data automation through pattern recognition and module automation, solve the inexhaustible mode of labor and excavate the effective value that is hard to obtain by labor.

(7) Data visualization. Visualization is an important expression of data uses. It can convey information clearly and effectively by the aid of graphical means. Data visualization is closely related to information graphics, information visualization, statistical graphics, etc., and transmits key information and feature of data directly.

Chapter 2
Sources of Data

第 2 章
数据的来源

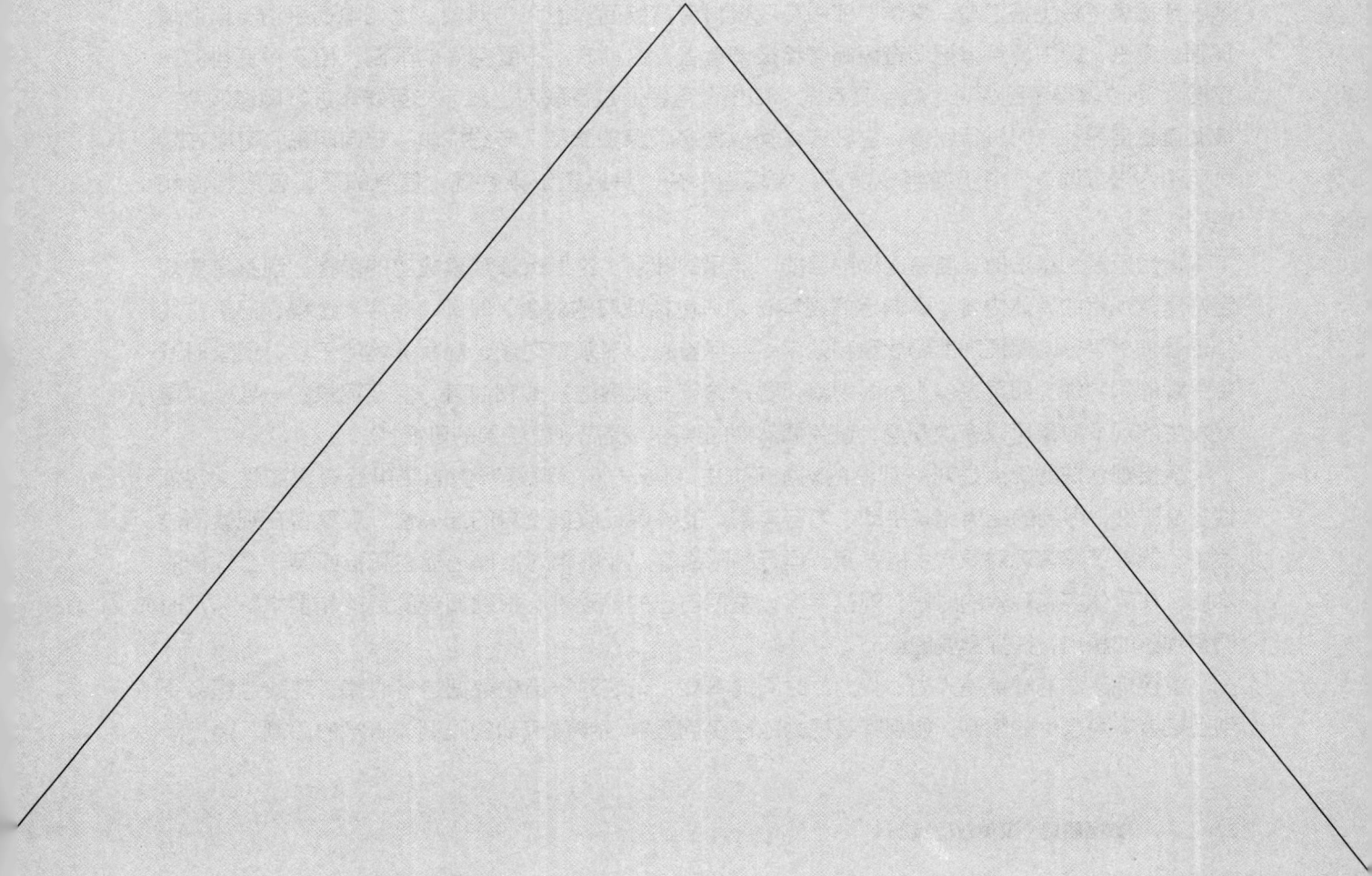

2.1　观察法

——

2.1.1 观察法的分类

数据的来源有很多方式，通过观察而获得数据是比较古老的方法，由于其直观的特点，至今还在使用。观察法主要分为直接观察法和间接观察法两大类。

1. 直接观察法

直接观察法是指人们凭借自己的眼睛、耳朵、手脚等器官和肢体、感观等，对研究对象进行观察的方法，观察的过程中没有任何中介媒质，观察人员（调查人员）亲临调查现场进行相应的观察工作。观察的内容包括对于被调查对象数量的清点、测定、计量等，然后登记入册，取得第一手资料。直接观察法在日常工作中运用较多，比如：人们盘点库存、观测当天的气温、记录空调出风口的温度、衡量量杯里面的液体容积、清点一个班级的学生数量、高速公路上发生事故的里程标记、洪水到了什么水位、发生事件的具体时间、消毒的次数记录等。

直接观察法可以分为几种类型：简单观察，就是进行日常的观察、记录，比如针对员工的行为进行相应的观察与记录。报告法，即由一个层级把观察资料上报给另一层，比如下级的资料根据统计报表逐级上报汇总。采访，即由获取资料者直接面对面采访对象，比如电话采访、自助填报等。通讯，即由调查者把调查表邮寄给被调查者，这种方法不受空间的限制，但获得资料的速度慢，回收率得不到保证。数字交互式，即由调查者通过互联网、数字交互媒体进行网络调查，被调查者自愿、及时进行回复，这种方法可以匿名，保护隐私，节约成本，定向明确，可以在较大范围内进行调查；由于使用多媒体，传播速率快，其调查结果客观，信息的质量容易检验和控制。

运用直接观察法时，要有明确的目的、详尽的计划，这是保证观察成功的前提。观察者要规定好需要观察的具体内容。要制定好观察计划，包括观察的对象、时间、顺序、过程、记录方式和记录表格等，以保证观察的质量和效率。要尽量减少对观察对象、观察环境的干扰，记录时避免主观的倾向性。如果是多人同时观察，要预先统一观察的方式和内容，保证记录的一致。对重复发生的观察对象可以多次观察，避免偶然性的误差，提高观察结果的可靠性。

直接观察法的优点是对被观察的对象不加控制或干涉，观察调查的结果具有真实性、客观性、及时性，得到的结果比较生动、内容丰富，能够保证收集资料的准确性。观察者在观察的过程中，能够了解客观对象发生的背景、运行的脉络等，能够相对准确地描述和推断某一个现象的本质。在具体实施观察的时候，观察者可以发挥自己的能动性，根据调查的目的和要求等，及时调整观察的重点，捕捉需要的细节。

直接观察法的缺点是人力、物力、时间成本高，无法对于历史数据进行收集。观察过程会受到观察者本身条件的限制，如观察者视力范围受到限制、肢体运动的限制、听觉的限制、捕捉瞬间发生事情的能力有限、人手不够等，因此会发生观察不够详细、不够精准等现象，而且会发生错记、漏记等情况。有时会受到时间的限制，特别是有些事情的发生是在一个特定时间段内，而观察者到来时已经过了这个时段。如果被研究的对象是人，当调查对象知道自己处于被观察的时候，有可能会产生影响正常行为的反应，影响观察结果的真实性和客观性。有时候，观察者自己的主观意识、先入为主的意识等会影响观察结果。

2. 间接观察法

间接观察法是指观察人员（调查人员）借助仪器设备等对被观察对象进行的观察、记录方法，通过对客观对象的观察，来记录正在发生的事情，追索、了解曾经发生的事情，也是一种实物观察方法。有时候，调查人员作为旁观者，以周围事物作为间接观察的对象，对被调查对象进行观察。调查者借助仪器设备进行间接观察的例子数不胜数，如测量体温、卫星定位等。调查者借助实物进行间接观察的案例也很多，比如查尔斯·巴林先生在芝加哥地区进行的街区垃圾调查，通过对居民垃圾的观察与记录，来分析得到当地居民的消费资料等。

间接观察法的类型主要有：借助仪器设备法，可以做大量人工所无法达到的观察结果，极大扩展人们获得信息的能力。间接物体观察法，即通过对其他对象的观察，来获取被调查对象的数据。创造情境法，即通过设定特定的情境，来获得相应的信息。谈话法，即通过明确目的的拟定谈话提纲，获得相应的信息，这种方法受到的干扰较多，比如谈话的情形、环境、心情等。痕迹与行为标志观察，即通过对被观察对象留下的痕迹、各种行为中设定的行为标志等获得有效信息。卫星遥感，这是一种新颖的方法，速度快、准确度高，现在的分辨率也高。

运用间接观察法时，要明确观察的目的、制定好详尽的观察计划。观察者要规定好需要观察的具体内容和间接观察的对象。制定的观察计划，包括间接观察的对象、时间、过程、使用的仪器设备、记录方式和记录表格等。如果是多人同时进行间接观察，要统一间接观察的方式和内容，保证记录的一致。

间接观察法的优点是观察的时间、空间不受限制，特别是被观察对象发生的时间转瞬即逝，或者被观察对象持续较长一段时间的时候，可以被真实记录下来。观察者不会因为自身生理条件的限制而无法完成观察，比如视力受到限制、高度受到限制、角度受到限制、温度受到限制等，这些问题不再干扰观察过程。如果研究对象是人的时候，调查对象不会知觉正处于被观察中，因此观察行为真实、可靠。观察者不会受到主观意识的影响，观察的结果客观、真实。

间接观察法的缺点是调查者没有亲身感受，涉及心理活动的数据难以准确把握。观察的结果会受到仪器设备受限的影响，有时会受到仪器设备操作不当的影响，导致观察结果失真。在观察行为表现时，所搜集的资料不能保证其效度。间接观察时得到的被观察对象的痕迹、标志等不能真正反映调查者所要研究的行为、现象，难以判断这些痕迹、标志是否具有普遍性。

2.1.2 观察法的其他分类

1. 定性观察与定量观察

为了对研究对象的变化程度进行描述，可以分为定性观察和定量观察。

定性观察主要指对研究对象的性质和特征方面进行观察，并用语言文字对客观对象进行描述。定性观察只要求对研究对象的组成、表现特征等有所了解，比如物体的外观、颜色、状态、所处位置、环境等，观察的结论除了对特征的直接描述外，还有就是其保持不变或存在变化。

定量观察是以具体的数值来衡量研究对象变化的幅度、过程等，定量观察往往要借助于仪器设备，以便有更精确的记录。定量观察的数据精度与观察时使用的仪器设备有关。

2. 实验室观察与实地观察

根据观察场所的不同、是否对观察对象进行过程中的控制等，可以分为实验室观察和实地观察。

实验室观察是指在实验室条件下进行观察活动，要求观察者要有明确的实验目的，观察前需要设定相应的实验条件、实验对象，确定各种变量，包括自变量、因变量、无关变量等，通过实验来获得相应数据。

实地观察是指观察者在调查对象处于自然状态下的一种观察方法，观察者要有明确的观察目的，制定观察计划，运用适当的观察工具（如人体本身的视觉听觉器官、照相机、摄像机、录音机、探测器），设计相应的观察表式，到调查对象的实地进行观察。根据观察者的角色，可分为参与观察和非参与观察；根据观察的内容和要求，可分为有结构观察和无结构观察。

3. 参与型观察与非参与型观察

根据观察者的角色，即观察者是否参与被观察者相关的活动，可以分为参与型观察和非参与型观察。严格来说，直接观察法包括了参与型观察和非参与型观察，而间接观察法只有非参与型观察。

参与型观察是指观察者直接参与到被观察者的活动之中，在大家共同进行的活动过程中进行内部的观察。参与型观察是调查研究的重要方法，起源于"田野工作法（field work）"，适合于社会科学的研究，特别是与人的活动相关的调查研究。参与型观察记录的数据可以更深入到研究对象的内部，但有时会干扰到研究结果。

非参与型观察是指观察者不参与被观察者的活动，是以局外人的身份进行观察，在观察的过程中保持客观性和独立性。非参与型观察不会造成对研究对象的干涉，观察结果的信度比较高。

4. 结构型观察与非结构型观察

根据观察的内容和要求，即观察过程是否有统一设计的、有一定构架的观察内容、要求等，可以分为结构型观察和非结构型观察。

结构型观察是观察者事先确定好观察的内容，包括观察的项目范围、观察清单、样本及抽样方法、指标、周期等，设计好记录的表式，对观察资料进行准确的分类、编码、记录。观察的形式有事件取样观察，即根据调查研究的目的从观察对象的行为中选出有代表性的样本进行观察；时间取样观察，即在特定的时间段内对选定的对象进行观察，记录发生的频率和持续的时间；等级量表观察，即采用等级量表的方法来观察调查对象的相关行为、结果，等级量表包括数字量表、图示量表、描述量表等类型。

非结构型观察是事先没有具体设计要求的一种观察类型，观察者有一个总的观察目的和要求，对于观察的项目范围、观察内容等不作详尽规定。在具体实施时，观察者可以根据当时的环境、条件变化等调整观察的角度和观察内容，方法比较灵活，实施简单。

5. 时间抽样观察与事件抽样观察

根据观察前是先设定观察的时间还是先设定观察的事件，可以分为时间抽样观察和事件抽样观察。

时间抽样观察是指在事先设定的时间段中，对调查对象进行的观察，它可以观察记录某一现象是否会发生、发生的频率、持续的时间等。在实施前，要先明确观察目的、观察对象、观察频率和观察时间，适用于发生频率比较高的对象。

事件抽样观察是指事先设定一个"靶子事件"，在事先规定的行为等内容下进行的观察。在实施前要定义好所要观察的事件，然后在确定的地点实施，可以对发生的行为与当时的情景等进行因果分析。由于记录的是与"事件"相关的信息，该观察更多的是定性分析。

2.1　Observation Method

___

2.1.1 Classification of observation methods

There are many ways to get data. It is an old method to get data through observation. Because of its intuitive characteristics, it is still used today. Observation method can be mainly divided into direct observation and indirect observation.

1. Direct observation method

Direct observation refers to that people observe the research objects with their own eyes, ears, hands, feet and other limbs, senses, etc. There is not any intermediary medium in the observation process. Observers (investigators) come to the investigation site to conduct relevant observation. The contents of observation include the counting, measurement and meterage of the number of respondents, then register and obtain first-hand information. The direct observation method is widely used in daily work. For example, people check inventory, observe the temperature, record the temperature of the air conditioner outlets, measure the liquid volume in the measuring cup, count the number of students in a class, set up the mileage mark of an accident on the highway, sign water level the flood reached, record the specific time of the event, and tagging the number of disinfection records.

The direct observation method can be divided into several types: One is simple observation, that is, daily observation and recording, such as corresponding observation and recording of employees' behaviors. The second is report method, which refers to the reporting of observation data from one level to another. For example, the data of lower level is reported and summarized level by level according to the statistical report. The third is interview, that is, the person who obtained the information directly interviewed the object face to face, such as telephone interview, self-service filling, etc. The fourth is correspondence. The investigator sends the questionnaire to the respondent. This method is not limited by space, but the speed of obtaining information is slow, and the recovery rate cannot be guaranteed. The fifth is digital interaction. The investigators conduct online surveys through the internet and digital interactive media, and the respondents respond voluntarily and timely. This method can be anonymous, protect privacy, save costs, and have a clear orientation. The survey can be conducted in a wide range. Due to the use of multimedia, the transmission rate is fast. The survey results are objective, and the quality of information is easy to check and control.

When using the direct observation method, there should be a clear purpose and detailed plan, which is the premise to ensure the success of observation. The observer shall specify the specific content to be observed. A good observation plan should be made. It includes the observed object, time, sequence, process, recording method, recording form, etc., so as to ensure the quality and efficiency of observation. The interference to the observation object and observation environment shall be minimized, and the subjective tendency shall be avoided when recording. If multiple people observe at the same time, the method and content of observation shall be unified in advance to ensure the consistency of records. Repeated observation objects can be observed for many times to avoid accidental errors and improve the reliability of observation results.

The advantage of the direct observation method is that it does not control or interfere with the observed objects, and the observation and investigation results are authentic, objective, and timely. The results obtained are vivid and rich in content, which can ensure the accuracy of the collected data. In the process of observation, the observer can understand the background of the objective object and the context of its operation. The essence of a phenomenon described and inferred is relatively accurate. In the specific implementation of observation, observers can play their own initiative, adjust the keynotes of observation in time according to the purpose and requirements of the investigation, and capture the details needed.

The disadvantage of the direct observation method is that it takes a lot of manpower, material resources and time. It cannot collect historical data. The observation process will be limited by the observer's own condition. For example, the scope of vision of the observer is limited, the movement of the body is limited, the hearing is limited, the ability to catch the instantaneous events is limited, and the hands are insufficient, etc. As a result, the observation is not detailed and accurate enough, and mistakes and omissions will occur. Sometimes it will be limited by time. In particular, some things happen in a specific period of time, which has already passed when the observer arrives. If the object of study is a person, when the person knows that he or she is being observed, there may be a reflection that affects his or her normal behavior and affects the authenticity and objectivity of the observation results. Sometimes, the subjective consciousness and preconceived consciousness of the observer will affect the observation results.

2. Indirect observation method

The indirect observation method refers to that the observer (investigator) observes and records the observed objects with the help of instruments and equipment, records what is happening through the observation of the objects, traces and understands what has happened. It is also a kind of physical observation. Sometimes, the investigators observe the investigated objects as bystanders, taking the surrounding objects as indirect observation objects. There are numerous examples of indirect observation by investigators with the help of instruments and equipment, such as temperature measurement, satellite positioning, etc. Investigators also have many cases of indirect observation with the aid of physical objects, such as the street garbage survey conducted by Charles Bahrain in Chicago, which analyzed and obtained the consumption materials of local residents through the observation and recording of residents' garbage.

The types of indirect observation methods mainly include the following items: The method with the help of instruments and equipment can obtain a lot of observation results which can not be achieved manually, and greatly expand people's ability to obtain information. The indirect object observation method is used to obtain the data of the investigated objects by observing other objects. Create situation method is to obtain corresponding information by setting specific situations. Conversation method is to draw up an outline of the conversation by specifying the purpose and obtain the relevant information. This method is subject to more interference, such as the situation, environment, mood, etc. of the conversation. Observation of traces and behavior signs is to obtain effective information through the traces left by the observed objects, behavior signs set in various behaviors, etc. Satellite remote sensing is now a method with high speed, high accuracy and high discrimination.

When using the indirect observation method, the purpose of observation should be clarified and a detailed observation plan should be formulated. The observer shall specify the specific content to be observed and the indirect observed object. The observation plan prepared includes the object, time, process, instruments and equipment used, recording method and form of indirect observation. If multiple people conduct indirect observation at the same time, the method and content of indirect observation shall be unified to ensure consistent records.

The advantage of indirect observation is that the time and space of observation are not limited. Especially when the time of the observed object is fleeting or the observed object lasts for a long time, it can be recorded truly. The observers will not be able to complete the observation due to the limitations of their own physiological conditions, such as limited vision, restricted height, limited angle, limited temperature, etc. These problems will not disturb the observation process. If the research target is a person, the target will not perceive that he is being observed, so the observation behavior is real and reliable. The observers will not be affected by the intuitious subjective consciousness. The observation result is objective and true.

The disadvantage of indirect observation is that the investigators have no personal experience, and it is difficult to accurately grasp the data related to psychological activities. The observation result may be affected by the limited instruments and equipment. Sometimes it will be affected by improper operation of instruments and equipment, resulting in distortion of observation results. When observing behavior, the data collected cannot guarantee its validity. The traces and marks of the observed object obtained through indirect observation cannot really reflect the behavior and phenomenon that the investigators want to study, and it is difficult to judge whether these traces and marks are universal.

2.1.2 Other classifications of observation

1. Qualitative observation and quantitative observation

To describe the change degree of the research object, it can be divided into qualitative observation and quantitative observation.

Qualitative observation mainly refers to observing the nature and characteristics of the research object, and describes the object with language and text. Qualitative observation only requires an understanding of the composition and performance characteristics of the research object, such as the appearance, color, status, location, environment, etc. The observation conclusion is not only a direct description of the characteristics, but also remains unchanged or changed.

Quantitative observation is to measure the change range and process of the research object with specific values. Quantitative observation often requires the help of instruments and equipment, so as to have more accurate records. The data accuracy of quantitative observation is related to the instruments and equipment used in observation.

2. Laboratory observation and field observation

According to the different observation places and whether the observation objects are controlled in the process, it can be divided into laboratory observation and field observation.

Laboratory observation refers to the observation activities under laboratory conditions. The observers have clear test purposes and set corresponding experimental conditions and objects before observation, determine various variables, including independent variable, dependent variable, irrelevant variable, etc., and obtain corresponding data through experiments.

Field observation refers to the observation method that the respondents are in a natural state. The observers should have clear observation purpose, make the observe plan, use appropriate observation tools (such as human visual and auditory organs, cameras, video cameras, tape recorders, detectors), design the corresponding observe tables, then go to the field of the object for observation. According to the role of the observer, it can be divided into participative observation and non participative observation. According to the contents and requirements of observation, it can be divided into structural observation and unstructured observation.

3. Participative observation and non participative observation

According to the role of the observer, that is, whether the observer participates in the activities related to the observed, it can be divided into participative observation and non participative observation. Strictly speaking, the direct observation includes participative observation and non participative observation, but the indirect observation has non participative observation only.

Participative observation refers to that the observer directly participates in the activities of the observed person and conducts internal observation in the process of activities jointly conducted by everyone. Participative observation is an important method of investigation and study, which originated from "field work method". It is suitable for the research of social sciences, especially the investigation and research related to human activities. The data recorded by participative observation can go deep into the interior of the research object, but sometimes it may interfere with the research results.

Non participative observation refers to the fact that the observer does not participate in the observed activity. The observer observes as an outsider, maintaining objectivity and independence in the process of observation. Non participative observation will not cause interference to the research object, and the reliability of the observation results is relatively high.

4. Structural observation and unstructured observation

According to the contents and requirements of observation, that is, whether there are uniformly designed and structured observation contents and requirements in the observation process, it can be divided into structural observation and unstructured observation.

Structural observation means that the observer determines the contents of observation in advance, including the scope of observation items, observation list, samples and sampling methods, indicators, period, etc., designs a table for recording, and accurately classifies, codes, and records the observation data. The forms of observation are as follows: Event sampling observation, that is, according to the purpose of investigation and research, select representative samples from the behavior of the observation object for observation. Time sampling observation is to observe the selected objects in a specific time period, and record the frequency and duration of occurrence. Grade scale observation refers to the observation of the relevant behaviors and results of the respondents by the method of grade scale, which includes digital scale, graphic scale, descriptive scale and other types.

Unstructured observation is an observation type without specific design requirements in advance. Observers have a general observation purpose and requirements, and do not specify the scope and content of observation items in detail. During the specific implementation, the observer can adjust the observation angle and content according to the current environment and condition changes. The method is flexible and the implementation is simple.

5. Time sampling observation and event sampling observation

According to whether the observation time or the observation event is set before the observation, it can be divided into time sampling observation and event sampling observation.

Time sampling observation refers to the observation of the survey object in a preset time period. It can observe and record whether a phenomenon will occur, the frequency of occurrence, the duration, etc. Before implementation, the purpose, object, frequency and time of observation shall be clarified. It is applicable to objects with high frequency.

Event sampling observation refers to setting a "target event" in advance and observing it under the content of predetermined behavior. Before implementation, the event to be observed should be defined, and then implemented in a certain place. Causal analysis can be conducted on the behavior and the current situation. Because the information recorded is related to "events", more qualitative analysis is used.

2.2 调查法

——

2.2.1 调查法的内涵

调查法是指调查者通过对于客观对象的调查过程，直接获得相关调查资料，然后对于调查资料进行分析并得出结论的一种数据获得方法。调查法在日常工作中比较常用，适用于描述客观事物、当前问题的分析、解释客观规律或现象、过去发生事情的研究、探索未知对象、不同事物的比较研究等情形。

通过调查，调查者可以获得相关主题的第一手资料和数据，解释客观事物的本质。调查者可以了解现状，发现问题，找到症结，提出解决问题的路径；可以发现好的做法与经验，以便今后的推广应用。通过调查，应用者可以提出解决问题的新思路，找到新方法，提出新理论等。

2.2.2 调查法的实施

调查法的实施比较灵活，不受时间和空间的限制。在具体执行时，可以针对客观对象进行全数调查（普查），也可以进行抽样调查。调查的方法有：问卷调查、访谈调查、电话调查、网络调查、个案调查、群体调查等。在开展调查工作时，先要进行调查设计，明确调查目的和调查对象。实施调查的周期可以根据工作的需要而进行调整。

1. 问卷调查

调查者开展问卷调查时需要事先设计调查所用的问卷，采用书面提问的方式，通过问卷的收集，得出定量和定性的分析结论。在具体实施时，要明确调查的主题，围绕主题来设计相应的问题。为了便于问卷调查的开展，不宜设计需要大篇幅回答的问题、不设计太笼统的问题。在问题的设计上，有选择题、是非题、简答题，在考察量度上可以采用计分法、等级排列法、比较法等。

2. 访谈调查

调查人员采用直接交谈的方式与被调查者进行沟通，从而获得相应的调查数据。访谈方法使用简单、针对性强，可以根据不同的情形灵活设定访谈的环境，有利于了解多重因素影响的各种问题。访谈的形式可以是个别访谈，也可以是群体访谈（如座谈会）；可以是正式访谈，也可以是非正式访谈。在访谈开始时，调查者必须说明访谈目的和基本要求，根据设计的问题逐一提问，注意倾听。必要时进行录音、摄像。访谈结束后对访谈材料进行整理、统计，然后得出相应结论，供后续管理使用。

3. 电话调查

调查人员通过电话的方式向被调查者进行问询，以获得相应的调查数据。电话调查适用于服务后的质量问询、产品的试用体验等。不适用于大量问题的调查。由于被调查者处于不同的工作状态，有时电话调查会被拒绝。

4. 网络调查

调查人员应用网络、APP 等软件等实时获得调查数据。网络调查不受时间和空间的限制，问题的形式可以多样化。调查结束后，对于数据的统计快速、便捷，容易得出调查结论。

5. 个案调查

个案调查是针对某一个特定的对象而开展的调查，特定的对象包括个体、单位、某个现象、某个主题等。调查者围绕特定对象，进行广泛的数据收集，了解被调查对象的产生、发展过程，分析各种内、外部因素及相关关系，对特定的对象形成全面的认识。在调查的方式上，可以采用观察、测验、文件分析、心里投射、全面检查等。个案调查可以针对某一个人，也可以针对某一个团队。

6. 群体调查

群体调查是调查者围绕一个主题对某个群体所开展的调查，可以通过调查获得该群体的某些特性或显著性的数据，在分析研究后，对该群体做出相应的判断，提出相应的后续管理措施。

群体的对象可以根据调查目的而设定，比如消费者群体、建筑施工人员群体、乘客群体等等。

2.2.3 调查设计

为了更好地实施调查，调查者应事先进行相关的调查设计，包括确定调查的时间、地点、对象、调查过程与方法、对调查资料进行分析的方法设想等。具体调查设计的内容如下：

一是明确调查目的并予以具体化。明确调查目的是整个调查环节中最核心的内容，如果没有确切的调查目的，调查将成为空洞的形式。调查者在确定调查目的时，要搞清楚为什么开展调查，对于调查预期要达成的目标及结果作预判，如调查将会解决什么问题，问题将会解决到什么程度；本次调查是为了理论的研究，还是为了解决一个具体的问题，或者是对某一个现象做出预测；是需要了解总体参数还是研究不同对象之间的相关联系；本次调研将在管理活动中起到什么样的作用。调查目的同时构成了一次调查活动的内在尺度，在调查目的的基础上，才可以展开具体的调查设计。在明确了调查目的之后，整个调查活动过程组成了一个有机的整体，每个调查阶段、调查步骤等相互连接、逻辑性缜密。在确定调查目的后，调查者要把调查目的的内容进行具体化，形成相应的调查指标，这些指标要经过慎重筛选，要体现客观性、精确性、灵敏度、量化等特征。

二是明确调查的时间与地点。调查的时间主要指一次完整调查活动的起讫时间，包括调查工作的筹备时间、调查目的及工作指标的确定时间、调查过程时间安排、调查过程的机动时间、调查结论及调查报告的完成时间等。调查地点是实施调查过程的具体地址，在抽样调查时，要确保获取样本的地点同样具有随机性质。

三是明确调查对象。调查对象又称调查总体，即需要进行调查的总体范围。调查者根据调查目的来确定调查对象，就是要确定调查总体及其同质的调查范围，明确总体的边界，防止调查过程中有重复或遗漏。调查观察的对象可以是人员、物体、群体、地区、项目等。

四是明确调查方法。调查者根据一次调查的目的和要求确定并选择调查方法，调查方法是确保调查成功的关键工作。调查方法有很多，从调查的形式上区分，有问卷调查、访谈调查、电话调查、网络调查等（在 2.2.2 中有详述）。 调查方法从样本数量来区分，有全数调查（普查），它是对总体中所有的观察单位进行调查，一般用于了解总体在某一特定"时点"上的情况；有抽样调查，它是通过随机抽样的方法，从总体中随机抽取一定数量具有代表性的观察单位，从而组成调查样本进行调查，然后根据样本信息来推断总体特征；有典型调查（案例调查），它是在对事物进行全面了解的基础上，有目的地选择典型的人和单位进行调查。

五是确定数据的收集与记录方式。数据的收集与记录方式是确定在调查过程中把相关数据予以收集与记录的具体载体，主要有：纸质文本、电脑、手机、网络、照片、有关仪器设备（包括摄像录音设备、记录仪）等。

六是确定调查项目和调查表。调查者根据调查目的，确定预期的分析指标和项目，并按照一定的逻辑关系制成相关的表格，即形成调查表。调查表通常包含以下一些内容：标题，说明本次调查的主体，要简洁明了，如果是问卷形式则要增强答题者的兴趣和责任；调查说明，简单交代本次调查的目的和意义，调查过程中的注意事项，标注调查者的名称、日期等信息；填写说明，讲清楚填写具体表式的要求；调查主体，它是本次调查主题的具体化，比如问卷调查，其问题和答案都是问卷的主体，从形式上看，主体可以分开放式和封闭式，从内容上看，主体可以分为事实、意见、判断、假设、敏感等形式；调查项目，一般包括背景资料、研究项目等内容；核查项目，主要指在调查过程、调查结束后可以复核的内容，它与调查目的无关，主要有调查者姓名、调查日期、复核结果、未能开展调查原因等；编码，当调查的数据量较多时，往往要借助计算机进行后续统计分析，因此相关的资料应予以编码。

如果是问卷式的调查，则调查表中问题的形式可分为开放式和封闭式两种。开放式问题是对答案不加限制，由调查对象针对问题进行自由回答。封闭式问题是根据问题的具体内容，提出两个或多个固定答案供调查对象从中选择。调查者在进行问题设计的时候要注意，尽量避免使用专业术语，要避免概念中的混淆现象，要回避双重标准或双重答案的问题，不能涉及诱导性或强制性的问题，所有的问题要适合全部调查对象进行回答，要符合逻辑。针对专业技术特别强的问题，要采用专门的方法进行调查。

七是制定调查的组织计划。组织计划包括组织架构和相关人员、负责人确定、调查的时间进度、调查工作的分工与联系、对参与调查人员的培训（让相关人员熟悉调查内容，明确调查态度，及时解决调查中突发问题等）、经费预算的确定（包括调查资料的印刷、相关物品的采购、通讯费、餐饮、资料收集过程中的费用）等。

八是制定调查数据的整理与分析计划。主要包括调查数据的核查、分组的设计、调查数据的录入（包括输入计算机等）、拟定数据整理表和数据分析表、对数据的归纳汇总、适宜的统计方法的选用等。

2.2.4 调查获取外部数据和内部数据

1. 外部数据的获取

直接获得外部数据的途径主要有：

（1）从政府相关部门、统计专业部门公布的有关资料，比如从各种统计年鉴等获得相关数据。

（2）从各类咨询机构、信息中心、专业调查机构等获得相关数据。

（3）从各类专业期刊、报纸、书籍等获得相关数据。

（4）从各类图书馆、科技情报中心获得相关数据。

（5）从行业协会、专业论坛等获得相关数据。

（6）从互联网获得相关数据。

（7）从国内外同行处获得相关数据。

2. 内部数据的获取

直接获得内部数据的途径主要有：

（1）各类组织内部的各种业务资料、生产经营活动的记录等相关数据。

（2）组织内部的所有统计报表。

2.2 Survey Method

2.2.1 Connotation of survey method

Survey method refers to a data acquisition method that the investigators directly obtain relevant investigation data through the investigation process of objective objects. The survey method is commonly used in daily work, and is applicable to describing objective things, analyzing current problems, explaining objective laws or phenomena, studying past events, exploring unknown objects, and comparing different things, etc.

Through investigation, investigators can obtain first-hand information and data on relevant topics and explain the nature of objective things. Investigators can understand the current situation, find problems, find the crux, and propose solutions to problems. They can find good practices and experiences for future promotion and application. Through investigation, users can put forward new ideas to solve problems, find new methods, and put forward new theories.

2.2.2 Implementation of survey method

The implementation of survey method is flexible. It will not be restricted by time and space. In specific implementation, we can conduct a full survey (census) or a sampling survey for objective objects. The methods of investigation are questionnaire investigation, interview investigation, telephone investigation, network investigation, case investigation, group investigation, etc. When carrying out the investigation, the investigation design shall be carried out first to clarify the purpose and object of the investigation. The investigation period can be adjusted according to the needs of the work.

1. Questionnaire investigation

When carrying out the questionnaire survey, the investigators need to design the questionnaire for the survey in advance, and use written questions to make quantitative and qualitative analysis conclusions through the collection of questionnaires. In specific implementation, it is necessary to define the theme of the survey and design corresponding questions around the theme. In order to facilitate the implementation of the questionnaire, it is not appropriate to design questions that need to be answered in large length and not to design too general questions. In the design of questions, there are multiple choice questions, right and wrong questions, and short answer questions. In terms of measurement, scoring method, ranking method, comparison method, etc. can be used.

2. Interview investigation

The investigators communicated with the respondents in a direct way, so as to get the corresponding survey data. The interview method is simple to use and highly targeted. The interview environment can be flexibly set according to different situations, which is conducive to understanding various problems affected by multiple factors. The forms of interview can be individual interview or group interview (such as symposium); It can be a formal interview or an informal interview. When the interview begins, the investigators must explain the purpose and basic requirements of the interview, ask questions one by one according to the designed questions, and pay attention to listening. If necessary, recording and video recording can be carried out. After the interview, the interview materials are collated and counted, and then corresponding conclusions are drawn for subsequent management.

3. Telephone investigation

Investigators asked the respondents by telephone to obtain the corresponding survey data. It is suitable for post service quality inquiry, product trial experience, etc. It is not applicable to the investigation of a large number of problems. Because the respondents are in different working status, sometimes the telephone survey will be rejected.

4. Network investigation

Investigators use the internet, App and other software to obtain survey data in real time. Network survey is not limited by time and space, and the forms of questions can be diversified. After the survey, the acquisition of the data is fast, convenient and easy to draw conclusions.

5. Case investigation

Case investigation refers to the investigation of a specific object, including an individual, a unit, a phenomenon, a theme, etc. Investigators collect extensive data around specific objects, understand the generation and development process of the investigated objects, analyze various internal and external factors and related relationships, and form a comprehensive understanding of specific objects. In the way of investigation, observation, test, document analysis, psychological projection and comprehensive examination can be used. Case investigation can be directed against a certain person or a certain team.

6. Group investigation

Group investigation is a survey conducted by investigators for a certain group around a theme. Some characteristics or significant data of the group can be obtained through the survey. After analysis and research, the group can make corresponding judgments and propose corresponding follow-up management measures. The objects of the group can be set according to the purpose of the survey, such as consumer group, construction personnel group, passenger group, etc.

2.2.3 Investigation design

In order to better implement the survey, investigators should make relevant survey designs in advance, including determining the time, place, object, process and method of the survey, and the method of analyzing the survey data. The specific investigation design is as follows.

First, clarify the purpose of the investigation and specify it. Clarifying the purpose of the investigation is the core content of the whole investigation process. Without an exact purpose, the investigation becomes an empty form. When determining the purpose of the investigation, investigators should make clear why the investigation is carried out, predict the expected objectives and results of the investigation, what problems will be solved by the investigation, and to what extent the problems will be solved; This survey is for theoretical research, or to solve a specific problem, or to predict a phenomenon; Whether it is necessary to understand the overall parameters or study the correlation between different objects; What role will this survey play in management activities. At the same time, the investigation purpose constitutes the internal scale of an investigation activity. Only on the basis of the investigation purpose can specific investigation design be carried out. After defining the purpose of the investigation, the whole process of investigation activities forms an organic whole, with each investigation stage and step connected and logical. After determining the purpose of the investigation, the investigators should specify the content of the purpose to form corresponding investigation indicators, which should be carefully screened to reflect the characteristics of objectivity, accuracy, sensitivity, quantification, etc.

Second, clarify the time and place of the investigation. The time of investigation mainly refers to the starting and ending time of a complete investigation, including the preparation time of the investigation, the determination time of the investigation purpose and work indicators, the time arrangement of the investigation process, the flexible time of the investigation process, the completion time of the investigation conclusion and the investigation report, etc. The place of investigation is the specific address where the investigation process is carried out. During sampling investigation, it is necessary to ensure that the place where samples are obtained is also random.

The third is to clarify the respondents. The survey object is also called the survey population, that is, the overall scope of the survey. Investigators determine the survey objects according to the purpose of the survey, which is to determine the population and its homogeneous survey scope, define the boundary of population, and prevent duplication or omission in the survey process. The objects of investigation and observation can be people, objects, groups, regions, projects, etc.

The fourth is to clarify the investigation method. Investigators determine and select investigation methods according to the purpose and requirements of an investigation. Investigation methods are the key work to ensure the success of the investigation. There are many survey methods, which are distinguished from the form of investigation, including questionnaire investigation, interview investigation, telephone investigation, network investigation, etc. (detailed in 2.2.2). The survey method is distinguished from the number of samples, including full survey (census), which is used to investigate all observation units in the population and generally understand the situation of the population at a specific "time point"; Sampling survey, which is to randomly select a certain number of representative observation units from the population through the method of random sampling, so as to form a survey sample for investigation, and then infer the overall characteristics according to the sample information; Typical investigation (case investigation), which is based on a comprehensive understanding of things and purposefully selects typical people and units for investigation.

The fifth is to determine the way of data collection and recording. The method of data collection and recording is to determine the specific carrier for collecting and recording relevant data during the investigation, mainly including paper text, computers, mobile phones, networks, photos, relevant instruments and equipment (including video recording equipment, recorders), etc.

Sixth, determine the investigation items and questionnaires. According to the purpose of the survey, the investigators determine the expected analysis indicators and items, and make relevant tables according to a certain logical relationship, that is, a questionnaire. The questionnaire usually contains the following contents: The title, which describes the main body of the survey, should be concise and clear. If it is in the form of a questionnaire, it should enhance the interest and responsibility of the respondents; Description of the survey, briefly explaining the purpose and significance of the survey, precautions during the survey, and marking the name, date and other information of the investigators; The instructions for filling in the form, the requirements for filling in specific forms shall be clarified; the subject of the survey, it is the embodiment of the subject of the survey. For example, in the questionnaire survey, the questions and answers are the subject of the questionnaire. From the perspective of form, the subject can be divided into open type and closed type. From the perspective of content, the subject can be divided into facts, opinions, judgments, assumptions, sensitivities and other forms; Investigation items, generally including background information, research items, etc.; Verification items mainly refer to the contents that can be reviewed during and after the investigation. They have nothing to do with the purpose of the investigation. They mainly include the name of the investigator, the date of the investigation, the results of the review, and the reasons for the failure to carry out the investigation; Coding. When there is a large amount of survey data, it is often necessary to use computers for subsequent statistical analysis, so relevant data should be coded.

If it is a questionnaire survey, the forms of questions in the questionnaire can be divided into two types: open type and closed type. Open type questions are open-ended answers that are freely answered by respondents. Closed questions are two or more fixed answers for respondents to choose from according to the specific content of the question. When designing questions, investigators should pay attention to avoiding the use of professional terms as much as possible, avoiding confusion in concepts, avoiding double standard or double answer questions, and not involving inductive or mandatory questions. All questions should be suitable for all respondents to answer, and should be logical. Special methods should be adopted to investigate problems with strong expertise.

The seventh is to formulate an organizational plan for the investigation. The organizational plan includes the determination of the organizational structure and relevant personnel, the person in charge, the time schedule of the investigation, the division and contact of the investigation work, the training of the personnel involved in the investigation (to make the relevant personnel familiar with the investigation content, clarify the attitude of the investigation, and solve the unexpected problems in the investigation in a timely manner, etc.), the determination of the budget (including the printing of the investigation materials, the purchase of related articles, communication expenses, catering expenses, and costs in the process of data collection), etc.

Eighth, the arrangement of survey data and the formulation of analysis plan. It mainly includes the verification of survey data, the design of grouping, the entry of survey data (including computer input), the preparation of data sorting tables and data analysis tables, the summary of data, and the selection of appropriate statistical methods.

2.2.4 Obtain external and internal data through surveys

1. Acquisition of external data

The main ways to directly obtain external data are as follows.

First, relevant information published from relevant government departments and statistical professional departments, such as various statistical yearbooks.

The second is to obtain relevant data from various consulting agencies, information centers, professional investigation institutions, etc.

Third, obtain relevant data from various professional journals, newspapers, books, etc.

Fourth, obtain relevant data from various libraries and science and technology information centers.

The fifth is to obtain relevant data from industry associations, professional forums, etc.

Sixth, obtain relevant data from the internet.

The seventh is to obtain relevant data from domestic and foreign counterparts.

2. Acquisition of internal data

The main ways to directly obtain internal data are as follows.

First, various business data, production and operation records and other relevant data within various organizations.

Second, all statistics reports within the organization.

2.3　测量法

——

2.3.1 测量的涵义

1. 测量的定义

测量是人类认识和揭示自然界物质运动的规律，用来定性区别和定量描述客观事物的一种重要手段。物理学家开尔文（Kelvins）说：测量就是认知。1963 年，美国标准局（NBS）的数理专家埃森哈特（Eisenhart）指出，测量是赋值给具体事物，以表示它们之间关于特定特性的关系。这个赋值的过程就是测量过程，事物被赋予的值则称为测量值。也可以这么理解：测量是按照特定的规则把数字、符号分配给目标、人物、状态、事件，并把它们的特性予以量化的过程，是一个分配数字的过程。

随着人类社会和科学技术的发展，人类认识自然的能力得到进一步提升，测量的对象从物理量拓展到化学量、工程量、生物量、电子信息量等，测量范围不断扩大。测量的方法也在不断延伸，有静态测量、动态测量、在线测量、综合测量、特殊环境下的特殊测量等。

2. 与测量有关的一些术语

测量仪器，指任何用来获得测量结果的装置，目的是获得被测对象的某些属性值。常见的测量仪器有：量具、长度测量仪、角度测量仪等。现在电子测量仪器已经遍及各行业。

测量系统，指用来对被测特性进行定量测量或定性评价的全套仪器、设备、量具、标准、操作、方法、夹具、软件、人员、环境和假设的集合，是用来获得测量结果的整个过程。测量系统中各要素对于测量结果的影响可能是独立的，也可能是相互影响的。

测量装备，是指固定安装好的测量系统。

基准量具，是指用来校对或调整计量器具，或作为标准尺寸进行相对测量的量具，如量块、标准线纹尺等。

通用计量器具，是指将被测量转换成可直接观测的指示值或等效信息的测量工具，有游标类量具（如游标卡尺等）、微动螺旋类量具（如内外径千分尺等）、机械类量具（如杠杆齿轮比较仪等）、光学类量具（如投影仪、激光干涉仪等）、电动量具（如容栅侧位仪等）、气动量具（如浮标式气动量仪等）、微机化量具（如电脑表面粗糙度测量仪等）。

极限量规，是指一种没有刻度（刻度线）的、用于检验被测量是否处于给定的极限偏差之内的专用检验工具，如位置量规等。

检验夹具，是一种专用检验工具，与相应的计量器具配套使用时，可方便地检验出被测件的各项参数。

分度值，是指计量器具刻度尺或刻度盘上相邻两个刻线所代表的量值之差。

刻度间距，是指量仪刻度尺或刻度盘上相邻刻线中心的距离。

示值范围，是指计量器具所指示或显示的最低值到最高值的范围。

测量范围，是指在允许误差限度内，计量器具所能测量的最低值到最高值的范围。

灵敏度，是指计量器具示值装置对被测量变化的反映能力。灵敏度 = 刻度间距/分度值。

灵敏限，是指能引起计量器具示值可觉察变化的、被测量的最小变化值。

测量力，是指在测量过程中，计量器具与被测表面之间的接触力。在接触测量中，希望有一定的恒定测量力，否则会使示值不稳定。但测量力太大会导致零件变形。

示值误差，是指计量器具示值与被测量真值之间的差值。

示值变动量，是指在测量条件不变的前提下，对同一被测量进行多测重复测量读数时，其读数的最大变动量。

回程误差，是指在相同测量条件下，对同一被测量进行往返两个方向测量时，测量仪的示值变化量。

修正值，是指为了清除或减少计量器具的系统误差，用代数法加到测量结果上的值。

测量系统分析，是指对于测量系统输出变差的分析，它主要用于判断一个测量系统是否可接受。在质量管理上，变差越小越好。测量观察到的总变差等于零件间的变差与测量系统误差之和，其中零件间的变差是指零件间客观存在的真实差异，由零件本身决定。测量系统误差是指由测量系统能力决定的测量偏差。测量系统误差等于测量系统的精确度与准确度之和。

基准值，是指认为规定的、可接受的属性值。

真值，是指客观对象的实际值。

偏倚，是指测量的平均观测值与基准值之间的差异。

线性，是指在整个测量操作范围内的偏倚值的差值。

分辨力，是指测量或测量仪器输出的最小刻度单位，又称分辨率、可读性。

精确度，是指测量变差的波动范围，它表示测量结果中随机误差大小的程度。精确度等于重复性与再现性之和。

准确度，是指测量变差离真值或参考值的差异，它表示测量结果中系统误差大小的程度。准确度等于偏倚、稳定性、线性三者之和。

稳定性，是指测量仪器保持其计量特性随时间恒定的能力。

重复性，是指由同一位评价人多次使用一种测量仪器，测量同一个对象的同一特性，从而获得的测量值的变差。

再现性，是指由不同的评价人使用同一种测量仪器，测量同一个对象的同一特性，从而获得的测量平均值的变差。

变差，是指测量仪器在上行程和下行程的测量过程中，同一被测量变量所得到结果之间的偏差。变差可以由测量仪器本身所引起，也可能由一些随机因素、试验条件变化而引起。变差是多次测量结果的变异程度，可理解为数据的波动，通常用标准差来表示。

误差，是指测量结果偏离真值的程度。误差可分为系统误差和随机误差。

精度，简单理解是指测量结果与真值接近的程度，是与误差相对的概念。精度是评价测量误差大小的量，与误差大小相对应，误差大则精度低，误差小则精度高。由于存在系统误差和随机误差，因此，精度是准确度与精确度的综合。

2.3.2 测量四要素

完成一次测量工作，有四个要素需要明确：

（1）明确测量的客体，也就是测量对象。在确定对象的基础上，要明确测量对象的特性、被测参数的定义、相关的标准要求等。

（2）明确计量单位。必须使用法定计量单位。比如在长度计量中，单位为米（m），其他常用单位有毫米（mm）和微米（μm）等；在角度测量中以度、分、秒为单位。

（3）明确测量方法。测量方法是测量时所采用的测量器具、测量原理、测量条件等的总和。一般根据被测量参数的特点，如公差值、大小、轻重、材质、数量等，分析研究该参数与其他参数的关系，最后确定对该参数如何进行测量的操作方法。

（4）明确测量的准确度。就是明确测量结果与真值的一致程度。由于任何测量过程都会出现测量误差，因此，必须明确测量的准确度。由于存在测量误差，测量结果以近似值来表示。

2.3.3 测量的原则与程序

1. 工程上测量的原则

在工程的测量原则，也适用于作为其他测量的参考，主要包括：

（1）测量布局上，遵循"从整体到局部"，即任何局部的测量工作要服从全局的测量需要。

（2）测量精度上，遵循"从高级到低级"，即先布设高精度的控制点，再逐步发展布设低一级的交会点，并进行碎部测量。

（3）测量次序上，遵循"先控制后碎部"，即先选择一些有控制意义的点（控制点），把它们的平面位置、高程等精确地测定出来，再根据这些控制点测定出附近碎部点地位置，以减少误差地积累。

（4）测量校核上，遵循"未校核不往下"，即上一个测量工作未做校核，不能进入下一步的测量工作。

（5）测量检查上，遵循"随时检查，杜绝错误"，不能把错误延伸到下一个步骤。

2. 机械上测量的原则

机械上测量工作遵循的测量原则，可以作为其他测量的参考，主要有：

（1）阿贝原则，又称布线原则、串联原则，是长度计量中的重要原则。它是指如果要使测量仪器得到准确结果，必须把仪器的标尺安装在被测件测量中心线的延长线上。

（2）最小变形原则。它是指为了使测量结果准确可靠，在测量中要做到使测量链中硬件部分各环节所引起的变形最小。其中的变形包括：测量力引起的接触变形、自重变形、热变形。

（3）最短测量链原则。它是指整个测量仪器由测量链组成，测量链中每个构成部分在制造和装配中都会存在误差，因此，测量链越短，仪器的误差越小。

（4）封闭原则。它是指封闭性连锁测量要遵循闭合原则，即最后的累计误差为0。因为圆分度具有封闭特性，在测量中满足封闭条件，则其间隔误差的总和为0。

（5）基准统一原则。对于加工零件上尽可能多的表面，要采用同一组的基准定位。这样可以简化工艺规程的制定工作，减少夹具设计与制造的工作量和成本，缩短生产准备周期。通过减少基准转换，保证加工表面相互位置的精度。

3. 测量的程序

测量程序是根据一种或多种测量原理、给定的测量方法，在测量模型和获得测量结果所需要计算的基础上，所开展测量的全部过程。下面简要介绍工程测量的程序。

工程测量的程序主要有：

（1）测量的准备工作，主要有：与建设单位办理控制点移交，交接引测点位，编制控制点记录表。根据设计意图、工程特点等确定控制轴线。配置测量人员和测量仪器。

（2）进行工程测量放线，主要有：平面测量控制网的建立和校核，包括首级及二级测量控制网的建立与校核。

（3）进行建筑物高程的控制，建立高程控制网，确定标高测量方法。

（4）测量精度的控制，包括主轴线测距精度、建筑物竖向垂直度、标高控制网闭合差等。

（5）设立工程测量最重要预控点，包括建筑物总体垂直度控制、模板标高控制、电梯井施工测量控制等。

（6）竣工测量及变形观测。

（7）确定其他相关注意事项。

4. 测量的分类

（1）直接测量与间接测量。这是按照是否直接测量被测参数进行的分类。直接测量是指在测量时，直接从测量仪器上读出被测几何量的大小数值。间接测量是指被测几何量无法直接测量得到时，先测出与被测几何量相关的其他几何量，再通过一定的数学关系式进行计算，从而求得被测几何量的数值。

（2）绝对测量与相对测量。这是按照测量器具的读数值是否直接表示被测尺寸进行的分类。绝对测量也称为全值测量，是指测量器具的读数值直接表示被测的尺寸。相对测量也称为比较测量，是指测量器具的读数值表示被测尺寸相对于标准量的微差值或偏差。

（3）接触测量与不接触测量。这是按照被测物的表面是否与测量器具的测量头有机械接触进行的分类。接触测量是指测量器具的测量头与被测物的表面以机械测量力接触。不接触测量是指测量器具的测量头与被测物表面不接触，不存在机械测量力。

（4）单项测量与综合测量。这是按照同时测量参数的多少进行的分类。单项测量是指单独测量物体的每一个参数。综合测量是指测量物体两个及两个以上相关参数的综合效应、综合指标。

（5）被动测量与主动测量。这是按照测量在加工过程中所起的作用不同进行的分类。被动测量是指零件加工后进行测量。主动测量是指在零件加工过程中进行测量。

（6）静态测量与动态测量。这是按照被测零件或测量头在测量过程中所处的状态进行的分类。静态测量是指测量时，被测物表面与测量头处于相对静止状态。动态测量是指测量时，被测物表面与测量头处于工作（或模拟）过程中的相对运动状态。

2.3　Measurement Method

———

2.3.1 Connotation of measurement

1. Definition of measurement

Measurement is an important means for human beings to understand and reveal the laws of material movement in nature and to qualitatively distinguish and quantitatively describe objective things. Physicist Kelvin said: Measurement is cognition. In 1963, Eisenhart, a mathematical expert of the National Bureau of Standards (NBS), pointed out that measurement is the assignment of values to specific things to express the relationship between them with respect to specific characteristics. This assignment process is a measurement process, and the value assigned to a thing is called a measurement value. It can also be understood that measurement is a process of allocating numbers and symbols to targets, people, status and events according to specific rules, and quantifying their characteristics. It is a process of allocating numbers. With the development of the human society, science and technology, the ability of human beings to understand nature has been further improved. The objects of measurement have been expanded from physical quantities to chemical quantities, engineering quantities, biomass, electronic information, etc., and the measurement scope has been expanding. The measurement methods are also extended, including static measurement, dynamic measurement, online measurement, comprehensive measurement, special measurement under special environment, etc.

2. Some terminology related to measurement

Measuring instrument refers to any device used to obtain measurement results in order to obtain some attribute values of the measured object. Common measuring instruments include measuring tools, length measuring instruments, angle measuring instruments, etc. Now electronic measuring instruments have been widely used in various industries.

Measurement system refers to the collection of a complete set of instruments, equipment, measuring tools, standards, operations, methods, fixtures, software, personnel, environment and assumptions used for quantitative measurement or qualitative evaluation of the measured characteristics. It is the whole process used to obtain measurement results. The influence of each element in the measurement system on the measurement results may be independent or mutual.

Measuring equipment refers to the fixed and installed measurement system.

Benchmark measuring tools refer to measuring tools used to check or adjust measuring instruments, or used as standard dimensions for relative measurement, such as measuring blocks, standard linear rulers, etc.

General measuring instruments refer to the measuring tools that convert the measured values into directly observable indication values or equivalent information, including vernier measuring tools (such as vernier calipers, etc.), inching screw measuring tools (such as internal and external micrometers, etc.), mechanical measuring tools (such as lever gear comparators, etc.), optical measuring tools (such as projectors, laser interferometers, etc.), electric measuring tools (such as capacitive grating side aligners, etc.), pneumatic measuring tools (such as buoy type air momentum meter, etc.), computerized measuring tools (such as computer surface roughness meter, etc.).

Limit gauge refers to a special inspection tool without scale (scale line), such as position gauge, used to check whether the measured value is within the given limit deviation.

Inspection fixture is a special inspection tool. When it is used together with the corresponding measuring instruments, it can easily inspect the parameters of the tested piece.

Graduation value refers to the difference between the measurement values represented by two adjacent scribes on the scale or dial of the measuring instrument.

Scale spacing refers to the distance between the centers of adjacent scribed lines on the gauge scale or dial.

The indication range refers to the range from the lowest value to the highest value indicated or displayed by the measuring instrument.

The measuring range refers to the range from the lowest value to the highest value that can be measured by measuring instruments within the allowable error limit.

Sensitivity refers to the ability of the indicating device of the measuring instrument to reflect the measured changes. Sensitivity is equal to scale spacing divided by division value.

The sensitivity limit refers to the minimum change value measured that can cause perceptible changes in the indication of measuring instruments.

The measuring force refers to the contact force between the measuring instrument and the measured surface during the measurement process. In contact measurement, it is desirable to have a certain constant measuring force, otherwise the indication will be unstable. But too much measuring force will cause part deformation.

The indication error refers to the difference between the indicated value of the measuring instrument and the measured true value.

The indication variation refers to the maximum variation of the reading when multiple measurements and repeated readings are made for the same measured object under the same measurement conditions.

Return error refers to the indicated value change of the measuring instrument when the same measured object is measured in two directions under the same measurement conditions.

Correction value refers to the value added to the measurement results by algebraic method to eliminate or reduce the systematic error of measuring instruments.

Measurement system analysis refers to the analysis of the output variation of the measurement system, which is mainly used to judge whether a measurement system is acceptable. In terms of quality management, the smaller the difference, the better. The total variation observed in the measurement is equal to the sum of the variation between parts and the measurement system error. The variation between parts refers to the real difference objectively existing between parts, which is determined by the parts themselves. Measurement system error refers to the measurement deviation determined by the capability of the measurement system. The measurement system error is equal to the sum of the accuracy and precision of the measurement system.

Reference value refers to the specified and acceptable attribute value.

The true value refers to the actual value of the objective object.

Bias refers to the difference between the measured average observation value and the reference value.

Linearity refers to the difference of bias value within the whole measurement operation range.

Resolving power refers to the smallest scale unit of measurement or measuring instrument output, also known as resolution and readability.

Accuracy refers to the fluctuation range of measurement variation, which indicates the degree of random error in measurement results. Accuracy is equal to the sum of repeatability and reproducibility.

Precision refers to the difference between the measurement variation and the true value or reference value, which indicates the degree of systematic error in the measurement results. Accuracy is equal to the sum of bias, stability and linearity.

Stability refers to the ability of measuring instruments to keep their metrological characteristics constant over time.

Repeatability refers to the variation of the measured value obtained by the same evaluator using a measuring instrument for many times to measure the same characteristics of the same object.

Reproducibility refers to the variation of the measured average value obtained by different evaluators using the same measuring instrument to measure the same characteristics of the same object.

Variation refers to the deviation between the results of the same measured variable during the measurement of up stroke and down stroke of the measuring instrument. The variation may be caused by the measuring instrument itself, or by some random factors and changes in test conditions. Variation is the variability degree of multiple measurement results, which can be understood as the fluctuation of data, and is usually expressed by standard deviation.

Error refers to the degree to which the measurement result deviates from the true value. Error can be divided into systematic error and random error.

Degree of accuracy, which is simply understood as the degree to which the measurement result is close to the true value, is a concept relative to error. Degree of accuracy is the quantity to evaluate the size of measurement error, which corresponds to the size of error. The larger the error, the lower the accuracy, and the smaller the error, the higher the accuracy. Due to the existence of systematic error and random error, degree of accuracy is a combination of accuracy and precision.

2.3.2 Four elements of measurement

To complete a survey, four elements need to be clarified.

(1)Specify the object of measurement. On the basis of determining the object, the characteristics of the object to be measured, the definition of the parameter to be measured, and the relevant standard requirements should be clarified.

(2) Specify the unit of measurement. The legal unit of measurement must be used. For example, in length measurement, the unit is meter (m), and other commonly used units are millimeter (mm) and micrometer (μm), etc. In angle measurement, it is measured in degrees, minutes and seconds.

(3) Specify the measurement method. The measuring method is the sum of measuring instruments, measuring principles, measuring conditions, etc. used in measurement. Generally, according to the characteristics of the measured parameter, such as tolerance value, size, weight, material, quantity, etc., analyze and study the relationship between this parameter and other parameters, and finally determine how to measure this parameter.

(4) Determine the accuracy of measurement. It is to determine the consistency between the measurement result and the true value. Since any measurement process will lead to measurement errors, the accuracy of measurement must be clear. Because of the measurement error, the measurement results are expressed by approximate values.

2.3.3 Principles and procedures of measurement

1. Principles of engineering measurement

The principle of engineering measurement is also applicable to other measurements for reference, mainly including the following items.

(1) In terms of measurement layout, follow the principle of "from the whole to the part", that is, any local measurement work should be subject to the whole situation measurement needs.

(2) In terms of measurement accuracy, follow the principle of "from high to low", that is, first lay out high-precision control points, then gradually develop and lay lower level intersection points, and carry out detailed measurement.

(3) In the measurement sequence, follow the principle of "control before detail", that is, select some points with control significance (control points), accurately measure their plane position, elevation, etc., and then measure the location of nearby detail points according to these control points to reduce the accumulation of errors.

(4) In terms of measurement verification, the principle of "do not go down without verification" is followed, that is, the previous measurement work has not been verified, and the next measurement work cannot be started.

(5) In terms of measurement inspection, the principle of "check at any time to eliminate errors" should be followed, and errors should not be extended to the next step.

2. Principles of mechanical measurement

The measurement principles followed by mechanical measurement can be used as reference for other measurements, mainly including the following items.

(1) Abbe's principle, also known as wiring arrangement principle and series principle, is an important principle in length measurement. It means that if the measuring instrument is to get accurate results, the scale of the instrument must be installed on the extension line of the measuring center line of the measured piece.

（2）Principle of minimum deformation. It means that in order to make the measurement results accurate and reliable, the deformation caused by each segment of the hardware in the measurement chain should be minimized. The deformation includes contact deformation caused by measuring force, self weight deformation and thermal deformation.

（3）The principle of the shortest measuring chain. It means that the whole measuring instrument is composed of a measuring chain, and each component of the measuring chain will have errors in manufacturing and assembly. Therefore, the shorter the measuring chain, the smaller the instrument error.

（4）Closure principle. It means that the closed interlocking measurement should follow the closed principle, that is, the final cumulative error is 0. Because the circular division has a closed characteristic, if the closed condition is met in the measurement, the sum of its interval errors is 0.

（5）The principle of datum unification. For as many surfaces as possible on the machined parts, the same group of datum positioning shall be adopted. This can simplify the process of planning, reduce the workload and cost of fixture design and manufacturing, and shorten the production preparation cycle. By reducing datum conversion, the precision of mutual position of machined surfaces is ensured.

3. Procedure of measurement

The measurement procedure is the whole process of measurement based on one or more measurement principles, given measurement methods, measurement models and calculations required to obtain measurement results. The procedure of engineering survey is briefly introduced below.

The procedures of engineering survey mainly include the following parts.

First, preparation for survey, mainly including: handing over control points with the construction unit, handing over survey points, and preparing control point record table. Determine the control axis according to the design intent and engineering characteristics. Allocate measuring personnel and measuring instruments.

The second is to carry out engineering survey, mainly including the establishment and verification of the horizontal surveying control network, including the establishment and verification of the primary and secondary surveying control networks.

The third is to control the building elevation, establish the elevation control network and determine the elevation measurement method.

The fourth is the control of measurement accuracy, including the ranging accuracy of the main axis, the vertical verticality of the building, the closure error of the elevation control network, etc.

Fifthly, the most important pre-control points for engineering survey shall be set, including the control of overall perpendicularity of buildings, formwork elevation, and elevator shaft construction survey.

Sixth, completion survey and deformation observation.

The seventh is to determine other relevant precautions.

4. Classification of measurement

(1) Direct measurement and indirect measurement. This is classified according to whether the measured parameters are directly measured. Direct measurement refers to reading the size value of the measured geometric quantity directly from the measuring instrument during measurement. Indirect measurement means that when the measured geometric quantity cannot be directly measured, other geometric quantities related to the measured geometric quantity shall be measured first, and then calculated through a certain mathematical formula to obtain the value of the measured geometric quantity.

(2) Absolute measurement and relative measurement. This is the classification according to whether the reading value of the measuring instrument directly represents the measured size. Absolute measurement is also called full value measurement, which means that the reading value of the measuring instrument directly represents the measured size. Relative measurement, also known as comparative measurement, refers to the reading value of the measuring instrument indicating the difference or deviation between the measured dimension and the standard quantity.

(3) Contact measurement and non-contact measurement. This is classified according to whether the surface of the object to be measured has mechanical contact with the measuring head of the measuring instrument. Contact measurement means that the measuring head of the measuring instrument contacts the surface of the object to be measured with mechanical measuring force. Non-contact measurement means that the measuring head of the measuring instrument does not contact the surface of the object to be measured, and there is no mechanical measuring force.

(4) Single measurement and comprehensive measurement. This is classified according to how many parameters are measured at the same time. Single measurement refers to measuring each parameter of an object separately. Comprehensive measurement refers to measuring the comprehensive effect and comprehensive index of two or more related parameters of an object.

(5) Passive measurement and active measurement. This is classified according to the role of measurement in the processing. Passive measurement refers to the measurement after part processing. Active measurement refers to measurement during part processing.

(6) Static measurement and dynamic measurement. This is classified according to the state of the measured part or measuring head in the measurement process. Static measurement means that the measured object surface and the measuring head are in a relatively static state during measurement. Dynamic measurement refers to the relative motion between the measured object surface and the measuring head during the working (or simulation) process.

2.4　抽样法

────

2.4.1 抽样的概念

抽样就是从总体中随机抽取出样品而组成样本的活动过程。抽样的基本要求是要保证所抽取的样品对总体具有充分的代表性。抽样可以分为随机抽样和非随机抽样两种。随机抽样，就是要使总体中的每一个个体（产品）都有同等的机会被抽取出来，而组成样本的活动过程。非随机抽样是调查者根据主观分析来抽取样本的方法，大数定律在非随机抽样上不适用，因此不能说明样本的统计值在多大的程度上适合于总体。

2.4.2 抽样量的确定

抽样问题主要分两大类，一类是用于检验判断的抽样，比如用于判断产品是否合格的离散型问题抽样，以及用于判断产品是否达到某项数值型指标（如均值）的连续型问题抽样，以上两种抽样问题称为简单抽样问题。还有一类是用于参数估计的抽样，比如针对正态分布、泊松分布等情形，对它们的均值、标准差等参数进行估计，直至样本量达到一定的程度才能使估计量达到给定的精度要求，这种抽样问题称为简单估计问题。

实践中所遇到的数据往往比较复杂，比如有的是可靠性指标，抽样后得到的是删失数据；有的数据并不服从正态分布，而是服从威布尔分布、指数分布、对数分布等寿命分布。因此，针对这一类抽样问题，称为可靠性抽样验收问题和可靠性参数估计问题。

在管理活动实践中，很多人对于抽样量取多少是心中没底的，一会儿取个很大的样本数量，一会儿取个较少的样本数量。而适当的样本数量将是管理活动质量的重要保证。通常抽样可以根据国家标准（如 GB 2828）来制定精细的抽样方案，也可以应用统计软件（如 Minitab）来确定抽样量。

2.4.3 随机抽样的方法

随机抽样有多种方法，主要有简单随机抽样法、系统抽样法、分层抽样法、整群抽样法等。

简单随机抽样法是从总体的全部个体中随机抽取样品，总体中的每个个体被抽取出来的概率相等。简单随机抽样法对总体不作任何处理，适用于总体个数较少的情形。

系统抽样法又叫等距抽样法，它是把总体分成均衡的几个部分，然后按照事先设定的规则，每隔一定的间隔抽取一个或若干个样本，组成所需要的样本进行检测。系统抽样法操作简便，适用于总体个数比较多的情形，在生产实际中人们经常使用它。例如在手机的生产流水线上，每隔一定时间就抽取一部手机进行检测，这就是系统抽样的一个实例。在实践中，系统抽样法还可以分为随机起点等距抽样、半距起点等距抽样和对称等距抽样等方法。

分层抽样法，是先把总体按照不同的性质分成互不交叉的层，再按照规定的比例从不同层中随机抽取样本的方法。分层抽样法适用于总体数量比较多、而且总体内部有比较明显的层次可分。分层抽样法的特点是样本的代表性比较好，抽样误差较小；但它对于如何分层有比较高的要求。如果分层不清楚，则抽取样本的代表性就差。比如，甲、乙、丙三人分别在不同的时间段用同一台机器生产同一种产品，三个人加工的产品分别堆放在三个不同的地方，现在要对他们加工的产品进行检测，要求总共抽取其中 30 件产品组成样本。现在采用分层抽样法，抽取者从三个堆放产品的地方分别随机抽取 10 件产品，总共 30 件产品组成样本加以检测，这样的样本代表性较好。假设在生产过程中，三个人所生产的产品混淆在一起，而且没有混放均匀，从中随机抽取 30 件产品组成样本进行检测，其结果的代表性就差。

整群抽样法又叫集团抽样法、聚类抽样法。它是把总体分成许多群，每个群由个体按一定的方式结合而成，群与群之间互不交叉、互不重复，然后随机抽取若干群，并由这些群中的所有个体组成样本。比如某集团公司在质量管理活动中，以群（工厂、公司、项目、班组、工序或一段时间内施工的项目等）为单位进行抽样，凡是抽到的群体要全面进行检查。现在集团在质量检查中，在下属 5 个子公司中抽取一家公司进行检查，该公司所有的产品线都构成了样本。这种抽样方法的特点是抽样实施方便、节约费用，但由于样本只来自于个别群体，因此要求群体有较好的代表性，即群体之间的差异要小。

2.4.4 非随机抽样的方法

非随机抽样的方法主要有：便利抽样法、判断抽样法、配额抽样法、滚雪球抽样法、图像抽样法、空间抽样法等。

便利抽样法，又称偶遇抽样法、随意抽样法，它是调查者在某个特定的事件、特定的位置上，遇到某个对象即作为样本所采取的方法。它的抽样准确性相对不高。

判断抽样法，又称立意抽样法，调查者根据本次研究的目标和自己的主观分析，来选择和确定调查对象的方法。这种抽样方法主观性比较强，客观性不足。

配额抽样法，又称定额抽样法。调查者对总体进行分层，分层的依据是对研究对象有影响的各种因素，找出具有不同特征的因素在总体中所占的比例，把这种分层划分及各种不同因素的比例组成配额，然后根据配额选择符合要求的样本进行分析。

滚雪球抽样法。调查者对于总体的情况不是很了解，就先从总体中的少数人群样本入手，向他们询问符合条件的其他人员样本，接着根据符合条件的人员样本信息，继续询问扩大符合条件的人群样本，如此像滚雪球似地不断扩大样本范围，完成抽样分析。

图像抽样法。在整个图像区域选取一些特定的区域位置，取出该特定位置的图像作为样本进行分析，通常是分析图像的亮度值、色度值。

空间抽样法。它是指调查者把非静止的、暂时性的、空间相邻的群体作为样本的抽样方法。比如参加一次城市马拉松比赛的选手，参赛人员从一个地方到另一个地方，有些人是参加全程比赛，有些人参加半程比赛，还有的人参加 10 千米比赛等，人员不断在发生着变化，但全程整体的比赛路线、范围是确定的。调查者可以从某一个起始点开始，先调查离他最近的选手，然后每隔 500 米再调查其他选手，由此完成抽样分析。

2.4 Sampling Method

2.4.1 Concept of sampling

Sampling is the process of randomly taking samples from the population to form a sample. The basic requirement of sampling is to ensure that the samples taken are fully representative of the population. Sampling can be divided into random sampling and non-random sampling. Random sampling is the activity process to make every individual (product) in the population have the same opportunity to be selected and form a sample. Non-random sampling is a method for investigators to sample according to subjective analysis. The law of large numbers is not applicable to non-random sampling, so it cannot be explained to what extent the statistical value of the sample is suitable for the population.

2.4.2 Confirmation of sample size

Sampling problems mainly fall into two categories: one is sampling for inspection and judgment, such as discrete problem sampling for judging whether a product is qualified, and the other is continuous problem sampling for judging whether a product reaches a certain numerical index (such as the mean value). The above two sampling problems are called simple sampling problems. The other is sampling for parameter estimation. For example, in the case of normal distribution and Poisson distribution, the mean value, standard deviation and other parameters are estimated. Only when the sample size reaches a certain level can the estimator meet the given accuracy requirements. This sampling problem is called simple estimation problem.

The data encountered in practice are often complex. For example, some data are reliability indicators, and the censored data are obtained after sampling. Some data do not obey normal distribution, but obey life distribution, such as Weibull distribution, exponential distribution, logarithmic distribution, etc. Therefore, this kind of sampling problem is called reliability sampling acceptance problem and reliability parameter estimation problem.

In the practice of management activities, many people have no idea how much to sample. Sometimes they take a large sample size and sometimes a small sample size. The appropriate sample size will be an important guarantee for the quality of management activities. Generally, fine sampling plans can be formulated according to national standards (such as GB/T 2828). Statistical software, such as Minitab, can be used to determine the sampling size.

2.4.3 Methods of random sampling

There are many methods for random sampling, including simple random sampling, systematic sampling, stratified sampling, cluster sampling, etc.

Simple random sampling method is to randomly select samples from all individuals of the population, and the probability of each individual in the population being selected is equal. The simple random sampling method does not deal with the population, and is suitable for the case of small population.

Systematic sampling, also known as equidistant sampling, divides the whole population into several balanced parts, and then draws out one or several samples every certain interval according to the preset rules to form the required samples for testing. The systematic sampling method is easy to operate and suitable for the situation with a large number of population. It is often used in production practice. For example, on the production line of mobile phones, one mobile phone is sampled every certain time for testing, which is an example of system sampling. In practice, systematic sampling can also be divided into random starting point equidistant sampling, half distance starting point equidistant sampling and symmetric equidistant sampling.

Stratified sampling is a method that divides the population into different layers according to different properties, and then randomly samples from different layers according to the specified proportion. The stratified sampling method is applicable to a large number of population with obvious levels within the population. The characteristics of stratified sampling method are that the sample is representative and the sampling error is small. But it has high requirements on how to layer. If the layering is not clear, the representativeness of the samples will be poor. For example, there are three people, Party A, Party B and Party C, who use the same machine to produce the same product in different periods of time. The products processed by the three people are stacked in three different places. Now we need to test the products processed by them. It is required to sample a total of 30 products. Now, stratified sampling method is adopted. The sampler randomly selects 10 products from three places where products are stacked. A total of 30 products are sampled for testing. Such samples are representative. Suppose that the products produced by three people are mixed together in the production process, and they are not evenly mixed, and 30 products are randomly selected from them to form a sample for testing, and the representativeness of the results is poor.

Cluster sampling is also known as group sampling. It divides the population into several groups. The individuals in each group should be combined in a certain way. Groups do not intersect or duplicate each other, then select several groups randomly, and all the individuals in these groups will be composed as a sample. For example, in the quality management activities of a group company, the group (factory, company, project, team, process or project constructed in a period of time, etc.) is taken as the unit for sampling, and all groups taken shall be comprehensively inspected. At present, the group managers select one of its five subsidiaries for quality inspection, and all its product lines constitute a sample. This sampling method is characterized by convenient sampling and cost saving. However, since the samples only come from individual groups, the groups are required to be quite good representative, that is, the differences between groups should be small.

2.4.4 Method of non-random sampling

Non-random sampling methods mainly include convenient sampling, judgment sampling, quota sampling, snowball sampling, image sampling, spatial sampling, etc.

Convenient sampling, also known as accidental sampling and random sampling, is a method that investigators take as samples when they meet a certain object at a specific event or location. Its sampling accuracy is relatively low.

Judgment sampling method, also known as purposive sampling method, is used by investigators to select and determine the method of survey objects according to the objectives of this study and their own subjective analysis. This sampling method is highly subjective and not objective enough.

Quota sampling method refers to that investigators layer the whole population and the basis of layering is to find out the proportion of factors with different characteristics in the whole population based on various factors that have influence on the research object. The quota is composed of this hierarchical division and the proportion of various factors, and then select samples that meet the requirements according to the quota for analysis.

Snowball sampling. Investigators do not know much about the overall situation, so they start with the sample of a small number of people in the population, ask them about the sample of other qualified people, and then continue to ask and expand the sample of qualified people according to the sample information of qualified people, so as to continue to expand the sample range like a snowball and complete the sampling analysis.

Image sampling method. Select some specific regional locations in the whole image area, and take the map image of this specific location as a sample for analysis, usually to analyze the brightness value and chromaticity value of the image.

Spatial sampling method. It refers to the sampling method in which the investigators take non-static, temporary and spatially adjacent groups as samples. For example, the participants in an urban marathon race move from one place to another. Some take part in the whole race, some take part in the half race, and some take part in the 10 km race. The personnel are constantly changing, but the overall competition route and scope are determined. Investigators can start from a certain starting point, first investigate the contestants closest to them, and then investigate other contestants every 500 meters to complete sampling analysis.

2.5　试验法

———

2.5.1 试验法的涵义

　　试验法是试验者根据其试验目的，利用科学仪器、试验设备及工具等，有意识地控制或改变试验条件、试验环境等，对所关心的对象进行研究的方法。试验法适用于在已规定标准、已规定置信水平的情况下做出评价，也适用于对两个或多个体系进行比较。在调查一个过程的多个因素的影响时，试验设计显得更为高效和经济；试验设计能够识别因素间的交互效应，提高过程的质量；试验设计还可以识别出一些偶尔影响因素，有利于准确判断。

　　试验法在各类管理活动中具有重要的作用。试验法的特点是：试验者可以根据研究的目的和任务，预先设计相应的试验方案，干预和控制研究对象，排除一些非必要因素的干扰，观察研究对象的变化；试验者可以在相同的条件下、根据相同的方式进行多次重复试验，以验证试验结果的信度和效度，使试验结果真实、可靠；试验者可以通过试验完成不同因素之间的相关关系的验证，如验证因果关系、相关关系等。但试验法也会受到一定的制约，比如试验技术只能处于当前的技术状态；试验对象存在丰富性和复杂性，而试验很可能只完成其中若干项的研究等。

2.5.2 与试验法有关的几个概念

　　因子（factor），又称因素，指可能影响试验结果，而且在试验中被考察的可控原因或其组合。在试验中，它就是被选中安排试验的因素。

　　水平（level），又称位级，指因子的一个给定值，或一种特定的措施，或一种特定的状态。

　　处理（treatment），指在试验中实施的因子水平的一个组合。

　　主效应（main effect），反映一个因子各水平的 平均响应之间差异的一种度量。一个因子第 i 个水平上的所有处理的响应之平均，与全部处理的响应平均之差，称为该因子第 i 个水平的主效应。

　　交互效应（interaction），由若干个因子之间水平的搭配而产生效应的一种度量。两个因子之间的交互效应称为"二因子交互效应"或"一级交互效应"；三个因子之间的交互效应称为"三因子交互效应"或"二级交互效应"。

　　真值，是指在某一时刻、某一状态下，某量的客观值或实际值。真值通常是未知的。在某些情况下，真值是已知的，比如平面三角形内角之和等于 180°、国际公认的计量值等。

2.5.3 试验法的作用

　　试验法的作用可以简单概括为三个方面：

　　一是简化和纯化试验步骤的作用。简化是指试验者可以在众多因子中缩小因子范围，把整个试验过程予以简单操作；纯化是指试验者可以摆脱干扰因子的影响。由于试验者可以控制整个试验的条件、试验环境等，只对所关心的因子进行试验研究，这样就摆脱了其他因子的干扰，使研究对象的某种属性得到相对纯粹的展现，这样容易得到试验结果，发现内在规律。

二是缩小样本数量的作用。在相对确保试验结果准确度的前提下，试验者可以根据试验的要求，从总体中选取样本来进行相关试验。通过样本的缩小，可以大量节约试验者的资源，包括试验时间、人财物的投入等。据称，盖洛普公司（一家专业调查公司）通过科学抽样的方式，抽取4000人所得到的结果与抽取40 000人所得到的结果基本一致。

三是实证的作用。试验者可以通过相关的科学试验，对于一些理论、假设、推断等进行实证，从而建立起科学的理论、体系、方法等，得到科学的结论。

2.5.4 试验法的分类

试验法可以根据不同的试验性质、目的、作用等进行相应的分类，主要有以下几类。

1. 定性试验与定量试验

这是根据试验过程中量与质的关系而进行的分类。

定性试验是指通过试验来确定某个因素或若干个因素是否存在、这些因素之间是否有联系、这些因素是否会对试验效果产生作用和影响等。定性试验是定量试验的基础。

定量试验是指通过试验来确定某些因素之间的内在联系，比如数量关系、因果关系、相关关系等，确定某些因素的数值，确定一个因素对其他因素的影响程度，可以深入了解试验对象的性质。

2. 探索性试验与验证性试验

这是根据试验的目的而进行的分类。

探索性试验是指试验者为了探索未知事物或现象的性质、规律等而进行的试验活动，比如新能源汽车刚出来时，它的电池续航能力究竟达到多少千米就需要进行探索性试验。

验证性试验是指试验者对于研究对象有一定的了解、形成一定的认识或已提出某种假设，现在为了验证这种认识或假设是否正确而进行的试验活动；有时结论已知，不存在假设问题，对已知的结论再做的试验也是验证性试验。比如现在知道某款新能源电池的续航达到500千米，通过验证性试验，再次验证它是否真的达到了500千米。

3. 析因试验、对照试验、空白试验、模拟试验

这是根据试验在科学认识中的作用而进行的分类。

析因试验是指通过试验来确定某一个因素或若干因素是否对现有的结果起作用，它可以去发现某个现象的原因，故名析因。

对照试验也称为比较试验，它是把两个及以上不同组别进行试验以确定某个因素的影响结果，也可以把一个需要研究的对象与一个已知结果的事物进行对比试验，以确定某个因素的影响。对照试验可以用来鉴别有无处理因素的影响效果差异，或者鉴别若干个处理因素的影响效果差异。对照试验时，不同试验组（对照组）之间在试验对象、试验条件、操作步骤等方面要尽量保持一致。

空白试验是对照试验中的特例，它是指在不增加某因子的情况下，按照与事先制定的分析方案完全一致的操作条件和步骤进行的试验，这样得到的值为空白值。然后在增加该因子的情况下

进行试验，从分析结果中扣除空白值。空白试验可以减小系统误差，特别是可以减小由背景、环境等引起的误差。

模拟试验是指针对现有条件尚不能进行实际试验，运用替代物来进行模拟的方法，在具体的模拟方法上，有数学建模后模拟、计算机模拟、物理模拟等，从而揭示试验对象的客观规律和性质。

4. 实验室试验与现场试验

这是根据试验的场所而进行的分类。

实验室试验是指试验开展的场所为一种人造的隔离环境，比如实验室。实验室试验包括生物试验、物理试验、化学试验等。

现场试验，也称为田野试验（field experiment），是指研究者不能控制某种事件，为了确定这种事件在什么样的条件下才会发生，或者在给定的条件下这种事件是否会发生、怎样发生，从而进行的一种试验活动。如现场训练试验等。

5. 预备性试验、中间试验、正式试验

这是根据试验的步骤而进行的分类。

预备性试验是指限于试验的条件，试验者先进行小规模的简单试验，根据试验结果进行一定的修正和完善，以便后续试验的展开。预备性试验通常可以保证正式试验的成功率。

中间试验是指研究者已经完成实验室的工作，但还没有进入正式的生产，需要通过一定的试验来验证技术的可行性、生产的合理性，就此而开展的试验活动。通过中间试验以后，可以对原来的设计、工艺、参数、材料等进行改进和完善，以便大规模生产和推广。有些中间试验可以通过计算机模拟来完成。

正式试验是指试验者根据试验方案，以最佳试验材料、符合要求的试验环境、试验条件下而开展的试验。正式试验在预备性试验之后，在完善了试验条件、试验材料、试验环境等基础上进行，可以是大规模的试验。

6. 单因子试验、多因子试验、正交试验、拉丁方试验

这是根据试验过程中因子的数量而进行的分类。

单因子试验，就是假设在管理活动中只存在一种影响因素，或只考虑对目标影响最大的因子而其他因素保持不变，通过设计出相应的试验方法来求得最优解，解决这个影响因子以达到管理目标。

多因子试验，就是在管理活动中存在多种因子的影响，通过设计出相应的试验方法来求得影响结果，从中获得最优解。

正交试验，是多因子试验设计中的简易方法，它的原理是针对预设的目标情况，利用正交表安排尽量少的试验次数，经济、快捷地求得满意解。正交表是一种把研究的因子、因子间的交互作用等进行了合理安排、且试验次数最少的一种表式。

拉丁方试验，是指在 n 个因子影响下，选择 n 个试验对象，进行 n 次试验，以检验不同因素的影响结果，它是多因子试验的特例。

2.5.5 提高试验结果准确度的方法

1. 提高试验结果准确度的方法

提高试验结果准确度的方法主要有：增加平行试验的次数，以减少随机误差；选择适当的方法，如选用灵敏度恰当的仪器等；开展对比试验，可以检查系统误差，比如用标准样品来检验新方法的准确度、安排不同的人员同时进行试验（内检）、试样送其他实验室分析（外检）；另外还有空白试验、校正仪器、应用控制图、校正分析结果等。

2. 评价试验结果的准确度

每一种新的试验方法都可以评价其准确度（正确度与精密度）。国家标准中有关于测试方法准确度的规定，如 GB/T6379 系列《测量方法与结果的准确度》。通常对于要求高或非标准试验方法准确度的评价，需要做 10 次以上的试验；对于一般的试验方法准确度的评价，需要做 5 次以上的试验。评价试验准确度通常有两种办法：

（1）TPI 评定法

TPI 是测试性能指标（Test Performance Index），它是根据美国 ASTMCS94 制定，其评价公式为

$$\text{TPI} = S/Sr$$

其中，S 是试验数据的标准差，Sr 是试验方法的准确度，即重复性标准差，$Sr = r/2.8$，r 是试验方法的重复性限（重复性限是指在重复性条件下，两个测试结果的绝对差小于或等于此数的概率为 95%）。判断时，$\text{TPI} \leqslant 1.33$，说明实验室准确度符合统计要求；$\text{TPI} \leqslant 0.7$，说明实验室准确度统计评估好于试验方法的准确度；$0.7 < \text{TPI} < 1.0$，说明实验室准确度接近试验方法的准确度；$\text{TPI} \geqslant 1.33$，说明试验室准确度没有达到试验方法准确度规定的最低限度。

（2）实验室重复性标准差 SD 和试验变异系数 CV 值评定法

该方法是美国分析化学家协会（AOAC）常用的一种方法。同样，试验的重复性标准差：

$$Sr = r/2.8$$

试验方法的相对标准差：

$$\text{RSD} = Sr / \bar{x} \times 100\%$$

评定时，按照试验方法进行 10 次测试，计算出试验的标准差 S，看 S 是否接近 Sr，接近者为符合规定。

2.5　Test Method

___

2.5.1 Connotation of test method

Test method is a method that the experimenter uses scientific instruments, test equipment and tools to consciously control or change the test conditions, test environment, etc. according to the purpose of the test to study the object of interest. The test method is applicable to the e-valuation under the condition of specified standards and confidence levels, and also to the comparison of two or more systems. When investigating the influence of multiple factors in a process, the experimental design is more efficient and economical; The experiment design can identify the interaction between factors and improve the quality of the process; The test design can also identify some occasional influencing factors, which is conducive to accurate judgment.

Test method plays an important role in various management activities. The characteristics of the experimental method are: according to the research purpose and task, the experimenter can design the corresponding experimental scheme in advance, intervene and control the re-search object, eliminate the interference of some unnecessary factors, and observe the change of the research object. Investigators can conduct multiple repeated tests under the same condi-tions and in the same way to verify the reliability and validity of the test results and make the test results true and reliable. Investigators can verify the correlation between different factors through experiments, such as verifying causality and correlation. However, the test method will also be subject to certain restrictions. For example, the test technology can only be in the current tech-nical state; The test object is rich and complex, and the test is likely to only complete some of the studies.

2.5.2 Several concepts related to test method

Factor refers to the controllable causes or their combinations that may affect the test results and are investigated in the test. In the experiment, it is the factor selected to arrange the experi-ment.

Level, also known as bit level, refers to a given value of a factor, or a specific measure, or a specific state.

Treatment refers to a combination of factor levels implemented in the experiment.

Main effect is a measure that reflects the difference between the average responses of each level of a factor. The difference between the average response of all treatments at the ith level of a factor and the average response of all treatments is called the main effect of the ith level of the factor.

Interaction is a measure of the effect produced by the horizontal collocation of several factors. The interaction effect between two factors is called "two factor interaction effect" or "primary interaction effect"; The interaction effect between the three factors is called "three factor interaction effect" or "secondary interaction effect".

The true value refers to the objective value or actual value of a certain quantity at a certain time and under a certain state. The truth value is usually unknown. In some cases, the true value is known. For example, the sum of the internal angles of a plane triangle equals 180°, the internationally recognized measurement value, etc.

2.5.3 Function of test method

The function of test method can be summarized in three aspects.

The first is to simplify and purify the test steps. Simplification means that the experimenter can narrow the range of factors among many factors and simply operate the whole test process; Purification means that the experimenter can get rid of the influence of interference factors. Because the experimenter can control the entire test conditions, test environment, etc., and only carry out experimental research on the factors concerned, this will get rid of the interference of other factors, so that a certain attribute of the research object can be displayed relatively purely, so that it is easy to get the test results and find the internal laws.

The second is to reduce the number of samples. On the premise of relatively ensuring the accuracy of the test results, the tester can select samples from the population for relevant tests according to the requirements of the test. By reducing the sample size, we can save a lot of resources of the experimenter, including test time, human resources and property investment, etc. It is said that Gallup, a professional survey company, through scientific sampling, obtained the results of 4000 people, which are basically consistent with the results of 40 000 people.

The third is the role of demonstration. The experimenter can demonstrate some theories, assumptions, inferences, etc. through relevant scientific experiments, so as to establish scientific theories, systems, methods, etc., and get scientific conclusions.

2.5.4 Classification of test method

Test methods can be classified according to different test properties, purposes, functions, etc., mainly including the following categories.

1. Qualitative test and quantitative test

This is classified according to the relationship between quantity and quality during the test.

Qualitative test refers to the test to determine whether a certain factor or several factors exist, whether these factors are related, and whether these factors will affect the test effect. Qualitative test is the basis of quantitative test.

Quantitative test refers to determining the internal relationship between some factors through experiments, such as quantitative relationship, causality, correlation, etc., determining the value of some factors, and determining the degree of influence of one factor on other factors, so as to deeply understand the nature of the test object.

2. Exploratory test and confirmatory test

This is classified according to the purpose of the test.

Exploratory test refers to the test activities carried out by the tester to explore the nature and laws of unknown things or phenomena. For example, when a new energy vehicle is just coming out, its battery endurance should be explored.

Confirmatory test refers to the test activity conducted by the experimenter to verify whether the cognition or hypothesis is correct after the experimenter has a certain understanding of the research object, has formed a certain understanding or has put forward a certain hypothesis. Sometimes the conclusion is known and there is no hypothesis problem. The test on the known conclusion is also a confirmatory test. For example, we now know that a certain new energy battery can travel 500km, and through the validation test, we will again verify whether it really has reached 500km.

3. Factorial test, control test, blank test, simulation test

This classification is based on the role of experiments in scientific understanding.

Factorial test refers to the test to determine whether a certain factor or several factors play a role in the existing results. It can find the cause of a phenomenon, so it is called factorial test.

The control test is also called comparative test. It refers to the test of two or more different groups to determine the effect of a factor, or the comparative test of an object to be studied and a thing with known results to determine the effect of a factor. The control test can be used to identify whether there are differences in the effects of treatment factors, or to identify differences in the effects of several treatment factors. During the control test, different test groups (control groups) should be consistent in terms of test objects, test conditions, operation steps, etc.

Blank test is a special case in the control test. It refers to the test conducted under the operating conditions and steps that are completely consistent with the previously developed analysis scheme without adding a factor. The value thus obtained is a blank value. Then the test is carried out with the factor increased, and the blank value is deducted from the analysis result. The blank test can reduce the system error, especially the error caused by background and environment.

Simulation test refers to the method of using substitutes to simulate when the existing conditions cannot be tested actually. In terms of specific simulation methods, there are mathematical modeling simulation, computer simulation, physical simulation, etc., so as to reveal the objective laws and properties of the test object.

4. Laboratory test and field test

This is classified according to the test site.

Laboratory test refers to the place where the test is carried out in an artificial isolation environment, such as a laboratory. Laboratory tests include biological tests, physical tests, chemical tests, etc.

Field test, also known as field experiment, refers to an experimental activity conducted by researchers who cannot control an event to determine under what conditions it will occur, or whether and how it will occur under given conditions, such as on-site training test.

5. Preparatory test, intermediate test and formal test

This is the classification according to the steps of the test.

Preparatory test refers to the limited conditions of the test. The experimenter first conducts a small-scale simple test, and then makes certain corrections and improvements according to the test results, so as to facilitate the subsequent test. The preparatory test can usually guarantee the success rate of the formal test.

Intermediate test refers to the test activities carried out by researchers who have completed the work in the laboratory but have not yet entered into formal production, and need to pass certain tests to verify the feasibility of technology and the rationality of production. After the intermediate test, the original design, process, parameters, materials, etc. can be improved and perfected for large-scale production and promotion. Some intermediate tests can be completed by computer simulation.

Formal test refers to the test conducted by the tester under the best test materials, qualified test environment and test conditions according to the test plan. The formal test can be a large-scale test after the preparatory test and on the basis of improving the test conditions, test materials, test environment, etc.

6. Single factor test, multi factor test, orthogonal test, Latin square test

This classification is based on the number of factors in the test process.

Single factor test is to assume that there is only one influencing factor in the management activities, or only consider the factor that has the greatest impact on the objectives while other factors remain unchanged, and to find the optimal solution by designing the corresponding test method to solve this influencing factor to achieve the management objectives.

Multi factor test means that there are many factors influencing the management activities, and the influence results can be obtained by designing corresponding experimental methods, from which the optimal solution can be obtained.

Orthogonal test is a simple method in multi factor test. Its principle is to use orthogonal table to arrange as few times as possible to obtain satisfactory solution economically and quickly according to the preset target situation. Orthogonal table is a kind of table that arranges the factors and interactions between factors reasonably, and has the least number of experiments.

Latin square test is a special case of multi factor test, which refers to selecting n test objects and conducting n tests under the influence of n factors to test the effects of different factors.

2.5.5 Accuracy of test result

1. Methods to improve the accuracy of test results

It mainly includes: increasing the number of parallel tests to reduce random errors; Select appropriate methods, such as selecting instruments with appropriate sensitivity; The comparative test can check the systematic error, such as using standard samples to check the accuracy of the new method, arranging different personnel to conduct the test at the same time (internal inspection), and sending samples to other laboratories for analysis (external inspection); In addition, there are blank tests, calibration instruments, application control charts, calibration analysis results, etc.

2. Methods to evaluate the accuracy of test results

Each new test method can evaluate its accuracy (precision). The national standards have provisions on the accuracy of test methods, such as GB/T 6379 Series Accuracy of Measurement Methods and Results. Generally, more than 10 tests are required to evaluate the accuracy of high or non-standard test methods; For the evaluation of the accuracy of general test methods, more than 5 tests are required. There are usually two methods to evaluate the test accuracy:

(1) TPI evaluation method

TPI refers to Test Performance Index. It is formulated according to ASTM CS94 of the United States, and its evaluation formula is

$$TPI = S/Sr$$

In the formula, S is the standard deviation of the test data and Sr is the accuracy of the test method, i.e. repeatability standard deviation, $Sr = r/2.8$ while is the repeatability limit of the test method (the repeatability limit means that under the repeatability condition, the probability that the absolute difference between two test results is less than or equal to this number is 95%). When TPI $\leqslant$ 1.33, it indicates that the accuracy of the laboratory meets the statistical requirements; When TPI $\leqslant$ 0.7, it indicates that the statistical evaluation of laboratory accuracy is better than the accuracy of test methods; When 0.7 < TPI < 1.0, it explains that the laboratory accuracy is close to the accuracy of the test method; When TPI $\geqslant$ 1.33, it indicates that the accuracy of the laboratory does not reach the minimum specified by the accuracy of the test method.

(2) Evaluation method of laboratory repeatability standard deviation SD and test coefficient of variation CV

This method is commonly used by the Association of Official Analytical Chemists (AOAC). Similarly, the repeatability standard deviation of the test is

$$Sr = r/2.8$$

The relative standard deviation of test method is

$$RSD = Sr / \bar{x} \times 100\%$$

During evaluation, conduct 10 tests according to the test method, and calculate the standard deviation S of the test, then check whether S is close to Sr, and the data that is close to it is qualified.

2.6 模拟法

这里的模拟法是指通过计算机进行模拟，从而获得相应数据的方法。

2.6.1 计算机模拟的过程

计算机模拟的过程主要分三个步骤：（1）建模，（2）产生随机数，（3）计算机模拟。

1. 建立数学模型

（1）定义。数学模型是指出于某个特定的目的，运用适当的数学工具，对于某个特定对象进行必要的简化和假设，然后得到的一个数学结构，这个结构能够解释特定对象的现实性态，或者预测对象的未来，或者能提供处理对象的最优决策等。这个数学结构的表现形式为：用字母、数字、其他数学符号等建立起来的等式或不等式，以及用图表、图像、框图等来描述客观对象的特征及内在联系。

（2）建立数学模型的方法。具体建立数学模型时，通常有三种方法。第一种是分析法，又称为演绎法、理论建模、机理建模，即白箱问题，它是根据系统的工作原理，运用已知的定理、原理、前提、规则等推导出系统的数学模型。第二种是测试法，又称为归纳法、实验建模、系统辨识，即黑箱问题，它是根据系统输入输出的记录，加以必要的数据处理和数学运算，估计出系统的数学模型。第三种是把上述两种方法综合起来运用，用分析法列出系统的理论模型，用测试法确定模型中的参数。

（3）数学模型的分类。数学模型有很多分类方法，主要有：

按照模型的特征分类，有静态模型与动态模型、确定性模型与随机模型、离散模型与连续模型、线性模型与非线性模型等。

按照模型应用的数学方法分类，有几何模型、图论模型、微分方程模型、概率模型、最优控制模型、规划论模型、马氏链（马尔科夫）模型等。

按照对模型的结构所了解程度分类，有白箱模型、灰箱模型、黑箱模型等。

按照模型的应用方向分类，有经济模型、生态模型、人口模型、交通模型、环境模型、资源模型、工程模型、航空模型等。

按照建模的目的分类，有预测模型、优化模型、决策模型、控制模型等。

（4）建立数学模型的步骤

一是观察。建模前对于实际对象（问题）的背景做全面、深入、细致的观察，明确要解决的问题，按要求收集必要的数据，所收集的数据必须符合要求。

二是提出模型假设。鉴于实际对象（问题）的复杂性，需要把对象（问题）进行理想化、简单化处理，抓住主要因素，暂不考虑次要因素，厘清变量之间的关系，提出必要的假设。如果假设合理，则后续模型与实际对象比较吻合；如果假设不合理，则需要修改假设。

三是建立数学模型。在建模的过程中，要分清变量的类型，选用恰当的数学工具；要抓住问题本质，模型尽量简单化；要有严密的数学推理，确保模型的正确；建模要有足够的精度，在剔除次要因素时不能影响反应对象的真实程度。

四是模型求解。不同的模型要用不同的数学知识求解，往往要借助于计算机求解。

五是模型的分析、验证。可以用已有的数据来验证，如果用模型计算出来的理论数值与实际数值比较吻合，则模型成功；若两者相差太大，则要修改模型。

六是模型的修改。要分析假设的合理性，保留合理部分，修改不合理部分，增加对于次要因素的分析，并在新的建模中考虑曾经忽略的内容。有时还需要去掉一些变量，或者改变变量的性质，或者改变变量之间的函数关系等。

2. 产生随机数

任何一个带有随机成分的系统或过程，对它们的模拟都需要一种方法来产生或得到随机数。因此，怎样方便而有效地从所需的概率分布中产生随机值，以便用于数学模型的执行是非常重要的。现在可以借助计算机来随机产生一系列的随机数，这些随机数的出现服从一定的概率分布。现在的编程语言、仿真模拟开发工具等都提供均匀分布随机数发生器，这些随机数的产生是由程序来决定的，实质上也符合一定的规律，因此称为"伪随机数"。但这些伪随机数恰恰是应用于模拟过程中的随机数。

比较简单的一个方法是可以借助 Excel 等软件来实现随机数的发生。在 Excel 中点击"数据分析"，选择其中的"随机数发生器"，然后根据需要选择变量个数、随机数个数等，就可以得到均匀分布的随机数。

均匀分布的随机数，可以通过变换得到按要求分布的随机数。在 Excel 的随机数发生器菜单中，选择不同的分布，就可以得到服从正态分布、二项分布、泊松分布等的一些随机数。

3. 计算机模拟

在完成数学建模、随机数产生以后，接下来就是按照各种算法来进行模拟。模拟的算法有很多，比如蒙特卡洛法、图论、穷举法、数值分析法、图像处理法、线性规划等。通过模拟，得到所模拟对象的某一个特征量 A 的统计特征值，如 A 的分布类型、均值、方差、分布函数等。

由于手工进行模拟计算工作量太大，现在的模拟工作基本上都由计算机来完成。市场上的计算机模拟软件有很多，常用的模拟软件包有：Arena，ExtendSim，Simio，SIMPLE++，Cradle CFD，Matlab，Simulink 及其 mode Frontier 等。

4. 模拟的发展

建立一个有效的、可信度高的模拟模型是有难度的，现在有了计算机软件来进行模拟模型的建立，模拟的发展越来越快，在软件设计中广泛采用了面向对象的思想和方法。今后模拟主要的发展有：发展"建模方法学"，重点是利用数学模型来求解，还能够利用计算机进行新的建模，使复杂的建模过程简单化。二是"面向对象仿真模拟"，把系统视为一个个独立的对象组成，通过对象之间互相发送消息来执行仿真，其建模过程接近人的思维。三是分布交互仿真，通过电子手段把不同地点的软硬件设备、人员联系起来，在人工合成的电子环境中交互进行模拟试验，适用于局域网、异地远程网络仿真模拟。四是人工智能模拟，利用知识库进行建模与模拟，实现智能化模拟。五是虚拟现实模拟，为用户创造一个实时反映实体对象变化与相互作用的三维图形世界。

2.6.2 蒙特卡洛法

1. 蒙特卡洛法的基本介绍

蒙特卡洛法，又称蒙特卡洛模拟法、随机抽样法、统计试验法，它是众多模拟算法中的一个重要方法。在 20 世纪 40 年代中期，传统的经验方法不能逼近真实的物理过程，而蒙特卡洛法能够模拟真实的物理过程，结果也比较理想。由于该方法的概率统计特征，人们借用摩纳哥的城市蒙特卡洛来命名。严格来说，蒙特卡洛法是以概率和统计理论方法为基础的一种计算方法，用随机数来解决很多计算问题，可以使用计算机来实现统计模拟，从而获得相应的输出。当采样越多，蒙特卡洛法的输出越接近最优解。蒙特卡洛法现状得到了广泛应用，包括金融、工程项目管理、医学、量子热力学计算、空气动力学计算等。由于计算机技术的发展，蒙特卡洛法得到了快速普及和发展。

2. 蒙特卡洛法的基本原理

（1）蒙特卡洛法的基本原理是：事件的概率可以用大量试验中发生的频率来估计，当样本的容量足够大，则可以认为该事件发生的频率就是该事件的概率。在运用蒙特卡洛法时，可以先对影响某个事件可靠度的随机变量进行大量的随机抽样，然后把抽样值依次输入到相应的函数式子中，来判断该函数式子的结构是否失效，最后得到函数式子结构的失效概率。蒙塔卡洛法的特点是通过大量重复试验，通过统计频率来估计概率，从而得到求解。

（2）举例说明蒙特卡洛法的思路

假设部件 C＝零件 A＋零件 B，

A 是一个随机变量，服从正态分布 $N(\mu_A, \sigma_A^2)$，

B 是一个随机变量，服从正态分布 $N(\mu_B, \sigma_B^2)$，

根据解析法的思路，则 C 服从正态分布 $N(\mu_A + \mu_B, \sigma_A^2 + \sigma_B^2)$。

现在按照蒙特卡洛法的思路，取一组部件（A_1，B_1），可以测得该部件的尺寸数据为 $c_1 = (a_1, b_1)$；接着再取一组部件（A_2，B_2），得到 $c_2 = (a_2, b_2)$，再继续下去，一直到 $c_n = (a_n, b_n)$。于是：

根据 c_1，c_2，$\cdots$，c_n 的数据，制作直方图，可以得到部件 C 分布的大致形态；

根据 $C = \sum_{i=1}^{n} c_i/n$，可以得到 μ_c 的估计值；

根据 $S_c = \left(\sum_{i=1}^{n} (c_i - \bar{c})^2/n - 1 \right)^{1/2}$，可以得到 σ_c 的估计值；

根据（$C_i < x$ 的部件数 $/n$），可以得到分布函数 $F(X < x)$ 的估计值。

由此得到了部件 C 的分布类型、均值、方差、分布函数的估计。

2.6.3 数字孪生

1. 数字孪生的概念

数字孪生是数字时代建模与模拟应用的重要形式。2002 年美国密歇根大学格利斯夫博士提出了数字孪生的概念。它是一种多物理场、多尺度、多概率的模型，能够利用物理模型、传感器数据、历史数据等来反映与模型对应的实体的功能、实时状态、演变趋势等。现在基本认同的是：关于物理系统的数字信息构造，可以作为一个独立的实体创建，这种数字信息是嵌入在物理系统内的信息的孪生，并在整个系统生命周期内与该物理系统相连接。

2. 数字孪生的元素

包括真实的客观对象、虚拟空间、从真实空间到虚拟空间的数据流连接、从虚拟空间到真实空间、虚拟子空间的信息连接。

3. 数字孪生的作用

包括资产优化、竞争差异化、改善用户体验，提高组织效率，改善企业决策，理解设备或系统的状态，并响应各种变化，改变操作和增加价值。可应用于产品研发、工艺规划、精益制造、设备维护、智慧城市建设等。

2.6.4 数字工程战略

数字工程战略最早由美国国防部研究与工程副部长迈克尔·格里芬于 2018 年 6 月在《国防部数字工程战略》中提出。该战略指出，数字技术已经彻底改变了多数主要行业的业务以及我们的个人生活。通过提高计算速度、存储能力和处理能力，数字工程赋予了从传统的"设计、构建、测试"方法到"模型、分析、构建"方法的范式转变。这种方法可以使国防部的项目在交付之前，能够在虚拟环境中构建原型，进行实验和测试，支撑决策和确定解决方案。数字工程需要新的方法、过程和工具，改变原来工程社区的运作方式，对研究、需求、采集、测试、成本、维持和情报社区产生了影响。

数字工程的愿景是使设计、开发、交付、运作和维护系统全面现代化。在整个生命周期中，使用的模型以数字方式代表虚拟世界中感兴趣的系统，包括过程、设备、产品、零件等。数字工程将整合先进计算、大数据分析、人工智能、自主系统、机器人技术等来改进工程实践。

数字工程倡议的五大战略目标：

（1）规范模型的开发、集成与使用。实现设计模型、制造模型、审核与验证模型、系统模型、生产支撑模型、特种工程模型、管理模型之间标准化数据的无缝流通，服务于企业和项目决策。

（2）提供持久、权威的可信源（即基础数据与模型）。让访问、管理、分析、使用和分发来自一组通用数字模型和数据的信息成为可能。授权的利益相关者能在生命周期内使用当前的、权威的、一致的信息。

（3）把技术创新融入工程实践的改进。利用大数据与分析、认知技术、先进计算技术、数字与物理融合技术，以及新方法和人机协作，推动向端到端的数字企业转型。

（4）构建数字工程的支撑架构和环境。包括工具、流程、方法、软硬件、网络、安全等，实现各利益相关者之间的互动、协作与沟通。

（5）完成文化与团队的数字转型。面向对全生命周期数字工程的适应与支撑，通过沟通与接触、领导方式、战略与实施、训练与教育的持续性改进，实现整个采办部门数字工程的制度化。

2.6.5 数字时代建模与模拟的机遇与挑战

数字孪生的发展和数字工程战略的提出，意味着建模与模拟技术将迎来全新的发展，同时也面临很多挑战。建模与模拟的应用将更加普及，从实验室进一步走进工厂、办公室、校园、社区，成为生产生活越来越不可缺少的工具。

1. 机遇

（1）建模与模拟的常态化应用。物联网、大数据使建模与模拟在线化、泛在化、常态化成为可能。数字时代的典型特征是通过无所不在的传感器网络构建的物联网，使用 5G、6G 通信技术乃至更高阶的通信技术，广泛搜集现实世界的海量数据，并通过大数据分析技术来提升决策与控制能力。以数字孪生为代表，数字时代的建模与模拟可以与真实世界建立永久、实时、交互的链接，建模与模拟不再是离线的、独立的、特定阶段的存在，而是向在线化、泛在化、常态化的服务发展。

（2）建模与模拟向纵深发展。数字工程战略的提出，有助于建模与模拟向科学化、规范化、智能化发展。科学化体现在建模与模拟的科学基础更加稳健深入；规范化体现在标准统一，流程更加科学；智能化体现在新一代的人工智能和机器学习在建模与模拟中广泛应用。这些变化和发展将促进数字工程战略的全面落实。数字工程战略是对传统工程思想的一次变革，建模与模拟作为数字工程战略的核心支撑技术也必将进一步得到发展。

（3）算力的极大提升。云计算、超级计算、量子计算的发展使数字时代建模与模拟不再受到计算能力的局限，从而为数字孪生与数字工程发展提供无限可能。芯片技术和软件技术的不断发展，使超级计算机的能力不断提升。云计算的发展使计算、存储、通信能力变成一种可以按需获取、可海量增长的资源。量子计算为彻底解决建模与模拟所需的计算能力提供支持。算力的提升将进一步提高模型的精细化程度，更好满足模拟的需要。

（4）模拟的产业化发展。数字时代的模拟产业将得到极大发展。数字孪生的成功应用、数字工程战略的提出将促进建模与模拟的现实应用得到井喷式增长。建模与模拟将进入各行各业，推动社会生产力的极大提高，推进各行业高质量发展。

2. 挑战

（1）理论上的挑战。大数据和复杂系统模拟相结合的"数字孪生"系统可以兼顾"过程"与"结果"，对"过去和未来"的全面情况进行深入学习，实现更加智能的建模与模拟。数据时代的模拟的理论重心从模型转向数据，模拟的方法从抽样和发现因果关系转到整体把握和发现相关关系。模拟计算研究与大数据分析为需要更加融合发展，建立坚实的数字时代建模与模拟理论基础。

（2）可信度挑战。可信度是建模与模拟的生命力之源。数字孪生与数字工程更强调把建模与模拟的结果作为权威可信数据源，这对模型的可信度和精度都提出了更高的要求。数字孪生与数据工程是复杂性、实时性、交互性、多样性极强的模拟应用，对其进行可信度评估往往超出现有的能力，必须深入研究更加有效的评估理论与方法。

（3）管理挑战。数字孪生与数字工程都需要进一步集中管理和积极利用模型、数据、支撑环境等模拟资源，这给数字时代的建模与模拟管理带来新挑战。需要在全生命周期中为模型创建、管理、集成以及相关的项目和工程活动制定正式计划，描述在执行工作活动时如何以连贯有效的方式实现模型、支持分析和决策。在管理上需要建立一个规范的体系，确定模型开发应遵循的基本质量标准和规则。维护好模型的来源和谱系，建立可信度、准确性和判断模型重用的基础。

（4）安全挑战。数字孪生和数字工程使建模与模拟的资源、基础设施越来越开放，其受到攻击的可能性越来越高。建模与模拟工具存在的漏洞将更多暴露于开放的互联环境，使其安全性面临巨大挑战。

（5）文化挑战。数字孪生与数字工程的实施所面临的根本性挑战是变革已有的组织文化。需要各组织采用深思熟虑的系统方法来规划、实施和支持数字化转型。这种转型必须超越技术，以解决人员和文化等方面的挑战，包括组织的价值观、行为。

2.6   Simulation Method

The simulation method here refers to the simulation by computer to obtain the corresponding data.

2.6.1 The process of computer simulation

The process of computer simulation is mainly divided into three steps: first, modeling; second, generating random numbers; and third, computer simulation.

1. Establish mathematical model

(1) Definition. Mathematical model refers to a mathematical structure obtained by using appropriate mathematical tools to make necessary simplifications and assumptions for a specific object for a specific purpose, which can explain the reality of a specific object, predict the future of the object, or provide the optimal decision to deal with the object. The expression form of this mathematical structure is: equality or inequality established with letters, numbers, other mathematical symbols, etc., and the characteristics and internal relations of objective objects described with charts, images, block diagrams, etc.

(2) Method of establishing mathematical model. There are usually three methods for establishing mathematical models. The first is the analytical method, also known as the deductive method, theoretical modeling, and mechanism modeling, that is, the white box problem. It is based on the working principle of the system, and uses known theorems, principles, prerequisites, rules, etc. to derive the mathematical model of the system. The second is the test method, also known as inductive method, experimental modeling, and system identification, that is, black box problem. It estimates the mathematical model of the system by necessary data processing and mathematical operations based on the records of system input and output. The third is to use the above two methods together, list the theoretical model of the system by analysis, and determine the parameters in the model by testing.

(3) Classification of mathematical models. There are many classification methods for mathematical models, which is listed as follows.

According to the characteristics of models, there are static models and dynamic models, deterministic models and stochastic models, discrete models and continuous models, linear models and nonlinear models.

According to the mathematical method of model application, there are geometric model, graph theory model, differential equation model, probability model, optimal control model, programming theory model, Markov chain model, etc.

According to the degree of understanding of model structure, there are white box model, gray box model, black box model, etc.

According to the application direction of models, there are economic models, ecological models, population models, traffic models, environmental models, resource models, engineering models, aviation models, etc.

According to the purpose of modeling, there are prediction models, optimization models, decision models, control models, etc.

(4) Steps of establishing mathematical model

The first is observation. Before modeling, make a comprehensive, in-depth and detailed observation of the background of the actual object (problem), identify the problem to be solved, collect necessary data as required, and the data collected must meet the requirements.

The second is to propose model assumptions. In view of the complexity of the actual object (problem), it is necessary to idealize and simplify the object (problem), grasp the main factors, temporarily ignore the secondary factors, clarify the relationship between variables, and put forward necessary assumptions. If the assumption is reasonable, the subsequent model is more consistent with the actual object. If the assumption is unreasonable, it needs to be modified.

Third, establish a mathematical model. In the process of modeling, it is necessary to distinguish the types of variables and select appropriate mathematical tools. We should grasp the essence of the problem and simplify the model as much as possible. There should be strict mathematical reasoning to ensure the correctness of the model. The modeling should have sufficient accuracy, and the authenticity of the reaction object should not be affected when removing secondary factors.

The fourth is to solve the model. Different models need to be solved with different mathematical knowledge, often with the help of computers.

The fifth is the analysis and verification of the model. It can be verified by existing data. If the theoretical value calculated by the model is consistent with the actual value, the model is successful; If the difference is too large, modify the model.

The sixth is the modification of the model. It is necessary to analyze the rationality of assumptions, retain the reasonable part, modify the unreasonable part, increase the analysis of secondary factors, and consider the content that has been ignored in the new modeling. Sometimes it is necessary to remove some variables, or change the nature of variables, or change the functional relationship between variables.

2. Generate random number

The simulation of any system or process with random components needs a method to generate or obtain random numbers. Therefore, how to conveniently and effectively generate random values from the required probability distribution for the execution of mathematical models is very important. Now we can use computer to generate a series of random numbers randomly. The occurrence of these random numbers follows a certain probability distribution. Today's programming languages, simulation development tools, etc. all provide uniformly distributed random number generators. The generation of these random numbers is determined by the program, which also conforms to certain laws in essence, so they are called "pseudo random numbers". But these pseudo random numbers are exactly random numbers applied in the simulation process.

A relatively simple method is to use Excel and other software to realize the occurrence of random numbers. Click "Data Analysis" in Excel, select "Random Number Generator", and then select the number of variables, random numbers, etc. as required to obtain the uniformly distributed random number.

Uniform distribution of random numbers can be obtained by transformation. In the random number generator menu of Excel, select different distributions to get some random numbers that obey normal distribution, binomial distribution, poisson distribution, etc.

3. Computer simulation

After completing mathematical modeling and generating random numbers, the next step is to simulate according to various algorithms. There are many simulation algorithms, such as Monte Carlo method, graph theory, exhaustive method, numerical analysis method, image processing method, linear programming, etc. Through simulation, the statistical characteristic value of a certain characteristic quantity A of the simulated object is obtained, such as the distribution type, mean, variance, distribution function, etc. of A.

Due to the heavy workload of manual simulation calculation, the current simulation work is basically completed by computers. There are many pieces of computer simulation software in the market, and the commonly used simulation software packages are Arena, ExtendSim, Simio, SIMPLE++, Cradle CFD, Matlab and its Simulink, mode Frontier, etc.

4. Development of simulation

It is difficult to establish an effective and reliable simulation model. Now there is computer software to establish the simulation model. Simulation is developing faster and faster. Object oriented ideas and methods are widely used in software design. The main development of simulation in the future is to develop "modeling methodology", which focus on solving problems by using mathematical models, and can also use computers for new modeling to simplify the complex modeling process. The second is "object-oriented simulation", which regards the system as an independent object and executes simulation by sending messages between objects. Its modeling process is close to human thinking. The third is distributed interactive simulation, which connects software and hardware equipment and personnel in different places through electronic means, and conducts simulation tests interactively in a synthetic electronic environment. It is suitable for LAN and remote network simulation in different places. The fourth is artificial intelligence simulation, which uses knowledge base for modeling and simulation to achieve intelligent simulation. The fifth is virtual reality simulation, which creates a three-dimensional graphic world for users to reflect the changes and interactions of solid objects in real time.

2.6.2 Monte Carlo method

1. Introduction to Monte Carlo method

Monte Carlo method, also called Monte Carlo simulation method, random sampling method, statistical test method, is an important method in the numerous simulation algorithms. In the mid 1940s, the traditional empirical methods could not approach the real physical process, while the Monte Carlo method could simulate the real physical process, and the results were relatively ideal. Because of the probability and statistical characteristics of this method, people borrowed the name of Monte Carlo, the city of Monaco. Strictly speaking, Monte Carlo method is a calculate method based on probability and statistical theory. Many calculation problems can be solved by using random numbers, and statistical simulation can be realized by using computers to obtain corresponding outputs. The more samples, the closer the output of Monte Carlo method is to the optimal solution. Monte Carlo method has been widely used in finance, project management, medicine, quantum thermodynamic calculation, aerodynamics calculation, etc. Due to the development of computer technology, Monte Carlo method has been rapidly popularized and developed.

2. Basic principle of Monte Carlo method

(1) The basic principle of Monte Carlo method is as follows: The probability of the event can be estimated by the frequency of a large number of experiments. When the sample size is large enough, the frequency of the event can be considered as the probability of the event. When using the Monte Carlo method, a large number of random samples can be taken of the random variables that affect the reliability of an event, and then the sampling values can be input into the corresponding function formula in turn to judge whether the structure of the function formula fails, and finally the failure probability of the function formula substructure can be obtained. The characteristic of Monte Carlo method is to estimate the probability through a large number of repeated tests and statistical frequency, so as to get the solution.

(2) Illustrate the idea of Monte Carlo method with examples.

Assume that component C = component A + component B,

A is a random variable and obeys normal distribution $N(\mu_A, \sigma_A^2)$,

B is a random variable and obeys normal distribution $N(\mu_B, \sigma_B^2)$,

According to the idea of analytic method, C obeys normal distribution $N(\mu_A + \mu_B, \sigma_A^2 + \sigma_B^2)$.

Now according to the idea of Monte Carlo method, take a set of components (A_1, B_1) and the dimension data of the component can be measured as $c_1 = (a_1, b_1)$; Then take another set of components (A_2, B_2), we get $c_2 = (a_2, b_2)$. Keep going, until $c_n = (a_n, b_n)$. Therefore:

Make histogram according to the data of $c_1, c_2, \cdots, c_n$, the approximate shape of component C distribution can be obtained;

According to C $= \sum_{i=1}^{n} c_i/n$, the estimated value of μ_c can be obtained;

According to $S_c = (\sum_{i=1}^{n} (c_i - \bar{c})^2/n - 1)^{1/2}$, the estimated value σ_c can be obtained.

According to ($C_i <$ Number of parts of x/n), the estimated value of distribution function $F(X < x)$ can be obtained.

Therefore, the estimation of distribution type, mean, variance and distribution function of component C is obtained.

2.6.3 Digital twins

1. The concept of digital twins

Digital twins are an important form of modeling and simulation applications in the digital age. The concept was proposed by Dr. Michael Grieves of the University of Michigan in 2002. It is a multi physical field, multi-scale, multi probability model, which can use physical models, sensor data, historical data, etc. to reflect the function, real-time status, evolution trend, etc. of the entities corresponding to the model. Now it is basically agreed that the digital information structure of the physical system can be created as an independent entity. This digital information is the twin of the information embedded in the physical system and connected with the physical system throughout the system life cycle.

2. The elements of digital twins

It includes real objective objects, virtual spaces, data flow connections from real spaces to virtual spaces, information connections from virtual spaces to real spaces, and virtual sub-spaces.

3. The role of digital twins

It includes asset optimization, competition differentiation, improving user experience, improving organizational efficiency, improving enterprise decision-making, understanding the status of equipment or systems, responding to various changes, changing operations and adding value. It can be applied to product research and development, process planning, lean manufacturing, equipment maintenance, smart city construction, etc.

2.6.4 Digital engineering strategy

The digital engineering strategy was first proposed by Michael Griffin, Deputy Secretary of Research and Engineering in the Digital Engineering Strategy of the Department of Defense in June 2018. The strategy points out that digital technology has completely changed the business of most major industries and our personal life activities. By improving the computing speed, storage capacity and processing capacity, digital engineering has endowed a paradigm shift from the traditional "design, build, test" method to the "model, analysis, build" method. This method can enable projects of the Department of Defense to build prototypes, conduct experiments and tests, support decisions and determine solutions in a virtual environment before delivery. Digital engineering needs new methods, processes and tools, changes the operation mode of the original engineering community, and has an impact on the research, demand, collection, testing, cost, maintenance and intelligence communities.

The vision of digital engineering is to fully modernize the design, development, delivery, operation and maintenance systems. In the whole life cycle, the models used represent the systems of interest in the virtual world in a digital way, including processes, equipment, products, parts, etc. Digital engineering will integrate advanced computing, big data analysis, artificial intelligence, autonomous systems, robot technology, etc. to improve engineering practice.

Five strategic goals of digital engineering initiative:

(1) Formalize the development, integration and use of normative models. Realize the seamless flow of standardized data among design model, manufacturing model, audit and verification model, system model, production support model, special engineering model and management model, and serve for enterprise and project decision-making.

(2) Provide durable, authoritative and trustworthy sources (i.e., basic data and models). Make it possible to access, manage, analyze, use, and distribute information from a common set of digital models and data. Authorized stakeholders can use current, authoritative, and consistent information throughout the lifecycle.

(3) Integrate technological innovation into the improvement of engineering practice. Use big data and analysis, cognitive technology, advanced computing technology, digital and physical integration technology, as well as new methods and human-computer collaboration to promote the transformation to an end-to-end digital enterprise.

(4) Build the supporting structure and environment of digital engineering. They include tools, processes, methods, software and hardware, networks, security, etc., to achieve interaction, collaboration and communication among stakeholders.

(5) Complete the digital transformation of culture and team. For the adaptation and support of digital engineering in the whole life cycle, through continuous improvement of communication and contact, leadership, strategy and implementation, training and education, the digital engineering of the entire procurement department is institutionalized.

2.6.5 Opportunities and challenges of modeling and simulation in the digital age

The development of digital twins and the proposal of digital engineering strategy mean that modeling and simulation technology will usher in new development, but also face many challenges. The application of modeling and simulation will become more and more popular. From the laboratory to the factory, office, campus and community, it will become an increasingly indispensable tool for production and life.

1. Opportunities

(1) Normalized application of modeling and simulation. The Internet of things and big data make modeling and simulation online, ubiquitous and normalized possible. The typical feature of the digital age is the Internet of things built through ubiquitous sensor networks, which uses 5G, 6G communication technologies and even higher level communication technologies to widely collect massive data in the real world, and improves decision-making and control capabilities through big data analysis technology. Represented by the digital twins, modeling and simulation in the digital era can establish a permanent, real-time and interactive link with the real world. Modeling and simulation is no longer offline, independent and at a specific stage, but develops into online, ubiquitous and normalized services.

(2) Modeling and simulation develops in depth. The proposal of digital engineering strategy is conducive to the scientific, standardized and intelligent development of modeling and simulation. Scientification is reflected in the more stable and in-depth scientific foundation of modeling and simulation; Standardization is embodied in the unification of standards and more scientific

processes; Intelligence is embodied in the new generation of artificial intelligence and machine learning, which are widely used in modeling and simulation. These changes and developments will promote the comprehensive implementation of the digital engineering strategy. Digital engineering strategy is a change of traditional engineering thought. Modeling and simulation, as the core supporting technology of digital engineering strategy, will also be further developed.

(3) Great improvement of computing power. With the development of cloud computing, supercomputing and quantum computing, modeling and simulation in the digital age are no longer limited by computing power, thus providing unlimited possibilities for the development of digital twins and digital engineering. With the continuous development of chip technology and software technology, the ability of supercomputers has been constantly improved. The development of cloud computing makes computing, storage and communication capabilities become a kind of resources that can be obtained on demand and can be massively increased. Quantum computing provides support for completely solving the computing power required for modeling and simulation. The improvement of calculation force will further improve the refinement of the model and better meet the needs of simulation.

(4) The industrialized development of simulation. The simulation industry in the digital era will be greatly developed. The successful application of digital twins and the proposal of digital engineering strategy will promote the blowout growth of practical applications of modeling and simulation. Modeling and simulation will enter all walks of life, promote the great improvement of social productivity, and promote the high-quality development of all industries.

2. Challenge

(1) Theoretical challenge. The "Digital Twin" system, which combines big data and complex system simulation, can give consideration to both "process" and "result", conduct in-depth study on the overall situation of "past and future", and achieve more intelligent modeling and simulation. In the data age, the theoretical focus of simulation has shifted from model to data, and the simulation method has shifted from sampling and finding causality to global and finding correlation. Analog computing research and big data analysis need to be more integrated and developed to establish a solid theoretical foundation for modeling and simulation in the digital era.

(2) Credibility challenge. Credibility is the source of vitality of modeling and simulation. Digital twins and Digital Engineering emphasize more on taking the results of modeling and simulation as the authoritative and reliable data source, which puts forward higher requirements for the reliability and accuracy of the model. Digital twins and data engineering are simulation applications with strong complexity, real-time, interactivity and diversity. It is often beyond the existing ability to evaluate their credibility. Therefore, more effective evaluation theories and methods must be studied in depth.

(3) Management challenge. Both digital twins and digital engineering need to further centralize management and actively use analog resources such as models, data and support environments, which brings new challenges to modeling and analog management in the digital age. A formal plan for model creation, management, integration and related project and engineering

activities needs to be formulated in the whole life cycle, describing how to implement the model, support analysis and decision-making in a coherent and effective way when implementing work activities. In terms of management, a standardized system should be established to determine the basic quality standards and rules that should be followed in model development. Maintain the source and pedigree of the model, and establish the basis for credibility, accuracy and judgment model reuse.

(4) Safety challenge. Digital twins and Digital Engineering make the resources and infrastructure of modeling and simulation more and more open, and the possibility of being attacked is higher and higher. The loopholes in the modeling and simulation tools will be more exposed to the open Internet environment, making its security face huge challenges.

(5) Culture challenge. The fundamental challenge facing the implementation of digital twins and digital engineering is to change the existing organizational culture. Organizations need to adopt a deliberate and systematic approach to planning, implementing and supporting digital transformation. This transformation must go beyond technology to solve the challenges of people and culture, including the values and behaviors of the organization.

2.7　计算法

───

2.7.1 主要计算方法

计算方法，也可以理解为数值分析，它是指为各种问题的数值解答提供有效的算法。在日常的工作生活中有很多计算方法，如数学计算方法（包括加、减、乘、除，计算长度、面积、体积、效率等）、物理计算方法（根据物理公式进行物理计算）、化学计算方法（根据各种化学原理进行化学计算）等。

这里介绍几种常用的、反映一组数据特性的数学计算方法（实质上也归于统计范畴）。

1. 平均数

平均数有算术平均数、几何平均数、调和平均数三种。

（1）算术平均数 $\bar{x}$

用 x_1，x_2，x_3，$\cdots$，x_n 表示一组数据，它们的平均值（也称均值）通常用 $\bar{x}$ 表示，则

$$\bar{x} = \frac{1}{n} \sum_{i=1}^{n} x_i$$

其中，x_i 表示第 i 个数据值，n 表示一组数据的数量（也称为样本量）。

（2）几何平均数

计算几何平均数的数据必须都大于零，则几何平均数是所有数据值之积的 n 次方根，记为 X_G：

$$X_G = \sqrt[N]{X_1 X_2 \cdots X_n}$$

（3）调和平均数

一组数据都不等于零，则先把每个数据求倒数，然后求出这些倒数的算术平均数，再求出该算术平均数的倒数，记为 X_H：

$$X_H = \frac{1}{\dfrac{\dfrac{1}{X_1} + \dfrac{1}{X_2} + \cdots + \dfrac{1}{X_n}}{n}}$$

2. 中位数 $\tilde{x}$

中位数是求得一组数据均值的简单、快捷的方法，通常用 $\tilde{x}$ 表示。计算中位数的步骤是：把一组数据按照由大到小的顺序排列，或按照由小到大的顺序排列，取最中间的一位数值，即为该组数据的中位数。如果一组数据共有偶数个数值，则取最中间两个数值的平均值为该组数据的中位数。中位数不是整组数据的平均值，但它可以客观地反映一组数据的均值状态。

3. 标准差 σ 和样本标准差 s

标准差是反映数据离散程度的一种方法，通常用 σ 表示。标准差数值越小，那么数据的离散度就越小。标准差公式为

$$\sigma = \sqrt{\frac{1}{n}\sum_{i=1}^{n}(x_i - \mu)^2}$$

其中，数值 $X_1, \ldots, X_n$ 皆为实数，其平均值为 μ，标准差为 σ。

在日常各种活动中，有些时候采集的一组数据是总体的数据，那么可以用标准差 σ 来反映这组数据的离散程度；有时候采集的一组数据是抽取样本的数据，那么就需用使用样本标准差来反映数据的离散程度。样本标准差通常用 S 表示，其计算公式为

$$s = \sqrt{\frac{1}{n-1}\sum_{i=1}^{n}(x_i - \bar{x})^2}$$

其中，x_i 表示第 i 个数据值，$\bar{x}$ 为平均值，n 是样本量。

4. 极差

极差也是用来反映一组数据的离散程度，它是指一组数据中的最大值与最小值之差，通常用 R 表示，其计算公式为

$$R = X_{\max} - X_{\min}$$

其中，$X_{\max}$ 表示一组数据中的最大值，$X_{\min}$ 表示一组数据中的最小值。

2.7.2 其他计算方法

1. 函数逼近论

函数逼近论涉及的基本问题是函数的近似表示问题，它是指在选定的一个函数中，寻找某个函数 g，使得 g 是已知函数 f 在一定意义下的近似表示，同时求出这个近似表示所产生的误差。函数逼近论现在发展与应用很快，比如在人工智能的深度学习上有很好的应用。

2. 微分与积分

微分是指有函数 $y = f(x)$，在 X 的数集中，当 dx 靠近时，函数在 dx 处的极限叫作函数在 dx 处的微分。

积分是把微分后无数无限小的东西重新集合为一个整体，它分为不定积分和定积分。不定积分是指由函数的已知导数或微分，去求出原来的函数，这个原函数有无穷多个。定积分是指函数 $f(x)$ 在区间 $[a,b]$ 上有定义，将该区间分成 n 个小区间，当 n 无限增大时，总和 S_n 的极限存在（$S_n = \sum_{i=1}^{n} f(\xi_i)\Delta x_i$，其中 ξ_i 表示任意一点），则把此极限值称为函数 $f(x)$ 在区间 $[a,b]$ 上的定积分。

3. 迭代法

迭代法是一种利用递推公式或循环算法，通过构造序列来求问题近似解的方法，它是一种不断用变量的旧值递推新值的过程，也称为辗转法。迭代法是计算机解决问题的一种基本方法。

4. 差分法

差分法是微分方程的一种近似数值解法，它是用有限差分代替微分，用有限差商代替导数，把求解微分方程的问题换成求解代数方程的问题。

5. 插值法

插值法是离散函数逼近的方法，它是在离散数据的基础上补插连续函数，并使这条连续曲线通过所有的离散数据点，然后利用函数在有限个点处的取值情况，来估算函数在其他点处的近似值。

6. 有限元素法

有限元素法是指将结构用网格划分为计算模型的一种结构分析数值方法。被网格所划分的若干小块称为有限元或有限元素，它可以是三角形、四边形、六面体等各种形状。这种方法已成为解数学物理方程的一种近似方法，极大提高工作的效率和可靠性。

2.7 Calculation Method

——

2.7.1 The main calculation methods

The calculation method can also be understood as numerical analysis, which refers to providing effective algorithms for numerical solutions of various problems. There are many calculation methods in daily work and life, such as mathematical calculation methods (including addition, subtraction, multiplication, division, calculation length, area, volume, efficiency, etc.), physical calculation methods (physical calculation based on physical formulas), chemical calculation methods (chemical calculation based on various chemical principles), etc.

Here we introduce several commonly used mathematical calculation methods that reflect the characteristics of a group of data (in essence, they also belong to the statistical category).

1. Average

There are three kinds of averages: arithmetic average, geometric average and harmonic average.

(1) Arithmetic average $\bar{x}$

Use x_1, x_2, x_3, $\cdots$, x_n to represent a set of data, and their average (also called mean value) usually be expressed by $\bar{x}$, then

$$\bar{x} = \frac{1}{n} \sum_{i=1}^{n} x_i$$

In the formula, x_i represents the ith data value, and n represents the number of a group of data or the sample size.

(2) Geometric mean

The data for calculating the geometric mean must be greater than zero, then the geometric mean is the nth root of the product of all data values, which is recorded as X_G,

$$X_G = \sqrt[N]{X_1 X_2 \cdots X_n}$$

(3) Harmonic mean

If a group of data is not equal to zero, first calculate the reciprocal of each data, then calculate the arithmetic mean of these reciprocals, and then calculate the reciprocal of the arithmetic mean, which is recorded as X_H,

$$X_H = \cfrac{1}{\cfrac{\dfrac{1}{X_1} + \dfrac{1}{X_2} + \cdots + \dfrac{1}{X_n}}{n}}$$

2. Median $\tilde{x}$

Median is a simple and fast method to obtain the mean value of a group of data, usually expressed in $\tilde{x}$. The procedure for calculating the median is to arrange a group of data in the order from the largest to the smallest, or from the smallest to the largest, and take the middle digit as the median of the group of data. If a group of data has an even number of values, the average of the middle two values is taken as the median of the group of data. The median is not the average of the whole group of data, but it can objectively reflect the average state of a group of data.

3. Standard deviation σ and sample standard deviation s

Standard deviation is a method to reflect the degree of data dispersion, usually using σ express. The smaller the standard deviation, the smaller the dispersion of the data. The standard deviation formula is

$$\sigma = \sqrt{\frac{1}{n} \sum_{i=1}^{n} (x_i - \mu)^2}$$

In the formula, value $X_1, \ldots, X_N$ are all real numbers, their average value is μ, and their standard deviation is σ.

In daily activities, sometimes a group of data collected is overall data, so standard deviation can be used σ to reflect the degree of dispersion of this group of data; Sometimes a group of data collected is sampled data, so the sample standard deviation is used to reflect the dispersion of data. The sample standard deviation is usually expressed by S, and its calculation formula is

$$s = \sqrt{\frac{1}{n-1} \sum_{i=1}^{n} (x_i - \bar{x})^2}$$

In the formula, x_i represents the ith data value, $\bar{x}$ is the mean value, n is the sample size.

4. Range

Range is also used to reflect the degree of dispersion of a group of data. It refers to the difference between the maximum value and the minimum value in a group of data, usually expressed in R, and its calculation formula is

$$R = X_{\max} - X_{\min}$$

In the formula, $X_{\max}$ represents the maximum value in a set of data, $X_{\min}$ represents the minimum value in a set of data。

2.7.2 Other calculation methods

1. Function approximation theory

The basic problem involved in the function approximation theory is the approximate representation of functions. It refers to finding a function g in a selected function so that g is the approximate representation of a known function f in a certain sense, and at the same time finding out the error caused by the approximate representation. The theory of function approximation has been developed and applied rapidly. For example, it has a good application in the deep learning of artificial intelligence.

2. Differential and integral

Differential refers to a function $y = f(x)$. In the number set of X, when dx is close, the limit of the function at dx is called the differential of the function at dx.

Integration is to reassemble innumerable infinitesimal things after differentiation into a whole, which is divided into indefinite integral and definite integral. Indefinite integral refers to finding the original function from the known derivative or differential of the function. There are infinitely many original functions. Definite integral means that the function $f(x)$ is defined on the interval $[a, b]$, and the interval is divided into n cells. When n increases infinitely, the limit of the sum S_n exists, ($S_n = \sum_{i=1}^{n} f(\xi_i) \Delta x_i$, ξ_i stands for any point), then this limit value is called the definite integral of function $f(x)$ on the interval $[a, b]$.

3. Iterative method

Iterative method is a method to obtain the approximate solution of a problem by constructing a sequence using recursive formula or cyclic algorithm. It is a process of continuously using the old value of a variable to recurse new value, also known as the rolling method. Iterative method is a basic method for computer to solve problems.

4. Difference method

The difference method is an approximate numerical solution of differential equations. It uses finite difference instead of differential, uses finite difference quotient instead of derivative, and replaces the problem of solving differential equations with that of solving algebraic equations.

5. Interpolation

The interpolation method is a method of approximating discrete functions. It is to interpolate continuous functions on the basis of discrete data, make this continuous curve pass through all discrete data points, and then use the value of the function at a limited number of points to estimate the approximate value of the function at other points.

6. Finite element method

The finite element method is a numerical method for structural analysis, which divides the structure into computational models by grids. The small pieces divided by the mesh are called finite element or finite element, which can be triangle, quadrilateral, hexahedron and other shapes. This method has become an approximate method for solving mathematical and physical equations, greatly improving the efficiency and reliability of work.

2.8 统计法

———

数据无处不在，要想用好数据，就需要会使用统计方法。统计的过程其实是对数据进行收集、整理、分析、解释的过程。统计的方法有很多，主要可以分为描述统计方法和推论统计方法。

2.8.1 描述统计方法

描述统计方法是对数据进行整理分析，并对数据的分布状态、数字特征、随机变量之间关系进行估计和描述的方法，它是通过文字、图示等描述性的方式对数据的特性进行处理、显示和分析。描述统计方法可以分为图示技术描述统计和语言文字描述统计两大类。

1. 图示技术描述统计

图示技术描述统计是根据收集到的统计数据，用简单的图形、精炼的文字等对数据的特性进行表示，具有直观、形象、生动、具体、易理解等特点。图示技术描述统计主要有排列图、饼分图、直方图、茎叶图、散布图、趋势图、柱状图、控制图等。

2. 语言文字描述统计

语言文字描述统计是以语言文字等非数字资料对数据的特性进行表示的方法，具有层次清晰、逻辑性强、通俗易懂等特点，主要有调查表、分层法、因果图、关联图、亲和图、流程图、系统图等。

2.8.2 推论统计

推论统计通常是指利用小样本的数据来推论总体的特征。推论统计的理论假设是概率论，主要有参数估计、假设检验、独立性检验。

1. 参数估计

参数估计就是根据抽样结果，科学地估计总体参数值的大小和范围。参数估计主要有两种方法：第一个是点估计，它是利用样本统计单值来估计未知总体参数的方法，有以点代面的估计特征，对总体参数的估计比较粗略，准确度和可靠度比较小。第二个是区间估计，它是在一定的程度上对总体参数可能落入的一个数值范围做出估计的方法，它对总体参数的估计比较准确。但由于存在抽样误差，区间估计会其影响而发生波动，因此估计准确的程度要用显著性水平（置信度）来表示。

显著性水平是估计总体参数落在某一个区间内可能犯错的概率，是根据概率计算的当样本与总体没有真实差异时却出现误差的最大可能性。比如对某项目部的工人重复抽样 100 次，其中有 95 次所作的区间估计显示工人的平均年龄在 40~50 岁之间，则进行一次估计成功的概率为 95%，即显著水平达到 5%，意味着估计错误的可能性不超过 5%。

按一定显著性水平求得的估计区间称为置信区间。

2. 假设检验

假设检验是指在规定的风险水平上，确定一组数据是否符合已经给定假设的一种统计方法，它是根据抽样调查的结果在一定可靠性基础上对原来的假设作出接受或拒绝的判断，或者说通过研究样本对事先作出的有关总体特征的假设进行检验的基本过程。

在假设检验中，显著性水平是指当原假设为正确时人们却把它拒绝了的概率或风险，它是公认的小概率事件的概率值，必须在每一次统计检验之前确定，通常取 $\alpha = 0.05$ 或 $\alpha = 0.01$，表明当作出接受原假设的决定时，其正确的可能性为 95%或 99%。

2.8.3 正态分布、t 分布、威布尔分布

推论统计的理论假设是概率论。概率论研究发现，当样本总体的样本容量在不同特定值的时候，样本总体的分布形状是不一样的。当样本容量在 50 以上时，样本总体分布的形状为标准正态分布，又称 Z 分布；当样本容量为 15 ~ 50 时，样本总体分布形状为 t 分布；当样本容量为 15 以下时，样本总体分布形状为 P 分布或概率分布。从样本总体中随机抽出一个样本，这个样本落在这个样本总体分布的中心区域的可能性较大，落在边缘区域的可能性较小，掉出了某一区域的可能性就更小。在可靠性分析中广泛应用的有威布尔分布。

1. 正态分布

正态分布，又称高斯分布，它的定义如下：

假设随机变量 x 服从一个位置参数为 μ、尺度参数为 σ 的概率分布，且它的概率密度函数为

$$f(x) = \frac{1}{\sigma\sqrt{2\pi}} \times \exp\left(-\frac{(x-\mu)^2}{2\sigma^2}\right)$$

则称这个随机变量为正态随机变量，该正态随机变量服从的分布为正态分布，记为 $X \sim N(\mu, \sigma^2)$。在上述其中，μ 为随机变量的平均值，σ 为它的标准差。当 $\mu = 0$，$\sigma = 1$ 的时候，正态分布成为标准正态分布。

2. t 分布

t 分布，因威廉·戈塞（William Sealy Gosset）于 1908 年发表相关论文时使用"学生"而得名"学生分布"，它主要用于根据小样本来估计呈正态分布但是方差未知的总体的均值。取名 t 的由来，主要是在实际工作中，标准差 σ 是未知的，σ 用 s 来作为估计；平均值 μ 也必须做变换，为了区别，使用 t 来变换。

t 分布的定义如下：假设随机变量 X 服从标准正态分布 $N(0, 1)$，Y 服从卡方分布 $X^2(n)$，那么

$$Z = \frac{X}{\sqrt{Y/n}}$$

的分布，成为自由度为 n 的 t 分布。

t 分布的曲线形态与自由度 n 大小有关。与标准正态分布曲线相比，自由度 n 越小，t 分布曲线愈平坦，曲线中间愈低，曲线双侧尾部翘得愈高；自由度 n 越大，t 分布曲线越接近正态分布，当自由度趋向无穷大时，t 分布曲线成为标准正态分布曲线。

3. 威布尔分布

由于局部失效而导致整体机能失效的串联式模型所采用的分布函数称为威布尔分布，它是由瑞典的威布尔在研究链的强度时所构造的一种分布函数。威布尔分布是可靠性分析和产品寿命检验的理论基础，具有普遍的意义，现在广泛应用于天气预报、雷达系统、拟合度、工业制造、生产过程和运输时间的关系、量化寿险模型的重复索赔、预测技术变革、描述风速分布等。

威布尔分布函数的形式为

$$F(t) = 1 - \exp\left(-\frac{(t-t_0)^\beta}{\theta}\right)$$

威布尔分布概率密度函数为

$$f(t;\theta,\beta) = \frac{\beta}{\theta}\left(\frac{t-t_0}{\theta}\right)^{\beta-1} \exp\left[-\left(\frac{t-t_0}{\theta}\right)^\beta\right]$$

其中，β 为形状参数；t_0 是位置参数，用于表征分布曲线的起始位置，在实践中常取值为0；β 是尺度参数（比例参数）。

当形状参数 $\beta =1$ 时，成为指数分布；$\beta =2$ 时，成为瑞利分布（Rayleigh distribution）。

2.8.4 其他统计方法

统计方法非常之多，常用的还有分层分析、比较分析、时间序列分析、指数分析、回归分析、相关分析、平衡分析等方法。有兴趣的读者可以参阅《质量管理统计应用》和其他相关文献。

2.8   Statistical Method

Data is everywhere. If you want to use data well, you will use statistical methods. The process of statistics is actually the process of collecting, sorting, analyzing and interpreting data. There are many statistical methods, mainly including descriptive statistical methods and inferential statistical methods.

2.8.1 Descriptive statistical methods

Descriptive statistics is a method to sort out and analyze data, and estimate and describe the distribution state, numerical characteristics, and the relationship between random variables of data. It processes, displays, and analyzes the characteristics of data in descriptive ways such as text, diagrams, etc. Descriptive statistics methods can be divided into graphic technology description statistics and language description statistics.

1. Graphic technology description statistics

Graphic technology description statistics is based on the collected statistical data, using simple graphics, refined text, etc. to express the characteristics of the data. It is intuitive, vivid, specific, and easy to understand. Graphic technology description statistics mainly include Pareto chart, pie chart, histogram, stem and leaf chart, scatter chart, trend chart, histogram, control chart, etc.

2. Language description statistics

Language description statistics is a method to express the characteristics of data with non numeral data such as verbal text. It is characterized by clear hierarchy, strong logic, and easy to understand. It mainly includes questionnaire, hierarchical method, causality diagram, correlation diagram, affinity diagram, flow diagram, system diagram, etc.

2.8.2 Inferential statistics

Inferential statistics usually refers to the use of small sample data to infer the characteristics of the population. The theoretical hypothesis of inferential statistics is probability theory, mainly including parameter estimation, hypothesis test and independence test.

1. Parameter estimation

Parameter estimation is to scientifically estimate the size and range of population parameter values according to the sampling results. There are two main methods for parameter estimation: the first is point estimation, which is a method to estimate unknown population parameters by using the single value of sample statistics. It has the estimation feature of point instead of surface. The estimation of population parameters is relatively rough, and the accuracy and reliability are relatively small. The second is interval estimation, which is a method to estimate a numerical range that the population parameters may fall into to a certain extent, and it is more accurate in estimating the population parameters. However, due to sampling error, interval estimation will fluctuate due to its influence, so the accuracy of estimation should be expressed by significance level (confidence level).

The significance level is to estimate the probability that the population parameters may make mistakes in a certain interval, and it is the maximum probability of error when there is no real difference between the sample and the population calculated according to the probability. For example, if the workers of a project department are sampled 100 times repeatedly, and 95 of them have made interval estimates indicating that the average age of the workers is between 40~50 years old, the probability of success in one estimate is 95%, that is, the significance level reaches 5%, which means that the probability of error in estimation is not more than 5%.

The estimation interval obtained according to a certain significance level is called confidence interval.

2. Hypothesis test

Hypothesis test refers to a statistical method to determine whether a group of data conforms to a given assumption at a specified risk level. It is a basic process to accept or reject the original assumption on the basis of certain reliability based on the results of sampling survey, or to test the assumptions made in advance about the overall characteristics through research samples.

In hypothesis testing, significance level refers to the probability or risk that people reject the null hypothesis when it is correct. It is the recognized probability value of small probability events, which must be determined before each statistical test, usually taking $\alpha = 0.05$ or $\alpha = 0.01$, indicating that the correct probability is 95% or 99% when the decision to accept the original hypothesis is made.

2.8.3 Normal distribution, t-distribution, Weibull distribution

The theoretical assumption of inferential statistics is probability theory. Probabilistic research shows that when the sample size of the sample population is at different specific values, the distribution shape of the sample population is different. When the sample size is more than 50, the shape of the sample population distribution is standard normal distribution, also known as Z distribution; When the sample size is 15~50, the overall distribution shape of the sample is t distribution; When the sample size is below 15, the overall distribution shape of the sample is P distribution or probability distribution. Randomly select a sample from the sample population. The sample is more likely to fall in the center area of the sample population distribution, less likely to fall in the edge area, and less likely to fall out of a certain area. Weibull distribution is widely used in reliability analysis.

1. Normal distribution

Normal distribution is also called Gauss distribution, which is defined as follows:

Suppose that the random variable x obeys a probability distribution with a position parameter of μ and a scale parameter of σ, and its probability density function is

$$f(x) = \frac{1}{\sigma\sqrt{2\pi}} \times \exp(-\frac{(x-\mu)^2}{2\sigma^2})$$

Then the random variable is called normal random variable, and the distribution that the normal random variable obeys is normal distribution. It is recorded as $X \sim N(\mu, \sigma^2)$. In the above formula, μ is the average value of the random variable, and σ is its standard deviation. When $\mu = 0$, $\sigma = 1$, normal distribution becomes standard normal distribution.

2. *t* distribution

t distribution, named "student distribution" because William Sealy Gosset used "students" when he published a related paper in 1908, is mainly used to estimate the mean value of the population with normal distribution but unknown variance according to small samples. The name t comes from the standard deviation σ is unknown, and σ use s as an estimate in practical work; The average value μ must also be transformed. To distinguish, use *t* to transform.

The definition of *t* distribution is as follows: suppose that the random variable X obeys standard normal distribution $N(0,1)$, Y obeys χ^2 distribution $\chi^2(n)$, then the distribution of

$$Z = \frac{X}{\sqrt{Y/n}}$$

is called the *t* distribution with the n degree of freedom.

The curve shape of *t* distribution is related to the degree of freedom *n*. Compare with the curve of standard normal distribution, the smaller the degree of freedom *n*, the flatter the distribution curve, the lower the middle of the curve, and the higher the tail of both sides of the curve. The greater the degree of freedom *n*, the closer the *t* distribution curve is to the normal distribution. When the degree of freedom tends to infinity, the *t* distribution curve becomes the standard normal distribution curve.

3. Weibull distribution

The distribution function used in the series model of global functional failure caused by local failure is called Weibull distribution, which is a distribution function constructed by Sweden's Weibull when studying the strength of the chain. Weibull distribution is the theoretical basis of reliability analysis and product life inspection, which is of universal significance. It is now widely used in weather forecast, radar system, fitness, industrial manufacturing, the relationship between production process and transportation time, repeated claims of quantitative life insurance model, prediction technology innovation, description of wind speed distribution, etc.

The form of the Weibull distribution function is:

$$F(t) = 1 - \exp\left(-\frac{(t-t_0)^\beta}{\theta}\right)$$

TheWeibull distribution probability density function is:

$$f(t;\theta,\beta) = \frac{\beta}{\theta}\left(\frac{t-t_0}{\theta}\right)^{\beta-1} \exp\left[-\left(\frac{t-t_0}{\theta}\right)^\beta\right],$$

In the formula, β is a shape parameter. t_0 is a positional parameter that is used to characterize the starting position of the distribution curve, in practice, the value is usually 0. θ is a scale parameter (scale parameter).

When the shape parameter $\beta = 1$, it becomes exponential distribution; when $\beta = 2$, it becomes Rayleigh distribution.

2.8.4 Other statistical methods

There are many statistical methods, including hierarchical analysis, comparative analysis, time series analysis, exponential analysis, regression analysis, correlation analysis, balance analysis, etc. Interested readers can refer to the Statistical Applicationin in Quality Control and other relevant literature.

Chapter 3
Quality Representation and
Assurance of Data

第 3 章
数据的质量体现和保证

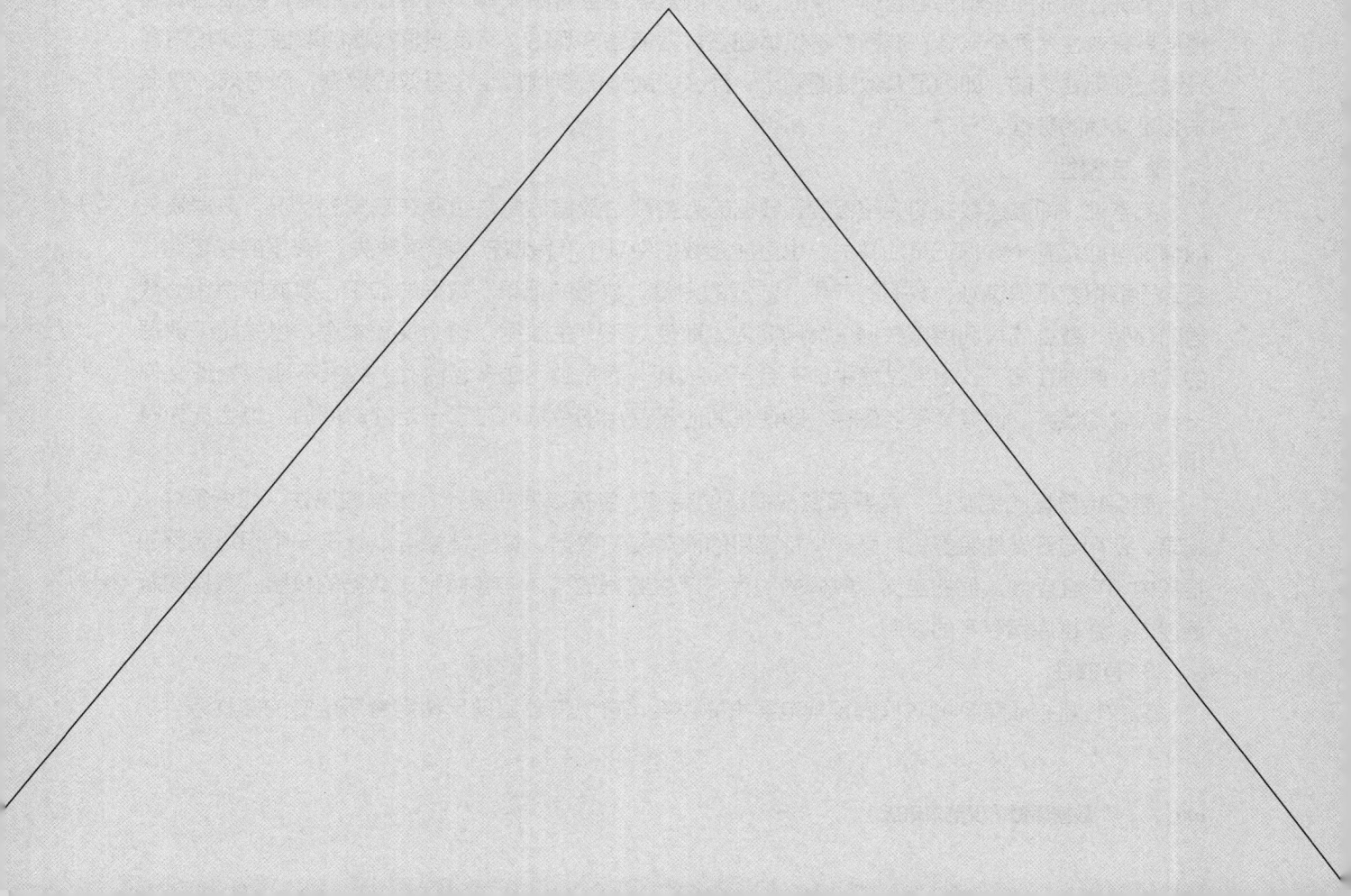

3.1 真实反映客观对象

——

高质量的数据，应该能够真实、正确、完整地反映客观对象的特性。被记录的数据在读取时不会发生误读、失真的情况。人们要使用数据时，是希望对客观对象有一个完整、真实的解读。

数据质量能够真实反映客观对象，通常具有以下几个特性：

1. 准确性

准确性，即一个记录值与它的真实值之间的接近程度，用来描述数据与其对应的客观实体的特征相一致的程度。确保数据的准确性是数据质量中的根本原则。数据的准确性主要体现在四个方面，一是数据的产生是否正常，对异常数据的产生是否有预警监控和处理，数据产生的过程是否受到干扰等。数据的名称、定义、指标、运行公式、逻辑等必须规范、标准。二是获取数据的准确性，体现在测量系统是否规范、测量人员是否具备专业技能、测量的环境是否收到干扰等。数据的记录格式要有明确规定，不产生歧义。比如"05/12/20"，是表示 2020 年 5 月 12 日，还是 2005 年 12 月 20 日，应当清晰标注。三是数据的读取的准确性，读取数据的人可以准确、清晰地解读所读到数据的真实含义，在数据传递的全过程中，数据的含义不会受到影响。如果发生有争议的数据解读，应当有相应说明。四是数据使用过程的准确性，在数据使用过程中，数据是否进行了清洗，清洗的逻辑是否合理、规范；收集的数据是否有缺失等。需要注意的是，数据的准确性与其有效性是两个概念，准确性不仅体现在数据值是客观的，而且要求数据的取值范围和内容是满足规范要求的，即满足有效性的要求，那么，准确的数据肯定是有效的数据，而有效的数据未必是准确的数据。

2. 完整性

完整性，即描述数据的完整程度。数据的完整性是数据信息是否存在缺失的状况，数据缺失的情况可能是整个数据记录的缺失，也可能是数据中某个字段信息的记录缺失。数据的完整性，贯穿于取得数据的策划、数据的产生、数据的测量、数据的记录、数据的加工、数据的使用、数据的存储、数据的再利用等数据生命周期内。即使"不产生数据"这个现象本身，也是属于数据的范畴。数据在加工、使用过程中，不会产生损耗，而是越用越多。因此，数据不能像处理文件一样被轻易废弃，而是要妥当保存。即使是暂时不使用的数据，过了一定的时间后，也会成为有用的数据。

要保持数据的完整性，需要保证数据链的完整，包括数据的设计、数据的属性、约束条件、记录、使用与开发的完整。就某一个特定目的而需要的数据，其完整性可以处于一个相对完整的框架内，它会受到数据的定义、数据的约束、产生数据的实体与参照物、数据的校验、数据的解读能力、数据库储存等的限制。

3. 有效性

有效性是指收集到的数据满足规范要求的程度，它也指数据满足使用者所定义的各种条件。

在数据的收集过程中，数据产生于调查者真正要调研的对象，试验的对象没有发生偏差或更换，测量的对象的确是测量人员或仪器所要测量的客观事物，测量的仪器等不受测量系统误差和随机误差的约束。从使用的角度，数据的有效性也是数据的规范性，是数据对于数据标准的满足程度，这些规范通常从数据的命名、数据类型、数据的长度、数据的取值范围、数据内容的标准、取得数据的日期等方面进行约束。对于各种不符合规范要求的数据，可以视为无效数据，但这些数据不是无用数据，如果规范进行了改变，曾经的无效数据也可以转变为有效数据。

4. 唯一性

唯一性是指在一个数据集里面没有重复、冗余的数据记录。数据的唯一性体现在数据的录入、导出、使用过程中是不发生重复现象，除非是要求重复录入、导出和使用。严格来说，即使是规定的重复录入、导出、使用，也是属于新的数据范畴，只是数据使用的目的改变了。举个数据唯一性例子，某咖啡店在一个时段真实卖出咖啡 1000 杯，咖啡店的数据表显示也是卖出 1000 杯，则数据是唯一的。假设从咖啡店的相关数据表中，看到有 10 条售出信息重复，显示卖出 1010 杯，那么该数据表就不具有唯一性。可以说，数据的唯一性是保持业务协同开展、流程可以追溯的重要因素，保持数据唯一性是数据开发、使用中的基本环节。

5. 时效性

时效性是指数据及时得到更新的程度。时效性可以体现在两个方面，一是数据采集的即时性，当数据产生以后，数据在采集的时间节点、传输的时间节点、存储的时间节点、加工使用的时间节点上，都是即时反映或动作，没有时滞或时滞几乎忽略。如果数据在经过一段时间之后再予以采集、传输、存储等，则数据的即时性就差。二是数据应用的及时性，从数据获得之后到数据应用之间的时间间隔在符合规定的要求，数据及时性好；如果时间间隔超过规定的要求，则及时性不理想。数据应用的及时性与使用者处理数据的速度、效率等有着很大的关联性，要确保使用者在需要数据的时候能够及时获得数据。

6. 一致性

一致性是指用来描述同一个客观对象的数据，在不同的数据集中保持信息属性相同。当有多个使用者同时访问、使用一个数据库时，他们所用到的数据在各项属性上保持一致，关联数据之间的逻辑关系保持正确和完整，数据存储保持一致。数据的属性有很多，包括数据对于客观对象描述的详细程度、数据与客观对象之间的差异、两个可测量数值之间最小的可辨识率、单位、格式等，数据的属性一致，有利于数据应用的开发和管理。

3.1 Real Reflection of Objects

High quality data should be able to truly, correctly and completely reflect the characteristics of objective objects. The recorded data will not be misread or distorted during reading. When people want to use data, they want to have a complete and true interpretation of the objective object.

Data quality can truly reflect the objective object, and usually has the following characteristics:

1. Accuracy

Accuracy, that is, the closeness between a recorded value and its true value, which is used to describe the degree to which the data is consistent with the characteristics of its corresponding objective entity. Ensuring the accuracy of data is the fundamental principle in data quality. The accuracy of data is mainly reflected in four aspects. The first is whether the data generation is normal, whether there is early warning monitoring and processing for the generation of abnormal data, and whether the data generation process is disturbed. The name, definition, indicator, operation formula and logic of data must be standardized. The second is the accuracy of data acquisition, which is reflected in whether the measurement system is standardized, whether the measurement personnel have professional skills, and whether the measurement environment is disturbed. The recording format of data shall be clearly specified without ambiguity. For example, whether "05/12/20" means May 12, 2020 or December 20, 2005 should be clearly marked. The third is the accuracy of data reading. Data readers can accurately and clearly interpret the true meaning of the data they read. The meaning of the data will not be affected in the whole process of data transmission. In case of controversial data interpretation, corresponding explanation shall be provided. The fourth is the accuracy of the data use process. It matters, whether the data has been cleaned, and whether the cleaning logic is reasonable and standardized, whether the collected data is missing, etc. It should be noted that the accuracy of data and its validity are two concepts. The accuracy is not only reflected in the fact that the data value is objective, but also requires that the value range and content of the data meet the requirements of the specification, that is, meet the requirements of validity. Then, accurate data must be effective data, while effective data may not be accurate data.

2. Integrity

Integrity is the description of the data completeness. Data integrity refers to whether data information is missing. Data missing may be the missing of the whole data record, or the missing of a field information record in the data. The integrity of data runs through the data life cycle of data acquisition planning, data generation, data measurement, data recording, data processing, data use, data storage, data reuse, etc. Even the phenomenon of "not producing data" itself belongs to the category of data. Data will not be lost during processing and use, but will be used more and more. Therefore, data cannot be discarded as easily as files, but should be properly saved. Even data that is not used temporarily will become useful after a certain period of time.

To maintain the integrity of data, it is necessary to ensure the integrity of the data chain, including the integrity of data design, data attributes, constraints, records, use and development. The integrity of data required for a specific purpose can be within a relatively complete framework, which will be limited by the definition of data, data constraints, entities and references that generate data, data verification, data interpretation ability, database storage, etc.

3. Validity

Validity refers to the extent to which the collected data meet the requirements of the specification, and it also refers to the extent to which the data meet various conditions defined by users. In the process of data collection, data are generated from the objects that the investigators really want to investigate. The test objects have not been deviated or replaced. The measured objects are indeed the objective things that the measuring personnel or instruments want to measure. The measuring instruments are not constrained by the measurement system error and random error. In view of usefulness, data validity is the normalization of data, which refers to the satisfaction of data to data standards. These specifications are usually constrained from data naming, data type, data length, data value range, data content standard, data acquisition date, etc. All kinds of data that do not meet the requirements of the specification can be regarded as invalid data, but these data are not useless data. If the specification is changed, the invalid data can also be transformed into valid data.

4. Uniqueness

Uniqueness means that there are no duplicate or redundant data records in a dataset. The uniqueness of data is reflected in the fact that there is no duplication in the process of data entry, export and use, unless repeated entry, export and use are required. Strictly speaking, even repeated entry, export and use of data are new data categories, but the purpose of data use has changed. For example, if a coffee shop actually sells 1000 cups of coffee in a period of time, and the coffee shop's data table shows that 1000 cups are also sold, then the data is unique. Suppose that from the relevant data table of the coffee shop, there are 10 pieces of duplicate sales information, showing 1010 cups sold, then the data table is not unique. It can be said that the uniqueness of data is an important factor to maintain business collaboration and traceability of processes. Maintaining data uniqueness is a basic link in data development and use.

5. Timeliness

Timeliness refers to the extent to which data is updated in a timely manner. Timeliness can be reflected in two aspects. First, the immediacy of data collection. When data is generated, it will be reflected or acted on the collected time nodes, transmitted time nodes, stored time nodes, and processing time nodes in real time. There is no time delay or time delay is almost ignored. If the data is collected, transmitted and stored after a period of time, the timeliness of the data is poor. Second, the timeliness of data application. The time interval between data acquisition and data application meets the specified requirements, and the data timeliness is good; If the time interval exceeds the specified requirements, the timeliness is not ideal. The timeliness of data application is closely related to the speed and efficiency of users' data processing. It is necessary to ensure that users can obtain data in time when they need it.

6. Consistency

Consistency refers to the data used to describe the same objective object and maintain the same information attributes in different data sets. When multiple users access and use a database at the same time, the data they use is consistent in various attributes, the logical relationship between associated data is correct and complete, and the data storage is consistent. There are many attributes of data, including the level of detail of data description for objective objects, the difference between data and objective objects, the minimum distinguishability, unit, format, etc. between two measurable values. The attributes of data are consistent, which is conducive to the development and management of data applications.

3.2 由抽样反映全貌

———

通过抽样而获得的数据，能够真实反应客观对象的全貌，是数据质量的重要体现。一次成功的抽样，要有设计完善的抽样方案，其中又需要至少两个方面的保证：一是抽样方法的正确性（在第二章第四节已经有介绍），主要是体现抽样的随机性；二是样本容量的确定。以下重点介绍样本容量的确定。

3.2.1 样本容量的确定

关于样本容量的确定，这里以简单随机抽样为例。

1. 估计平均值的样本容量

估计平均值是在管理活动中经常遇到的问题，估计平均值的样本容量公式为

$$n = \frac{Z^2 \sigma^2}{E^2}$$

其中，n 是指样本容量；

Z 是标准误差的置信水平；

σ 是总体标准差；

E 是可接受的抽样误差范围，即允许误差。

关于标准误差的置信水平 Z 和总体标准差 σ 的确定，需要根据抽样方的实际情况而定，比如要考虑统计原则、财务成本的投入、管理过程的执行情况、人员的安排等，因此，管理者需要在抽样结果的精确度、置信水平、财务成本、管理执行等各方面予以综合权衡。

关于可接受的抽样误差范围 E 的确定，是要有一些经验值作为参考，毕竟抽样这个实际过程还没有发生。那么在不抽样的前提下估计出总体的标准差，通常可以参考以下做法：

一是参考以往的调研结果，把过去所做的调研总体的标准差作为本次标准差的估计值。

二是开展小规模的试点调查，在调查完成后计算出标准差，以此作为总体标准差的估计值。

三是利用二手的标准差数据，可以向其他单位、同行等咨询。

四是在没有任何可借鉴数据的情况下，只能依赖资深的、与统计相关的管理人员，让他们根据自己的经验来判断标准差的估计值。

2. 估计比例的样本容量

估计比例在管理活动中遇到的概率也相当大。估计比例的样本容量公式为

$$n = \frac{Z^2 \left[P(1-P) \right]}{E^2}$$

其中，n 是指样本容量；

Z 是标准误差的置信水平；

P 是抽样中的样本比例，通常 $P \leqslant 5\%$；

E 是可接受的抽样误差范围，即允许误差。

3. 估计大比例的样本容量

当样本容量小于或等于总体的 5% 的时候，通常认为样本之间是相互独立的。如果样本容量超过总体的 5%，那就属于大比例的样本容量，样本之间就不再认为是相互独立的。因此估计大比例的样本容量公式为

$$n' = \frac{nN}{N + n - 1}$$

其中，n' 是修正后的样本容量；

n 是原来的样本量；

N 是总量。

3.2.2 举例说明样本容量的确定

现在有一个调查者需要估计某市每个月的网上购物次数。下面分别从平均值的样本容量、比例的样本容量和大比例的样本容量三个方面进行估计。

1. 估计平均值的样本容量

在估计平均值的样本容量中，几个参数的设置如下：

（1）考虑到调查者对于管理精确度的要求，把允许误差 E 设定为 0.10（代表估计值不超过实际值的 1/10）；

（2）把总体平均值在区间内的置信度设定为 95%，那么考虑 2 倍标准误差的范围，所以 Z 取值为 2（严格来说，应当为 1.96）；

（3）根据以往的资料，调查者曾经在相类似的城市所作的调查，在最近的一个月中网上购物的平均次数标准差为 1.35，即 $\sigma = 1.35$。

因此，

$$n = \frac{Z^2 \sigma^2}{E^2}$$

$$= \frac{2^2 \cdot (1.35)^2}{(0.10)^2}$$

$$= 729$$

说明，当样本容量达到 729 时，能够满足要求。

2. 估计比例的样本容量

现在估计最近的一个月中，中年人在网上购物所占的百分比。在估计比例的样本容量中，几个参数的设置如下：

（1）允许误差（可接受的误差范围）E 设定为 0.02。该允许误差显然小于对平均值估计时的允许误差，要求更高；

（2）抽样估计在实际总体比例的 ±1.96% 范围内，置信度为 95%，为计算方便，取 $Z = 2$；

（3）查阅资料标明，在相类似的城市中最近的一个月中，中年人网上购物的比例为 4.5%，即 $P = 0.045$。

因此，

$$n = \frac{Z^2 \left[P(1-P) \right]}{E^2}$$

$$= \frac{2^2 \left[0.045(1-0.045) \right]}{(0.02)^2}$$

$$= 430$$

说明，当样本量达到 430 时，能够满足要求。

3. 估计大比例的样本容量

当样本量>5%的时候，由于样本之间不再考虑为相互独立，因此，样本容量的计算公式做了调整。现在假设总量为 3000，原来的样本量为 400，则有

$$n' = \frac{nN}{N+n-1}$$

$$= \frac{400 \cdot 3000}{3000+400-1}$$

$$= 353$$

说明，当样本容量达到 353 时，能够满足要求。

3.2 Reflection of the Overall Perspective by Sampling

The data obtained through sampling can truly reflect the panorama of the objective object, which is an important manifestation of data quality. A successful sampling requires a well-designed sampling plan, which requires at least two guarantees: first, the correctness of the sampling method (introduced in Section 4 of Chapter 2), mainly reflecting the randomness of sampling; the second is the determination of sample size. The following focus on the determination of sample size.

3.2.1 Determination of sample size

For the determination of sample size, simple random sampling is taken as an example.

1. Sample size of estimated average

Estimating the average value is a common problem in management activities. The formula of the sample size of estimated average is

$$n = \frac{Z^2 \sigma^2}{E^2}$$

In the formula, n refers to the sample size;

Z is the confidence level of the standard deviation;

σ is the standard deviation of population;

E is the acceptable sampling error range, that is, the allowable error.

On the determination of the confidence level Z of standard error and the population standard deviation σ, it should be determined according to the actual situation of the sampling party, such as considering statistical principles, financial cost input, management process implementation, personnel arrangement, etc. Therefore, managers need to make a comprehensive balance in accuracy, confidence level, financial cost, management implementation and other aspects of the sampling results.

As for the determination of the acceptable sampling error range E, it is necessary to have some empirical values as a reference. After all, the actual process of sampling has not yet occurred. Then the standard deviation of the population can be estimated without sampling, and the following practices can generally be referred to:

First, refer to previous research results, and take the standard deviation of the past research population as the estimate of the standard deviation of this time.

Second, carry out small-scale pilot survey, and calculate the standard deviation after the survey is completed, which is used as the estimate of the overall standard deviation.

Third, use second-hand standard deviation data and you can consult with other units and peers.

Fourth, in the absence of any data for reference, we can only rely on senior management personnel related to statistics to judge the estimated value of standard deviation according to their own experience.

2. Sample size of estimated proportion

The probability of the estimated proportion in the management activities is quite large. The formula of the sample size of the estimated proportion is

$$n = \frac{Z^2 [P(1 - P)]}{E^2}$$

In the formula, n is the sample size;

Z is the confidence level of the standard deviation;

P is the sample proportion in sampling, and usually $P \leqslant 5\%$;

E is the acceptable sampling error range, that is, the allowable error.

3. Sample size of estimated large proportion

When the sample size is less than or equal to 5% of the population, it is generally considered that the samples are mutually independent. If the sample size exceeds 5% of the population, it belongs to a large proportion of sample size, and samples are no longer considered to be independent of each other. Therefore, the formula for estimating the large proportion of sample size is:

$$n' = \frac{nN}{N + n - 1}$$

In the formula, n' is the revised sample size;

n is the original sample size;

N is the total amount.

3.2.2 Illustration of the determination of sample size

Now a researcher needs to estimate the number of online shopping in a city every month. The following three aspects are estimated respectively from the sample size of the average value, the proportion of the sample size and the large proportion of the sample size.

1. Sample size of estimated average

In estimating the sample size of the average value, several parameters are set as follows:

(1) Considering the requirements of investigators for management accuracy, the allowable error E is set to 0.10 (representing that the estimated value does not exceed 1/10 of the actual value).

(2) Set the confidence level of the population average value within the interval to 95%, then consider the range of twice the standard error, so Z is taken as 2 (strictly speaking, it should be 1.96).

(3) According to previous data, the standard deviation of the average number of online shopping in the recent month was 1.35 according to the survey conducted by the investigators in similar cities, which means $\sigma = 1.35$.

So,

$$n = \frac{Z^2 \sigma^2}{E^2}$$

$$= \frac{2^2 \cdot (1.35)^2}{(0.10)^2}$$

$$= 729$$

It shows that when the sample size reaches 729, the requirements can be met.

2. Sample size of estimated proportion

Now we estimate the percentage of middle-aged people shopping online in the recent month. In the sample size of the estimated proportion, several parameters are set as follows:

(1) The allowable error (acceptable error range) E is set to 0.02. The allowable error is obviously less than the allowable error for average value estimation, so it is more demanding.

(2) The sampling estimation is within ± 1.96% of the actual population proportion, and the confidence level is 95%. For the convenience of calculation, $Z = 2$ is taken.

(3) According to the reference data, in the recent month in similar cities, the proportion of middle-aged people shopping online was 4.5%, that is, $P = 0.045$.

So,

$$n = \frac{Z^2 [P(1 - P)]}{E^2}$$

$$= \frac{2^2 [0.045(1 - 0.045)]}{(0.02)^2}$$

$$= 430$$

It shows that when the sample size reaches 430, the requirements can be met.

3. Sample size of estimated large proportion

When the sample size is >5%, the calculation formula of sample size is adjusted because the samples are no longer considered to be independent. Now suppose that the total amount is 3000 and the original sample size is 400, then:

$$n' = \frac{nN}{N + n - 1}$$

$$= \frac{400 \cdot 3000}{3000 + 400 - 1}$$

$$= 353$$

It shows that when the sample size reaches 353, the requirements can be met.

3.3　统计及计算过程准确

统计及计算过程准确是数据质量的保证。计算过程正确容易理解，关键在于计算公式使用正确，每个计算步骤遵循计算公式的内在逻辑，不要出错；遇到数据有单位的，关注单位的统一等。如果是使用计算机进行计算，要注意程序和输入数据的正确。统计过程正确，涉及的内容比较广，有统计设计、收集相关数据、统计数据的整理、统计分析等。收集数据的方法有很多，包括直接观察、试验、测量等（在第 2 章中有介绍）。统计数据的整理是指按照规定的方法、流程等对于收集到的数据进行归类、汇总，使数据能基本显示分布特征。统计分析是对上述数据的进一步运算、处理，得到相应的统计结果。这里重点介绍应用统计方法的适宜性及正确性、应用分布曲线的正确性。

3.3.1 应用统计方法的适宜性

应用统计方法的适宜性，是指在各项管理活动中所使用的统计方法与该活动过程所要达到的目的相匹配，能够方便使用者做出准确判断。统计方法是管理活动中使用的工具，不是活动的目的，应用统计方法重在提高管理的有效性和效率。不同的统计方法所体现出的功能与作用是不同的，唯有适宜的统计方法，才能让枯燥的数据发出最合适的声音。每一项统计方法都有其应用的前提和条件。比如：当质量管理人员要分析存在的质量问题时，可以应用图示技术描述方法，如直方图、散布图、趋势图、排列图等，也可应用语言文字描述方法，如因果图、关联图、分层法等；如果要监控生产和测量过程，则可以应用控制图；如果要确定哪些因素对过程、产品性能等有显著影响，或者是确定最优搭配，那么可应用试验设计方法，如 0.618 法、正交试验法等；当生产、设计条件发生变化时，为了预测和控制产品的特性，可以应用回归分析法来提供定量模型；当生产过程受到多因素影响，若干总体的分布都是正态分布，且它们的方差相等，可以应用方差分析来估计各因素的影响程度，如单因子方差分析、双因子方差分析；在检验时，经常会用到抽样方法，如随机抽样、系统抽样、整群抽样、多级抽样等，还可以通过假设检验来判断在规定的风险水平上，一组数据是否符合给定的假设；还有利用计算机软件进行仿真模拟，既安全又经济地获得所模仿系统的运行结果。统计方法非常多，要关注在管理活动中使用适宜的统计方法以达到最佳的管理效果。

3.3.2 应用统计方法的正确性

运用统计方法不仅需要关注适宜性，还需要关注应用统计方法的正确性。应用统计方法的正确性是指相关的人员所运用的某个具体统计方法是规范的，符合该统计方法的基本要求、应用条件、应用过程等，其应用结果能够真实反映该统计方法所要达到的目的。应用的正确性反映在以下几个方面：

（1）应用规范。每个统计方法都有其特定的内涵，在应用的时候，必须遵照每个统计方法的内在要求来开展。比如应用直方图，在收集到数据以后，必须计算数据的极差、根据数据数量的多少来确定相应的组数，然后用组数去除极差得到组距，不然，这些数据的波动特征就难以准确表达。比如应用排列图，左边数轴的单位是频数，右边数轴的单位是累计频率百分比，这样才能清晰地显示各项数据所占的比例，并快速从中找到占比最大的数据项。

（2）数据准确。只有准确的数据才能确保统计方法呈现的结果正确。在各项管理活动中所采集到的数据最好是来自于全部样本，即总体的数据一个不漏，显然这样的工作量是非常巨大的，假如把某个大型工厂的生产过程进行全部取样来开展质量检验，由于样本在不断地产生中，那么这样的大数据分析就不是人工能完成的。在这样的情况下，工厂的质检人员需要通过抽样的方法来获取样本。但如果质检人员的抽样方法不准确、或者抽到的样本数据不能真实反应总体的状态，就会导致数据出现偏差，这样应用统计方法也就失去了准星。

（3）保证应用条件。有些统计方法在应用时是有一定的前提条件的。比如运用控制图有一个重要的前提是生产产品的过程能力必须足够，如果过程能力大于 1.33，那么运用控制图来进行判断数据状况是有足够说服力的；如果过程能力小于 1，则数据离散程度大，运用控制图来判断质量状况会有失偏颇。另外在运用控制图时需要确定所控制的数据是计量值还是计数值，因为两者所选用的控制图是有区别的。有些统计方法在应用时，对于数据量也有一定的要求。比如运用直方图，如果数据量少于 50 个，则图形波动大，只有数据量足够多，直方图才能准确传递样本信息。在运用散布图时，数据组数应当在 30 对以上，全部数据才能较好地反映相关性或无相关性；如果少于 30 对数据，那么数据间的相关性很可能会失真，还需要借助假设检验来进行分析。

（4）计算过程准确。有些统计方法是需要计算过程的，虽然有现成的公式，但必须清楚如何计算来保证计算质量。比如在计算标准差时会面临标准差的计算公式，但其中究竟是除以" n "，还是除以" $n-1$ "，就必须搞清楚是计算总体的标准差还是样本的标准差。在绘制排列图时，首先需要计算不同对象所占有的频率和累计频率，如果相应的频率和累计频率计算错误，则绘制的排列图就不准确。在应用网络图时，需要计算具体每个工作节点上紧前、紧后的作业时间后才能正确绘制；如果紧前、紧后的时间计算错误，那么网络图绘制失败，寻找的关键路径会出现偏差。

（5）内在逻辑清晰。统计方法在应用时，需要分析每个步骤的逻辑性，只有确保逻辑正确，才能有正确的分析结果。比如运用因果图，在具体分析导致问题的原因时，必须确保两者的因果关系清晰、逻辑关系紧密。如果针对问题随意地进行分析原因，那么最后得到的结果是没有使用价值的，不能真正地解决质量问题。

（6）归类准确。有些统计方法应用时，需要把相关信息按照属性进行归类。比如应用亲和图，其主要特点就是把收集到有关某一个特定主题的观点、问题、产品、材料等，按照它们的相互亲近程度来整理、归类。如果在应用时，图中的具体内容不具有亲和性，那么亲和图的内容就不亲和，亲和图就失去了意义。

（7）关键内涵正确。有些统计方法在表述上有使用记号的，记号的内涵理解正确才能保证整个统计过程的应用正确。比如应用正交试验时，有些人对于正交符号的含义理解不清楚，特别是对于正交试验可安排的因素数和因素的水平数理解出偏差，也容易把两者混淆、颠倒，这样所引用的具体试验内容和结果就相差甚远。

（8）图示规范。有些统计方法是通过图示的方式来表达，很多图示有规范性的应用，有些是有约定成俗的应用，应用者要确保这些图示的规范性。比如应用流程图时，缺少"流程开始"和"流程结束"的图示，使局外人看不清该流程是完整的流程还是流程的一部分；有些人在应用流程图中，遇到"判断"时还是用矩形方框表示，不清楚国际规范是应用"菱形框"来表示"判断"的节点。有的人在绘制折线图、散布图时，数轴上没有单位名称，图中也缺少图例，那么这些图只能作为示意图。

3.3.3 应用分布曲线的正确性

在统计方法中有很多的分布曲线，比如正态分布、威布尔分布、贝塔分布、二项分布等，不同的分布曲线适用于不同的分析过程中（严格来说，不同的数据特性呈现出不同的分布曲线），因此，正确地运用分布曲线来辅助分析是十分重要的。

1. 应用正态分布的情形

正态分布以均值为中心，图形左右对称，且呈现钟形。正态分布曲线下的面积总和等于1。理论上，$\mu \pm 1\sigma$（μ 为均值，为标准差）范围内曲线下的面积占总面积的 68.27%，$\mu \pm 1.96\sigma$ 范围内曲线下的面积占总面积的 95.44%，$\mu \pm 2.58\sigma$ 范围内曲线下的面积占总面积的 99.74%。

举例：

某电商小店销售蔬菜包，假设蔬菜包的品种一定、重量一定、价格一定。现有市场检查人员抽查该店铺的蔬菜包重量情况，数据如下：每个蔬菜包的平均重量在 1008 克，标准差为 15 克。现在需要估计蔬菜包重量小于 980 克所占的比例。

由于每个蔬菜包 X 服从非标准正态分布，因此，先进行均值的标准化变换，得到：

$$u = \frac{x - \mu}{\sigma}$$

$$= \frac{980 - 1008}{15}$$

$$= -1.87$$

查正态分布表，得到 $\Phi(-1.8667) = 1 - 0.9693 = 0.0307$

说明在标准正态分布曲线下，从 $-\infty$ 到 $x = 980$ 的比例为 3.07%。

因此，检查人员估计蔬菜包重量小于 980 克的比例为 3.07%。

2. 应用二项分布的情形

二项分布适用于结果只有两种且呈相互对立情形的数据，如果一个结果为 x，则其对立的结果为 $1 - x$，n 个数据之间均相互独立。二项分布是重复 n 次的伯努利试验，其一次随机试验的概率为

$$P(X = k) = \binom{n}{k} p^k (1 - p)^{n-k}, \ k = 0, 1 \cdots, n$$

$$\binom{n}{k} = \frac{n!}{k! (n-k)!}$$

举例：

某公司新开发的一台新机器的有效率为 0.98，现在有 10 台新机器售出，那么，至少有 9 台机器运转正常的概率是多少？

$$P(X \geqslant 9) = P(X = 9) + P(X = 10)$$

$$= \binom{10}{9} 0.98^9 (1 - 0.98)^{10-9} + \binom{10}{10} 0.98^{10}$$

$$= 0.1667 + 0.8171$$

$$= 0.9838$$

因此，售出的 9 台机器运转正常的概率为 0.9838。

3. 应用威布尔分布的情形

威布尔分布是瑞典物理学家 Weibull 为解释疲劳试验的结果而建立的一种分布形式，主要应用于产品的可靠性分析和寿命检验，它的应用场合相当多，比如机电产品的磨损累计失效的分布形式、生产过程和运输时间的关系、天气预测、无线电信号的衰减模型、寿险中的重复索赔模型等，可适用威布尔分布曲线。威布尔分布的概率密度函数为

$$f(t;\theta,\beta) = \frac{\beta}{\theta} \left(\frac{t-t_0}{\theta}\right)^{\beta-1} \exp\left[-\left(\frac{t-t_0}{\theta}\right)^{\beta}\right], t \geqslant 0$$

其中，θ 是比例参数；

β 是形状参数，亦称威布尔斜率；

t_0 是位置参数，是产品的最低寿命，表示在此之前，产品的可靠度为 100%。

当 $\beta < 1$ 的时候，$f(t)$ 是以 $t = \theta_0$ 为渐近线，失效率 $\lambda(t)$ 是递减函数，可用来描述产品早期的失效期。

当 $\beta = 1$ 的时候，$f(t)$ 为指数分布曲线，失效率 $\lambda(t)$ 是常数，可用来描述产品的偶然失效期。

当 $\beta > 1$ 的时候，$f(t)$ 是单峰曲线，当 $2.7 < \beta < 3.7$ 的时候，为近似的正态分布曲线；当 $\beta = 3.313$ 的时候，为正态分布曲线。失效率 $\lambda(t)$ 是递增函数，可用来描述产品的损耗失效期。

举例：

某零件的寿命服从威布尔分布，其形状参数 $\beta = 4$，比例参数 $\theta = 3000$，位置参数 $t_0 = 1000$ 小时。现在求这个零件工作到 2000 小时的可靠度和失效率。

可靠度为：$R(2000) = e^{\wedge} - \left(\frac{2000 - 1000}{3000}\right)^4 = e^{-1/81} = 0.9877$

失效率为：$\lambda(2000) = \frac{4 \cdot (2000 - 1000)^{4-1}}{3000^4} = 0.000\ 049\ 4$

3.3　Accurate Statistics and Calculation Process

The accuracy of statistics and calculation process is the guarantee of data quality. The calculation process is correct and easy to understand. The key lies in the correct use of the calculation formula. Each calculation step follows the inherent logic of the calculation formula without making mistakes; If the data has units, focus on the unification of units. If a computer is used for calculation, pay attention to the correctness of the program and input data. The statistical process is correct and involves a wide range of contents, including statistical design, collection of relevant data, collation of statistical data, statistical analysis, etc. There are many methods to collect data, including direct observation, experiment, measurement, etc. (introduced in Chapter 2). The sorting of statistical data refers to classifying and summarizing the collected data according to the specified methods and processes, so that the data can basically show the distribution characteristics. Statistical analysis is the further operation and processing of the above numbers to obtain the corresponding statistical results. This paper focus on the suitability and correctness of statistical methods and the correctness of distribution curves.

3.3.1 Suitability of applying statistical methods

The suitability of the application of statistical methods means that the statistical methods used in various management activities match the purpose of the activity process and can facilitate users to make accurate judgments. Statistical methods are tools used in management activities, not the purpose of activities. The application of statistical methods focus on improving the effectiveness and efficiency of management. Different statistical methods have different functions. Only appropriate statistical methods can make boring data sound the most appropriate. Every statistical method has its premise and conditions of application. For example, when the quality management personnel want to analyze the existing quality problems, they can use graphic technology description methods, such as histogram, scatter chart, trend chart, pareto diagram, etc., as well as verbal description methods, such as cause and effect diagram, correlation diagram, hierarchical method, etc; If we want to monitor the production and measurement process, we can use the control chart; If we want to determine which factors have a significant impact on the process and product performance, or determine the optimal mix, we can use experimental design methods, such as 0.618 method, orthogonal test method, etc.; When production and design conditions change, in order to predict and control product characteristics, regression analysis can be used to provide quantitative models; When the production process is affected by multiple factors, the distribution of several populations is normal distribution, and their variances are equal. Variance analysis can be used to estimate the impact of various factors, such as single factor analysis of variance and double factor analysis of variance; In inspection, sampling methods are often used, such as random sampling, systematic sampling, cluster sampling, multistage sampling, etc. We can also judge whether a group of data conforms to the given assumptions at the specified risk level through hypothesis testing; In addition, computer software is used for simulation to obtain the operation results of the simulated system safely and economically. There are many statistical methods, so we should pay attention to the use of appropriate statistical methods in management activities to achieve the best management effect.

3.3.2 The correctness of applying statistical methods

The application of statistical methods requires attention not only to the suitability, but also to the correctness of the application of statistical methods. The correctness of the applied statistical method means that a specific statistical method used by the relevant personnel is standardized and conforms to the basic requirements, application conditions, application process, etc. of the statistical method, and its application results can truly reflect the purpose of the statistical method. The correctness of the application is reflected in the following aspects:

(1) The application is standard. Each statistical method has its own specific connotation. When it is applied, it must comply with the inherent requirements of each statistical method. For example, in the application of histogram, after the data is collected, the range of data must be calculated, the corresponding number of groups must be determined according to the number of data, and then the group spacing can be obtained by removing the range with the number of groups, otherwise, the fluctuation characteristics of these data will be difficult to accurately express. For example, in the application of the Pareto Chart, the unit of the left number axis is frequency, and the unit of the right number axis is cumulative frequency percentage. In this way, the proportion of each data item can be clearly displayed, and the data item with the largest proportion can be quickly found.

(2) The data is accurate. Only accurate data can ensure that the results presented by statistical methods are correct. The data collected in various management activities should preferably come from all samples, that is, the overall data should not be missed. Obviously, such workload is very huge. If the production process of a large factory is sampled to carry out quality inspection, because the samples are constantly generated, then such big data analysis cannot be completed manually. In this case, the factory's quality inspectors need to obtain samples by sampling. However, if the sampling method of the quality inspection personnel is not accurate, or the sampled sample data cannot truly reflect the overall state, the data will be biased, so the application of statistical methods will lose the benchmark.

(3) Guarantee the application conditions. Some statistical methods have certain preconditions when applied. For example, an important prerequisite for using control charts is that the process capability of producing products must be sufficient. If the process capability is greater than 1.33, it is persuasive to use control charts to judge data conditions; If the process capability is less than 1, the data dispersion is large, and it is biased to use the control chart to judge the quality status. In addition, when using the control chart, it is necessary to determine whether the controlled data is the measured value or the counted value, because there is a difference between the two control charts. Some statistical methods have certain requirements for data volume when they are applied. For example, if the amount of data is less than 50, the graph will fluctuate greatly. Only when the amount of data is sufficient, the histogram can accu-

rately transmit sample information. When using the scatter chart, the number of data sets should be more than 30 pairs, so that all data can better reflect the correlation or non correlation; If there are less than 30 pairs of data, the correlation between the data is likely to be distorted, and hypothesis testing is also needed for analysis.

(4) The calculation process is accurate. Some statistical methods require calculation process. Although there are ready-made formulas, it must be clear how to calculate to ensure the calculation quality. For example, when calculating the standard deviation, there will be a formula for calculating the standard deviation. However, whether the formula is divided by " n " or " $n-1$ " must be clarified to calculate the standard deviation of the population or the sample. When drawing a Pareto diagram, you first need to calculate the frequency and cumulative frequency occupied by different objects. If the corresponding frequency and cumulative frequency are calculated incorrectly, the Pareto diagram is not accurate. When applying the network diagram, it is necessary to calculate the operation time before and after the time of each specific work node before it can be correctly drawn; If the time before and after is calculated incorrectly, the network diagram fails to be drawn, and the critical path will be deviated.

(5) The internal logic is clear. In the application of statistical methods, it is necessary to analyze the logic of each step. Only by ensuring that the logic is correct, can correct analysis results be obtained. For example, when using the cause and effect diagram to specifically analyze the causes of problems, we must ensure that the causal relationship between the two is clear and the logical relationship is close. If we randomly analyze the cause of the problem, the final result is of no use value and can not really solve the quality problem.

(6) The classification is accurate. When some statistical methods are applied, relevant information needs to be classified by attributes. For example, the main feature of application affinity map is to collect opinions, problems, products, materials, etc. on a specific topic and sort them according to their closeness. If the specific content in the graph is not compatible during application, the content of the affinity graph will be incompatible and the affinity graph will lose its meaning.

(7) The key connotation is correct. Some statistical methods use marks in their expressions. The correct understanding of the connotation of the marks can ensure the correct application of the whole statistical process. For example, when using orthogonal test, some people do not understand the meaning of orthogonal symbol clearly, especially for the number of factors that can be arranged in orthogonal test and the deviation of factor level mathematical solution. It is also easy to confuse and reverse the two, so the specific test contents and results quoted are far from each other.

(8) The graphic is normative. Some statistical methods are expressed in the form of diagrams. Many diagrams have normative applications, and some have conventional applications. Users should ensure that these diagrams are normative. For example, when using a flowchart, there is a lack of diagrams for "process start" and "process end", which makes it difficult for outsiders to see whether the process is a complete process or a part of the process; In the application flow chart, some people still use rectangular boxes to represent "judgments". It is unclear that the international standard is to use "diamond boxes" to represent "judgments" nodes. When drawing broken line charts and scatter charts, some people do not have unit names on the number axis, nor do they have legends, so these charts can only be used as

schematic diagrams.

3.3.3 Correct application of distribution curve

There are many distribution curves in the statistical methods, such as normal distribution, Weibull distribution, Beta distribution, binomial distribution, etc. Different distribution curves are applicable to different analysis processes (strictly speaking, different data characteristics show different distribution curves). Therefore, it is very important to correctly use distribution curves to assist analysis.

1. Application of normal distribution

The normal distribution takes the mean value as the center, the graph is symmetrical on the left and right, and presents a bell shape. The sum of the areas under the normal distribution curve is equal to 1. μ means the average value and σ means the standard deviation. Theoretically, the area under the curve within the range of $\mu \pm 1 \sigma$ accounts for 68.27% of the total area, the area under the curve within the range of $\mu \pm 1.96 \sigma$ accounts for 95.44% of the total area, and the area under the curve within the range of $\mu \pm 2.58 \sigma$ accounts for 99.74% of the total area.

Case:

A small e-commerce store sells vegetable bags, assuming that the variety, weight and price of vegetable bags are certain. The existing market inspectors spot check the weight of the vegetable bags in the shop. The data are as follows: the average weight of each vegetable bag is 1008g, and the standard deviation is 15g. Now we need to estimate the proportion of vegetable bags weighing less than 980g.

Because each vegetable bag obeys non-standard normal distribution, therefore, the standardized transformation of the mean value is performed first, and the following results are obtained:

$$u = \frac{x - \mu}{\sigma}$$
$$= \frac{980 - 1008}{15}$$
$$= -1.87$$

Check the normal distribution table, we can get $\Phi(-1.8667) = 1 - 0.9693 = 0.0307$, which shows that under the standard normal distribution curve, the ratio from $-\infty$ to $x = 980$ is 3.07%.

So, the inspector estimates that the proportion of vegetable bags weighing less than 980g is 3.07%.

2. Application of binomial distribution

Binomial distribution is applicable to data with only two opposite results. If one result is x, the opposite result is $1 - x$, and n data are independent of each other. The binomial distribution is a Bernoulli test repeated n times, and the probability of one random test is

$$P(X = k) = \binom{n}{k} p^k (1 - p)^{n-k}, k = 0, 1 \cdots, n$$

$$\binom{n}{k} = \frac{n!}{k! (n - k)!}$$

Case:

The efficiency of a new machine developed by a company is 0.98. Now there are 10 new machines sold. What is the probability that at least 9 machines will operate normally?

$$P(X \geqslant 9) = P(X = 9) + P(X = 10)$$
$$= \binom{10}{9} 0.98^9 (1 - 0.98)^{10-9} + \binom{10}{10} 0.98^{10}$$
$$= 0.1667 + 0.8171$$
$$= 0.9838$$

Therefore, the probability of the 9 sold machines operating normally is 0.9838.

3. Application of Weibull distribution

Weibull distribution is a distribution form established by the Swedish physicist Weibull to explain the results of fatigue tests. It is mainly used in the reliability analysis and life inspection of products. It has been applied in many fields, such as the distribution form of wear cumulative failure of mechanical and electrical products, the relationship between production process and transportation time, weather prediction, the attenuation model of radio signals, the repeated claim model in life insurance, etc, Applicable to Weibull distribution curve. The probability density function of Weibull distribution is:

$$f(t;\theta,\beta) = \frac{\beta}{\theta} \left(\frac{t - t_0}{\theta}\right)^{\beta-1} \exp\left[-\left(\frac{t - t_0}{\theta}\right)^{\beta}\right], t \geqslant 0$$

In the formula, θ is a sacale parameter.

β is a shape parameter, also called Weibull slope.

t_0 is a positional parameter, which is the minimum life of the product, indicating that the reliability of the product is 100% before that.

When $\beta < 1$, $f(t)$ is the asymptote of $t = \theta_0$. The failure rate $\lambda(t)$ is a decreasing function, which can be used to describe the early failure period of the product.

When $\beta = 1$, $f(t)$ is the exponential distribution curve. The failure rate $\lambda(t)$ is a constant, which can be used to describe the accidental expiration period of the product.

When $\beta > 1$, $f(t)$ is a unimodal curve. When $2.7 < \beta < 3.7$, it is an approximate normal distribution curve; When $\beta = 3.313$, it is a normal distribution curve. The failure rate $\lambda(t)$ is an increasing function, which can be used to describe the loss and expiration period of products.

Case:

The life of a part follows Weibull distribution, its shape parameter $\beta = 4$, scale parameter $\theta = 3000$, the positional parameter $t_0 = 1000$ hours. Now we calculate the reliability and failure rate of this part when it works for 2000 hours.

The reliability is: $R(2000) = e\char`^ -\left(\dfrac{2000-1000}{3000}\right)^4 = e^{-1/81} = 0.9877$

The failure rate is: $\lambda(2000) = \dfrac{4 \cdot (2000-1000)^{4-1}}{3000^4} = 0.000\,049\,4$

3.4 推论过程正确

——

推论，也可以称之为推理，是指由一个或者若干个已知的命题（即前提）得出新命题（即结论）的逻辑思维过程或逻辑思维形式。推论在日常管理工作中普遍存在。推论过程的正确，是数据质量的重要保证。推论主要包括逻辑推理和统计推论。

3.4.1 逻辑推理

逻辑推理是指从某些已知条件出发，推出合理结论的过程。逻辑推理中的已知条件和结论都是可以判断真假的命题。如果只研究命题推理的规律，就是命题逻辑。把命题细分为谓词、量词，就得到谓词逻辑。用符号表示命题、谓词、量词，就得到符号逻辑。符号逻辑通常用来研究数学中的推理，因此也叫作数理逻辑。数理逻辑有四个主要分支，分别是集合论、模型论、递归论、证明论。现在，各种应用逻辑发展很快，已有量子逻辑、时态逻辑、概率逻辑、模态逻辑、模糊逻辑等各种现代逻辑。这里主要介绍逻辑推理中的三个主要形式：演绎推理、归纳推理和类比推理。

1. 演绎推理

演绎推理是从一般性的前提出发，通过推导（也就是演绎）的方式，得出具体结论的过程。它是一种确实性推理，推理的前提与结论之间的联系是必然的。关于演绎推理，还有几种定义的方式：演绎推理是从一般到特殊的推理；演绎推理是前提中蕴涵结论的推理；演绎推理是前提和结论之间具有必然联系的推理；演绎推理是前提与结论之间具有充分条件或充分必要条件联系的必然性推理。

演绎推理主要有三段论、假言推理、选言推理、关系推理等几种形式。

（1）三段论

三段论是演绎推理的一般模式，它是由两个含有一个共同项的性质判断作为前提，得出一个新的性质判断为结论的演绎推理过程。三段论包含了三个部分，即大前提，指已知的一般原理；小前提，指所研究的特殊情况；结论，指根据一般原理，对特殊情况做出判断。关于三段论有以下几个概念：

小项：结论的主项叫作"小项"，用"S"表示；

大项：结论中的谓项叫作"大项"，用"P"表示；

中项：两个前提中共有的项叫作"中项"，用"M"表示；

大前提：含有大项的前提叫作大前提；

小前提：含有小项的前提叫作小前提。

结论：三段论的推理过程，就是根据两个前提中所表明的中项与大项、小项之间的关系，通过中项的媒介作用，推导出确定小项与大项之间的关系。

举例：

所有的水果都含有大量水分，苹果是水果，因此，苹果含有大量水分。

在例子中，"苹果"是小项，"含有水分"是大项，"水果"是"中项"。"所有的水果都含有大量水分"是大前提，"苹果是水果"是小前提，"苹果含有大量水分"是结论。

（2）假言推理

假言推理是以假言判断为前提的推理。它分为充分条件假言推理和必要条件假言推理两种。

充分条件假言推理的原则是：如果小前提肯定大前提的前件，结论就肯定大前提的后件；如果小前提否定大前提的后件，结论就否定大前提的前件。

举例：

如果一个数的末位数是 0，那么这个数能够被 5 整除；30 这个数的末位为 0，所以 30 这个数能被 5 整除。

如果一个图形是正五边形，那么它的五条边相等；这个图形的五条边不相等，所以它不是正五边形。

必要条件假言推理的原则是：如果小前提肯定大前提的后件，结论就要肯定大前提的前件；如果小前提否定大前提的前件，结论就要否定大前提的后件。

举例：

只有认真复习，才能考试成绩好；小明的考试成绩好，所以小明认真复习了。

只有认真复习，才能考试成绩好；小明没有认真复习，所以小明考试成绩不好。

（3）选言推理

选言推理是以选言判断为前提的推理。它分为相容的选言推理和不相容的选言推理两种。

相容的选言推理的原则是：大前提是一个相容的选言判断，小前提否定了其中一个（或一部分）选言支，结论就要肯定剩下的选言支。

举例：

一个地方环境不好，或者是空气质量不好导致，或者是水污染导致；这个地方的空气质量好，所以这个地方的环境不好是水污染导致。

一个地方环境不好，要么是空气质量不好导致，要么是水污染导致，要么是乱堆垃圾导致；这个地方的环境不好是水污染导致，所以这个地方的环境不好不是空气质量不好导致，也不是乱堆垃圾导致。

一个地方环境的不好，要么是空气质量不好导致，要么是水污染导致，要么是乱堆垃圾导致；这个地方的环境不好不是空气质量不好导致，也不是水污染导致，所以这个地方的环境不好是乱堆垃圾导致。

（4）关系推理

关系推理是前提中至少有一个是关系命题的推理。通常有：对称性关系推理、反对称性关系推理、传递性关系推理。

举例：

对称性关系推理：因为 $a = b$，所以 $b = a$.

反对称性关系推理：因为 $a > b$，所以 $b < a$.

传递性关系推理：因为 $a > b$，$b > c$，所以 $a > c$.

2. 归纳推理

归纳推理是指由个别到一般的推理，是从个别对象推知一类对象，从个别的知识推知中，概括出一般原理或规律的推理形式和思维方法。归纳推理包括完全归纳法和不完全归纳法。

（1）完全归纳法

完全归纳法是指根据某类事物中的每一个对象都具有某种属性，从而推出该类事物都具有该种属性的结论。完全归纳法适用于数量不多的事物。

完全归纳法的逻辑形式是：

A_1 是 B，A_2 是 B，…，A_n 是 B；

A_1，A_2，…，A_n 是 A 类别的全部对象；

所以，所有的 A 都是 B。

完全归纳法要获得正确的结论，必须满足两个条件：一是在前提中考察了一类事物的全部对象；二是前提中对该类事物中每一个对象所作的断定都是真的。

完全归纳法有两个作用：一是认识作用，它使人的认识从个别上升到了一般。二是论证作用，它的前提和结论之间的联系是必然的。

（2）不完全归纳法

不完全归纳法是根据某类事物中部分对象都具有某种属性，从而推出该类事物都具有该种属性的结论。不完全归纳推理的结论，不能作为演绎推理的大前提。不完全归纳法还可以分为简单枚举归纳、科学归纳两种。

简单枚举归纳是指在一类事物种，根据已经观察到部分对象都具有某种属性，并且没有反例，从而推出该类事物都具有该种属性的结论。

简单枚举法的逻辑形式是：

A_1 是 B，A_2 是 B，…，A_n 是 B；

A_1，A_2，…，A_n 是 A 类别的部分对象，并且其中没有 A 不是 B；

所以，所有的 A 都是 B。

要提高简单枚举归纳的可靠性，要注意两个方面：一是枚举的数量要足够多，考察的范围要足够广。二是考察有无反例。否则会出现"以偏概全"或"轻率概括"的现象。

科学归纳是指根据某类事物中部分对象与某种属性之间因果关系的分析，推出该类事物具有该种属性的推理。

科学归纳法的逻辑形式是：

A_1 是 B，A_2 是 B，…，A_n 是 B；

A_1，A_2，…，A_n 是 A 类别的部分对象，其中没有 A_i（$1 \leqslant i \leqslant n$）不是 B；并且，科学研究表

明，A 和 B 之间有因果关系。

所以，所有的 A 都是 B。

科学归纳的可靠性较高，因为它考察了对象与属性之间的因果关系。

3. 类比推理

类比推理是根据两个或两类事物在某些属性上有相同或相似之处，而且已知其中一个事物具有某种属性，由此推知另一个事物也可能具有这种属性的推理。类比推理是从观察个别现象开始的，有点近似于归纳推理；但类比推理不是从特殊推到一般，而是从特殊推到特殊，所以与归纳推理存在区别。

举例：

"杂志对于编辑"，相当于"蔬菜对于农民"。

上面这句话就是类比推理。其中，"杂志与编辑"两者是"产品和生产者"的关系，"蔬菜与农民"两者也是"产品和生产者"的关系。

4. 推论的有效性

推论的有效性，从演绎的角度是指当一个演绎推论的所有前提为真时，其结论必然为真，那么具有这种性质的推论就是有效的。推论的有效性取决于推论的形式，意味着一个推论是有效的，当且仅当这个推论是一个有效推论形式的替换。

5. 推论的可靠性

推论的可靠性是指当且仅当一个推论是有效的，并且它的所有前提都是真的。因为推论有四种方式：一是有效推论所有的前提为真而且结论也为真。二是反例，所有的前提为真，而结论为假。三是有效推论至少有一个前提为假，而结论为真。四是有效推论中的另一个情形：至少有一个前提为假，而结论也假。可以看出，一个有效的推论，当它的前提为真时，结论必真；而它的前提为假时，结论可以是真，也可以是假。由此得出推论可靠性的定义。可以看出，推论的可靠性是推论有效性的充分条件，推论有效性是推论可靠性的必要条件。

3.4.2 统计推论

统计推论是指调查者收集数据，然后对数据进行分析，从抽样调查的局部情况来推断总体的情形，从而做出相应的决策。统计推论主要有参数估计、假设检验两种。参数估计是根据抽样的结果，来估计总体参值的大小和范围，它主要有两种方法，一是点估计，是利用样本的统计单值来直接估计未知总体参数；二是区间估计，它是在有一定把握的程度上，对于总体参数可能落入的一个区间范围做出估计。按照一定的显著性水平求得的估计区间为置信区间。

下面主要介绍正态分布中未知参数的置信区间和正态分布中未知参数的假设检验。

1. 正态分布中未知参数的置信区间

一个正态总体的情形：

有 $X_1, \cdots, X_n$ 是正态总体 $N(\mu, \sigma^2)$ 的大小为 n 的样本，在置信水平 $1-\alpha$ 下，求得未知参

数的置信区间。

第一种情况，μ 未知，σ^2 已知，则 μ 的一个双侧置信区间为

$$(\bar{X} - \mu_{1-\frac{\alpha}{2}}\frac{\sigma}{\sqrt{n}}, \ \bar{X} + \mu_{1-\frac{\alpha}{2}}\frac{\sigma}{\sqrt{n}})$$

μ 的一个单侧置信区间上限为

$$\bar{X} + \mu_{1-\alpha}\frac{\sigma}{\sqrt{n}}$$

μ 的一个单侧置信区间下限为

$$\bar{X} - \mu_{1-\alpha}\frac{\sigma}{\sqrt{n}}$$

第二种情况，μ 与 σ^2 均未知，则 μ 的一个双侧置信区间为

$$(\bar{X} - t_{1-\frac{\alpha}{2}}(n-1)\frac{S}{\sqrt{n}}, \ \bar{X} + t_{1-\frac{\alpha}{2}}(n-1)\frac{S}{\sqrt{n}})$$

μ 的一个单侧置信区间上限为

$$\bar{X} + t_{1-\alpha}(n-1)\frac{S}{\sqrt{n}}$$

μ 的一个单侧置信区间下限为

$$\bar{X} - t_{1-\alpha}(n-1)\frac{S}{\sqrt{n}}$$

σ^2 的一个双侧置信区间为

$$(\frac{nS_n^2}{\mathsf{X}^2_{1-\frac{\alpha}{2}}(n-1)}, \ \frac{nS_n^2}{\mathsf{X}^2_{\frac{\alpha}{2}}(n-1)})$$

2. 正态分布中未知参数的假设检验

一个正态分布的情形：

（1）u 检验

$(X_1, \cdots, X_n)$ 是取自正态分布 $N(\mu, \sigma^2)$ 的一个样本，其中，$-\infty < \mu < \infty$，$\sigma^2 > 0$。取显著性水平为 α。

假定 μ 未知，σ^2 已知，要检验

$$H_0: \ \mu = \mu_0 \ (H_1: \ \mu \neq \mu_0)$$

取检验统计量

$$U = u(X_1, \cdots, X_n) = \sqrt{n}\frac{\bar{x} - u_0}{\sigma}$$

当 H_0 成立，即 $\mu = \mu_0$ 时，$U \sim N(0, 1)$。

拒绝域 $W_1 = \{(X_1, \cdots, X_n): \ |u(X_1, \cdots, X_n)| > c\}$，

对于给定的显著性水平 α，当 $\mu = \mu_0$ 时，

$$P(|U| > u_{1-\frac{\alpha}{2}}) = \alpha$$

临界值 $c = u_{1-\alpha/2}$

（2） t 检验

假定 μ 和 σ^2 均未知，要检验原假设

$$H_0 : \mu = \mu_0 (H_1 : \mu \neq \mu_0)$$

取检验统计量

$$T = \sqrt{n}\, \frac{\bar{X} - \mu_0}{S}$$

对于给定的显著性水平 α ，一个检验的拒绝域为

$$W_1 = \left\{ (X_1, \cdots, X_m) : \sqrt{n}\, \frac{|\bar{X} - \mu_0|}{S} > t_{1-\frac{\alpha}{2}}(n-1) \right\}$$

这个检验，通常称为 t 检验（也称为学生检验 student′s t -test）。

3.4 Correct Inference Process

Inference, also known as reasoning, refers to the logical thinking process or form of logical thinking in which a new proposition (i.e. a conclusion) is obtained from one or more known propositions (i.e. premises). Inference is common in daily management. The correctness of inference process is an important guarantee of data quality. Inference mainly includes logical inference and statistical inference.

3.4.1 Logical inference

Logical reasoning refers to the process of deriving reasonable conclusions from certain known conditions. The known conditions and conclusions in logical reasoning are all propositions that can judge whether they are true or false. If we only study the laws of propositional reasoning, it is propositional logic. The proposition is subdivided into predicate and quantifier, and predicate logic is obtained. Using symbols to express propositions, predicates, quantifiers, we get symbolic logic. Symbolic logic is usually used to study reasoning in mathematics, so it is also called mathematical logic. Mathematical logic has four main branches, namely, set theory, model theory, recursion theory and proof theory. Now, various applied logics have developed rapidly, including quantum logic, temporal logic, probability logic, modal logic, fuzzy logic and other modern logics. This paper mainly introduces three main forms of logical reasoning: deductive reasoning, inductive reasoning and analogical reasoning.

1. Deductive reasoning

Deductive reasoning is the process of drawing specific conclusions from the general premise through deduction (that is, deduction). It is a kind of certainty reasoning, and the connection between the premise and conclusion of reasoning is inevitable. There are several ways to define deductive reasoning: deductive reasoning is from general to special; Deductive inference is the inference of the conclusion contained in the premise; Deductive inference is a kind of inference that has an inevitable connection between premise and conclusion; Deductive reasoning is a kind of inevitable reasoning with sufficient or necessary conditions between premises and conclusions.

Deductive reasoning mainly includes syllogism, hypothetical reasoning, disjunctive reasoning, relational reasoning, etc.

(1) Syllogism

Syllogism is a general mode of deductive reasoning. It is a deductive reasoning process that takes two property judgments with a common term as the premise and draws a new property judgment as the conclusion. Syllogism includes three parts, namely, the major premise, which refers to the known general principle; the minor premise, which refers to the special situation under study; conclusion, which refers to making judgments on special situations according to general principles. There are several concepts about syllogism.

Minor term: The main term of the conclusion is called "minor term", which is represented by "S";

Big term: The predicate in the conclusion is called "big term", which is represented by "P";

Middle term: The common term of the two premises is called "middle term", which is represented by "M";

Big premise: The premise with big items is called "big premise";

Minor premise: The premise with minor items is called "minor premise".

Conclusion: The reasoning process of syllogism is to deduce and determine the relationship between minor items and major items through the intermediary role of the medium term according to the relationship between the medium term and major items and minor items indicated in the two premises.

Case:

All fruits contain a lot of water. Apples are fruits. Therefore, apples contain a lot of water.

In the above case, "apple" is the minor term, "contain a lot of water" is the big term and "fruit" is the middle term. "All fruits contain a lot of water" is the big premise, "Apples are fruits" is the minor premise and "Apples contain a lot of water" is the conclusion.

(2) Hypothetical reasoning

Hypothetical reasoning is based on hypothetical judgment. It is divided into two types: sufficient condition hypothetical reasoning and necessary condition hypothetical reasoning.

The rule of the sufficient condition hypothetical reasoning is: If the minor premise affirms the antecedent of the major premise, the conclusion affirms the consequent of the major premise; If the minor premise negates the consequent of the major premise, the conclusion negates the antecedent of the major premise.

Case:

If the last digit of a number is 0, the number can be divided by 5; The last digit of the number 30 is 0, so the number 30 can be divided by 5.

If a figure is a regular pentagon, its five sides are equal; The five sides of this figure are not equal, so it is not a regular pentagon.

The rule of the necessary condition hypothetical reasoning is: if the minor premise affirms the consequent of the major premise, the conclusion must affirm the antecedent of the major premise; if the minor premise negates the antecedent of the major premise, the conclusion will negate the consequent of the major premise.

Case:

Only if a person review carefully can the person get good exam results; Xiao Ming did well in the exam, so he reviewed carefully.

Only if a person review carefully can the person get good exam results; Xiao Ming did not review carefully, so he did not do well in the exam.

(3) Disjunctive reasoning

Disjunctive reasoning is based on disjunctive judgment. It is divided into compatible disjunctive reasoning and incompatible disjunctive reasoning.

The rule of the compatible disjunctive reasoning is as follows: the major premise is a compatible disjunctive judgment. The minor premise negates one (or part of) disjunctive branch, and the conclusion must affirm the remaining disjunctive branch.

Case:

The poor environment of a place is caused by poor air quality or water pollution; The air quality in this place is good, so the bad environment in this place is caused by water pollution.

The bad environment of a place is either caused by poor air quality, water pollution, or littering; The bad environment of this place is caused by water pollution, so the bad environment of this place is not caused by poor air quality, nor by littering.

The bad environment of a place is either caused by poor air quality, water pollution, or littering; The bad environment in this place is not caused by poor air quality or water pollution, so the bad environment in this place is caused by littering.

(4) Relational reasoning

Relational reasoning is the reasoning that at least one of the premises is a relational proposition. Generally, it includes symmetric relational reasoning, anti symmetric relational reasoning, and transitive relational reasoning.

Case:

Symmetrical relation reasoning: because $a=b$, so, $b=a$.

Antisymmetric relational reasoning: because $a>b$, so, $b<a$.

Transitive relational reasoning: because $a>b$ and $b>c$, so, $a>c$.

2. Inductive reasoning

Inductive reasoning refers to the reasoning from individual to general, which infers a class of objects from individual objects and generalizes the reasoning form and thinking method of general principles or laws from individual knowledge inference. Inductive reasoning includes complete induction and incomplete induction.

(1) Complete induction

Complete induction refers to the conclusion that every object in a certain kind of thing has a certain attribute, so that this kind of thing has this attribute. Complete induction applies to a small number of things.

The logical form of complete induction is:

A_1 is B, A_2 is B, $\cdots$, A_n is B;

A_1, A_2, $\cdots$, A_n are all objects of category A;

So, all A's are B.

To get a correct conclusion, complete induction must satisfy two conditions: the first is to investigate all objects of a class of things in the premise; the second is that the judgment made on each object in the premise is true.

Complete induction has two functions: one is cognitive function, which makes people's understanding rise from individual to general. The second is the role of argumentation. The connection between its premise and conclusion is inevitable.

（2）Incomplete induction

Incomplete induction is based on the fact that some objects in a certain kind of things have certain attributes, so it can be concluded that all such things have such attributes. The conclusion of incomplete inductive reasoning cannot be the major premise of deductive reasoning. Incomplete induction can also be divided into simple enumeration induction and scientific induction.

Simple enumeration induction refers to the conclusion that some objects of a kind of things have certain attributes according to the observation that some objects have certain attributes and there are no counterexamples.

The logical form of the simple enumeration induction is:

A_1 is B, A_2 is B, ..., A_n is B;

$A_1, A_2, ..., A_n$ are some objects of category A, and none of A is not B;

So, all A's are B.

To improve the reliability of simple enumeration induction, we should pay attention to two aspects: first, the number of enumerations should be enough and the scope of investigation should be wide enough; second, check whether there are counterexamples, otherwise, the phenomenon of "generalizing the whole from a part" or "summarizing carelessly" will occur.

Scientific induction refers to the inference that a certain kind of thing has this kind of attribute based on the analysis of the causal relationship between some objects and some attributes.

The logical form of the scientific induction is:

A_1 is B, A_2 is B, ..., A_n is B;

$A_1, A_2, ..., A_n$ are some objects of category A, no A_i ($1 \leq i \leq n$) is not B; Moreover, scientific research shows that there is a causal relationship between A and B.

So, all A's are B.

Scientific induction is highly reliable because it examines the causal relationship between objects and attributes.

3. Analogical reasoning

Analogical reasoning is based on the fact that two or two kinds of things have the same or similar attributes, and that one of them has a certain attribute, thus inferring that another thing may also have this attribute. Analogical reasoning starts from observing individual phenomena, which is somewhat similar to inductive reasoning; However, analogical reasoning is different from inductive reasoning because it is not from special reasoning to general reasoning, but from special reasoning to special reasoning.

Case:

"Magazine to editor" is equivalent to "vegetable to farmer".

The above sentence is analogical reasoning. Among them, the relationship between "magazine and editor" is "product and producer", and the relationship between "vegetable and farmer" is also "product and producer".

4. Validity of inference

The validity of inference, from the perspective of deduction, means that when all the premises of a deductive inference are true, its conclusion must be true, then the inference with this property is valid. The validity of inference depends on the form of inference, which means that a inference is valid if and only if the inference is a replacement of an effective inference form.

5. Reliability of inference

The reliability of inference means that if and only if a inference is valid and all its premises are true. Because there are four ways of inference: first, effective inference. All the premises are true and the conclusion is true. The second is the counterexample, where all the premises are true and the conclusion is false. Third, effective inference, at least one premise is false and the conclusion is true. The fourth is another situation in effective inference: at least one premise is false, and the conclusion is also false. It can be seen that when the premise of an effective inference is true, the conclusion must be true; When its premise is false, the conclusion can be true or false. The definition of inferential reliability is obtained. It can be seen that the reliability of inference is a sufficient condition for the validity of inference, and the validity of inference is a necessary condition for the reliability of inference.

3.4.2 Statistical inference

Statistical inference refers to that investigators collect data, analyze the data, infer the population situation from the local situation of the sampling survey, and then make corresponding decisions. Statistical inference mainly includes parameter estimation and hypothesis testing. Parameter estimation is to estimate the size and range of population parameters according to the sampling results. There are two main methods: one is point estimation, which is to directly estimate unknown population parameters by using the statistical single value of samples; The second is interval estimation, which estimates an interval range that the overall parameters may fall into to a certain extent. The estimation interval obtained according to a certain significance level is the confidence interval.

The confidence intervals of unknown parameters in normal distribution and hypothesis test of unknown parameters in normal distribution are mainly introduced below.

1. Confidence intervals of unknown parameters in normal distribution

The case of a normal population:

$X_1, \cdots, X_n$ are samples of normal population $N(\mu, \sigma^2)$ with size n, at the confidence level $1-\alpha$. Obtain the confidence interval of the unknown parameter.

In the first case, μ is unknown. σ^2 is known and a two-sided confidence interval of μ is

$$\left(\bar{X} - \mu_{1-\frac{\alpha}{2}} \frac{\sigma}{\sqrt{n}}, \bar{X} + \mu_{1-\frac{\alpha}{2}} \frac{\sigma}{\sqrt{n}} \right)$$

An upper limit of one side confidence interval of μ is

$$\bar{X} + \mu_{1-\alpha} \frac{\sigma}{\sqrt{n}}$$

A lower limit of one side confidence interval of μ is

$$\bar{X} - \mu_{1-\alpha} \frac{\sigma}{\sqrt{n}}$$

In the second case, μ and σ^2 are unknown. A two-sided confidence interval of μ is

$$\left(\bar{X} - t_{1-\frac{\alpha}{2}}(n-1) \frac{S}{\sqrt{n}}, \bar{X} + t_{1-\frac{\alpha}{2}}(n-1) \frac{S}{\sqrt{n}} \right)$$

An upper limit of one side confidence interval of μ is

$$\bar{X} + t_{1-\alpha}(n-1)\frac{S}{\sqrt{n}}$$

A lower limit of one side confidence interval of μ is

$$\bar{X} - t_{1-\alpha}(n-1)\frac{S}{\sqrt{n}}$$

A two-sided confidence interval of σ^2 is

$$\left(\frac{nS_n^2}{\chi^2_{1-\frac{\alpha}{2}}(n-1)}, \frac{nS_n^2}{\chi^2_{\frac{\alpha}{2}}(n-1)}\right)$$

2. Hypothesis test of unknown parameters in normal distribution

The case of a normal population:

(1) u -test

$(X_1,\cdots,X_n)$ is a sample of normal distribution $N(\mu,\sigma^2)$, $-\infty < \mu < \infty$, $\sigma^2 > 0$. Take the significance level as α.

Assume μ is unknown and σ^2 is known, to test

$$H_0:\mu = \mu_0(H_1:\mu \neq \mu_0)$$

Take test statistics

$$U = u(X_1,\cdots,X_n) = \sqrt{n}\frac{\bar{x} - u_0}{\sigma}$$

When H_0 is established, i.e. when $\mu = \mu_0$, $U \sim N(0,1)$.

The critical region $W_1 = \{(X_1,\cdots,X_n): |u(X_1,\cdots,X_n)| > c\}$,

For a given significance level α, when $\mu = \mu_0$,

$$P(|U| > u_{1-\frac{\alpha}{2}}) = \alpha$$

The critical vlue $c = u_{1-\frac{\alpha}{2}}$

(2) t -test

Suppose that μ and σ^2 are all unknown, we need to test the null hypothesis

$$H_0:\mu = \mu_0(H_1:\mu \neq \mu_0)$$

Take test statistics

$$T = \sqrt{n}\frac{\bar{X} - \mu_0}{S}$$

For a given significance level α, the rejection field of an inspection is

$$W_1 = \left\{(X_1,\cdots,X_m): \sqrt{n}\frac{|\bar{X} - \mu_0|}{S} > t_{1-\frac{\alpha}{2}}(n-1)\right\}$$

This test is usually called t -test (also called student's t -test).

3.5　正确观察

——

观察是获取数据的重要手段，是一项有目的、有计划的知觉活动，是人本身知觉主动及能动体现的重要形式。观察者利用自己的感官，借助必要的观察仪器设备等，对研究的对象进行相应数据获取的过程。观，主要指人的感观、知觉行为，充分利用各项器官进行视、听、嗅、触、尝等各种感知行为。察，主要指积极思考、分析、逻辑推理的过程。由于人的感觉器官具有一定的局限性，观察者往往要借助各种现代化的仪器和手段，如照相机、摄像机等来辅助观察。除了运用"观察"这个方法外，人们在操作各种仪器设备时，对于仪器设备的正确使用也需要观察来辅助完成。因此，正确观察是确保数据质量的重要途径。

3.5.1 保证观察质量

要保证一项观察活动的质量，需要从策划、执行过程、数据收集等各方面予以关注，主要有：

（1）制订观察计划。包括设定观察的目的，明确观察的对象，确认观察的内容，确定观察范围，确认观察的地点，确认观察的周期和起讫时间，明确观察的取样安排，核定观察记录的表式、确定观察的预算，准备观察的相关辅助设备、仪器等。

（2）选择适宜的观察方法。在控制成本的前提下，采用适宜、正确的观察方法。要善于抓住主要观察点，不忽视次要观察内容，留意细微观察对象的变化。要注意有序观察，按照事物的先后排序进行、按照时间的先后顺序开展。有必要执行周期观察时，遵从一个产品的生命周期进行动态观察，关注过程内容。关注动态观察，对于变化着的事物，关注其变化和发展趋势。必要时进行分类分层观察，区分不同的事物类别，从整体到部分，或者由部分到整体进行完整观察。观察方法上运用远近观察，由近及远、由远至近观察。根据需要进行对照观察，获得不同样本的数据等。

（3）按照观察计划实施观察。观察过程要注意系统性，不遗漏观察对象和内容。要注意多角度观察，可以采取不同位置、不同方向进行观察。必要时可以安排多人次观察。观察的过程中，要有积极的观察思维，观察者要有应付突发状况的准备，应时而变、应势而变。观察时要集中注意力，减少及避免无关事物的影响。

（4）如实记录观察内容。根据事先设计的表式进行记录事实。记录要注重描述过程和观察到的现象。不能记录带有虚构、推理的内容，如带有传闻色彩的内容，不记录对于各种现象、问题的解释。

（5）整理和分析观察资料。对于观察记录进行初步的整理，剔除错误的数据，纠正和修复相关记录资料，注重观察细节材料的完善。对于有特殊情形、重点关注的对象，要进行专项整理和分析。

（6）观察人员要培养观察能力。包括培养专注力、辨别事物细微差别及细小特征的能力。要有良好的职业精神，尊重客观事实，记录以事实为依据，对每一个环节做认真详细的记录。分析数据要忠实于记录，还原事物现状。平时要注重积累一定的知识、经验和技能，提升专业素养。

3.5.2 不可忽视的观察盲区

观察法的应用有着很多优点，比如观察者能够通过观察直接获得第一手资料，避免很多中间环节，观察到的资料真实、可靠，所以有"眼见为实"的说法。观察具有及时性，观察者可以捕捉到正在发生的现象。观察的记录可以得到生动的资料，能搜集到语言文字所无法细腻表达的材料。但正如一枚硬币有正反两个面，观察法也有自身的缺陷，于是产生了观察盲区。观察盲区的例子有：

观察一张激光 CD 片，仅仅用肉眼观察，是平坦的，刻录面会折射五彩阳光；如果使用放大镜看，会看到一个个圆圈纹路，甚至会感到有些凹凸不平；但如果借助于电子显微镜观察，会发现这些圆圈纹路并不是连续的，而是由大量二进制的细微块状连接而成。正是有了这些细微的二进制块状符号，通过激光的照射，才能解读出丰富多彩的信息。

我们观察蜜蜂采蜜，以为蜜蜂像人类一样看到了花朵才去采蜜的，但事实是蜜蜂根据花朵的紫外光线所呈现的内容去采蜜。

据一位物理学家介绍，每个人在自然环境下，每天大约有 3 至 4 吨的暗物质穿过了人体，而这些暗物质不仅人类的肉眼、身体无法感知，连当今最高级的仪器设备也无法探知，只能通过相关的计算才能确认暗物质的存在。

归纳起来，主要的观察盲区有：

（1）受到时间上的限制。某些事情的发生是在特定时间内完成或发生的，观察者必须在这个时间段进行观察，否则坐失良机。

（2）受到空间上的限制。观察者由于自身空间上的局限性，或者借助于仪器设备也受到限制，导致观察缺失。

（3）受到屏蔽。由于特定的原因，观察对象受到屏蔽，无法进行观察。

（4）被观察对象不配合。由于被观察对象的不配合态度，导致观察失败。

（5）受到仪器设备观察能力的限制。虽然可以借助仪器设备来辅助观察，但仪器设备也有自身的局限性，无法完成观察过程，或者观察精度不高。

（6）观察者本身能力不够或观察意识不强。由于观察者自己没有受到专业训练或缺少专业训练，无法保证观察的成功。

（7）观察的结果只能限于表象、结构，不能直接观察到本质。

（8）受到时代中科技发展的能力限制。比如无法看到暗物质。

（9）如果是人工观察，完成大规模调查有难度。现在借助于当代科技，可以实现。

3.5.3 六西格玛管理思想在观察中的应用

正确观察的一个重要前提是要确保观察的总体思路清晰。观察者可以引入六西格玛管理的思想。六西格玛的管理目标是通过减小过程输出变差来提高过程表现，那么观察者可以通过分析每次观察的数据输出，通过努力来缩小后续相同观察内容的数据变差。观察者应该对本次需要观察数据的定义、哪些数据是属于当前过程中满足使用者的需求、所观察的数据对整体的业务开展有什么样的影响等有清晰的认知。

在观察过程中要注意的有：

（1）衡量。衡量包括：提出数据的收集计划，明确本次观察的内容、收集什么样的数据。明确怎样衡量、收集数据的方式，判断所提出的衡量系统是否适当。在执行观察过程中，要明确收集数据的量、收集数据的层次、分类、时间、执行人员等。要及时判断整个观察过程是否稳定，观察过程是否满足相关规范要求。计算观察数据的极差、标准差、方差等，了解数据变差的最大来源。

（2）分析。观察者要分析哪些因素对于观察过程输出的影响最大，分析观察数据输出的波动规律，找到的影响因素是真实的原因还是疑似的原因，便于后续观察的开展。

（3）改进。了解当前状况与目标指标之间的差距，识别改进方向。根据分析的结论，不断改进观察的计划，修正观察的过程和其他内容，促使观察过程的改进，提高输出数据的质量。

（4）控制。作为观察者要通过完整的管理流程，确认观察的过程是否持续保持改进水平；观察所需的全部目标是否都实现。

3.5.4 机器观察

1. 机器观察的优势

观察，以调研行为、现象等显性的内容居多，各种人工的观察，通过培训，可以做到客观的记录，但有时会受到个人情绪的影响。人工观察还要受到身体感官的生理限制，会出现疲劳、无法触及到一定的空间等。为此，人们的观察就需要借助各种仪器设备，以不断拓展观察能力。在这样的情形下，机器观察顺应了这个需求。现在人们使用摄像、视听设备、卫星、自动记录仪、智能感应设备等各种机器进行观察，观察的过程和结果更客观、更精确、更稳定、更详细，不会出现疲劳等人工现象（暂且不考虑机器疲劳），基本不会受到时间干扰和空间干扰，在很多的情形下不会出现环境干扰等。

机器观察由于其独特的优势，现在已经得到了广泛应用，尤其是电子（数字）观察设备的大量应用，大大提升了观察的质量和效率。机器观察的应用场景有：机场的乘客自助通关系统；工地上的人脸识别系统；很多企业的网络考勤系统；一些搜索系统的三维实景地图；网络上的关键词等搜索功能；CIM+BIM（City Information Model and Building Information Model）的开发与应用；高速公路上的 ETC（电子不停车收费系统）；企业的工作绩效自动考核系统；安检中的太赫兹波检测；工业机器人、农场机器人等。随着数字化、信息化不断融合进社区、企业、各类机构，机器观察的作用将大放异彩。

2. 流量观察

流量观察是机器观察中的一项重要功能，现在已经得到了大量推广。各种人流、物流、信息流等普遍用到了流量观察。比如：进入机场门口，就有电子人流量的统计；在高铁车站刷身份证进入站台时，是人流量的统计，而且精确到具体个人及相应车厢、座位号；邮政系统普遍应用物流自动统计；网上购物后的物品驿站（临时存放处）是物流统计；网上购物平台是信息流统计；搜索平台是信息流统计；文档的关键词纠错是信息流统计；火车票、机票的手机订票系统是信息流统计；道路上的电子监控系统是人流、车流、物流、信息流的合成统计；医院的 ICU 是医务信息流统计等。流量观察无处不在，尤其是手机、电脑、电子摄像器等成为机器观察的载体以后，流量观察的功能也得到了进一步提升，主要有：

（1）流量记录功能。流量记录是流量观察的基础性功能。各种流量的记录，帮助决策者进行相应的管理判断。比如：某个十字路口的交通流量统计，可以为红绿灯的时间间隔设计提供数据。城市内外的交通流量统计，可以为自动导航设备提供实时数据，方便驾驶人员避开拥堵路段、临时事故路段。

（2）工作行为提醒功能。流量的观察，可以判断工作者的工作状况，提醒工作者做出工作行为的改善。比如：车载观察系统在卫星和人工智能的支持下，及时判断驾驶员的驾驶行为，发出疲劳驾驶提醒、超速提醒、大客车经过提醒、转弯提醒、前方车辆紧急处理提醒等，提高了驾驶安全。道路口的观察系统可以实时提醒行人（尤其是盲人）什么时候可以穿过马路，确保行人安全。电脑上的观察系统，可以评估操作者的工作状态，提醒主人及时调整工作行为、注意休息等。护理院的监护系统，可以随时提醒医务人员对患者做出急救行为，保障生命安全。

（3）营销功能。流量的观察，可以帮助商家实现精准营销。现在互联网已经遍及城乡的每个角落，消费者在网上购物行为的记录、门店购物的清单等，让商家通过汇总分析后及时了解不同消费群体的购物习惯，快速分析哪些商品紧俏、哪些商品滞销等信息，然后在不同的季节、假期等做出相应的营销策略。通过对于广告效果的评估，调整广告的播放时段、播放内容，可以评价不同广告明星的广告效应。餐饮店等可以通过流量观察实现菜品调整、配送调整等，帮助实现销售额和服务质量的提升。

（4）改善供应链功能。通过流量的观察，可以实现对于各种供应商供应能力的评估。比如通过不同渠道销售记录的分析，来及时调整商品采购渠道和物流渠道，改善供应链管理。通过终端对于供应链的评价信息收集，可以及时向供应商提出服务提升的要求。当某个地区发生突发事件、事故时，及时做出某些供应链环节调整的预判等，保障了企业的生产运营工作。

（5）拉动欠发达地区经济功能。流量的观察，可以让决策者获取不同地区物资信息，判断不同地区的不同物品的生产能力、不同商品的消费能力等。比如，相关部门可以通过互联网信息、各地区门店的实时货物旺销信息、某些地区货物的质量高价格低等信息，及时带动相关的市场，尤其可以撬动农村市场、欠发达地区的消费，提升当地的经济。

（6）机器智能管理功能。借助于流量的观察，可以在各种网络平台上实现相关的机器智能管理。比如：通过收集人们点击手机网页的习惯，掌握不同人群的消费爱好，然后向这些人群实行机器智能推送，并进一步评估机器智能推送的效果，进一步确认不同群体的消费爱好、兴趣，强化智能推送的实现。网络上的智能互动功能通过流量观察得以实现，机器评价人员的爱好，人员藉此评价增加与机器的互动，互动式购物、自助式消费等日益普及。

（7）虚拟商店货品自动更新功能。通过流量的观察，管理者可以通过大量商品游览记录的分析，及时掌握网络虚拟商店的货品布置，实现自动货品的更新，以巩固现有的消费人群，开发新兴的消费群体。网络上可以设置虚拟人物，通过流量观察，实现虚拟任务的换装、换肤及相应升级，带动消费。

（8）调配资源的功能。通过流量的观察，管理者可以及时掌握本企业在不同时间、不同地点的生产运营状况，及时感知不同时点的突发情况、事故及相应需求，准确预测、发现不同地区的资源缺少、富余等信息，对于相关资源进行事先调配或紧急调配等，避免不必要的损失。

（9）预测功能。通过流量的观察，管理者可以对相关的管理要素需求与供应进行预测。跨地区管理者可以针对某些流量行为的改变而做出判断。比如：在某地突然出现某种药品的热销，大致可以判断某些紧急病情的出现。通过对搜索关键词变动的分析，可以判断不同商品的流行趋势，比如服装的色彩、样式等。通过不同的视频点击等分析，可以催生专项热销产品。

（10）人力资源管理功能。通过流量观察，管理者可以提升人力资源管理质量。比如：记录上班人员的工作状况，掌握不同岗位工作人员的操作习惯，记录相关人员的工作效率。通过实时掌握有缺陷产品的生成与处理，评价操作人员的生产过程质量和效率。通过分析送货及时率、准确率、好评率、价格波动情况等信息，相关商家可以评价工作人员的工作质量，提升管理水平。

（11）生理监测功能。比如有些观察系统可以监测人员的生理变动情况，当一个人在看了某些广告以后，会激活人体内部的活化作用水平，然后用脑电仪监测脑电波的频率变动情况，用测瞳仪获得眼睛的关注焦点，用声音分析仪器来分析个人的情感与喜好；用监测系统来观察皮肤出汗、体温上升、血流及心跳加快等状况，以此来作为决策的依据。

（12）辅助管理决策功能。数字化时代，数据是决策的重要依据。通过各种流量数据的分析，实现对竞争对手的竞争策略判断，追踪竞争对手的新品情况。在数据流量的分析过程中，寻找自己的数据伙伴，找到新的识别途径，界定合作方的绩效及改进机会，强化战略合作。通过数据分析，实现销量的增加途径，追求利润最大化。在数据分析中，总结经验得失，改善产品策略，制定促销方案，量化不同的机会，辅助快速行动，寻找新的商机。

流量观察的功能还有很多，它必将在实践中得到不断的提升。

3.5　Correct Observation

———

Observation is an important means of obtaining data, a purposeful and planned perceptual activity, and an important form of reflecting the initiative and initiative of human perception. Observers use their own senses, with the help of necessary observation equipment, etc., to conduct the corresponding data acquisition process for the research object. The concept mainly refers to people's sense, perception and behavior, making full use of various organs to see, hear, smell, touch, taste and other perceptual behaviors. Observation mainly refers to the process of positive thinking, analysis and logical reasoning. Due to the limitations of human sensory organs, observers often have to use various modern instruments and means, such as cameras, video cameras, to assist observation. In addition to using the method of "observation", people also need observation to assist in the correct use of instruments and equipment when operating various instruments and equipment. Therefore, correct observation is an important way to ensure data quality.

3.5.1 Ensure observation quality

To ensure the quality of an observation activity, attention should be paid to the planning, implementation process, data collection and other aspects, mainly including the following items.

(1)Draw up an observation plan. It includes setting the purpose of observation, specifying the object of observation, confirming the content of observation, determining the scope of observation, confirming the location of observation, confirming the period and starting and ending time of observation, specifying the sampling arrangement of observation, verifying the form of observation records, determining the budget of observation, preparing relevant auxiliary equipment and instruments for observation, etc.

(2)Select an appropriate observation method. On the premise of cost control, appropriate and correct observation methods shall be adopted. We should be good at grasping the main observation points, not ignoring the secondary observation contents, and pay attention to the changes of subtle observation objects. Pay attention to orderly observation, and carry out according to the order of things and time. When it is necessary to perform cycle observation, follow the life cycle of a product for dynamic observation and focus on process content. Pay attention to dynamic observation, and the change and development trend of things that are changing. When necessary, conduct classified and layered observation to distinguish different categories of things, from whole to part, or from part to whole. The observation method is from near to far, from far to near. Carry out comparative observation as required to obtain data of different samples.

(3)Observe according to the observation plan. The observation process shall be systematic, and the objects and contents shall not be omitted. Attention shall be paid to multi angle observation, which can be conducted in different positions and directions. If necessary, multiple people can be arranged for observation. In the process of observation, there should be positive observation thinking. Observers should be prepared to deal with unexpected situations and change from time to time. Focus on observation, reduce and avoid the influence of irrelevant things.

(4) Record the observation truthfully. Record the facts according to the table designed in advance. Records should focus on describing the process and observed phenomena. It is not allowed to record the content with fiction and reasoning, such as the content with hearsay color, and it is not allowed to record the explanation of various phenomena and problems.

(5) Sort out and analyze the observed data. The observation records shall be preliminarily sorted out, the wrong data shall be eliminated, the relevant records shall be corrected and repaired, and the improvement of observation details shall be emphasized. Special sorting and analysis shall be carried out for objects with special circumstances and focus.

(6) Cultivate their observation abilities. It includes the ability to cultivate concentration and distinguish the subtle differences and characteristics of things. Have good professionalism, respect objective facts, record based on facts, and make careful and detailed records of each link. Data analysis should be faithful to records and restore the status quo of things. At ordinary times, we should focus on accumulating certain knowledge, experience and skills to improve our professional quality.

3.5.2 Non-negligible blind area

The application of observation method has many advantages. For example, observers can directly obtain first-hand data through observation, avoiding many intermediate links, and the observed data is true and reliable, so there is a saying that "seeing is believing". The observation is timely, and the observer can catch what is happening. The observation record can get vivid data and collect materials that cannot be expressed in detail by language. But just as a coin has both positive and negative aspects, the observation method also has its own defects, thus creating a blind area for observation. Examples of blind areas are as follows.

When observing a laser CD, it is flat with the naked eye, and the recording surface will refract colorful sunlight; If you look with a magnifying glass, you will see circle lines, and even feel a little uneven; However, with the aid of electron microscope, it can be found that these circles are not continuous, but are connected by a large number of binary fine blocks. It is with these subtle binary block symbols that colorful information can be interpreted through laser irradiation.

We observe that bees collect honey, thinking that bees, like humans, only collect honey when they see flowers, but the fact is that bees collect honey according to the content presented by the ultraviolet light of flowers.

According to a physicist, every person in the natural environment, about 3 to 4 tons of dark matter pass through the human body every day. This dark matter is not only imperceptible to the human eye and body, but also impossible to detect even the most advanced instruments and equipment. Only through relevant calculations can we confirm the existence of dark matter. To sum up, the main blind areas are as follows:

(1) Limited by time. Some things happen in a specific time. The observer must observe them in this time period, or he will miss the opportunity.

(2) Limited by space. Observers are limited due to their own space limitations, or by means of instruments and equipment, resulting in lack of observation.

(3) Shielded. Due to specific reasons, the observation object is shielded and cannot be observed.

(4) The observed object does not cooperate. The observation failed due to the uncooperative attitude of the observation object.

(5) Limited by the observation ability of instruments and equipment. Although instruments and equipment can be used to assist observation, they also have their own limitations, unable to complete the observation process, or the observation accuracy is not high.

(6) The observer's own ability is not enough or the observation consciousness is not strong. Because the observers themselves have not received professional training or lack of professional training, the success of observation cannot be guaranteed.

(7) The observation results can only be limited to the appearance and structure, and cannot directly observe the essence.

(8) Limited by the ability to develop science and technology in the times. For example, we can't see dark matter.

(9) If it is manual observation, it is difficult to complete large-scale investigation. Now, with the help of modern science and technology, the large-scale investigation can be realized.

3.5.3 Application of Six Sigma management thought in observation

An important prerequisite for correct observation is to ensure that the overall idea of observation is clear. Observers can introduce the idea of Six Sigma management. The management goal of Six Sigma is to improve process performance by reducing the variation of process output. Then observers can reduce the variation of subsequent data of the same observation content by analyzing the data output of each observation. Observers should have a clear understanding of the definition of the data to be observed this time, which data belongs to the needs of users in the current process, and what impact the observed data has on the overall business development.

The contents to be noted during observation include the following items.

(1) Measurement. The measurement includes: proposing the data collection plan, specifying the content of this observation and what data to collect. Specify how to measure and collect data, and judge whether the proposed measurement system is appropriate. In the process of observation, the amount of data collected, the level of data collected, the classification, the time, and the executors should be clear. Timely judge whether the whole observation process is stable and whether the observation process meets the requirements of relevant specifications. Calculate the range, standard deviation, variance, etc. of the observed data, and understand the largest source of data variation.

(2) Analysis. The observer should analyze which factors have the greatest impact on the output of the observation process, analyze the fluctuation rule of the observation data output,

and find out whether the influencing factors are real or suspected, so as to facilitate the follow-up observation.

(3) Improvement. Understand the gap between the current situation and the target indicators, and identify the direction for improvement. According to the analysis conclusion, continuously improve the observation plan, revise the observation process and other contents, promote the improvement of the observation process, and improve the quality of the output data.

(4) Control. As an observer, it is necessary to confirm whether the observation process continues to maintain the improvement level through a complete management process; Observe whether all the required goals have been achieved.

3.5.4 Machine observation

1. The advantages of machine observation

Observation is dominated by research behaviors, phenomena and other explicit contents. Various manual observations can be objectively recorded through training, but sometimes they are affected by personal emotions. Manual observation is also subject to physiological limitations of body senses, which may lead to fatigue and inability to reach a certain space. Therefore, people need to use various instruments and equipment to constantly expand their observation ability. In this case, machine observation complies with this requirement. Now people use cameras, audio-visual equipment, satellites, automatic recorders, intelligent induction equipment and other machines to observe. The observation process and results are more objective, more accurate, more stable, and more detailed. There will be no fatigue and other artificial phenomena (machine fatigue is not considered for the time being), basically no time and space interference, and in many cases no environmental interference.

Because of its unique advantages, machine observation has been widely used, especially the extensive application of electronic (digital) observation equipment, which greatly improves the quality and efficiency of observation. The application scenarios of machine observation include: airport passenger self-service customs clearance system; Face recognition system on the construction site; Network attendance system of many enterprises; 3D real map of some search systems; Search functions such as keywords on the network; The development and application of CIM+BIM (City Information Model and Building Information Model) ETC (Electronic Toll Collection System) on expressway; Automatic performance appraisal system; Terahertz wave detection in security inspection; Industrial robots, farm robots, etc. With the continuous integration of digitalization and informatization into communities, enterprises and various institutions, the role of machine observation will shine.

2. Flow observation

Flow observation is an important function of machine observation, which has been widely promoted. Flow observation is widely used in various flows of people, materials and information. For example, when entering the airport gate, there will be electronic people flow statistics; When people swipe their ID cards to enter the platform at the high-speed railway station, it is the statistics of the flow of people, and it is accurate to the specific individual and the corresponding compartment and seat number; Logistics automatic statistics is widely used in postal system; The goods post (temporary storage) after online shopping is logistics statistics; Online shopping platform is information flow statistics; The search platform is information flow statistics; Keyword error correction of documents is information flow statistics; The mobile booking system of train tickets and air tickets is information flow statistics; The electronic monitoring system on the road is the composite statistics of the flow of people, vehicles, logistics and information; The ICU of the hospital is medical information flow statistics, etc. Traffic observation is everywhere, especially after mobile phones, computers, and electronic cameras have become the carriers of machine observation. The function of traffic observation has also been further improved, mainly including the following items.

(1) The flow recording function. Flow recording is the basic function of flow observation. Records of various flows help decision-makers make corresponding management judgments. For example, the traffic flow statistics of an intersection can provide data for the time interval design of traffic lights. Traffic flow statistics inside and outside the city can provide real-time data for automatic navigation equipment, so that drivers can avoid congested roads and temporary accident roads.

(2) Work behavior reminder function. Through the observation of work flow, we can judge the working conditions of workers and remind them to improve their working behaviors. For example, with the support of satellites and artificial intelligence, the on-board observation system can timely judge the driver's driving behavior, and send out fatigue driving reminders, over-speed reminders, bus passing reminders, turning reminders, and emergency handling reminders for vehicles ahead, which improves driving safety. The observation system at the intersection can remind pedestrians (especially blind people) when they can cross the road in real time to ensure pedestrian safety. The observation system on the computer can evaluate the working state of the operator, remind the owner to adjust the working behavior in time, pay attention to rest, etc. The monitoring system of the nursing home can remind medical personnel to give first aid to patients at any time to ensure life safety.

(3) Marketing function. Through the observation of customer flow, businesses can achieve precision marketing. Now the Internet has spread in every corner of urban and rural areas. The records of consumers' online shopping behavior and the list of store shopping, etc., enable businesses to timely understand the shopping habits of different consumer groups through summary and analysis, quickly analyze which goods are in demand, which goods are unsalable, and other information, and then make corresponding marketing strategies in different seasons, holidays, etc. Through the evaluation of advertising effect, the advertising effect of different advertising stars can be evaluated by adjusting the advertising time and content. Restaurants and others can adjust dishes and distribution through flow observation to help improve sales and service quality.

(4) The function of improving the supply chain. By observing the data flow, we can evaluate the supply capacity of various suppliers. For example, through the analysis of sales records of different channels, we can timely adjust the commodity procurement channels and logistics channels to improve supply chain management. Through the collection of evaluation information on the supply chain by the terminal, the requirements for service improvement can be put forward to the supplier in a timely manner. When an emergency or accident occurs in a cer-

tain region, it timely makes a prediction on the adjustment of certain supply chain links, which ensures the production and operation of the enterprise.

(5) The function of stimulating the economy of underdeveloped regions. The observation of flow can enable decision-makers to obtain material information in different regions and judge the production capacity and consumption capacity of different goods in different regions. For example, relevant departments can drive relevant markets in a timely manner through Internet information, real-time information about strong sales of goods in stores in various regions, and information about the high quality and low price of goods in some regions. In particular, they can leverage the consumption of rural markets and underdeveloped areas to improve the local economy.

(6) The function of machine intelligent management. With the help of net flow observation, relevant machine intelligent management can be realized on various network platforms. For example, by collecting people's habits of clicking on mobile web pages, we can master the consumption preferences of different groups, and then implement machine intelligence push to these groups, further evaluate the effect of machine intelligence push, further confirm the consumption preferences and interests of different groups, and strengthen the realization of intelligent push. The intelligent interaction function on the network can be realized through flow observation. Machines evaluate the hobby of personnel, so that personnel can increase their interaction with machines. Interactive shopping, self-service consumption, etc. are increasingly popular.

(7) The automatic updating function of virtual stores. Through the observation of flow, managers can timely grasp the layout of goods in online virtual stores through the analysis of a large number of goods tour records, and realize the automatic updating of goods, so as to consolidate the existing consumer groups and develop new consumer groups. Virtual characters can be set up on the network. Through traffic observation, virtual tasks can be changed and upgraded to drive consumption.

(8) The function of resource allocation. Through the observation of flow, managers can timely grasp the production and operation status of the enterprise at different times and places, timely perceive emergencies, accidents and corresponding needs at different time points, accurately predict and find the lack of resources, surplus and other information in different regions, and make advance deployment or emergency deployment of related resources to avoid unnecessary losses.

(9) Prediction function. By observing the flow, managers can forecast the demand and supply of related management elements. Cross regional managers can make judgments about changes in certain traffic behaviors. For example, the sudden hot sale of a certain drug in a certain place can roughly judge the emergence of some emergency conditions. By analyzing the changes of search keywords, we can judge the trend of different commodities, such as the color and style of clothing. Through the analysis of different video clicks, special hot selling products can be expedited.

(10) Human resource management function. Through flow observation, managers can improve the quality of human resource management. For example, record the working conditions of the staff, master the operating habits of the staff at different posts, and record the work effi-

ciency of relevant personnel. Evaluate the production process quality and efficiency of operators by mastering the generation and processing of defective products in real time. By analyzing the timeliness, accuracy, praise rate, price fluctuation and other information, relevant businesses can evaluate the work quality of staff and improve the management level.

（11）Physiological monitoring function. Some observation systems can monitor the physiological changes of personnel. When a person reads some advertisements, he or she will activate the activation level in the human body. Then, he or she will monitor the frequency changes of brain waves with an electroencephalograph, obtain the focus of attention of the eyes with a pupil detector, and analyze personal feelings and preferences with a sound analysis instrument; The monitoring system is used to observe the skin sweating, temperature rise, blood flow and rapid heartbeat, which can be used as the basis for decision-making.

（12）Auxiliary management decision-making function. In the digital era, data is an important basis for decision-making. Through the analysis of various traffic data, we can judge the competitors' competitive strategies and track their new products. In the process of data flow analysis, find our own data partners and new identification ways, define the performance and improvement opportunities of partners, and strengthen strategic cooperation. Through data analysis, we can increase sales and maximize profits. In the data analysis, summarize the experience, improve the product strategy, formulate promotion plans, quantify different opportunities, assist in rapid action, and find new business opportunities.

There are still many functions of data flow observation, which will be constantly improved in practice.

3.6　正确的逻辑关系

———

逻辑关系是指有两个相连的活动之间，其中一个活动的变化会影响到另一个活动变化的一种关系。表述为：设前导活动为 A，后续活动为 B，当 A 发生变化，那么 B 也发生了变化；或者当 B 发生了变化，A 也发生了变化。逻辑关系存在于各种概念之间、命题之间、事物之间、时空之间等，而且通常以数据（包括数字、文字）的形式予以反映或表达。因此，数据之间正确的逻辑关系，是数据高质量的基石，同样数据的高质量可以在数据的逻辑关系上得到直接体现。

常见的逻辑关系主要有：相关关系、因果关系、并列关系、递进关系、转折关系、让步关系、列举关系、举例关系、总结关系、主次关系、总分关系、相对关系、对立关系等十三项。

3.6.1 相关关系

相关关系是指一个变量与另一个变量之间存在的一种非确定的相互依存关系，相对于自变量的每一个取值，因变量因依存关系而受到影响，但因变量的变化是随机的，与它所对应的数值是非确定的。在相关关系中，自变量与因变量之间可以互换，不存在严格的区别。

1. 相关关系的类别

当两个变量之间存在相关关系时，当一个变量取一定的数值时，与之对应的另一个变量的数值虽然不确定，但它仍然会按照某种规律在一定的范围内发生变化。两个变量之间的相关关系可以根据不同的类别进行区分：

（1）按照相关的程度区分。如果一个变量的数量变化由另一个变量的数量变化所唯一确定，那么两者存在着函数关系，那就是完全相关。如果两个变量之间彼此的数量变化互相独立，那么两者之间没有关系，那就是不相关。当两个变量之间的关系介于完全相关和不相关之间，存在着非确定的依存关系，那就是不完全相关。

（2）按照相关的方向区分。如果两个变量的变化趋势基本趋同，当一个变量不断增大时，另一个变量也基本在不断增大，则两个变量呈正相关。当两个变量的变化趋势相反，当一个变量不断增大时，另一个变量基本在不断变小，则两个变量呈负相关。

（3）按照相关的形式区分。如果一个变量在均匀地发生变动时，另一个变量也相应地发生均匀的变动，则两个变量之间呈线性相关，或称为直线相关。如果一个变量在均匀地发生变动时，另一个变量相应地发生不均匀的变动，则两个变量之间呈非线性相关，或称为曲线相关。

2. 相关性分析

相关性分析（亦称相关分析）是反映两个变量之间的相关性，以衡量两个变量之间的相关密切程度。相关性分析在各种管理活动中起着非常重要的作用。它可以建立两个变量之间的数学关系来分析相关性，也可以在没有明确数学关系的情况下分析两个变量之间的相关性。比如在工程质量调查中，发现有一个地区的大楼存在施工质量问题，而且这些工程主要是由 A 公司承建，那就是这家 A 公司的施工管理与产品质量之间产生了相关性。

（1）最小二乘法

在进行相关分析时，两个变量之间的关系可以通过最小二乘法计算来确定。最小二乘法的原理是：在散点图中，有 n 个点分布在几个象限中，这 n 个点的坐标为（x_1，y_1），…，（x_n，y_n）。现在有一条任意的直线 l：

用 $y = a + bx$ 表示，

观察点（x_i，y_i）沿着平行于 y 轴方向到直线 l 的铅直距离（注意：不是垂直于直线 l 的距离）。用这个铅直距离来表示该点（x_i，y_i）偏离直线 l 的程度，可以用数量

$$| y_i - (a + bx_i) |$$

表示点（x_i，y_i）到直线 l 的远近程度。把所有的绝对误差相加，可以定量地刻画直线 l 与 n 个点之间总的远近程度。但它有两个不足：一是由于绝对值进行运算不方便；二是对于大小不等的绝对误差没有区别对待。因为当某个点离开这条直线越远时，其绝对误差越大。

为了使回归直线与所有的已知点配合更好，需要寻找一种方法来处理较大的绝对误差。思路为：在单项绝对误差求和之前，先把它们中的每一个都加以平方，这样可以有两个优势：一是去掉绝对值符号，便于后续的运算；二是让大的绝对误差得到放大，使之处于更不利的地位。

这种利用"误差的平方和为最小"来求取回归直线的方法，称为"最小平方法"，又称为"最小二乘法"。

上述"误差的平方和"，可用下面公式表示：

$$\sum_{n=1}^{n} (y_i - (a + bx_i))^2$$

该公式定量地刻画了直线 l 与 n 个点（x_i，y_i）的总的远近程度。当这 n 个点已知时，这个量是因直线的不同而变化的，也就是随着 a、b 的变化而变化。接下来需要求取一条回归直线，使该直线总的来说"最接近"这 n 个点。利用微积分学中的极值原理，可以求得 $\hat{a}$ 和 $\hat{b}$ 的值，从而得到回归直线。

（2）相关系数

两个变量之间的线性关系是否密切，可以用相关系数来表示。通过最小二乘法来求得回归直线，通过计算，可以求得两个变量之间的线性关系密切程度，一般用相关系数 r 来表示。

相关系数 r 的含义是：描述两个变量之间线性关系密切程度的一个数量指标，它的计算公式为：

$$r = \frac{\sum_{i=1}^{n} (x_i - \bar{x})(y_i - \bar{y})}{\sqrt{\sum_{i=1}^{n} (x_i - \bar{x})^2 \sum_{i=1}^{n} (y_i - \bar{y})^2}}$$

相关系数具有的性质为：$0 \leq |r| \leq 1$。

当 $r = 0$ 时，$\sum_{i=1}^{n} (x_i - \bar{x})(y_i - \bar{y}) = 0$，回归直线的斜率为 0，这时的回归直线与 x 轴平行，说明变量 x 与变量 y 之间无线性关系。但需要注意的是，当 $r = 0$ 时，虽然 x 与 y 之间无线性关系，但可能存在非线性关系，所以不能轻易下结论。但 $|r|$ 越接近于 0 时，x 与 y 之间的线性相关程度越小，$|r|$ 为 0.0~0.2 为极弱相关或无相关。

当 $0<|r|<1$ 时，变量 x 与变量 y 之间存在一定的线性关系。$|r|$ 越接近于 1 时，散点越靠近回归直线。$r>0$ 时，回归直线的斜率大于 0，称 x 与 y 正相关；$r<0$ 时，回归直线的斜率小于 0，x 与 y 负相关。$|r|$ 在 0.8~1.0 为极强相关，0.6~0.8 为强相关，0.4~0.6 为中等程度相关，0.2~0.4 为弱相关。

当 $|r|=1$ 时，所有的散点都在回归直线上，回归直线的斜率为 1，这时变量 x 与变量 y 之间存在确定的线性关系，即 x 与 y 完全线性相关。当 $r=1$ 时，随着 x 的增加或减少，y 同时增加或减少；当 $r=-1$ 时，随着 x 的增加，y 同时减少。

（3）相关性的显著性检验

从统计的角度分析，通常两个变量之间的相关系数是根据两个变量的样本数据计算而得到。由于抽样的随机性和样本数量的关系，样本的相关系数是不能直接用来说明两个变量是否具有显著的相关关系，因此，需要进行显著性检验。这里以两个变量的线性相关关系为例。

先假设两个 X、Y 变量之间没有显著性线性相关关系，即两个变量的相关系数为 0，则构建统计量 t：

$$t = \frac{r \cdot \sqrt{n-2}}{\sqrt{1-r^2}}$$

当变量 X、Y 都服从正态分布时，统计量 t 服从自由度为 $n-2$ 的 t 分布。

接着计算统计量 t，并查询 t 分布中所对应的概率 P 值。通常取显著性水平 $\alpha=0.05$.

判断时，如果 $P<\alpha$，则否定原假设，说明变量 X、Y 之间存在显著的线性相关关系；反之，如果如果 $P>\alpha$，则接受原假设，说明变量 X、Y 之间不存在显著的线性相关关系。

3. 相关性的应用

基于相关性数据的例子很多，在科学发现上、管理活动中等都有体现。现举例如下：

（1）基于相关性的营销

沃尔玛超市平时注重把各门店的数据收集起来分析，发现了一个有规律的现象：每当飓风来临的前期，超市里的手电筒销量猛增，这个容易理解，因为很多家庭担心飓风来临后会造成临时的断电。但有趣的是，在手电筒销量上去的同时，蛋挞的销售项也出现激增。为此，沃尔玛超市决定，每当飓风到来之前，把蛋挞与飓风的用品摆放在一起，既方便了顾客的购买，又让蛋挞的销量得到很大的提升。

（2）基于大数据相关性的预测

大型零售公司塔吉特（Target）一直注重收集销售数据，以进行顾客采购行为的分析。塔吉特公司的分析人员注意到，婴儿礼品登记簿上的妇女在怀孕三个月的时候会采购较多的无香乳液，再过一段日子，她们会采购一些营养品等。就这样，塔吉特公司建立起孕期妇女预产期的分析数据，并适时向她们送去优惠券以促销。

有一天，一个父亲冲进一家塔吉特的门店投诉，说自己的女儿尚在高中读书，而塔吉特公司却给她邮寄婴儿用品的优惠券，这位父亲对此气愤异常。但过了一段日子，当塔吉特公司的负责人向这位父亲打电话表示道歉时，父亲却说，根据他和女儿的沟通，得知女儿的预产期的确快要到了。大数据的相关性预测在这里得到了验证。

（3）量子纠缠的发现

量子纠缠是相关关系的一个证明。量子纠缠是指在量子力学里，当几个粒子在彼此相互作用后，各个粒子所拥有的特性已经综合称为整体性质，无法单独描述各粒子的性质，只能描述整体系统的性质。普通人难以理解量子纠缠，那么按照常人的思路去解释量子纠缠的现象，即两个暂时耦合的粒子，当它们不再耦合之后彼此之间仍然维持着关联性。打个比方，有两个量子，一个在月球上，另一个在地球上。当月球上的量子发生变化时，地球上的量子会发生相同的变化，这是一种超距离的相关关系。如今，量子密码、墨子号量子科学试验卫星、量子计算机等是量子纠缠的最佳实践。

3.6.2 因果关系

1. 因果关系的涵义

因果关系是指前一个事件"因"与后一个事件"果"之间的关系，对某个结果产生影响的任何一个事件都是这个结果的"因"，因果关系普遍存在。在数学上，因果关系可以存在于函数中，比如"面积 $S = \pi R^2$"，R 为半径，因为 R 发生了变化，所以面积 S 也变化。对于一些函数中的因果关系是可以证明的。在统计学上，人们用统计规律来反映客观对象的偶然性与必然性在因果关系上的统一，可以用置信度来大致推论出因果关系。在物理学中，因果关系的例子有很多，如在经典力学中，只要知道力学体系内所有物体的初始位置和动量，又知道它们之间的相互作用力，就可以通过牛顿定律来确定这个体系在某一个时刻的状态，这是经典力学中的因果关系。化学反应中的因果关系相当多，比如因为金属锌的活动性比铁强，所以在远洋轮船上的外壳装锌块可以减缓对轮船外壳的腐蚀；MgO 的熔点很高，所以 Mg（OH）$_2$ 可以用作阻燃物。

因果关系可以用一些词语来表达，如："因为、从此、根据、随后、由于、导致、太……而……、既然……就……、之所以……是因为……、因为……所以……、因此、因而、正因为……才……"等。

2. 因果关系的证明

要证明两个变量之间的因果关系，其实是要证明一种变量的变化是否引起了另一种变量的预见性变化。如果单纯用数学的角度去证明两个变量之间的因果关系是几乎不可能的，因为数学中的推理有点类似对数学对象的"观察"，因此，只能是从逻辑的角度来证明因果关系。

（1）证明两个变量之间存在因果关系的条件

设有 A、B 两个变量，为了证明 A 的变化引起了 B 的变化，那么就需要证明 A 与 B 之间存在以下三个条件：

第一个条件，A 与 B 之间存在相关关系，也可以称 A、B 为共生变量，即两者有紧密的相互生存关系，当然实际上指两者的紧密度高。

第二个条件，A 与 B 在发生时存在适当的时间顺序。

第三个条件，在 A、B 之外不存在其他可能的原因性因素。

需要说明三个情况：

一是这里的因果关系并不是指 A、B 之间是唯一存在的因果关系，或者说 A 是引起 B 发生变化的唯一原因，而是指 A 是引起 B 产生变化的众多因素中的一个因素。

二是因果关系并不意味着 A 是 B 的完全决定关系，即如果 A 是 B 的原因，那么 A 必须导致 B。因果关系意味着一种可能性关系，即 A 的变化（或产生）可能导致 B 的变化（或产生），所以 A 是 B 的原因。

三是因果关系无法通过证明来体现，而是通过判断而得到。

综合以上三点，才能推断出 A、B 之间的因果关系。

（2）两个变量之间相关关系的证明

为了证明 A 的变化引起了 B 的某些变化，那么需要先说明 A、B 两者之间存在的相关关系，即它们是按照一些可以预见的方式一起发生变化，这些变化表现为正相关、负相关、指数相关等。

A、B 之间存在相关关系，是不能证明两者之间存在因果关系的。因为存在这样的事实：两个变量碰巧以某些可预见的方式一起发生了变化，但这不能证明 A 的变化引起了 B 的变化。比如：一个城市的汽车销量上升（称为变量 A）与另一个城市餐饮零售额的上升（称为变量 B）有着高度的相关性，这说明 A、B 两种变量碰巧以上升的形式一起存在，但后续的调研发现，A、B 两个变量之间的变化其实不存在真正的联系，是偶然现象而已。

（3）两个变量之间存在发生变化的时间顺序

为了证明 A 的变化引起了 B，那么还需要证明 A 在 B 的之前发生，两者之间存在着发生时间的先后，无论这个时间间隔是多么长或多么短。比如：这次的车辆撞击事故（A）导致了车辆的严重变形（B），虽然事故几乎是在一刹那发生，但必须是 A 在前，B 在后；或者说，在 B 发生之前，A 已经发生。

（4）排除其他因素的干扰

由于是证明 A 的变化导致了 B 的变化，因此，有必要排除其他因素的干扰导致了 B 变化的发生。现在假定当 A 发生变化的时候，B 也发生了变化；A 与 B 之间的相关关系存在；A、B 之间的前后顺序也存在，这时还不能证明 A 是导致 B 的原因。因为还存在着一种可能，当 A 发生变化的时候，C 也发生了变化，而恰恰是 C 的变化导致了 B 的变化。因此，有必要排除所有 C 类因素的干扰，这样才能只针对 A、B 之间的关系来进行论证。

3.6.3 并列关系

并列关系是指在同一属概念之中存在同层次的种概念，这些种概念之间存在着并列的关系。具有并列关系的变量，在相互之间不相隶属又相对独立。并列关系可以分为相容并列关系和不相容并列关系。

并列关系可以用一些词语表达，比如"和、也、像、那就是说、是……也是……、既不是……也不是……、相等的、同样的、与……一样、即……又……、一边……一边……、有时……有时……、一会儿……一会儿……"等。

1. 属种的涵义

属概念是指在具有真包含于和真包含关系的一对概念中外延较大的概念；种概念则是其中外延较小的概念。通俗地理解，属是客观事物中的大类，种是客观事物中的小类。真包含关系又叫作属种关系；真包含于关系又叫作种属关系；通常把真包含关系和真包含于关系统称为"属种关系"。所以说，属种关系是概念外延间的一种关系，它反映了大类的概念与反映该大类下某一小类的概念之间的关系。比如："学校"与"中学"就是属与种的关系。

2. 相容并列关系

它是指如果在同一个属概念之中有若干个同层次的种概念，而且这些种概念的外延相互交叉，也就是有一部分是相同的，那么这几个概念之间为相容并列关系。比如，属概念为数学科目，种概念为几何、代数、微分等，则这些种概念之间为相容并列关系。

3. 不相容并列关系

它是指如果在同一个属概念之中有若干个层次的种概念，而且这些种概念的外延相互排斥，那么这几个概念之间为不相容并列关系。比如属概念为自然资源，种概念为湖泊、矿山、森林等，则这些种概念之间为不相容并列关系。

3.6.4 递进关系

递进关系是指两个不同的客观事物，根据相应的衡量尺度，其中一个客观事物比另一个事物在尺度上更进一层，这样两个事物之间的逻辑构成了递进关系。递进可以表现为：

（1）结构层次的递进，比如大数据的体系构建分了不同的层级，从"数据基础平台"到"数据报表分析与可视化"，到"精细化业务分析"再到"战略分析与决策"，这四个层级之间是递进的关系。

（2）语义上的递进，比如小张不仅数学等各学科考得好，而且特别有礼貌。

（3）重要性的递进，比如车辆要做好平时的保养，轮胎的胎压要保持在标准的区域，尤其是刹车系统一定要保持好的状态。

递进关系可以用一些词语来表达，比如"更、而且、不但……而且……、甚至、那么、除了、更多的是、另外、不仅……还……、不只……也……、也……、别说……连……、不但没……反倒……、尚且……何况……、特别是、尤其"等。

3.6.5 转折关系

转折是表示某个事物的转变、变化。转折关系是指两个不同的客观事物，其中一个客观事物呈现为另一个客观事物变化后的状态。转折关系可以表现为：

（1）结构层次上的转折，比如小张虽然是基层一线的管理人员，但由于他自身的努力，马上要调至总部担任部门负责人。

（2）语义上的转折，比如小张做事情很低调，事实上他是一所著名高校的毕业生。

（3）重要性的转折，比如在传统的工业社会，生产要素主要有人、财、物、土地等，但是到了信息社会，信息成为了更重要的生产要素。

（4）运行趋势的转折，比如当前的股票指数在低位运行，但马上要迎来拐点向上运行了。

转折关系可以用一些词语来表达，比如"虽然……但是……、但是、尽管……可是……、然而、而、相反、另一方面、事实上、不像、还不如、用……替代、只是、不过、倒是、却"等。

3.6.6 让步关系

让步关系是指某一个客观对象先退了一步，然后进入它的一个状态。让步通常表现在重要性、语义、策略、沟通等方面。

让步关系可以用一些词语来表达，比如"虽然、不管怎样、尽管、甚至于、哪怕、就算、纵然、即使……也……"等。

3.6.7 列举关系

列举是人们在揭示某种概念的外延时，予以一一罗列。列举是明确了某个客观事物的全部元素或部分元素的一种逻辑方法，通常是纲目式的罗列，形式上比较笼统。因此，列举关系是指一个母项 A 罗列了它的全部元素或部分元素，另外的子项 B、C 等是其中罗列的内容，它们之间构成了列举关系。列举可以分为几个类型：

（1）在构成形式上，分为线性列举（又称一次列举），链式列举（又称多次列举）。线性列举母项在前，链式列举母项在后。

（2）在数量上，分为穷尽性列举和非穷尽性列举。

（3）在子项性质上，分为事物性列举和事件性列举。

（4）在表义上，分为主观列举和客观列举。

列举关系可以用一些词语来表达，比如"首先、其次、再次，从……开始，接着，一方面……另一方面……，其他仍然……，等等，之类"等。

3.6.8 举例关系

举例是明确某种概念的部分外延、揭示某些事理的逻辑方法。举例通常用具体的、容易理解的实例来阐述抽象化的事物或道理。举例关系是指为了解释、说明、阐述客观对象（母项）A，用事例（子项）B 来进行描述，这样所形成的一种逻辑关系。

举例是列举的一种特殊形式。列举与举例是有区别的，列举是清单式的罗列，举例用典型事物来阐述；列举强调数量，举例强调代表性。列举的子项是母项的直接组成部分，举例的子项不一定是母项的直接组成部分，可以用隐喻来替代。

举例关系可以用一些词语来表达，比如"例如、比如、比方、在……中、举例来说、更确切地说、可命名为……"等。

3.6.9 总结关系

总结是对于某个过程、若干个客观对象等予以总的归结，得到一个客观的结论，有时是做出带有规律性的结论。总结通常是最后的结果。总结关系是指针对客观对象、过程、内容等（A）予以归结，得到客观的结论（B），B 与 A 之间的逻辑关系。

总结关系可以用一些词语来表达，比如"最后、概括来说、总之、总结一下、言而总之、归纳为、归结为"等。

3.6.10 主次关系

主次关系是指两个客观对象之间，重点（A）与一般（B）之间的关系，两者具有对立统一的性质，有时因为条件的发展变化，两者的支配地位是可以互换的。主要与次要的二者之间没有隶属关系。

3.6.11 总分关系

总分关系是指客观对象 A 与 B 之间存在着纲与目的相互关系，纲统领目，目受纲支配，二者不能并列，也不能颠倒。

3.6.12 相对关系

相对关系是指两个客观对象之间，因为它们各自的内容、属性等是相比较而存在。如果没有了其中的一个，那么另一个也不存在；脱离了其中的一个，另一个无从比较。相对是有条件的、暂时的、有限的、特殊的。

3.6.13 对立关系

对立关系是指两个客观事物或一个客观事物中的两个方面之间，存在着相互排斥、相互矛盾、相互斗争、相互竞争、相互冲突、相互抵消、相互抑制的关系。

3.6　Correct Logical Relationship

———

Logical relationship refers to the relationship between two connected activities, in which the change of one activity will affect the change of the other. It is expressed as follows: Let the leading activity be A and the subsequent activity be B. When A changes, B also changes; Or when B changes, A also changes. Logical relations exist among various concepts, propositions, things, time and space, and are usually reflected or expressed in the form of data (including numbers and words). Therefore, the correct logical relationship between data is the cornerstone of high-quality data, and the same high-quality data can be directly reflected in the logical relationship of data.

Common logical relationships mainly include: correlation, causality, juxtaposition, progressive relationship, transition relationship, concession relationship, enumeration relationship, example relationship, summary relationship, primary and secondary relationship, general and separate relationship, relative relationship, opposite relationship, etc.

3.6.1 Correlation

Correlation refers to an uncertain interdependence between a variable and another variable. Relative to each value of an independent variable, the dependent variable is affected by the dependency, but the change of the dependent variable is random and its corresponding value is uncertain. In the correlation relationship, independent variables and dependent variables can be interchanged, and there is no strict difference.

1. Category of correlation

When there is a correlation between two variables and one variable takes a certain value, the value of the corresponding other variable is uncertain, but it will still change in a certain range according to a certain rule. The correlation between the two variables can be distinguished according to different categories.

(1) Distinguish according to the degree of correlation. If the quantity change of one variable is uniquely determined by the quantity change of another variable, then there is a functional relationship between the two, that is, complete correlation. If the quantity changes between two variables are independent of each other, then there is no relationship between them, which is irrelevant. When the relationship between two variables is between complete correlation and non correlation, there is an uncertain dependency relationship, that is incomplete correlation.

(2) Distinguish according to relevant directions. If the change trends of the two variables are basically similar, and when one variable is increasing, the other variable is also basically increasing, then the two variables are positively correlated. When the change trend of the two variables is opposite, and one variable is increasing while the other is basically decreasing, the two variables are negatively correlated.

（3）Distinguish according to relevant forms. If one variable changes uniformly, and the other variable changes uniformly, then the two variables are linearly related, or called linear correlation. If one variable changes uniformly and the other variable changes unevenly, the two variables are nonlinear or curve related.

2. Correlation analysis

Correlation analysis reflects the correlation between two variables to measure the closeness of correlation between two variables. Correlation analysis plays a very important role in various management activities. It can establish the mathematical relationship between two variables to analyze the correlation, or analyze the correlation between two variables without a clear mathematical relationship. For example, in the project quality survey, it is found that there is a construction quality problem in a building in a region, and these projects are mainly constructed by Company A. That is, there is a correlation between the construction management and product quality of Company A.

（1）Least square method

During correlation analysis, the relationship between the two variables can be determined by least square calculation. The principle of the least squares method is: In the scatter plot, there are n points distributed in several quadrants, the coordinates of these n points are（x_1, y_1）,$\cdots$,（x_n, y_n）. Now there is an arbitrary straight line l:

expressed by $y = a + bx$,

The vertical distance from the observation point（x_i,y_i）to the line l along the direction parallel to the y-axis（note: it is not the distance perpendicular to the line l）. This vertical distance is used to express the deviation degree of the point（x_i, y_i）from the straight line l. The distance from point（x_i, y_i）to line l can be expressed by the quantity

$$| y_i - (a + bx_i) |$$

The total distance between the line l and n points can be quantitatively described by adding all the absolute errors. But it has two shortcomings: one is that it is inconvenient to operate the absolute value; second, there is no distinction between absolute errors of different sizes. Because when a point is farther away from the straight line, its absolute error is greater.

In order to make the regression line match all known points better, it is necessary to find a method to deal with the larger absolute error. The idea is: before the sum of individual absolute errors, square each of them first, which has two advantages: first, remove the absolute value symbol for subsequent operations; the other is to magnify the large absolute error and make it more disadvantageous.

This method, which uses "the sum of squares of errors is the minimum" to obtain the regression line, is called "the least square method".

The above "sum of squares of errors" can be expressed by the following formula:

$$\sum_{n=1}^{n} (y_i - (a + b x_i))^2$$

The formula quantitatively depicts the total distance between line l and n points（x_i, y_i）. When the n points are known, the quantity changes with the difference of the straight line, that is, with the change of a and b. Next, we need to find a regression line to make it "closest" to the n points in general. Using the extreme value principle in calculus, the value of $\hat{a}$ and $\hat{b}$ can be obtaubed, and then the regression line can be obtained.

(2) Correlation coefficient

Whether the linear relationship between the two variables is close can be expressed by correlation coefficient. The regression line is obtained by the least square method. Through calculation, the closeness of the linear relationship between the two variables can be obtained, which is generally expressed by the correlation coefficient r.

The meaning of correlation coefficient r is a quantitative indicator that describes the closeness of the linear relationship between two variables. Its calculation formula is

$$r = \frac{\sum_{i=1}^{n} (x_i - \bar{x})(y_i - \bar{y})}{\sqrt{\sum_{i=1}^{n} (x_i - \bar{x})^2 \sum_{i=1}^{n} (y_i - \bar{y})^2}}$$

The property of correlation coefficient is: $0 \leqslant |r| \leqslant 1$。

When $r = 0$, $\sum_{i=1}^{n} (x_i - \bar{x})(y_i - \bar{y}) = 0$, the slope of the regression line is 0, at this time, the regression line is parallel to the x axis, indicating that there is no linear relationship between variable x and variable y. However, it should be noted that when $r = 0$, although there is no linear relationship between x and y, there may be a nonlinear relationship, so it is not easy to draw a conclusion. But when $|r|$ is closer to 0, the less linear correlation between x and y. When $|r|$ is between 0.0~0.2, it is extremely weak or irrelevant.

When $0 < |r| < 1$, there is a certain linear relationship between variable x and variable y. When $|r|$ is closer to 1, the scatter point is closer to the regression line. When $r > 0$, the slope of the regression line is greater than 0, it is said that x is positively correlated with y; When $r < 0$, the slope of the regression line is less than 0, at this time, x is negatively correlated with y. When $|r|$ is between 0.8~1.0, it is highly correlated; If $|r|$ is between 0.6~0.8, it is strongly correlated; If $|r|$ is between 0.4~0.6, it is medium correlated; If $|r|$ is between 0.2~0.4, it is weak correlated.

When $|r| = 1$, all scattered points are on the regression line, and the slope of the regression line is 1. At this time, there is a certain linear relationship between variables x and y, that is, x and y are completely linear. When $r = 1$, as x increases or decreases, y increases or decreases at the same time; When $r = -1$, as x increases, y decreases at the same time.

(3) Significance test of correlation

From the perspective of statistics, usually the correlation coefficient between two variables is calculated based on the sample data of two variables. Because of the relationship between the randomness of sampling and the number of samples, the correlation coefficient of samples cannot be directly used to indicate whether two variables have significant correlation. Therefore, a significance test is required. Take the linear correlation between two variables as an example.

First, suppose that there is no significant linear correlation between variable X and variable Y, that is, the correlation coefficient of the two variables is 0, then construct statistics t:

$$t = \frac{r \cdot \sqrt{n-2}}{\sqrt{1-r^2}}$$

When variables X abd Y all obey normal distribution, statistics t obeys t distribution with $n - 2$

degree of freedom.

Then calculate the statistic t and query the corresponding probability P value in the t distribution., usually take the significance level $\alpha = 0.05$.

When judging, if $P < \alpha$, then the null hypothesis is denied, indicating that there is a significant linear correlation between variables X and Y; Conversely, if $P > \alpha$, Then accept the original hypothesis, indicating that there is no significant linear correlation between variables X and Y.

3. Application of correlation

There are many examples based on relevant data, which are reflected in scientific discovery and management activities. Here are some examples:

（1）Marketing based on correlation

Wal Mart supermarket usually pays attention to collecting and analyzing the data of each store, and finds a regular phenomenon: every time in the early days of the hurricane, the sales of flashlights in the supermarket increase rapidly, which is easy to understand, because many families worry that the hurricane will cause temporary power failure. But interestingly, while the sales of flashlights increased, the sales of egg tarts also increased dramatically. To this end, Wal Mart decided to put egg tarts and hurricane supplies together before the hurricane, which not only facilitates customers' purchase, but also greatly improves the sales of egg tarts.

（2）Prediction based on the big data correlation

Target, a large retail company, has been focusing on collecting sales data to analyze customers' purchasing behavior. The analysts of Target Company noticed that women on the baby gift register would purchase more odorless lotion when they were three months pregnant, and in a few days, they would purchase some nutrition products. In this way, Target has established the analysis data of pregnant women's expected date of childbirth, and sent them coupons to promote sales.

One day, a father rushed into a Target store to complain that his daughter was still in high school, but the Target company sent her coupons for baby products. The father was very angry about this. However, some days later, when the person in charge of Target called the father to apologize, the father said that, according to his communication with his daughter, he knew that her due date was indeed approaching. The correlation prediction of big data is verified here.

（3）Discovery of quantum entanglement

Quantum entanglement is a proof of correlation. Quantum entanglement refers to that in quantum mechanics, when several particles interact with each other, the properties of each particle have been integrated into a whole property. It can not describe the properties of each particle alone, but can only describe the properties of the whole system. It is difficult for ordinary people to understand quantum entanglement, so we should explain the phenomenon of quantum entanglement according to ordinary people's ideas, that is, two temporarily coupled particles still maintain correlation with each other when they are no longer coupled. For example, there are two quanta, one on the moon and the other on the earth. When the quantum on the moon changes, the quantum on the earth will have the same change, which is a kind of long-distance correlation. Nowadays, quantum cryptography, Mozi quantum science experiment satellite and quantum computer are the best practices of quantum entanglement.

3.6.2 Causality relationship

1. Connotation of causality relationship

Causality refers to the relationship between the "cause" of the previous event and the "effect" of the next event. Any event that affects a result is the "cause" of the result, and causality is universal. Mathematically, causality can exist in functions, such as "area $S = \pi R^2$". R is the radius. Because R changes, so does the area S. The causality in some functions can be proved. In statistics, people use statistical laws to reflect the unity of causality between the contingency and inevitability of objective objects, and can use confidence to roughly infer causality. In physics, there are many examples of causality. For example, in classical mechanics, as long as you know the initial position and momentum of all objects in the mechanical system and the interaction force between them, you can determine the state of the system at a certain time through Newton's law. This is the causality in classical mechanics. There are quite a lot of causal relationships in chemical reactions. For example, since the mobility of metal zinc is stronger than iron, installing zinc blocks in the hull of ocean going ships can slow down the corrosion to the hull of ships; MgO has a high melting point, so $Mg(OH)_2$ can be used as a flame retardant.

Causality can be expressed by some words, such as "because, since, according to, after, cause, too...and..., since...is..., because...is..., therefore, because...".

2. Proof of causality

To prove the causal relationship between two variables is to prove whether the change of one variable causes the predictive change of another variable. It is almost impossible to prove the causal relationship between two variables from a purely mathematical perspective, because the reasoning in mathematics is somewhat similar to the "observation" of mathematical objects, so the causal relationship can only be proved from a logical perspective.

(1) Conditions for proving causality between two variables

There are two variables, A and B. In order to prove that the change of A causes the change of B, it is necessary to prove that the following three conditions exist between A and B.

The first condition is that there is a correlation between A and B. A and B can also be called symbiotic variables, that is, they have a close relationship with each other. Of course, it actually means that they are very close.

The second condition is that A and B have an appropriate time sequence when they occur.

Third, there are no other possible causal factors other than A and B.

Three situations need to be explained:

First, the causality here does not mean that A and B are the only causal relationship, or that A is the only reason that causes B to change, but that A is one of the many factors that cause B to change.

Second, causality does not mean that A is completely determined by B, that is, if A is the cause of B, then A must lead to B. Causality means a possible relationship, that is, the change (or generation) of A may lead to the change (or generation) of B, so A is the cause of B.

Third, causality cannot be demonstrated through proof, but can be obtained through judgment.

Based on the above three points, the causal relationship between A and B can be inferred.

(2) Proof of correlation between two variables

In order to prove that the change of A causes some changes of B, it is necessary to first explain the correlation between A and B, that is, they change together in some predictable ways. These changes are positive correlation, negative correlation, exponential correlation, etc.

There is a correlation between A and B, which cannot prove that there is a causal relationship between them. Because there is the fact that two variables happen to change together in some predictable way, but this cannot prove that the change of A causes the change of B. For example, there is a high correlation between the increase of car sales in one city (called variable A) and the increase of catering retail sales in another city (called variable B), which means that variables A and B happen to exist together in the form of increase, but subsequent research found that there is no real connection between the changes of variables A and B, which is just an accident.

(3) Time sequence of changes between two variables

In order to prove that the change of A causes B, it is also necessary to prove that A occurs before B, and there is a time sequence between them, no matter how long or how short the time interval is. For example, this vehicle collision accident (A) caused serious deformation of the vehicle (B). Although the accident occurred almost in a flash, A must be in front of B; In other words, before B occurs, A has already occurred.

(4) Elimination interference of other factors

Since it is proved that the change of A leads to the change of B, it is necessary to exclude the interference of other factors that lead to the change of B. Now suppose that when A changes, B also changes; The correlation between A and B exists; The sequence between A and B also exists, and it cannot be proved that A is the cause of B. Because there is a possibility that when A changes, C also changes, and it is precisely the change of C that causes the change of B. Therefore, it is necessary to eliminate the interference of all Class C factors, so that only the relationship between A and B can be demonstrated.

3.6.3 Juxtaposition relation

Parallelism refers to the existence of species concepts at the same level in the same generic concept, and there is a parallel relationship between these species concepts. Variables with parallel relationship are not subordinate to each other but independent. Juxtaposition can be divided into compatible and incompatible.

The juxtaposition relationship can be expressed by some words, such as "and, also, like, that is to say, is…is also…, is neither…nor…, is equal, is the same, is the same as…, is…and…, while, sometimes…and sometimes…, for a while…".

1. Meaning of genus and species

The concept of genus refers to a pair of concepts with the relation between true inclusion and true inclusion, which have a larger extension; The concept of species is the one with less extension. Commonly understood, genus is a big category of objective things, and species is a small category of objective things. True inclusion relation is also called genus species relation; True inclusion relation is also called species relation; Generally, the true inclusion relation and true inclusion relational system are called "genus species relation". Therefore, the genus species relationship is a relationship between the extension of concepts, which reflects the relationship between the concept of a large category and the concept of a small category under the large category. For example, "school" and "middle school" are the relationship between genus and species.

2. Compatible juxtaposition

It means that if there are several species concepts at the same level in the same generic concept, and the extensions of these species concepts cross each other, that is, some of them are the same, then these concepts are compatible and juxtaposed. For example, if the concept of genus is a mathematical subject, and the concept of species is geometry, algebra, differential, etc., then these concepts are compatible and juxtaposed.

3. Incompatible juxtaposition

It means that if there are several levels of species concepts in the same generic concept, and the extensions of these species concepts are mutually exclusive, then these concepts are incompatible and juxtaposed. For example, if the concept of genus is natural resources, and the concept of species is lakes, mines, forests, etc., then these concepts are incompatible and juxtaposed.

3.6.4 Progressive relation

Progressive relationship refers to that according to the corresponding measurement scale, one objective thing is more advanced than the other in terms of scale in two different objective things, so the logic between the two things constitutes a progressive relationship. Progression can be expressed as follows.

(1) Progression of structural levels. For example, the system construction of big data is divided into different levels, from "basic data platform" to "data report analysis and visualization", to "refined business analysis" to "strategic analysis and decision-making", and the relationship between these four levels is progressive.

(2) Progressive in semantics. For example, Xiao Zhang not only did well in mathematics and other subjects, but also was very polite.

(3) Progressive importance. For example, the vehicle should be well maintained at ordinary times. The tire pressure should be kept in a standard area, especially the brake system.

Progressive relationship can be expressed by some words, such as "more, and, not only...but also..., even, then, besides, more is, in addition, also..., let alone...even..., not to mention...especially...".

3.6.5 Turning relation

Turn is the change of something. The turning relationship refers to two different objective things, one of which is presented as the changed state of the other. The turning relationship can be shown as follows.

(1) A transition in the structure level. For example, although Xiao Zhang is a front-line manager at the grass-roots level, he will be transferred to the headquarters as the department head due to his own efforts.

（2）Semantic turn. For example, Xiao Zhang is very low-key when doing things. In fact, he is a graduate of a famous university.

（3）The turning point of importance. For example, in the traditional industrial society, the factors of production mainly include people, money, materials, land, etc. But in the information society, information has become a more important factor of production.

（4）The turning point of the operating trend. For example, the current stock index is operating at a low level, but it is about to usher in an inflection point for upward operation.

The turning relationship can be expressed by some words, such as "although...but..., but, however, on the contrary, on the other hand, in fact, not like, not as good as, replace with..., just, rather".

3.6.6 Concessional relationship

A concession relationship is a state in which an objective object takes a step back and then enters it. Concessions are usually expressed in terms of importance, semantics, strategies, communication, etc.

The concession relationship can be expressed by some words, such as "although, no matter what, despite, even if, even though, granted that, so much...so that..., anyway, whatever happens, in any case, even so, notwithstanding".

3.6.7 Enumerative relation

Enumeration is the listing of a concept when people reveal its extension. Enumeration is a logical method that clarifies all or part of the elements of an objective thing. It is usually a compendium listing with a general form. Therefore, the enumeration relationship refers to that a parent item A lists all or part of its elements, and the other sub items B, C, etc. are the contents listed therein, which constitute the enumeration relationship. Enumeration can be divided into several types.

（1）In the form of composition, it can be divided into linear enumeration (also known as one-time enumeration) and chain enumeration (also known as multiple enumeration). The parent term of linear enumeration comes first, and the parent term of chain enumeration comes last.

（2）In terms of quantity, it can be divided into exhaustive enumeration and non exhaustive enumeration.

（3）In terms of the nature of sub items, it can be divided into materialistic enumeration and eventful enumeration.

（4）In terms of meaning, it can be divided into subjective enumeration and objective enumeration.

The enumerative relationship can be expressed by some words, such as "first, second, third, starting from, then, on the one hand...on the other hand..., next, once more, once again, afterwards, for one thing, for another, others still..."

3.6.8 Exemplification relationship

Example is a logical method to clarify the partial extension of a concept and reveal some reasons. Examples usually use concrete and easy cases to understand examples to illustrate abstract things or principles. Example relation refers to a logical relation formed by using case (sub item) B to describe objective object (parent item) A in order to explain, explain and elaborate.

Example is a special form of enumeration. There is a difference between enumerating and giving examples. Enumerating is a list type listing, which is illustrated with typical things; Give examples to emphasize quantity and representativeness. The listed sub item is a direct component of the parent item. The cited sub item is not necessarily a direct component of the parent item, but can be replaced by metaphor.

The example relationship can be expressed by some words, such as "e.g., for instance, for example, such as, suppose, that is, in…, the case in point, more specifically, it can be named as…, to be precise, as a matter of fact, precisely speaking".

3.6.9 Summarized relationship

A summary is a general summary of a process, several objective objects, etc., to arrive at an objective conclusion, sometimes a conclusion with regularity. The summary is usually the final result. The summary relationship refers to the logical relationship between B and A, which refers to the conclusion of objective object, process, content, etc. (A) and objective conclusion (B).

The summary relationship can be expressed by some words, such as "finally, in summary, ultimate, posterior most, utmost, summarize, in words, summarization, generalize, in short, in brief, all in all, attribute to, come down to, to sum up, altogether".

3.6.10 Primary secondary relation

The primary and secondary relationship refers to the relationship between two objective objects, the key (A) and the general (B). They have the nature of unity of opposites. Sometimes, because of the development and change of conditions, the dominant positions of the two can be interchanged. There is no subordination between the primary and secondary.

3.6.11 General and partial relationship

General and partial relationship means that objective objects A and B are related to each other in terms of headrope and mesh. The headrope dominates the mesh and the mesh is dominated by the headrope. They cannot be parallel or reversed.

3.6.12 Relative relation

Relativity refers to the existence of two objective objects because their respective contents and attributes are compared. If one does not exist, the other does not exist; Without one, there is no comparison with the other. "Relative" is conditional, temporary, limited and special.

3.6.13 Antagonistic relationship

Antagonistic relationship refers to the relationship between two objective things or two aspects of an objective thing, which is mutually exclusive, contradictory, struggling, competing, conflicting, offsetting and restraining.

3.7　可用于预测趋势

——

在必要的时候，用数据预测是科学决策的重要方法之一，因此，数据质量的高低决定了预测的质量。

预测是根据研究对象发展变化地实际数据和历史资料，利用科学理论和方法，以及各种经验、判断和知识，对事物在未来一定时期内的可能变化进行推测、估计和分析。预测分析的实质是充分分析、理解事物变化发展的规律，根据事物过去和现在的情形，估计在未来的状态。预测的一个重要作用是通过对未知状态的预知，减少对未来出现的不确定性事物的干扰，使决策更科学。

由于预测的对象、时间、范围、性质等的不同，可以进行预测的分类。主要有：定性预测方法、时间序列分析预测方法、因果关系预测方法、回归分析预测方法等。

3.7.1 定性预测方法

定性预测方法是指人们根据经验和直觉，结合对象的过去和现状，对其进行分析、判断所做出的预测。

1. 定性预测方法的应用

定性预测方法主要应用于三个方面的预测：

一是对客观事物的发展性质的预测，在科研活动中，根据客观事物的过去表现、当前状况等，结合经验数据，参考一定的文献资料等，对于其未来的发展性质做出定性的预测。

二是对客观事物未来发展趋势的预测，根据客观事物过去的运行规律、当前的运行状况，结合经验数据，参考一定的文献资料等，对于其未来的运行趋势做出定性的预测。未来的发展趋势包括趋于稳定、趋于波动、走势向好、走势向下、乐观、悲观等内容。

三是对客观事物内在规律的预测，根据客观事物过去的内在规律、当前的运行状况，结合经验数据，参考一定的文献资料等，对其是否保持内在规律做出预测，通常的预测结果是延续已有的内在规律、终结已有的内在规律等。

2. 定性预测方法的优缺点

定性预测方法的优点主要是灵活性大，能够充分发挥人的主观能动性，特别是专家型的预测相对预测质量比较好。定性预测方法简单，受到的客观制约条件少，预测占用的时间短。

定性预测方法的缺点是受到预测者能力、知识、经验的影响大，预测结果的波动性大，如果不是专家型的预测，则其权威性差。定性预测基本上是基于个人的经验判断，往往还会受到当时的环境、情绪等影响。定性预测的结果更多的是作为决策的参考。

3. 做好定性预测的注意点

定性预测毕竟受到主观的影响大，因此，要做好定性预测，需要关注以下几个方面：

一是预测者尽量挑选有丰富经验、知识、能力的专业人员，非专业人员更多可以参加旁听，以逐步积累经验。

二是要加强对于前期相关资料的积累，特别是要做好一定的调研工作，想方设法掌握被预测对象的全景资料。

三是对于预测过程中可能遇到的各种不利因素等事先做出研判，尽量避免不必要的干扰。

四是在做前期背景调研时，尽量多采集定量化的数据，使定性预测人员在工作时能够有更多的依据作为参考，提高预测质量。

五是多收集与被预测对象具有相同性质、相同背景、类似发展过程、接近的各种环境影响等的同行资料，分析它们的运行趋势、发展规律等，作为重点参考。

3.7.2 时间序列分析预测方法

时间序列分析预测方法是指根据客观对象随时间变化的历史资料，包括关于客观对象的各类统计数据、试验数据、变化趋势等，考虑该对象随时间的推移而发生的变化规律，对其未来的运行趋势等作出预测的一种方法。

时间序列样本与简单随机样本有着很大的区别。简单随机样本中的 x_i 之间是相互独立的；而时间序列样本中的 x_i 之间并不相互独立，是呈现出时间序列的自相关性。

1. 时间序列分析预测方法的内在逻辑

时间序列分析预测方法仅仅依据当前为止所获得的、依照时间顺序记录的历史数据来进行分析，它的内在逻辑是：时间 T 综合了各种影响观察对象的因素，过去的时序数据反映了事物的变化规律，这种变化规律在各种因素的作用下而持续到未来的时间。研究者可以根据已经发生的历史数据，建立随时间变化的时序模型，不仅可以分析出事物已有的变化规律，还可以推断出未来的数值，也就是对未来数值的预测。

2. 时间序列分析预测的主要方法

（1）移动平均法

移动平均法是平滑法中的一种方法。平滑法的思路是采用修匀后的历史数据，即对历史数据中的偏大值、偏小值予以消除，使它们平滑化。移动平均法所采用的数据并不是全部历史数据，而是离预测期最近的 N 个观察值，取这些观察值的平均值来作为预测值。这些平均值的公式为

$$\widehat{x_{t+1}} = \frac{x_t + x_{t-1} + \cdots + x_{t-N+1}}{N} = \frac{1}{N} \sum_{i=0}^{N-1} x_{t-i}$$

其中，N 称为移动跨距。

随着时间 t 的推移，不断增加新的数据，相应的就不断截取最近的 N 个数据，并继续取得新的平均值，以此类推。

通过简单运算，可以得到移动平均法的递推预测公式为

$$\widehat{x_{t+1}} = \widehat{x_t} + \frac{1}{N} (x_t - x_{t-N})$$

举例：

某农产品加工厂生产农产品，现在有该厂 2020 年全年产量数据，生产人员做了数据收集，

并分别进行了 3 个月和 5 个月的移动平均值，来预测 2021 年 1 月的生产数据。具体见表 3-1。

表 3-1　三个月及五个月的移动平均值及预测值　　　　　　　　单位：吨

时间	观察值	三个月的移动平均值	五个月的移动平均值
2020.1	105	—	—
2020.2	114	—	—
2020.3	128	—	—
2020.4	126	115.7	—
2020.5	120	122.7	—
2020.6	100	124.7	118.6
2020.7	107	115.3	117.6
2020.8	130	109.0	116.2
2020.9	111	112.3	116.6
2020.10	131	116.0	113.6
2020.11	148	124.0	115.8
2020.12	157	130.0	125.4
2021.1	143	145.3	135.4

由表中数据可知，采用三个月移动平均法预测得到 2021 年 1 月的生产数据为 145.3 吨，采用五个月移动平均法预测得到的 1 月份生产数据为 135.4 吨，而该厂 1 月份的实际生产数据为 143 吨。

根据表 3-1 中数据，可以制作"农产品月产量及移动平均值曲线"，见图 3-1。

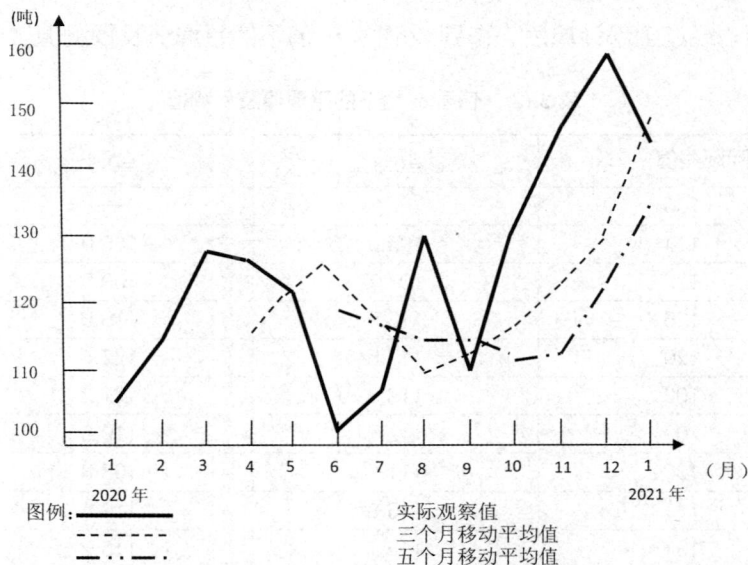

图 3-1　农产品月产量及移动平均值曲线

从图 3-1 中可以看出，随着移动平均值的计算从三个月到五个月，月与月之间的连线趋于平滑。一般来说，移动跨距 N 取较大值时，适合比较平稳的对象；N 取较小值时，适合波动大的对象。

（2）指数平滑法

从时间序列的自相关特点来说，未来的数据值受到现在和近期数据的影响较大，而受相隔时间远的历史数据影响小，因此有"厚近薄远"的说法，即对历史数据根据其离开预测期的近远程度而分别给予不同的权数，近期的数据给予大的权数，远期的历史数据给予小的权数，基于这样的思路，产生了指数平滑法。

指数平滑法的递推预测公式为

$$\hat{x}_t = \hat{x}_{t-1} + \alpha(x_t - \hat{x}_{t-1})$$

这个公式表示 t 期的平滑值 $\hat{x}_t$，是 $t-1$ 期的平滑值 $\hat{x}_{t-1}$ 加上部分误差 " $x_t - \hat{x}_{t-1}$ " 的修正。x_t 是指 t 时期的实测值。α 是平滑常数，取值通常在 0.1~0.9 之间，如果时间序列数值波动很大，α 取值大，以跟上这种波动剧烈的情况；如果时间序列数值波动小，则 α 取值小。

举例：

现在利用表 3-1 的数据，进行指数平滑，并计算 2021 年 1 月的指数平滑预测值。

递推公式为：$\hat{x}_t = \hat{x}_{t-1} + \alpha(x_t - \hat{x}_{t-1})$。

分别计算 α =0.3、α =0.5、α =0.7 时的平滑值。

以 α =0.3 计算，2020 年 1 月的平滑值 $\hat{x}_t$ 为空白。

2 月份的值套用 1 月份的实际观察值。

3 月份的平滑值 ＝ 105+0.3（114－105）＝107.7。

4 月份的平滑值 ＝ 107.7+0.3（128－107.7）＝ 113.8。

5 月份的平滑值 ＝ 113.8+0.3（126－113.8）＝ 117.5。

后面以此类推，经过全部计算后，得到"不同 α 值下的平滑值及预测值"，见表 3-2。

表 3-2　不同 α 值下的平滑值及预测值

时间	实际观察值 x_t／吨	α =0.3 $\hat{x}_t$	α =0.5 $\hat{x}_t$	α =0.7 $\hat{x}_t$
2020.1	105	—	—	—
2020.2	114	105.0	105.0	105.0
2020.3	128	107.7	109.5	111.3
2020.4	126	113.8	118.8	123.0
2020.5	120	117.5	122.4	125.1
2020.6	100	118.3	121.2	121.5
2020.7	107	112.8	110.6	106.5
2020.8	130	111.1	108.8	106.9
2020.9	111	116.8	119.4	123.1
2020.10	131	115.1	115.2	114.6
2020.11	148	119.9	123.1	126.1
2020.12	157	128.3	135.6	141.4
2021.1	143	136.9	146.3	152.3

根据表 3-2 的数据，可以制作"农产品月产量及指数平滑值曲线"，见图 3-2。

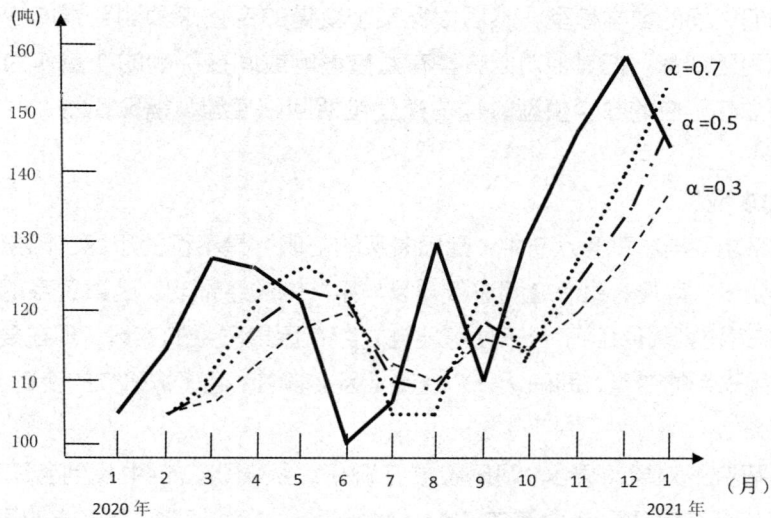

图 3-2　农产品月产量及指数平滑值曲线

由图 3-2 可以看出，当 α =0.3 时，2021 年 1 月的预测值为 136.9 吨；当 α =0.5 时，2021 年 1 月的预测值为 146.3 吨；当 α =0.7 时，2021 年 1 月的预测值为 152.3 吨。

（3）长期趋势分析预测

有时会遇到较长时期的时间序列，则需要进行长期趋势分析。通常可以把这个较长的时间序列分解成四个部分：

一是长期趋势项，用 T 表示；

二是周期变化项，用 C 表示；

三是季节波动项，用 S 表示，一般季节波动项 S 比周期变化项 C 更短；

四是随机干扰项，用 I 表示。

为了分析和讨论它们相互间的作用，通常把上述四个项进行复合，提出两种假设：

一种是叠加模型，表达为 X =T+ C+ S+ I；

一种是乘积模型，表达为 X = T×C×S ×I。

实践表明，乘积模型更能够反映时间序列的变动特征。

为了把时间序列中的趋势项 T 分解出来，可以使用一些趋势曲线，比如：

直线：　$X_t = a + bt$

二次曲线：$X_t = b_0 + b_1 t + b_2 t^2$

指数曲线：$X_t = a\,e^{bt}$

对数曲线：$X_t = a + b\ln t$

双曲线：　$X_t = a + b\dfrac{1}{t}$

皮尔曲线：$X_t = L/(1 + a\,e^{-bt})$

具体趋势曲线的选择，可以根据分析对象的性质、统计模型的识别等来确定，也可以用指数趋势线、皮尔曲线等进行拟合。拟合的方法可以是最小二乘法，或者用迭代法、分段估计法等来解决。

3.7.3 因果关系预测方法

因果关系预测方法是指变量之间存在着一些因果关系，通过找出影响某种结果的一个或若干个因素，建立起它们之间的数学模型，然后根据某个变量的变化来预测结果的变化。一般地，把需要预测的变量作为因变量，把其他对因变量有直接影响或间接影响的变量作为自变量。研究者根据实际情况，建立起相应的数学模型，这些模型包括回归模型、指数模型、经济计量模型、投入产出模型等。

1. 因果关系的求取

这里的因果关系求取与本章第六节中"因果关系的证明"是不同的思路。因果关系的证明是针对两个已经确认的变量（客观对象）之间进行"因与果"的关系证明。因果关系的求取是指针对需要研究的变量，现在通过调研找到了若干项可能与之存在因果联系的变量，现在要从中找到与被研究变量的确存在因果联系的变量，即确认一个变量（因）必然引起了被研究的变量（果）。

（1）求取的思路

第一步，针对研究的对象（果），调研者展开调研，从调研资料中找到若干项可能对研究对象有因果关联的变量，结合调研者已有的知识、经验等，先确认某几个可能的原因（因）。或者是针对研究的对象（因），调研者展开调研，从调研资料中找到若干项可能受研究对象影响的关联变量，结合调研者已有的知识、经验等，确认某几个可能的结果（果）。

第二步，采用适当的方法，来去除不太可能的几个变量。通常采用的方法有"穆勒五法"等。

第三步，确认没有被去除的变量为研究对象的最可能的原因（因）或结果（果）。

（2）穆勒五法

穆勒五法，又称穆勒氏方法，是英国心理学家穆勒在《逻辑学体系》中提出的五种实验推论的方法，用于研究自然界的因果关系。穆勒五法属于归纳推理的范畴，共有契合法、差异法、契合差异并用法、共变法、剩余法五种。

一是契合法，也称求同法。它是指针对研究对象，在可能的影响因素中，找到契合度最高的那一个因素为最可能的原因（或结果）。

举例：

现在有研究的对象（果）是"梁柱节点连接质量差"，通过调查，可能的原因是"结构位置偏差、焊缝不合格、板材生锈、焊条未烘焙"等四项。现在调查人员到不同的项目展开调查，重点挑选"梁柱节点连接质量不好"的几个项目，收集的资料见表3-3。

表3-3　梁柱节点连接质量问题调查表 I

项目名称	相关因素				被研对象
	结构位置偏差	焊缝质量	板材质量	焊条质量	
甲项目	A	B1	C1	D1	
乙项目	A	B2	C2	D2	
丙项目	A	B3	C3	D3	梁柱节点连接质量差
丁项目	A	B4	C4	D4	
……n 项目	A	Bn	Cn	Dn	

由表 3-3 可以看出，每一个项目中，"结构位置偏差"出现的偏差基本相同，而"焊缝质量、板材质量、焊条质量"等三项因素，在不同的项目上出现了不一样的程度，因此研究人员认为"结构位置偏差"的契合度最高。

结论："结构位置偏差"是"梁柱节点连接质量差"最有可能的原因。

二是差异法，也称求异法。它是指针对研究对象，在可能的影响因素中，采用对照组和实验组进行比较，在比较的过程中，分别锁定不同的相关因素不变，在仅剩一个变量的情况下，看被研究对象的变动情况。当某一个变量出现或不出现时，被研究对象也同步出现某种特性或不出现某种特性，则基本确认该变量为最可能的原因。差异法是使用相当普遍的方法。

举例：

现在有研究的对象是"梁柱节点连接质量差"这个问题中的"型钢柱连接偏位、柱轴线定位偏差、标高偏差"三项。现在调查人员分别在甲、乙两个项目上开展对照调研，收集的资料见表 3-4。

表 3-4　梁柱节点连接质量问题调查表 Ⅱ

| 项目名称 | 相关因素 | | | | 被研究对象 |
	结构位置	焊缝质量	板材质量	焊条质量	型钢柱连接偏位 a、柱轴线定位偏差 b、标高偏差 c
实验组 甲项目	结构位置偏差 A1	B	C	D	型钢柱连接偏位 a、柱轴线定位偏差 b、标高偏差 c
对照组 乙项目	结构位置正确 A2	B	C	D	柱轴线定位偏差 b、标高偏差 c

从调研资料得知，"焊缝质量、板材质量、焊条质量"三个因素的情况都基本一致，可以排除它们是产生不同结果的原因。而当"型钢柱连接偏位"这个结果出现或不出现时，"结构位置偏差"和"结构位置正确"两种情况也分别出现或不出现。

结论："结构位置偏差"是"型钢柱连接偏位"的最有可能的原因。

三是契合差异并用法，也称求同求异并用法。它是指针对研究对象，先用契合法分别找到可能的若干项原因。然后采用差异法，在可能的原因中找到最有可能的原因。

举例：

现在有研究的对象是"梁柱节点连接质量差"这个问题中的"型钢柱连接偏位、柱轴线定位偏差、标高偏差"三项。现在调查人员分别在甲、乙、丙、丁四个项目上开展调研，收集的资料见表 3-5。

表3-5　梁柱节点连接质量问题调查表Ⅲ

项目名称	相关因素				被研究对象出现情形
	结构位置	焊缝质量	板材质量	焊条质量	型钢柱连接偏位 a、柱轴线定位偏差 b、标高偏差 c
甲项目	结构位置正确 A1	B1	C1	D2	型钢柱连接正确、柱轴线定位偏差 b、标高偏差 c
乙项目	结构位置正确 A1	B2	C3	D4	型钢柱连接正确、柱轴线定位偏差 b、标高偏差 c
丙项目	结构位置偏差 A2	B1	C1	D2	型钢柱连接偏位 a、柱轴线定位偏差 b、标高偏差 c
丁项目	结构位置偏差 A2	B2	C3	D4	型钢柱连接偏位 a、柱轴线定位偏差 b、标高偏差 c

从调研资料得知，在甲、乙两个项目上，当"型钢柱连接正确"时，"焊缝质量、板材质量、焊条质量"三项各有不同，但"结构位置正确"始终存在，因此，"结构位置正确"是"型钢柱连接正确"的最有可能的原因。

在丙、丁两个项目上，当"型钢柱连接偏位"出现时，"焊缝质量、板材质量、焊条质量"三项各有不同，但"结构位置偏差"始终存在，因此，"结构位置偏差"是"型钢柱连接偏位"最有可能的原因。结论："结构位置偏差"是"型钢柱连接偏位"的最有可能的原因。

通过甲、乙、丙、丁四个项目的综合比较，得出结论："结构位置偏差"是"型钢柱连接偏位"的最有可能的原因。

四是共变法。它是指针对研究对象，对于各可能相关的若干项建立对照组进行比较。对于其中的一项因素给予不同的情况，其余项予以锁定不变，观察研究对象的变化情况。若该项因素与研究对象发生变化具有相关性，则该因素为最有可能的原因。

举例：

现在有研究的对象是"梁柱节点连接质量差"这个问题中的"型钢柱连接偏位"一项。现在调查人员分别在甲、乙、丙、丁四个项目上开展调研，收集的资料见表3-6。

表3-6　梁柱节点连接质量问题调查表Ⅳ

项目名称	相关因素				被研究对象出现情形
	结构位置	焊缝质量	板材质量	焊条质量	型钢柱连接偏位
甲项目	结构位置偏差 A1	B	C	D	型钢柱连接偏位 1
乙项目	结构位置偏差 A2	B	C	D	型钢柱连接偏位 2
丙项目	结构位置偏差 A3	B	C	D	型钢柱连接偏位 3
丁项目	结构位置偏差 A4	B	C	D	型钢柱连接偏位 4

从调研资料得知，在甲、乙、丙、丁四个项目上，共有"结构位置、焊缝质量、板材质量、焊条质量"四个因素，当"型钢柱连接偏差"不断发生变化的同时，"焊缝质量、板材质量、焊条质量"三项保持不变，只有"结构位置偏差"在发生不同程度的变化。

结论："结构位置偏差"是"型钢柱连接偏位"的最有可能的原因。

五是剩余法。它是指针对研究对象，存在着已知的复合原因 A、已知的复合结果 a，则构成了复合原因导致复合结果的关系（即 A 是 a 的原因）。经过调研分析后发现，A 由 A1、A2、A3 构成，a 由 a1、a2、a3 构成。进一步的调研发现，A1 是 a1 的原因，A2 是 a2 的原因。因此，A3 是 a3 的原因。剩余法还有一种思路是：先预计 A＝a，经过调研后发现 A＜a，所以，a ＝ A＋x。因此，x 是新的发现。

举例：

现在有研究的对象是"梁柱节点连接质量差"，这个对象有"型钢柱连接偏位 a1、柱轴线定位偏差 a2、标高偏差 a3"三个问题组成。

通过调研发现，导致"梁柱节点连接质量差"的原因有三种，分别是"结构位置偏差 A1""焊缝质量差 A2"和"板材质量差 A3"。

经过进一步的调研发现，"焊缝质量差 A2"是"柱轴线定位偏差 a2"的原因；"板材质量差 A3"是"标高偏差 a3"的原因。

结论："结构位置偏差 A1"是"型钢柱连接偏位 a1"的最有可能的原因。

2. 建立数学模型

建立数学模型的步骤，可参考第 2 章第 2.6 节的相关内容。

在建模过程中，必须关注模型的有效性，通常表现为三个级别：

一是复制有效（replicatively valid），把实际系统视为一个黑箱，仅在输入输出行为水平上识别系统，这是低水平的有效。

二是预测有效（predictively valid），所建立模型可以预测实际系统将来的状况和变化，其有效性高于复制有效，但对于实际系统内部的结构还不清晰。

三是结构有效（structurally valid），模型能反映实际系统内部的工作关系，能重复反映实际系统的行为，是模型中最高级别。

举例：

在沿海某市最东面有一个 A 小镇，根据天气预报，A 小镇的正东面有一个超强台风，台风中心距离 A 小镇有 200 千米。现在该台风正以每小时 40 千米的速度向西北方向移动。根据气象观测，距离台风中心 150 千米以内，将会受到台风影响。现在需要预测 A 小镇在多长时间后会受到台风的影响，并预计受到台风影响的持续时间。

具体建模及预测过程如下：

根据已知的资料，先建立解析几何模型。

把 A 小镇所在地设为原点，建立一个平面直角坐标系。

把台风中心设为 B 点，坐标为（200，0）。

那么，A 点的坐标为（0，0），台风中心的运动轨迹为从 B 点以 45 度角向 C 点移动。当台风中心运动至以 A 为圆心、半径为 150 千米的圆圈内时（也就是在线段 MN），A 小镇将受到台风的影响。

这样，就得到一个"台风移动及影响范围模型"，见图 3-3。

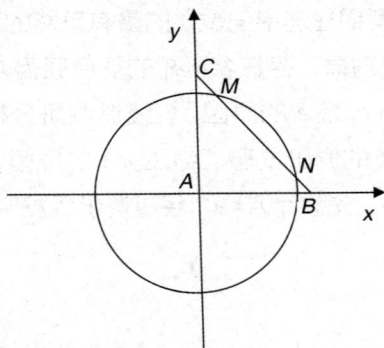

图 3-3　台风移动及影响范围模型

根据所有的资料，可知：

圆的方程为：$x^2 + y^2 = 150^2$

直线 BC 的方程为：$\begin{cases} x = 200 - 40t \cdot \cos 45° \\ y = 40t \cdot \sin 45° \end{cases}$

其中，t 表示时间，单位为小时。

当台风中心抵达圆圈上 N 点，到离开圆圈 M 点，可以建立数学模型如下：

$$\left(200 - 40 \cdot \sqrt{2}/2 \cdot t\right)^2 + \left(40 \cdot \sqrt{2}/2 \cdot t\right)^2 \leqslant 150^2$$

解这个方程（精确到 0.1），可以得到：$2.3 \leqslant t \leqslant 4.9$

$$4.9 - 2.3 = 2.6 \text{（小时）}$$

那么预计在 2.3 小时以后，A 小镇将受到台风影响。持续时间大致为 2.6 小时。

3.7.4 回归分析预测方法

回归分析预测方法是指利用回归综合移动平均数和相应的数学方程，建立变量（因变量）和影响变量因素（自变量）之间的关系，做出变量的回归趋势预测。回归分析预测方法是因果关系预测方法中的一种，在建模过程中，它把回归方程作为模型。回归分析预测在日常生活、工作中比较常见，如证券指数的回归、股票价格的回归、人群身高的回归、产品销量与价格的回归等。回归分析预测方法可以分为一元回归分析和多元回归分析两种。

1. 回归分析预测的基本步骤

回归分析预测的基本步骤主要如下：

一是根据预测的对象和相关要求，确定自变量和因变量。

二是建立回归分析方程，也就是回归分析预测的数学模型。通常使用最小二乘法来求取一元线性回归方程。关于最小二乘法、相关系数的内容，可参考本章第六节的相关内容。

三是对自变量和因变量进行回归分析，通常可求出相关系数来判断自变量和因变量之间的相关程度。

四是对回归分析预测模型进行统计检验，计算预测的误差。当检验后自变量和因变量之间呈线性相关，而且误差较小时，可以进行预测。

五是进行预测，确定预测值。

由于手工运用最小二乘法计算的工作量比较大，在实际操作中，可以借助于相关统计软件来快速获得回归方程、相关系数等，提高工作效率。

2. 一元回归预测方法

一元回归预测方法是根据自变量和因变量之间的相关关系，建立一元线性回归方程，然后进行预测的一种方法。它适用于影响预测对象的因素中，只有一个自变量因素是基本的，同时自变量与因变量之间有着线性分布的关系。

一元回归预测方法的预测模型是一元线性回归方程：

$$y_i = a + b x_i$$

其中，y_i 是因变量的值；x_i 是自变量的值；a、b 是一元线性回归方程的参数。

$$b = \frac{\sum_{i=1}^{n} (x_i - \bar{x})(y_i - \bar{y})}{\sum_{i=1}^{n} (x_i - \bar{x})^2}$$

$$a = \bar{y} - b \bar{x}$$

举例：

某化学品公司需要对某种化学物的投入与产品的产出之间构建一个一元线性回归方程，以开展后续投入与产出的预测工作。生产人员提交的过去 12 个月的资料见表 3-7。

表 3-7　化学物投入与产品产出资料表

月份	1 月	2 月	3 月	4 月	5 月	6 月
投入 x	20.0	15.5	20.9	16.1	17.8	16.5
产出 y	1.98	2.80	1.48	3.50	2.22	2.47
月份	7 月	8 月	9 月	10 月	11 月	12 月
投入 x	19.2	16.6	16.7	21.3	18.4	17.9
产出 y	1.46	2.75	3.83	1.14	1.79	2.01

具体分析如下：

（1）建立一元线性回归模型

$$\sum_{i=1}^{12} x_i = 216.9$$

$$\sum_{i=1}^{12} y_i = 27.43$$

$$\bar{x} = 18.08$$

$$\bar{y} = 2.29$$

$$\sum_{i=1}^{12} (x_i - \bar{x})^2 = 3.69+6.66+7.95+3.92+0.08+2.50+1.25+2.19+1.90+10.37+0.10+0.03$$

$$= 40.64$$

$$\sum_{i=1}^{12} (y_i - \bar{y})^2 = 0.10+0.26+0.66+1.46+0.00+0.03+0.69+0.21+2.37+1.32+0.25+0.08$$

$$= 7.43$$

$$\sum_{i=1}^{12} (x_i - \bar{x})(y_i - \bar{y}) = -0.60-1.32-2.28-2.40+0.02-0.28-0.93-0.68-2.13-3.70-0.16+0.05$$

$$= -14.41$$

因此

$$b = \frac{\sum_{i=1}^{12}(x_i - \bar{x})(y_i - \bar{y})}{\sum_{i=1}^{12}(x_i - \bar{x})^2} = \frac{-14.41}{40.64} = -0.355$$

$$a = \bar{y} - b\bar{x} = 2.29+0.355 \times 18.08 = 8.708$$

由此得到一元线性回归方程为

$$y = a + bx = 8.708 - 0.355x$$

（2）统计检验

使用 F 分布进行统计检验，取 $\alpha = 0.01$。

$$S_{xx} = \sum_{i=1}^{n}(x_i - \bar{x})^2$$

$$S_{yy} = \sum_{i=1}^{n}(y_i - \bar{y})^2$$

$$S_{xy} = \sum_{i=1}^{n}(x_i - \bar{x})(y_i - \bar{y})$$

$$S_{残} = \sum_{i=1}^{n}(y_i - \widehat{y_i})^2$$

其中，$\widehat{y_i} = a + bx_i$

经过计算，得到：

$S_{残} = 0.138+1.507+0.476+1.022+0.167+0.757+0.008+0.697+0.639+0.694+0.038+0.139$

$= 6.282$

建立线性模型为

$$y_i = \beta_0 + \beta_1 x_i + \varepsilon_i, i = 1, 2, \cdots, n$$

其中，假定 ε 服从正态分布 $N(0, \sigma^2)$。

现在原假设为：$H_0: \beta_1 = b_0, H_1: \beta_1 \neq b_0$

取 $b_0 = 0$ 时，为标准误差。

因此，F 检验统计量值为

$$F = T^2$$

$$= \frac{(b - b_0)^2 S_{xx}}{\dfrac{1}{n-2} S_{残}}$$

$$= \frac{(-0.355)^2 \cdot 40.64}{\frac{1}{12-2} \cdot 6.282}$$

$$= \frac{5.1227}{0.6282}$$

$$= 8.154$$

查 F 分布值表，自由度为（1，$n-2$），即（1，10），

当 $\alpha = 0.01$ 时，F 的临界值为 10.04。

显然，8.154 < 10.04，

因此，接受原假设，即可以认为 $\beta_1 = 0$。

那么，$y = 8.708 - 0.355x$ 满足线性模型的假设，可以进行后续的预测工作。

3. 多元回归预测方法

多元回归预测方法是指根据自变量和因变量之间的相关关系，建立多元线性回归方程，然后进行预测的一种方法。它适用于影响预测对象的因素是若干个，即自变量有若干个，且这些自变量是相互独立的。如果这些自变量之间有主次关系，那么次要因素的作用还是不能忽略。

（1）多元回归预测方法的预测模型

多元回归预测方法的预测模型是多元线性回归方程。其表达式为

$$y_i = b_0 + b_1 x_1 + b_2 x_2 + \cdots + b_i x_i + \varepsilon$$

其中，y_i 是因变量的值；

x_i 是自变量的值；

b_0 为常数；

b_i 为对应自变量的回归系数。

以二元线性回归方程为例，b_0，b_1，b_2 的求解方程为

$$\begin{cases} \sum y = n b_0 + b_1 \sum x_1 + b_2 \sum x_2 \\ \sum x_1 y = b_0 \sum x_1 + b_1 \sum x_1^2 + b_2 \sum x_1 x_2 \\ \sum x_2 y = b_0 \sum x_2 + b_2 \sum x_2^2 + b_1 \sum x_1 x_2 \end{cases}$$

（2）多元回归中自变量的确定

在多元回归预测中，对于自变量的确定要注意以下几个方面：

一是所选定的自变量应当对于需要预测的对象（因变量）有着显著的影响，而且这种影响不是偶然的，是内在的因果关系。

二是对于所选定的自变量，研究者应当有比较完整的历史数据，或者通过当前的调查掌握足够的第一手资料。

三是所选定的自变量，在运行的时候有着一定的内在规律，研究者可以对其进行定量的分析。

四是多个自变量之间应当是相互独立的，自变量之间基本不存在明显的相关关系。

五是自变量数量的确定，要尽可能减少，不仅方便计算，更方便确认预测的结果。要抓住主要矛盾，把影响程度大的自变量作为首要分析的变量。

六是当自变量的数量多于两个时，可以采取消元法、逐步回归法等，来减少自变量的数量。

（3）二元线性回归标准差的计算

计算二元线性回归标准差的公式，与计算一元线性回归标准差的公式相同（多元线性回归标准差的公式也是），为

$$S = \sqrt{\frac{\sum_{i=1}^{n} (y_i - \widehat{y_i})^2}{n-k}}$$

其中，y_i 是因变量的第 i 个观察值；

$\widehat{y_i}$ 是因变量在 i 的估计值；

n 为观察期的个数，即样本容量；

k 为自由度，即变量的个数（包括因变量和自变量）。

（4）显著性检验

二元线性回归使用 F 检验，用于检验自变量作为一个整体对于因变量的影响是否有显著的相关关系。

F 检验的计算公式为

$$F = \frac{\sum (\widehat{y} - \overline{y})^2 / (k-1)}{\sum (y - \widehat{y}) / (n-k)}$$

其中，y 是因变量的观察值；

$\overline{y}$ 是因变量观察值的平均值；

$\widehat{y}$ 是因变量在 i 的估计值；

n 是观察期的个数，也是样本容量；

k 是自由度，也是自变量的个数。

在检验时，原假设是 $H_0 : \beta_i = 0$，即所有的 β 都为 0，说明系数没有显著影响，呈线性相关；备择假设是 $H_1 : \beta_i$ 不全为 0，系数有显著影响，不呈明显线性相关。

根据给定的已知条件（即显著性水平），查 F 检验临界值表，得到临界值 $F_a (k, n-k-1)$，"k" 是自变量的个数。如果检验统计量 $F < F_a$，则接受原假设，说明系数没有显著影响，呈线性相关；如果 $F > F_a$，则拒绝原假设，说明系数有显著影响，不呈线性相关。

二元线性回归方程用手工计算工作量相当大，通常借助于统计软件来快速获得。

3.7.5 抽样调查预测方法

抽样调查预测方法是指通过抽样调查的方法，从总体中随机抽取样本，然后根据样本资料对总体进行预测（估计）的方法。

抽样调查的方法可参考第 2 章第 2.4 节的内容。样本容量的确定可参考第 3 章第 3.2 节的内容。统计推论的内容可参考第 3 章第 3.4 节的内容。

举例：

1. 比例估计

某城市想统计该市所有机动车数量（含流动车辆）。为此，调研人员随机在一条城市道路上进行了调查统计（该道路的通行量在全市处于平均水平），统计的方法是采用电子观察（即跟踪牌照），发现在调查当天的上午 10 点至 10 点 30 分，该道路上的通行车辆为 300 辆。在当天下午 3 点，调研人员在相隔比较远的道路上进行了电子观察，在半小时内通行了 200 辆车，其中与上午车辆牌照有重复的为 1 辆。现在估计车辆的全部数量。

该方法为样本比例估计。

由重复出现的车辆，可知样本比例为 $\hat{P} = 1/200 = 0.005$。

假设全市的车辆随机地在城市道路上通行，则总体中出现重复牌照的比例也为 0.005。

所以，$N = 300/0.005 = 60\,000$ 辆。

当然，比例估计的精度比较差。

2. 总体均值的区间估计

举例：

某型喷淋装置的出水量 X 服从 $N(\mu, \sigma^2)$。质量检测人员测量了 10 次，得到 $\bar{x} = 30.5L/S$，$s = 1.53L/S$。现在求 μ 的双侧 90% 置信区间。

具体分析：

μ 的双侧 90% 置信区间上限为

$$\bar{X} + t_{1-\frac{\alpha}{2}}(n-1)\frac{S}{\sqrt{n}} = 30.5 + t_{1-\frac{0.1}{2}}(10-1)\frac{1.53}{\sqrt{10}}$$

$$= 30.5 + 1.833 \times 0.484 = 31.39$$

μ 的双侧 90% 置信区间下限为 $\bar{X} - t_{1-\frac{\alpha}{2}}(n-1)\frac{S}{\sqrt{n}} = 30.5 - 1.833 \times 0.484 = 29.61$

因此，μ 的双侧 90% 置信区间为（29.61，31.29）。

3.7.6 德尔菲法

德尔菲法是定性预测方法中的一种，它不是一个人的定性预测，而是集合了众多人员而进行的一项预测活动。德尔菲方法的具体过程如下：

（1）由召集人召集行业内外的专家、资深人士、骨干等，就某一个具体的主题进行预测前的信息沟通。

（2）在预测人员了解全部相关信息后，根据每个人的经验、知识、能力等进行第一轮的预测。

（3）把第一轮的预测结果进行汇总后，进行预测结果的按序排列，把预测结果基本相同的、得票高的排序在前，得票低的排后。

（4）根据预测工作的需要，选取位列前几位的预测结果（比如是选取位列前 8 的结果）分发到所有的预测人士手中。

（5）进行第二轮的预测。把第二轮的预测结果，按照得票的高低进行排序。

（6）在第二轮的预测结果中，选取位列前茅的预测结果（比如是选取位列前 3 的结果）分发到所有的预测人士手中，进行第三轮的预测。

（7）把第三轮的预测结果，按照得票高低进行排序，选择得票最高的预测结果作为本次预测的最终结果。

德尔菲法简单易行，它的预测过程实质上是把大多数人的意见进行汇总后，挑出得票最高的意见再度进行预测。经过多轮次的预测后，可以快速得到最集中的预测结果。在具体执行的时候，可以给平时预测质量比较高的人员以更大的权重，以进一步提高德尔菲法预测的有效性。

3.7.7 大数据预测

1. 关于大数据的概念

大数据是指需要处理的信息量（数据）极大，超过了一般电脑在处理数据时所使用的内存量。因此，大数据是常规的数据处理软件在一定时间内无法进行处理的数据集合。大数据技术是指对各种类型的数据进行高速的处理，并从中获得有价值信息，并辅助决策（预测）、管理、应用的能力。大数据的数据体量巨大、数据类型繁多、价值密度低。有时候，大数据成为了一种方法论，即一切都可以被记录，一切都可以被数字化，然后从中挖掘价值。

用大数据进行预测工作，是大数据技术的核心应用内容。严格意义上的大数据是全部样本，从实际的应用情况看，大数据是占有了某个研究对象的大部分数据，它不再是传统的抽样数据，因此进行数据的挖掘和分析，对预测工作提供了很大的支持，也提高了预测的精度和效率。当然，由于大数据不是全部数据，因此，也会存在一定的系统性偏差。

对大数据的分析，要用到大量的统计学的内容，包括搜索、比较、判别、聚类、分类等很多分析、归纳等方法。大数据中不仅有大量的有效信息，也有大量的干扰信息。对大数据的分析非常复杂。现在主要使用分布式数据库、分布式计算集群进行分析和分类汇总，来支持大数据的统计分析。

2. 算法

算法是指解决问题的一种方法或一个过程。通常一个算法有以下几个特点：

一是输入，一个算法应该有 n 个初始的输入数据。

二是输出，一个算法的输出信息可以是一个或多个，也可以是没有输出。输出与输入的数据之间存在着特定的联系。

三是确定性，算法中的每一个步骤都必须具有确切的涵义，不能模糊。

四是可行性，算法中所描述的每一个步骤都是可执行的，即都可以通过计算机实现。

五是穷尽性，一个算法必须在有限个步骤之后能够正常结束，不能形成死循环。

3. 机器学习

（1）机器学习的涵义

机器学习是一种能够赋予机器学习的能力，并让机器完成直接编程所无法完成的功能。在实践中，机器学习是通过算法对大数据进行处理，然后"训练"出适当的模型，并应用该模型进行预测。"训练"和"预测"是机器学习的两个重要过程，"模型"是过程的中间输出结果。"训练"产生了"模型"，"模型"指导"训练"。训练过程类似于归纳的过程。机器学习的实质是计算机对人类在学习、成长中的一个模拟。机器学习并非基于编程形成的结果，所以它的处理过程不是因果的逻辑关系，而是通过归纳而得到相关性结论。

从范围上，机器学习与模式识别、统计学习、数据挖掘相类似；同时机器学习又与其他处理技术结合，形成了计算机视觉、语音识别、自然语言处理等交叉学科。机器学习重在对于大数据的处理，现在对于大数据的处理平台很多，如 Hadoop、Spark（Apache）、Atom、MongoDb、IBM PureData、Oracle Exadata、SAP Hana、Matlab 等。

机器学习分监督学习和无监督学习两类。监督学习是指机器（学习系统）通过对输入的信息进行如何组合的学习，对从来没有见过的数据做出预测。无监督学习是指机器不需要对输入的信息进行组合学习，而只是把数据根据某些特性或接近的特性进行分类组织。

（2）算法

机器学习中会用到很多算法，估计现在不下几百种。这里把一些常见的算法予以罗列：

贝叶斯理论（Bayes Theory），是指根据一个已经发生的事件的概率，来计算另一个事件的发生概率。

决策树（Decision Trees），它是一个决策支持工具，使用树形图、决策模型、序列可能性等，通过一个结构化的方式或系统化的方法，对一项业务的决定作出选择。

朴素贝叶斯分类（Naive Bayes），它是对于给出的待分类项，求解在该分类项出现的条件下各类别出现的概率，并把该待分类项归属于概率最大的类别。人脸识别就是这个方法的应用例子。

最小二乘法（Least Square Method），（可参考第三章第六节内容）可以计算线性回归，得到相关关系。而相关性的识别是大数据应用的重要方法，已经得到广泛应用。

线性回归（Linear Regression），与最小二乘法同理。它在自变量和因变量之间建立线性回归方程，并把相应的数学函数称之为模型，然后进行检验、预测。

逻辑回归（Logistic Regression），它属于线性回归算法的范畴，但又与线性回归有所不同。逻辑回归的结果只有两种情形：真（用"1"表示）或假（用"0"表示），常用于数据的分类。逻辑回归算法的拟合函数称为西格蒙（sigmond）函数，函数的输出值只有 0 和 1，该函数也被称为逻辑函数。

K –均值聚类算法（K -Means Clusterring），它是一种无监督机器学习算法，是一种划分聚类算法，用来对未标记的数据进行分类。它通过在数据中查找组来工作，组的数量由变量 k 来表示。在运算过程中，它根据提供的特征，把数据点相应分配给 K 组中的一个。它依据算法中"距离"的概念（在数学中，任何集合中任意两个元素之间距离的函数，为距离函数或度量），把数据予以聚类。之所以称为无监督学习，是因为它可以把数据集当中的、具有某些方面相似的数据进行分类组织，聚类的数目 K 可以由用户指定，在计算时，它根据某个距离函数反复把数据分入 K 个聚类中。

关于度量，可以分为两种类型。一种是欧几里得度量，它可以依据笛卡尔平面上的坐标点来计算两点间距离。另一种是出租车度量（形象化的称呼），它是指实际生活、工作种，两点之间的距离并不是直线，而是有许多的弯路要绕，因此，两点之间的距离有不同的算法和结果。

神经网络（Neural Network），也称人工神经网络，ANN。它的学习机理就是分解与整合，是模拟大脑来工作。比如，把一个正方形分解为四个折线，每个神经元处理一个折线；接着每个折线再继续分解为两条直线，交给神经元处理；每条直线继续分解为黑白两个面。这样，一个正方形变成了大量的细节进入神经元。神经元在处理以后再进行整合，还原成正方形。在神经网络中，每个处理单元就是一个逻辑回归模型。逻辑回归模型接受上层的输入，把模型的预测结果作为输出传输到下一个层次。这样，神经网络完成非常复杂的非线性分类。

支持向量机（Support Vector Machine），简称 SVM。它是一种强化的逻辑回归算法，条件更严格，获得的分类界线比逻辑回归更好。它与高斯"核"结合，表达出非常复杂的分类界线，从而又很好的分类效果。这里的 "核"是一种特殊的函数，最典型的特征是可以将低维的空间映射到高维的空间。比如把二维平面划分出的非线性分类界线，等价于三维平面的线性分类界线，然后通过在三维空间中简单的线性划分，达到在二维平面中的非线性划分效果。支持向量机并不会带来计算复杂性的提升。

降维算法（Dimension Reduction Algorithm），也是一种无监督学习算法，它是把数据从高维降低到低维层次。这里的维度是指数据特征量的大小。降维算法可以去除冗余信息，把数据从高维降低到低维，并最大程度地保留数据的信息。降维算法的主要作用是压缩数据，即把具有几千个特征的数据压缩至若干个特征，以提升其他算法的效率。通过降维算法，还可以实现数据的可视化。

推荐算法（Recommendation Algorithm），是指通过数据分析，可以自动向用户推荐他们最感兴趣的内容。推荐算法主要有两个类别：一类是基于物品内容的推荐，把与用户所需相近似的物品推荐给用户。另一类是基于用户相似度的推荐。有时是两类混合使用。

梯度下降法（Gradient Descent），它是是一个最优化算法，也称最速下降法。最速下降法是求解无约束优化问题最简单、最古老的方法之一。它是用负梯度方向为搜索方向进行迭代搜索，越接近目标值时，步长越小，前进越慢。

牛顿法（Newton's Method），又称牛顿逼近法。它是利用目标函数的泰勒展开式，把非线性函数的最小二乘法问题，化为每次迭代的线性函数的最小二乘法问题。牛顿法在二阶导数的作用下，从函数的凸性出发，直接搜索怎样到达极值点。从收敛速度看，梯度下降是线性收敛，而牛顿法是超线性收敛的。当目标函数不是凸函数时，可以将目标函数近似转化成凸函数。

4. 其他预测方法

主要有：

（1）AR 模型（Auto Regression Model），也称为自回归模型，是统计上一种处理时间序列的方法，用同一变数例如 X 的之前各期，亦即 x_1 至 x_{t-1} 来预测本期 x_t 的表现，并假设它们为线性关系。

（2）MA 模型（Moving Average Model），也称滑动平均模型，指 q 阶移动平均模型的自相关系数 q 阶截尾，偏自相关系数拖尾。

（3）ARMA 模型（Auto Regression Moving Average Model），也称自回归滑动平均模型，是研究平稳随机过程有理谱的典型方法。

3.7　Availability to Predict Trends

When necessary, data used for prediction is one of the important methods for scientific decision-making. Therefore, the quality of data determines the quality of prediction.

Prediction is to speculate, estimate and analyze the possible changes of things in a certain future period based on the actual data and historical data of the development and changes of the research object, using scientific theories and methods, as well as various experiences, judgments and knowledge. The essence of prediction analysis is to fully analyze and understand the laws of change and development of things, and estimate the future state according to the past and present situation of things. An important role of prediction is to reduce the interference of uncertain things in the future and make decisions more scientific by predicting the unknown state.

Due to different prediction objects, time, scope and nature, the prediction can be classified. It mainly includes qualitative prediction method, time series analysis prediction method, causal relationship prediction method, regression analysis prediction method, etc.

3.7.1 Qualitative prediction method

Qualitative prediction method refers to the prediction made by analyzing and judging the object based on people's experience and intuition, combined with its past and current situation.

1. Application of qualitative prediction method

Qualitative prediction methods are mainly applied to three aspects of prediction:

First, predict the development nature of objective things. In scientific research activities, make qualitative predictions about the future development nature of objective things according to their past performance, current situation, combined with empirical data, with reference to certain literature, etc.

The second is to predict the future development trend of the objective things. According to the past operation rules and current operation conditions of the objective things, combined with empirical data, with reference to certain literature, we can make qualitative predictions about the future operation trend of the objective things. The future development trend includes stability, fluctuation, good trend, downward trend, optimism and pessimism.

The third is to predict the internal laws of objective things. According to the past internal laws and current operating conditions of objective things, combined with empirical data, and referring to certain literature, we can predict whether they will maintain the internal laws. The usual prediction results are to continue the existing internal laws and end the existing internal laws.

2. Advantages and disadvantages of qualitative prediction method

The main advantage of qualitative prediction method is that it is flexible and can give full play to the subjective initiative of people, especially the expert prediction is better than the prediction quality. The qualitative prediction method is simple, subject to few objective constraints, and the prediction takes up short events.

The disadvantage of qualitative prediction method is that it is greatly affected by the ability, knowledge and experience of the predictor, and the prediction results are volatile. If it is not an expert prediction, its authority is poor. Qualitative prediction is basically based on personal experience, and is often affected by the environment and emotions at that time. The results of qualitative prediction are more used as a reference for decision-making.

3. Focus on qualitative prediction

After all, qualitative prediction is greatly affected by subjectivity. Therefore, to make a good qualitative prediction, we need to pay attention to the following aspects:

First, the forecaster should try to select professionals with rich experience, knowledge and ability, and more non professionals can participate in the audit to gradually accumulate experience.

Second, it is necessary to strengthen the accumulation of relevant data in the early stage, especially to do some research work, and try to master the panoramic data of the predicted object.

The third is to study and judge various unfavorable factors that may be encountered in the prediction process in advance, and try to avoid unnecessary interference.

The fourth is to collect as much quantitative data as possible during the preliminary background survey, so that qualitative forecasters can have more basis for reference and improve the quality of prediction.

Fifth, collect more peer data with the same nature, background, similar development process and close environmental impacts as the predicted object, and analyze their operating trends and development laws as key references.

3.7.2 Time series analysis and prediction method

Time series analysis and prediction method refers to a method to predict the future operation trend of an objective object based on its historical data of changes with time, including various statistical data, test data, change trend, etc. of the objective object, taking into account of the change rule of the object over time.

Time series samples are very different from simple random samples. x_i in simple random samples are independent of each other; However, x_i in the time series sample is not independent of each other, which shows the autocorrelation of the time series.

1. The inherent logic of time series analysis and prediction method

The time series analysis and prediction method is only based on the historical data obtained so far and recorded in chronological order. Its internal logic is that time T integrates various factors that affect the observation object, and the past time series data reflects the change rule of things, which will last until the future time under the action of various factors. Researchers can establish a time series model that changes with time based on the historical data that has occurred, which can not only analyze the existing change rules of things, but also infer future values, that is, predict future values.

2. Main methods of time series analysis and prediction

(1) Moving average method

Moving average method is one of the smoothing methods. The idea of smoothing method is to use the smoothed historical data, that is, eliminate the large and small values in the historical data to smooth them. The data used in the moving average method is not all historical data, but N observations closest to the prediction period. The average of these observations is taken as the prediction value. The formula for these averages is

$$\widehat{x}_{t+1} = \frac{x_{t+} x_{t-1+\cdots+} x_{t-N+1}}{N} = \frac{1}{N}\sum_{i=0}^{N-1} x_{t-i}$$

In the formula, N is called move span.

With the pass of time t, new data will be added continuously, and the corresponding N data will be intercepted continuously, and new average values will be obtained continuously, and so on.

Through simple operation, we can get the recursive prediction formula of moving average method as follows:

$$\widehat{x}_{t+1} = \widehat{x}_{t} + \frac{1}{N}(x_{t} - x_{t-N})$$

Case:

An agricultural product processing factory produces agricultural products. Now there is the annual output data of the factory in 2020. The production personnel have collected the data and carried out three and five-month moving average values respectively to forecast the production data in January 2021. See Table 3-1 for details.

Table 3-1　Three and five-month moving average and forecast values　　Unit: Tons

Time	Observed value	Three-month moving average	Five-month moving average
2020.1	105	—	—
2020.2	114	—	—
2020.3	128	—	—
2020.4	126	115.7	—
2020.5	120	122.7	—
2020.6	100	124.7	118.6
2020.7	107	115.3	117.6
2020.8	130	109.0	116.2
2020.9	111	112.3	116.6
2020.10	131	116.0	113.6
2020.11	148	124.0	115.8
2020.12	157	130.0	125.4
2021.1	143	145.3	135.4

It can be seen from the data in the table that the production data in January 2021 predicted by the three-month moving average method is 145.3 tons, the production data in January predicted by the five month moving average method is 135.4 tons, and the actual production data of the plant in January is 143 tons.

According to the data in Table 3-1, the "Monthly output and moving average curve of agricultural products" can be made, as shown in Figure 3-1.

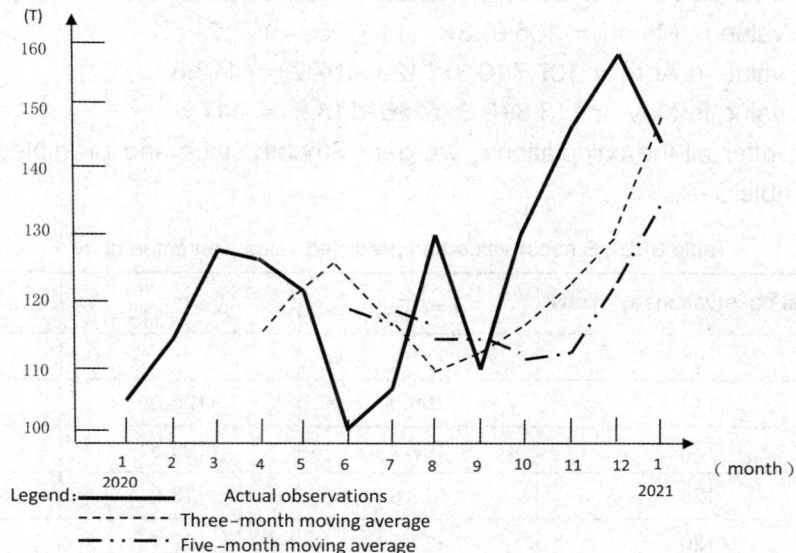

Figure 3-1   Monthly output and moving average curve of agricultural products

It can be seen from Figure 3-1 that with the calculation of the moving average from three months to five months, the line between months tends to smooth. Generally speaking, when the moving span N takes a larger value, it is suitable for relatively stable objects; When N is smaller, it is suitable for objects with large fluctuations.

(2) Exponential smoothing method

In terms of the autocorrelation characteristics of time series, the future data values are greatly affected by the current and recent data, while they are less affected by the historical data far apart. Therefore, there is a saying of "the near-term data weight is significant, while the long-term data weight is small", that is, historical data are given different weights according to their near distance from the prediction period. The recent data are given large weights, and long-term historical data are given small weights. Based on this idea, the exponential smoothing method is generated.

The recursive prediction formula of exponential smoothing method is

$$\widehat{x_t} = \widehat{x_{t-1}} + \alpha\,(x_t - \widehat{x_{t-1}})$$

This formula represents the smooth value $\widehat{x_t}$ of period t, which is the correction of the smoothing value $\widehat{x_{t-1}}$ in period "$t-1$" plus partial error "$x_t - \widehat{x_{t-1}}$". x_t refers to the measured value in period t. α is a smoothing constant, usually between 0.1 and 0.9. If the time series values fluctuate greatly, take the large value of α to keep up with such severe fluctuations; if the time series values fluctuate slightly, then the value of α is small.

Case:

Now we use the data in Table 3-1 to perform exponential smoothing and calculate the predicted value of exponential smoothing in January 2021.

The recursive formula is: $\widehat{x_t} = \widehat{x_{t-1}} + \alpha\,(x_t - \widehat{x_{t-1}})$.

Calculate the smoothing value that $\alpha = 0.3$, $\alpha = 0.5$, $\alpha = 0.7$ respectively.

With $\alpha = 0.3$, smooth value $\widehat{x_t}$ in January 2020 is blank.

The value in February is applied to the actual observation value in January.

The smooth value in March $= 105 + 0.3 \times (114 - 105) = 107.7$；

The smooth value in April $= 107.7 + 0.3 \times (128 - 107.7) = 113.8$；

The smooth value in May $= 113.8 + 0.3 \times (126 - 113.8) = 117.5$.

By analogy, after all the calculations, we get "Smooth value and predicted value with different α". See Table 3–2.

Table 3–2　Smooth value and predicted value with different α

Time	Actual observations x_t /Tons	$\alpha = 0.3$ $\widehat{x_t}$	$\alpha = 0.5$ $\widehat{x_t}$	$\alpha = 0.7$ $\widehat{x_t}$
2020.1	105	—	—	—
2020.2	114	105.0	105.0	105.0
2020.3	128	107.7	109.5	111.3
2020.4	126	113.8	118.8	123.0
2020.5	120	117.5	122.4	125.1
2020.6	100	118.3	121.2	121.5
2020.7	107	112.8	110.6	106.5
2020.8	130	111.1	108.8	106.9
2020.9	111	116.8	119.4	123.1
2020.10	131	115.1	115.2	114.6
2020.11	148	119.9	123.1	126.1
2020.12	157	128.3	135.6	141.4
2021.1	143	136.9	146.3	152.3

According to the data in Table 3–2, "Monthly output and exponential smoothing value curve of agricultural products" can be made, as shown in Figure 3–2.

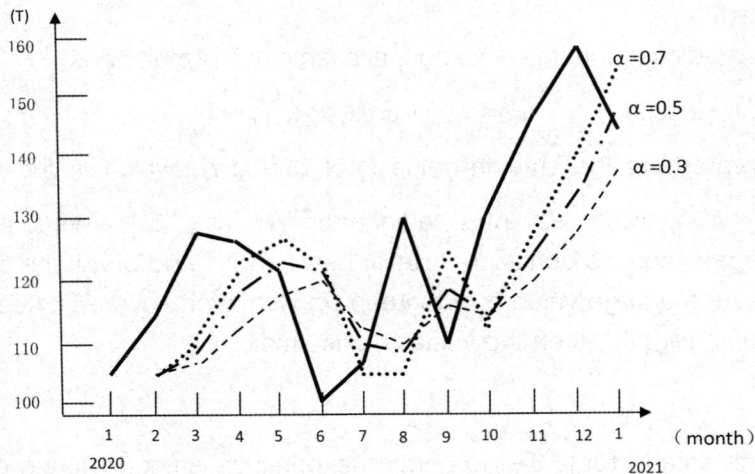

Figure 3–2　Monthly output and exponential smoothing value curve of agricultural products

It can be seen from Figure 3–2 that when $\alpha = 0.3$, the forecast value in January 2021 is

136.9 tons; When $\alpha = 0.5$, the forecast value in January 2021 is 146.3 tons; When $\alpha = 0.7$, the forecast value in January 2021 is 152.3 tons.

（3）Long term trend analysis and prediction

Sometimes long time series are encountered, so long-term trend analysis is required. The long time series can usually be decomposed into four parts:

First, long-term trend term, which can be expressed by T;

The second is the periodic vairation term, which can be expressed in C;

The third is the seasonal fluctuation term, which is expressed by S. Generally, the seasonal fluctuation term S is shorter than the periodic variation term C;

The fourth is random interference term, which is expressed by I.

In order to analyze and discuss the interaction between them, the above four terms are usually combined and two hypotheses are proposed:

One is the superposition model, which is expressed as X=T+C+S+I;

The other is the product model, which is expressed as X = T×C×S×I.

Practice shows that the product model can better reflect the change characteristics of time series.

In order to decompose the trend item T in the time series, some trend curves can be used, such as:

Straight line: $X_t = a + bt$

Conic: $X_t = b_0 + b_1 t + b_2 t^2$

Exponential curve: $X_t = a\, e^{bt}$

Logarithmic curve: $X_t = a + b\ln t$

Hyperbola: $X_t = a + b\,\dfrac{1}{t}$

Pearl curve: $X_t = L/(1 + a\, e^{-bt})$

The specific trend curve can be selected according to the nature of the analysis object and the identification of the statistical model, and can also be fitted with the exponential trend line and the Pearl curve. The fitting method can be least square method, or iterative method, subsection estimation method, etc.

3.7.3 Causality prediction method

Causality prediction method refers to that there are some causal relationships between variables. By finding out one or several factors that affect a certain result, a mathematical model is established between them, and then the change of the result is predicted according to the change of a variable. Generally, the variables to be predicted are regarded as dependent variables, and other variables that have direct or indirect influence on the dependent variables are regarded as independent variables. According to the actual situation, researchers have established corresponding mathematical models, including regression model, index model, econometric model, input and output model, etc.

1. Calculation of causality

The causality here is different from the "Proof of causality" in Section 6 of this chapter. The proof of causality refers to the proof of "cause and effect" relationship between two confirmed variables (objective objects). The calculation of causality refers to the variables to be studied. Now we have found several variables that may have causal links with them through research.

Now we need to find the variables that do have causal links with the variables to be studied, that is, to confirm that a variable (cause) must cause the variables to be studied (result).

(1) The idea of calculation

The first step is to investigate the research object (result), find out several variables that may be causal related to the research object from the research data, and identify some possible reasons (causes) based on the existing knowledge and experience of the investigators. Or for the research object (cause), the researcher conducts research, finds several related variables that may be affected by the research object from the research data, and confirms some possible results (results) based on the existing knowledge and experience of the researcher.

The second step is to use appropriate methods to divide the unlikely variables. The commonly used methods are "Mill's Methods", etc.

Step three is to confirm that the variables that have not been removed are the most likely causes or results of the research object.

(2) Mill's Methods

Mill's methods, also called Mill's five methods, is the method of five experimental inferences proposed by British psychologist Mill in *A System of Logic* which is used to study the causal relationship in nature. Mill's five methods belong to the category of inductive reasoning, including the method of agreement, the method of difference, the joint method, the method of concomitant variation and the method of residues.

The first is the method of agreement, also called the method of seeking the same, which refers to the most likely cause (or result) with the highest fit degree of the possible influencing factors of the research object.

Case:

At present, the object of study (result) is "poor connection quality of beam column joints". Through investigation, the possible causes are "structural position deviation, unqualified welds, rusty plates, and un-baked welding rods". Now the investigators are going to different projects to carry out investigation, focusing on several projects with "poor connection quality of beam column joints". The data collected are shown in Table 3-3.

Table 3-3　Poor connection quality of beam column joints Ⅰ

Project name	Relevant factors				Object of study
	Structural position deviation	Weld quality	Plates quality	Welding rod quality	
Project Ⅰ	A	B1	C1	D1	Poor connection quality of beam column joints
Project Ⅱ	A	B2	C2	D2	
Project Ⅲ	A	B3	C3	D3	
Project Ⅳ	A	B4	C4	D4	
Pjoject *n*	A	Bn	Cn	Dn	

It can be seen from Table 3-3 that in each project, the deviation of "structure position deviation" is basically the same, while the three factors of "weld quality, plates quality and welding rod quality" have different degrees in different projects, so researchers believe that "structure position deviation" has the highest degree of fit.

Conclusion: "structural position deviation" is the most likely cause of "poor connection quality of beam column joints".

The second is the method of difference, also called the method of seeking difference, which refers to the comparison between the control group and the experimental group among the possible influencing factors of the research object. In the process of comparison, different relevant factors remain unchanged, and when there is only one variable, the change of the research object can be seen. When a variable appears or does not appear, and a certain characteristic of the object under study appears or does not appear synchronously, it is basically confirmed that the variable is the most likely cause. The difference method is widely used.

Case:

Now, the research object is the "connection deviation of section steel column, positioning deviation of column axis and elevation deviation" in the problem of "poor connection quality of beam column joints". At present, the investigators have carried out comparative research on projects I and II, and the data collected are shown in Table 3-4.

Table 3-4　Poor connection quality of beam column joints II

Project name	Relevant factors				Object of study
	Structure position	Weld quality	Plates quality	Welding rod quality	a connection deviation of section steel column, b positioning deviation of column axis, c elevation deviation
Experience group Project I	Deviation of structure position A1	B	C	D	a connection deviation of section steel column, b positioning deviation of column axis, c elevation deviation
Control group Project II	Correct structure position A2	B	C	D	b positioning deviation of column axis, c elevation deviation

According to the survey data, the three factors of "weld quality, plates quality and welding rod quality" are basically the same, which can be ruled out as the cause of different results. When the result of "connection deviation of section steel column" appears or does not appear, "deviation of structure position" and "correct structure position" also appear or do not appear respectively.

Conclusion: "deviation of structure position" is the most likely cause of "connection deviation of section steel column".

The third is the joint method, which is also called seeking similarities and differences method. It refers to finding several possible reasons for the research object by the method of agreement, then use the difference method to find the most likely cause among the possible causes.

Case:

Now, the research object is the "connection deviation of section steel column, positioning deviation of column axis and elevation deviation" in the problem of "poor connection quality of beam column joints". Now the investigators have carried out research on the four projects I, II, III and IV respectively, and the data collected are shown in Table 3-5.

Table 3-5　Poor connection quality of beam column joints Ⅲ

Project name	Relevant factors				Situations of the object
	Structure position	Weld quality	Plates quality	Welding rod quality	A connection deviation of section steel column, b positioning deviation of column axis, c elevation deviation
Project Ⅰ	Correct structure position A1	B1	C1	D2	a connection deviation of section steel column, b positioning deviation of column axis, c elevation deviation
Project Ⅱ	Correct structure position A1	B2	C3	D4	a connection deviation of section steel column, b positioning deviation of column axis, c elevation deviation
Project Ⅲ	Deviation of structure position A2	B1	C1	D2	a connection deviation of section steel column, b positioning deviation of column axis, c elevation deviation
Project Ⅳ	Deviation of structure position A2	B2	C3	D4	a connection deviation of section steel column, b positioning deviation of column axis, c elevation deviation

It is known from the survey data that in the project Ⅰ and project Ⅱ, when the "section steel column connection is correct", the three items of "weld quality, plates quality and welding rod quality" are different, but "correct structure position" always exists. Therefore, "correct structure position" is the most likely reason for "correct connection of section steel column".

For project Ⅲ and project Ⅳ, when the "connection deviation of steel section column" occurs, the three items of "weld quality, plates quality and welding rod quality" are different, but the "structural position deviation" always exists. Therefore, the "structural position deviation" is the most possible reason for the "connection deviation of steel section column".

Conclusion: "structure position deviation" is the most likely cause of "section steel column connection deviation".

Through the comprehensive comparison of the four projects of Ⅰ, Ⅱ, Ⅲ and Ⅳ, it is concluded that "structure position deviation" is the most likely cause of "section steel column connection deviation".

The fourth is the method of concomitant variation, it refers to the establishment of a control group for comparison of several items that may be relevant. Give different conditions to one of the factors, and lock the rest to observe the changes of the research object. If this factor is related to the change of the research object, it is the most likely cause.

Case:

At present, the research object is the "section steel column connection deviation" in the problem of "poor connection quality of beam column joints". Now the investigators are conducting research on four projects, namely, Ⅰ, Ⅱ, Ⅲ and Ⅳ. The data collected are shown in Table 3-6.

Table 3-6　Poor connection quality of beam column joints Ⅳ

Project name	Relevant factors				Object of study
	Structure position	Weld quality	Plates quality	Welding rod quality	Connection deviation of section steel column
Project Ⅰ	Deviation of structure position A1	B	C	D	connection deviation of section steel column 1
Project Ⅱ	Deviation of structure position A2	B	C	D	connection deviation of section steel column 2
Project Ⅲ	Deviation of structure position A3	B	C	D	connection deviation of section steel column 3
Project Ⅳ	Deviation of structure position A4	B	C	D	connection deviation of section steel column 4

According to the survey data, there are four factors in the four projects of Ⅰ, Ⅰ, Ⅲ and Ⅳ, namely "structure position, weld quality, plates quality and welding rod quality". When the "connection deviation of section steel column" is constantly changing, the three items of "weld quality, plates quality and welding rod quality" remain unchanged, and only the "structure position deviation" is changing in varying degrees.

Conclusion: "structure position deviation" is the most likely cause of "section steel column connection deviation".

The fifth is the method of residues, which refers to the existence of a known composite cause A and a known composite result a, which constitute the relationship between composite causes and composite results (that is, A is the cause of a). After investigation and analysis, it is found that A is composed of A1, A2 and A3, and a is composed of a1, a2 and a3. Further investigation shows that A1 is the cause of a1 and A2 is the cause of a2. Therefore, A3 is the cause of a3. Another idea of the residual method is to estimate A = a first, and after investigation, it is found that A < a, so a = A + x. Therefore, x is a new discovery.

Case:

At present, the research object is "poor connection quality of beam column joints", which consists of three problems: "a1 connection deviation of section steel column, a2 positioning deviation of column axis, and a3 elevation deviation".

Through investigation, it is found that there are three reasons for "poor connection quality of beam column joints", namely, "A1 structure position deviation" "A2 poor weld quality" and "A3 poor plate quality".

Through further investigation, it is found that "A2 poor weld quality" is the cause of "a2 column axis positioning deviation"; "A3 poor plate quality" is the cause of "a3 elevation deviation".

Conclusion: "A1 structure position deviation" is the most likely cause of "a1 section steel column connection deviation".

2. Establish mathematical model

For the steps of establishing the mathematical model, please refer to the relevant contents in Section 2.6 of Chapter 2.

In the process of modeling, we must pay attention to the effectiveness of the model, which is usually manifested in three levels:

The first is replicatively valid. The actual system is regarded as a black box, and the system is identified only at the input/output behavior level, which is a low level of effectiveness.

The second is predictive validity. The established model can predict the future status and changes of the actual system. Its effectiveness is higher than that of the replicatively valid, but the internal structure of the actual system is not clear.

The third is structurally valid. The model can reflect the working relationship within the actual system, and can repeatedly reflect the behavior of the actual system. It is the highest level in the model.

Case:

There is a small town A in the east of a coastal city. According to the weather forecast, there is a super typhoon just east of A, and the typhoon center is 200 kilometers away from A. Now the typhoon is moving northwest at a speed of 40 kilometers per hour. According to meteorological observation, typhoon will affect the area within 150 km from the typhoon center. Now it is necessary to predict how long after Town A will be affected by the typhoon, and predict the duration of the impact of the typhoon.

The specific modeling and prediction process is as follows:

Based on the known data, the analytical geometric model is established first.

Set the location of Town A as the origin and establish a plane rectangular coordinate system.

Set the typhoon center as point B, with coordinates of $(200, 0)$.

Then, the coordinate of point A is $(0, 0)$, and the motion track of the typhoon center is moving from point B to point C at an angle of 45 degrees. When the typhoon center moves to a circle with A as the center and a radius of 150 kilometers (i.e., in the segment MN), Town A will be affected by the typhoon.

In this way, a "Typhoon movement and influence range model" is obtained, as shown in Figure 3-3.

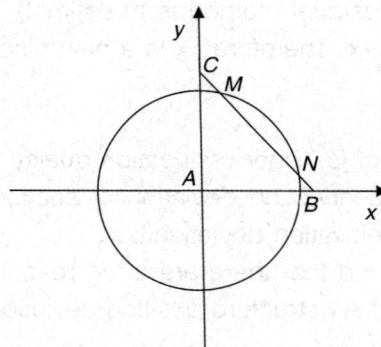

Figure 3-3 Typhoon movement and influence range model

According to all the data:

The equation of the circle is: $x^2 + y^2 = 150^2$

The equation of line BC is: $\begin{cases} x = 200 - 40t \cdot \cos 45° \\ y = 40t \cdot \sin 45° \end{cases}$

In the formula, t means time, and its unit is hour.

When the typhoon center arrives at point N on the circle and leaves point M on the circle, the mathematical model can be established as follows:

$$(200 - 40 \cdot \sqrt{2}/2 \cdot t)^2 + (40 \cdot \sqrt{2}/2 \cdot t)^2 \leqslant 150^2$$

By solving this equation (accurate to 0.1), we can get: $2.3 \leqslant t \leqslant 4.9$

$$4.9-2.3 = 2.6 \text{ (hours)}$$

It is estimated that after 2.3 hours, Town A will be affected by the typhoon. The duration is about 2.6 hours.

3.7.4 Regression analysis and prediction method

Regression analysis prediction method refers to establishing the relationship between variables (dependent variables) and influencing variable factors (independent variables) by using the regression comprehensive moving average and the corresponding mathematical equation to predict the regression trend of variables. Regression analysis prediction method is one of the causal relationship prediction methods. In the modeling process, it takes the regression equation as the model. Regression analysis and prediction are common in daily life and work, such as the regression of stock index, stock price, population height, product sales and price. Regression analysis and prediction methods can be divided into single regression analysis and multiple regression analysis.

1. Basic steps of regression analysis and prediction

The basic steps of regression analysis and prediction are as follows:

First, determine the independent variable and dependent variable according to the prediction object and relevant requirements.

The second is to establish the regression analysis equation, that is, the mathematical model of regression analysis and prediction. The least square method is usually used to solve the linear regression equation with one variable. For the contents of least squares method and correlation coefficient, refer to the relevant contents in Section 6 of this chapter.

The third is to conduct regression analysis on independent variable and dependent variable, and usually calculate correlation coefficient to judge the correlation degree between independent variable and dependent variable.

The fourth, the regression analysis prediction model is statistically tested to calculate the prediction error. When there is a linear correlation between the independent variable and the dependent variable after the test, and the error is small, it can be predicted.

The fifth, predict and determine the predicted value.

Due to the heavy workload of manual calculation with the least squares method, the regression equation and correlation coefficient can be quickly obtained with the help of the relevant statistical software in practical operation to improve the work efficiency.

2. One variable regression prediction method

One variable regression prediction method is a method to establish a linear regression equation based on the correlation between independent variables and dependent variables, and then predict. It is applicable to the factors that affect the prediction object. Only one independent variable factor is basic, and there is a linear distribution relationship between independent variables and dependent variables.

The predict model of the one variable regression prediction is theunary linear regression equation:

$$y_i = a + b x_i$$

In the formula, y_i is the value of the dependent variable; x_i is the value of the independent variable;

a and b are the parameters of the unary linear regression equation.

$$b = \frac{\sum_{i=1}^{n}(x_i - \bar{x})(y_i - \bar{y})}{\sum_{i=1}^{n}(x_i - \bar{x})^2}$$

$$a = \bar{y} - b\bar{x}$$

Case：

A chemical company needs to build a one variable linear regression equation between the input of a certain chemical and the output of the product to carry out the follow-up input and output prediction. The data submitted by the production personnel in the past 12 months are shown in Table 3-7.

Table 3-7　Data sheet of chemical input and product output

Month	January	February	March	April	May	June
Input x	20.0	15.5	20.9	16.1	17.8	16.5
Output y	1.98	2.80	1.48	3.50	2.22	2.47
Month	July	August	September	October	November	December
Input x	19.2	16.6	16.7	21.3	18.4	17.9
Output y	1.46	2.75	3.83	1.14	1.79	2.01

Make specific analysis.

（1）Establish a linear regression model with one variable.

$$\sum_{i=1}^{12} x_i = 216.9$$

$$\sum_{i=1}^{12} y_i = 27.43$$

$$\bar{x} = 18.08$$

$$\bar{y} = 2.29$$

$$\sum_{i=1}^{12}(x_i - \bar{x})^2 = 3.69+6.66+7.95+3.92+0.08+2.50+1.25+2.19+1.90+10.37+0.10+0.03$$
$$= 40.64$$

$$\sum_{i=1}^{12}(y_i - \bar{y})^2 = 0.10+0.26+0.66+1.46+0.00+0.03+0.69+0.21+2.37+1.32+0.25+0.08$$
$$= 7.43$$

$$\sum_{i=1}^{12}(x_i - \bar{x})(y_i - \bar{y}) = -0.60-1.32-2.28-2.40+0.02-0.28-0.93-0.68-2.13-3.70-0.16+0.05$$
$$= -14.41$$

so：

$$b = \frac{\sum_{i=1}^{12}(x_i - \bar{x})(y_i - \bar{y})}{\sum_{i=1}^{12}(x_i - \bar{x})^2} = \frac{-14.41}{40.64} = -0.355$$

$$a = \bar{y} - b\bar{x} = 2.29+0.355 \times 18.08 = 8.708$$

Thus, the linear regression equation of one variable is

$$y = a + bx = 8.708 - 0.355x$$

（2）Statistical test

Statistical test using F-distribution, take $\alpha = 0.01$.

$$S_{xx} = \sum_{i=1}^{n} (x_i - \bar{x})^2$$

$$S_{yy} = \sum_{i=1}^{n} (y_i - \bar{y})^2$$

$$S_{xy} = \sum_{i=1}^{n} (x_i - \bar{x})(y_i - \bar{y})$$

$$S_{残} = \sum_{i=1}^{n} (y_i - \hat{y}_i)^2$$

Among them, $\hat{y}_i = a + b x_i$.

After calculation, it is obtained that

$S_{残}$ =0.138+1.507+0.476+1.022+0.167+0.757+0.008+0.697+0.639+0.694+0.038+0.139

　　= 6.282

So we can establish the linear model as：

$$y_i = \beta_0 + \beta_1 x_i + \varepsilon_i, i = 1, 2, \cdots, n$$

In the formula, assume that ε obeys normal distribution $N(0, \sigma^2)$.

The null hypothesis is：$H_0 : \beta_1 = b_0$, $H_1 : \beta_1 \neq b_0$

When $b_0 = 0$, it is the standard error.

Therefore, the statistical value of the F test is

$$F = T^2$$

$$= \frac{(b - b_0)^2 S_{xx}}{\dfrac{1}{n-2} S_{残}}$$

$$= \frac{(-0.355)^2 \cdot 40.64}{\dfrac{1}{12-2} \cdot 6.282}$$

$$= \frac{5.1227}{0.6282}$$

$$= 8.154$$

Check the value table of F distribution, the degree of freedom is（1, $n-2$）, that is（1, 10）,

When $\alpha = 0.01$, the critical value of F is 10.04.

Obviously, 8.154 < 10.04,

So, the null hypothesis is accepted, that is to say, $\beta_1 = 0$.

Then, $y = 8.708 - 0.355x$ satisfies the assumptions of the linear model, and subsequent prediction can be carried out.

3. Multivariate regression prediction method

Multivariate regression prediction method refers to a method of establishing a multivariate linear regression equation according to the correlation between independent variables and dependent variables, and then forecasting. It is applicable to several factors that affect the prediction object, that is, there are several independent variables, and these independent variables are mutually independent. If there is a primary and secondary relationship between these independent variables, the role of secondary factors cannot be ignored.

（1）Prediction model of multivariate regression prediction method

The prediction model of multivariate regression prediction method is multiple linear regression equation. Its expression is

$$y_i = b_0 + b_1 x_1 + b_2 x_2 + \cdots + b_i x_i + \varepsilon$$

In the formula, y_i is the value of dependent variable;

x_i is the value of independent variable;

b_0 is a constant,

b_i is the regression coefficient of the corresponding independent variable.

Take the binary linear regression equation as an example, the equation for b_0 , b_1 , b_2 is

$$\begin{cases} \sum y = n b_0 + b_1 \sum x_1 + b_2 \sum x_2 \\ \sum x_1 y = b_0 \sum x_1 + b_1 \sum x_1^2 + b_2 \sum x_1 x_2 \\ \sum x_2 y = b_0 \sum x_2 + b_2 \sum x_2^2 + b_1 \sum x_1 x_2 \end{cases}$$

(2) Determination of independent variables in multivariate regression

In multivariate regression prediction, the following aspects should be paid attention to when determining independent variables.

First, the selected independent variable should have a significant impact on the object (dependent variable) to be predicted, and this impact is not accidental, but an internal causal relationship.

Second, for the selected independent variables, researchers should have relatively complete historical data, or grasp enough first-hand information through the current survey.

Third, the selected independent variables have certain internal laws when they are running, and researchers can conduct quantitative analysis on them.

Fourth, multiple independent variables should be independent of each other, and there is basically no obvious correlation between independent variables.

Fifth, the number of independent variables should be determined, which should be reduced as much as possible to facilitate not only the calculation, but also the confirmation of the prediction results. It is necessary to grasp the main contradiction and take the independent variable with great influence as the primary variable for analysis.

Sixth, when the number of independent variables is more than two, the elimination method and stepwise regression method can be used to reduce the number of independent variables.

(3) Calculation of standard deviation of binary linear regression

The formula for calculating the standard deviation of binary linear regression is the same as that for calculating the standard deviation of unitary linear regression (as is the formula for calculating the standard deviation of multivariate linear regression):

$$S = \sqrt{\frac{\sum_{i=1}^{n} (y_i - \widehat{y_i})^2}{n - k}}$$

In the formula, y_i is the ith observation value of the dependent variable;

$\widehat{y_i}$ is the estimated value of the dependent variable at i ;

N is the number of samples in the observation period, that is, the sample size;

k is the degree of freedom, that is, the number of variables (including dependent variables and independent variables).

(4) Significance test

Binary linear regression uses the F test to test whether the independent variable as a whole has a significant correlation with the dependent variable.

The calculation formula of F test is

$$F = \frac{\sum (\hat{y} - \bar{y})^2/(k-1)}{\sum (y - \hat{y})/(n-k)}$$

In the formula, y is the observed value of dependent variable;

$\bar{y}$ is the average of the dependent variable observation;

$\hat{y}$ is the estimated value of the dependent variable at i ;

n is the number of observation periods, and it is also the sample size;

k is the degree of freedom, which is the number of the independent variable.

In the test, the null hypothesis is $H_0 : \beta_i = 0$, that is, all β are 0, indicating that the coefficient has no significant effect and it is the linear correlation; the alternative hypothesis is $H_1 : \beta_i$ is no tall 0, which shows that the coefficient has significant influence, and it does not show significant linear correlation.

According to the given known conditions (i.e. significance level), check the critical value table of F test to get the critical value $F_a (k, n - k - 1)$. "k" is the number of the independent variable. If the test statistic $F < F_a$, then accept the null hypothesis, which indicates that the co-efficient has no significant effect, and the linear correlation is established; If $F > F_a$, then reject the null hypothesis, which indicates that the coefficient has significant effect, and the linear cor-relation is not existed.

The workload of manual calculation of binary linear regression equation is quite large, and it is usually obtained quickly with the help of statistical software.

3.7.5 Sampling survey prediction method

Sampling survey forecasting method refers to the method of random sampling from the population through sampling survey, and then forecasting (estimating) the population according to sample data.

For the method of sampling survey, please refer to Section 2.4 of Chapter 2. The determi-nation of sample size can refer to Section 3.2 of Chapter 3. For the content of statistical infer-ence, please refer to Section 3.4 of Chapter 3.

1. Proportional estimation

Case:

A city wants to count the number of all motor vehicles (including mobile vehicles) in the cit-y. For this reason, the investigators randomly conducted an investigation and statistics on an ur-ban road (the traffic volume of the road is at the average level in the city). The statistical meth-od is to use electronic observation (i.e. tracking license plate). It was found that there were 300 vehicles on the road from 10:00 a.m. to 10:30 a.m. on the day of investigation. At 3:00 p.m. on the same day, the investigators conducted an electronic observation on the road far away, and 200 vehicles passed in half an hour, including one with the same license plate as that in the morning. Now estimate the total number of vehicles.

This method is sample proportion estimation.

From the repeated vehicles, the sample proportion is $\hat{P} = 1/200 = 0.005$.

Assuming that vehicles in the whole city pass on urban roads at random, the proportion of duplicate license plates in the whole population is 0.005.

So, $N = 300/0.005 = 60\ 000$.

Obviously, the precision of proportional estimation is poor.

2. Interval estimation of population mean

Case:

The water output X of a certain type of sprinkler obeys $N(\mu, \sigma^2)$. The quality inspector measured for 10 times and got $\bar{x} = 30.5 \text{L/S}$, $s = 1.53 \text{L/S}$. Now solve the double 90% confidence interval of μ.

he specific analysis is as follows:

The upper limit of the double 90% confidence interval of μ is

$$\bar{X} + t_{1-\frac{\alpha}{2}}(n-1)\frac{S}{\sqrt{n}} = 30.5 + t_{1-\frac{0.1}{2}}(10-1)\frac{1.53}{\sqrt{10}}$$

$$= 30.5 + 1.833 \times 0.484 = 31.39$$

The lower limit of the double 90% confidence interval of μ is

$$\bar{X} - t_{1-\frac{\alpha}{2}}(n-1)\frac{S}{\sqrt{n}} = 30.5 - 1.833 \times 0.484 = 29.61$$

Therefore, the double 90% confidence interval of μ is $(29.61, 31.29)$.

3.7.6 Delphi method

Delphi method is one of the qualitative prediction methods. It is not a qualitative prediction of one person, but a prediction activity that gathers many people. The specific process of Delphi method is as follows.

(1) The convener shall convene experts, senior personages, backbones, etc. inside and outside the industry to communicate information on a specific topic before prediction.

(2) After all the relevant information is known to the forecasters, the first round of prediction is carried out according to everyone's experience, knowledge, ability, etc.

(3) After the first round of forecast results are summarized, the forecast results are sorted in order. The ones with the same forecast results and the highest votes are ranked first, and the ones with the lowest votes are ranked last.

(4) According to the needs of forecasting, select the top prediction results (such as the top 8 results) and distribute them to all the forecasters.

(5) Carry out the second round of prediction. Rank the second round of forecast results according to the number of votes.

(6) In the second round of prediction results, select the top three prediction results (for example, select the top three results) and distribute them to all forecasters for the third round of prediction.

(7) The prediction results of the third round are sorted according to the number of votes obtained, and the prediction result with the highest number of votes is selected as the final result of this prediction.

The Delphi method is simple and feasible. Its prediction process is essentially to summarize the opinions of most people and then jump out of the opinion with the highest vote to predict again. After multiple rounds of prediction, the most centralized prediction results can be quickly obtained. During specific implementation, personnel with high prediction quality can be given greater weight to further improve the effectiveness of the Delphi method.

3.7.7 Big data forecast

1. The concept of big data

Big data refers to the huge amount of information (data) to be processed, which exceeds the amount of memory used by ordinary computers when processing data. Therefore, big data is a data set that cannot be processed by conventional data processing software within a certain period of time. Big data technology refers to the ability to process various types of data at a high speed, obtain valuable information from them, and assist in decision-making (prediction), management and application. Big data has huge data volume, various data types and low value density. Sometimes, big data becomes a methodology, that is, everything can be recorded, everything can be digitized, and then the value can be mined from it.

Forecasting with big data is the core application of big data technology. Strictly speaking, big data is all samples. From the actual application, big data is the most data of a certain research object, and it is no longer the traditional sampling data. Therefore, data mining and analysis provide great support for the prediction work, and improve the accuracy and efficiency of prediction. Of course, because big data is not all, there will also be some systematic deviation.

The analysis of big data requires a lot of statistical content, including search, comparison, discrimination, clustering, classification and many other analysis and induction methods. There is not only a lot of effective information, but also a lot of interference information in big data. The analysis of big data is very complex. Now, distributed databases and distributed computing clusters are mainly used for analysis and classification to support the statistical analysis of big data.

2. Algorithm

Algorithm refers to a method or process of solving problems. Generally, an algorithm has the following characteristics:

One is input. An algorithm should have n initial input data.

The second is output. The output information of an algorithm can be one or more, or it can be no output. There is a specific relationship between the output and the input data.

The third is certainty. Every step in the algorithm must have a precise meaning and cannot be ambiguous.

Fourth, feasibility. Every step described in the algorithm is executable, that is, it can be realized by computer.

Fifth, it is exhaustive. An algorithm must end normally after a limited number of steps and cannot form an endless loop.

3. Machine learning

(1) The meaning of machine learning

Machine learning is a kind of ability that can endow machine with learning ability and enable the machine to complete functions that cannot be completed by direct programming. In practice, machine learning is to process big data through algorithms, then "train" an appropriate model, and apply the model to predict. "Training" and "prediction" are two important processes of machine learning, and "model" is the intermediate output of the process. "Training" produces "model", and "model" guides "training". The training process is similar to the induction process. The essence of machine learning is a computer simulation of human learning and growth. Machine learning is not based on the results of programming, so its processing process is not a logical relationship of cause and effect, but a conclusion of relevance through induction.

In terms of scope, machine learning is similar to pattern recognition, statistical learning and data mining; At the same time, machine learning is combined with other processing technologies to form computer vision, speech recognition, natural language processing and other cross disciplines. Machine learning focus on big data processing, and there are many big data processing platforms, such as Hadoop, Spark(Apache), Atom, MongoDb, IBM PureData, Oracle Exadata, SAP Hana, Matlab, etc.

Machine learning is divided into supervised learning and unsupervised learning. Supervised learning means that a machine (learning system) can predict the data that has never been seen by learning how to combine the input information. Unsupervised learning means that the machine does not need to combine the input information for learning, but only classifies and organizes the data according to some characteristics or similar characteristics.

(2) Algorithm

Many algorithms will be used in machine learning, and it is estimated that there are no fewer than hundreds of algorithms. Here are some common algorithms:

Bayes theory refers to calculating the probability of another event according to the probability of one event that has occurred.

Decision Trees, a decision support tool, use tree diagrams, decision models, sequence possibilities, etc. to make choices for a business decision in a structured or systematic way.

Naive Bayes classification, which is to solve the probability of occurrence of each category under the condition that the given classification item appears, and assign the classification item to the category with the highest probability. Face recognition is an example of this method.

The least square method (refers to Section 6 of Chapter 3) can calculate linear regression and obtain correlation. The identification of correlation is an important method of big data application, which has been widely used.

Linear regressionis the same as the least square method. It establishes a linear regression equation between the independent variable and the dependent variable, and calls the corresponding mathematical function a model, and then tests and predicts.

Logic regressionbelongs to the category of linear regression algorithm, but it is different from linear regression. There are only two cases of the results of logical regression: true (represented by "1") or false (represented by "0"), which are often used for data classification. The fitting function of the logic regression algorithm is called the Sigmond function. The output values of the function are only 0 and 1. This function is also called the logic function.

K-Means Clusterring, which is an unsupervised machine learning algorithm and a partition clustering algorithm used to classify unlabeled data. It works by looking for groups in the data, and the number of groups is represented by the variable k. During the operation, it assigns data points to one of the k groups according to the provided characteristics. It clusters data according to the concept of "distance" in the algorithm (in mathematics, the function of distance between any two elements in any set is distance function or metric). It is called unsupervised learning because it can classify and organize the data in the dataset that are similar in some aspects. The number of clusters k can be specified by the user. During calculation, it repeatedly divides the data into k clusters according to a distance function.

There are two types of metrics. One is Euclidean measurement, which can calculate the distance between two points according to the coordinate points on the Cartesian plane. The other is taxi metric (figurative name), which refers to the actual life and work. The distance between two points is not a straight line, but there are many detours to go around. Therefore, the distance between two points has different algorithms and results.

Neural Network is also called artificial neural network, ANN. Its learning mechanism is decomposition and integration, which simulates the brain to work. For example, a square is decomposed into four polylines, and each neuron processes one polyline; Then each polyline is further decomposed into two straight lines and handed over to neurons for processing; Each line continues to split into black and white faces. In this way, a square becomes a large number of details into neurons. After treatment, neurons were integrated and reduced to square. In neural networks, each processing unit is a logical regression model. The logistic regression model accepts the input from the upper level and transmits the prediction results of the model to the next level as the output. In this way, the neural network completes the very complex nonlinear classification.

Support Vector Machine is called SVM for short. It is an enhanced logical regression algorithm with more strict conditions and better classification boundaries than logical regression. It combines with Gaussian "kernel" to express a very complex classification boundary, which has a good classification effect. The "kernel" here is a special function, and the most typical feature is that it can map a low dimensional space to a high-dimensional space. For example, the nonlinear classification boundary divided by the two-dimensional plane is equivalent to the linear classification boundary of the three-dimensional plane, and then the nonlinear classification effect in the two-dimensional plane is achieved through simple linear division in the three-dimensional space. Support vector machine does not bring about the increase of computational complexity.

Dimension Reduction Algorithm is also an unsupervised learning algorithm, which reduces data from high dimension to low dimension. The dimension here refers to the size of the data characteristic quantity. Dimension reduction algorithm can remove redundant information, reduce data from high dimension to low dimension, and retain information of data to the greatest extent. The main function of dimension reduction algorithm is to compress data, that is, to compress data with thousands of features to several features, so as to improve the efficiency of other algorithms. Through the dimension reduction algorithm, data visualization can also be realized.

Recommendation Algorithm means that users can automatically recommend the content they are most interested in through data analysis. There are two main categories of recommendation algorithms: one is based on item content recommendation, which recommends items similar to users' needs to users. The other is recommendation based on user similarity. Sometimes the two types are mixed.

Gradient descent is an optimization algorithm, also known as the steepest descent method. The steepest descent method is one of the simplest and oldest methods for solving unconstrained optimization problems. It uses the negative gradient direction as the search direction for iterative search. The closer to the target value, the smaller the step size is and the slower the progress is.

Newton's Method is also known as Newton's approximation method. It uses Taylor expansion of objective function to transform the least squares problem of nonlinear function into the least squares problem of linear function in each iteration. Under the action of the second derivative, Newton's method directly searches how to reach the extreme point from the convexity of the function. In terms of convergence rate, gradient descent is linear convergence, while Newton's method is superlinear convergence. When the objective function is not convex, it can be approximately transformed into convex function.

4. Other prediction methods

It mainly includes the following methods.

(1) ARModel, also known as autoregression model, is a statistical method for processing time series, using the same variable, such as the previous periods of X, that is to say, use x_1 to x_{t-1} to forecast x_t in the current period, and assume that they have a linear relationship.

(2) MAModel, also known as the moving average model, refers to the q order truncation of the autocorrelation coefficient of the q order moving average model. It is biased towards autocorrelation coefficient tailing.

(3) ARMAModel, also known as autoregressive moving average model, is a typical method to study the rational spectrum of stationary random processes.

3.8 有测量系统保证

人们以为看明白了客观世界，但借助于各种测量仪器之后，又看到了不同的世界，而且也是客观存在的。比如：眼睛看到的是自然光线，但在仪器下是七种颜色的，乃至更多色彩；人们以为周边没有任何东西，除了眼睛所及，但现代物理学揭示，人的身边充满了暗物质，而且数量很多，是有重量的；人们看到了蜜蜂在花朵上采蜜，但从蜜蜂的角度，它看到的不是花朵，而是花朵所呈现的一种紫外的光；人的肉眼看地面没有什么的时候，小鸟可以轻松自然地在地面觅食；雄鹰可以在数百米高空轻松发现地面的猎物等。由此可见，测量仪器等对于获得真实数据、高质量数据的重要性。

测量中高质量的数据获取，离不开测量系统的三个基本要求：测量系统的分辨力、测量系统的统计稳定性、测量系统的线性。对于测量系统的可接受评价要关注其重复性、再现性、测量对象间的变差。

3.8.1 测量系统

1. 基本概念

测量系统是与确定测量对象、测量过程、测量输入、测量输出紧密联系在一起的一个大集合，它包括了测量仪器与设备、计量器具、测量标准、测量程序和方法、测量软件、测量环境、测量人员、测量假设等内容。通过一个完整的测量系统，可以获得测量结果，确定数据的可靠性，对测量方法进行评估和比较、评价测量仪器、确定并解决测量系统的误差、评价测量人员的能力等。测量的过程是一个数据制造与获取的过程，因此可以用 SPC（Statistical Process Control）控制图来分析测量系统。

测量所获得的数据质量有高有低，假设对某一个基准值做多次测量，如果测量的数据与基准值接近，那么数据的质量就高；如果测量的数据偏离基准值，则数据的质量低。一般可以用偏倚和变差来表示测量数据的质量。如果测量数据的偏倚小且变差小，则数据质量高；如果偏倚大且变差大，或偏倚小但变差大，或变差小但偏倚大，那么数据质量低。

2. 对测量系统的要求

对于一个测量系统来说，有三个基本的要求：

（1）测量系统的分辨力，它是指测量系统识别被测量特性中最小变化的能力，也可以理解为测量的精度。比如某测量系统能够分辨 0.1 mm 的变化，但不能分辨 0.01 mm 的变化，那么其分辨力就是 0.1 mm。每一个测量系统都必须有足够的分辨力，否则就不能准确表示被测对象的特性值，不能识别测量过程中的数据波动。

（2）测量系统的统计稳定性，它是指测量系统在重复测量的前提下，其计量特性测量值随着时间的推移保持稳定的能力，也可以指测量系统的偏倚随时间推移的变化范围。通常用"平均值–极差控制图"来评价测量系统的稳定性。比如现在有一个重量为 5 kg 的产品，可以设计一个固定的时间间隔来测量该产品的重量，如每天测量一次，每次测量 5 遍，一共连续测量 30 天。然后根据测得的数据来计算平均值和极差，制作控制图，根据控制图的判异准则来判断点子有无异常。如果没有异常，则测量系统是稳定的；如果有异常，则在消除由异常原因而产生的异常现象后，才能认定测量系统是稳定的。

（3）测量系统的线性，它是指测量数据的偏倚与基准值之间的线性关系，可以用两者之间的相关性来表示。如果偏倚与基准值之间相关系数接近 1，则线性关系好；如果相关系数越小（指相关系数 r 在 0 到 1 的范围内），则线性关系越差。由于每个测量系统都有自己的量程，比如有的磅秤只能称量 1 kg 以内的物品，有的磅秤能够称量 1000 kg 以内的物品，显然对两者提出相同的偏倚是不合理的。因此，对于量程较小的测量系统，偏倚应小一些；对于量程较大的测量系统，偏倚应大一些。

3. 测量系统的可接受评价

针对一个测量系统，可以做"可接受评价"。

（1）测量系统评估的内容

对于测量系统的评估内容，通常反映在三个方面：

一是测量系统必须具有充分的稳定性。它是指测量系统必须具有充分的分辨力，能够敏感地、有效地测量产品或流程地变化。

二是测量系统必须稳定。它是指测量系统能尽量排除特殊原因的干扰。

三是测量系统的误差在预期范围内要一致，而且测量系统对于所有测量的目标（产品或流程）来说是充分的。

（2）测量系统的变异

测量系统的变异（波动）主要由测量仪器、测量人员引起。测量系统的变异（波动）主要有：

一是由于测量仪器导致的测量数据的波动，可以用测量仪器的重复性 EV 来表示，也就是其重复测量值的变差，EV 越小，测量仪器的可接受度高。重复性是在相同操作者、相同设定、相同零件、相同环境条件、短期等前提下，对相同变量重复测量所产生的波动。

二是由于测量人员的因素而导致测量数据的波动，可以用测量系统的再现性 AV 来表示，也就是测量人员在测量技术上的变差，AV 越小，测量人员的可接受度高。再现性是在不同操作者、不同设定、不同测试零件、不同环境条件、长期等前提下，进行测量所产生的波动。

在实践中，是无法只针对一名操作人员进行 AV 考察的，而是需要考察 2 名及 2 名以上的操作人员进行变差估计。通常，测量系统的总方差小于 10% 时，测量系统没有问题；若大于 30%，则为不可靠。

（3）测量系统变异的影响

测量系统变异对于产品的决策、过程的判断等会产生影响，具体有：

一是可能影响对于产品的决策，比如不合格的测量系统导致产品错误接收或错误报废。

二是可能影响对于过程的决策，比如把产生变异的普通原因识别为特殊原因，或者把特殊原因识别为普通原因。普通原因是指平时一直客观存在的原因，对于过程有一定的影响，但不是很明显。特殊原因是指偶然出现的原因，对于过程的影响程度大。当过程只受到普通原因影响的时候，那么"过程"在控制图上的表现是受控的；而当过程受到特殊原因影响的时候，在控制图上会出现异常点。

三是可能影响对于过程能力的评价，因为观测到的过程变差等于实际变差加上测量变差。而过程能力判断出错，会直接影响到生产过程是否受控。

四是可能导致过程作业准备的错误，比如原来是合格的过程，因测量误差而对其做出了错误的调整，导致批量不合格。

（4）GR&R 评价

GR&R（Gauge Repeatability & Reproducibility），是测量系统的重复性和再现性，是评价测量系统的重要方法之一。

GR&R 的主要评判标准是：

如果 GR&R 小于所测零件公差的 10%，那么测量系统是没有问题的。

如果 GR&R 大于所测零件公差的 10%，但小于 20%，那么测量系统是可以接受的。

如果 GR&R 大于所测零件公差的 20%，但小于 30%，那么测量系统是否可以接受，应当依据数据的重要程度、改善测量系统所要花费的成本来决定。

如果 GR&R 大于所测零件公差的 30%，那么测量系统是不能接受的，需要进行改善。

3.8.2 测量的不确定度

测量的不确定度表示测量的能力。测量不确定度的概念最早由埃森哈特于 1963 年在研究"仪器校准系统的精密度和准确度估计"时提出的，意思是没有误差的测量结果是不存在的，根据规则所求得的误差的值就是测量的不确定度。不确定度的英语表达为"uncertainty"，意思是不确定、不稳定，在描述测量结果时，它由定性名词转为定量名词，表示对测量结果质量的定量表征，以确定测量结果的可信程度。

1980 年国际计量委员会要求国际计量局（BIPM）征求了 32 个国家计量研究院以及 5 个国际组织的意见，由国际计量局工作组发表了推荐采用测量不确定度来评定测量结果的建议书，该建议书为 INC-1（1980），同时向各国推荐了测量不确定度的表示原则。1993 年，在 INC-1（1980）的基础上，国际标准化组织（ISO）第四技术顾问组（TAG4）的第三工作组（WG3）起草，并以 7 个国际组织名义发布了《测量不确定度表示指南》（Guide to the Expression of Uncertainty in Measurement，简称 GUM），以及第二版的《国际通用计量学基本术语》（International Vocabulary of Basic and General Terms in Metrology，简称 VIM），为在全球统一采用测量结果的不确定度评定和表示奠定了基础。

1995 年 GUM 的修订版发布。7 个国际组织分别为国际计量局（BIPM）、国际电工委员会（IEC）、国际标准化组织（ISO）、国际法制计量组织（OIML）、国际纯粹和应用物理联合会（IUPAP）、国际纯粹和应用化学联合委员会（IUPAC）、国际临床化学联合会（IFCC）。国际实验室认可合作组织（ILAC）也承认 GUM。1998 年，我国发布了国家计量技术规范 JJF1001—1998《通用计量术语及定义》，其中前六章的内容与 VIM 的第二版完全对应。

不确定度的通常计算方法是《测量不确定度表示指南》GUM 法，其中包含因子（coverage factor）k 的含义是指"为求得扩展不确定度，对合成标准不确定度所承之数字因子"，包含因子数值上等于扩展不确定度与合成标准不确定度之比。在以长度测量为主的很多领域，都选择包含因子 k =2。不确定度乘以包含因子 k（k >1）则称为扩展不确定度。当 k =2 的时候，测量的可靠度水平达到 95%，相当于 ±2σ 的可靠区间，当 k =3 的时候，测量的可靠度水平达到 99%，相当于 ±3σ 的可靠区间。一个测量数据的提供者在测量结果中明示测量的不确定度，表明该提供者是在负责任地保证测量结果的质量。

3.8　Measurement System Guarantee

People think that they can see the objective world, but with the help of various measuring instruments, they can see different worlds, which also exist objectively. For example, the eye sees natural light, but under the instrument there are seven colors, or even more; People think that there is nothing around except the eyes, but modern physics reveals that people are full of dark matter, and there are a lot of dark matter, which is heavy; People see bees picking honey from flowers, but from the perspective of bees, what they saw was not flowers, but a kind of ultraviolet light presented by flowers; When human eyes see nothing on the ground, birds can easily and naturally find food on the ground; The eagle can easily find ground prey hundreds of meters high. It can be seen that measuring instruments are important for obtaining real data and high-quality data.

High quality data acquisition in measurement is inseparable from the three basic requirements of the measurement system: resolution, statistical stability and linearity of the measurement system. For the acceptable evaluation of the measurement system, attention should be paid to its repeatability, reproducibility, and variation between measurement objects.

3.8.1 Measurement system

1. Basic concept

The measurement system is a large collection closely related to the determination of measurement objects, measurement processes, measurement inputs and measurement outputs. It includes measurement instruments and equipment, measuring instruments, measurement standards, measurement procedures and methods, measurement software, measurement environment, measurement personnel, measurement assumptions, etc. Through a complete measurement system, the measurement results can be obtained, the reliability of data can be determined, the measurement methods can be evaluated and compared, the measurement instruments can be evaluated, the error of the measurement system can be determined and solved, and the ability of the measurement personnel can be evaluated. The measurement process is a process of data production and acquisition, so SPC (Statistical Process Control) control chart can be used to analyze the measurement system.

The quality of the data obtained by measurement varies from high to low. Suppose that a certain reference value is measured several times, if the measured data is close to the reference value, the quality of the data is high; If the measured data deviates from the reference value, the quality of the data is low. Generally, bias and variation can be used to express the quality of measurement data. If the bias of the measurement data is small and the variation is small, the data quality is high; If the bias is large and becomes large, or the bias is small but becomes large, or the bias is small but becomes large, the data quality is low.

2. Requirements for measurement system

There are three basic requirements for a measurement system.

(1) Resolution of measurement system refers to the ability of the measurement system to identify the minimum change in the measured characteristics, which can also be understood as the measurement accuracy. For example, if a measuring system can resolve a change of 0.1 mm, but cannot resolve a change of 0.01 mm, its resolution is 0.1 mm. Each measurement system must have sufficient resolution, otherwise it cannot accurately represent the characteristic value of the measured object and cannot identify the data fluctuation in the measurement process.

(2) Statistical stability of the measurement system refers to the ability of the measurement system to keep the measured value of its measurement characteristics stable over time on the premise of repeated measurement, and also refers to the variation range of the bias of the measurement system over time. The stability of the measurement system is usually evaluated by the "average-range control chart". For example, if there is a product with a weight of 5kg, a fixed time interval can be designed to measure the weight of the product, such as once a day, five times a time, and a total of 30 consecutive days. Then calculate the average value and range according to the measured data, make a control chart, and judge whether the points are abnormal according to the criteria of the control chart. If there is no abnormality, the measurement system is stable; If there is any abnormality, the measurement system can be determined to be stable only after the abnormal phenomena caused by the abnormal reasons are eliminated.

(3) Linearity of the measuring system refers to the linear relationship between the bias of measurement data and the reference value, which can be expressed by the correlation between them. If the correlation coefficient between bias and reference value is close to 1, the linear relationship is good; If the correlation coefficient is smaller (meaning that the phase relationship number r is in the range of 0 to 1), the linear relationship is worse. Since each measuring system has its own measuring range, for example, some scales can only weigh items less than 1 kg, and some scales can weigh items less than 1000 kg, it is obviously unreasonable to put forward the same bias between the two. Therefore, for a measuring system with a small measuring range, the bias should be smaller; For measuring systems with large measuring ranges, the bias should be larger.

3. Acceptable evaluation of measurement system

"Acceptable evaluation" can be made for a measurement system.

(1) Contents of measurement system evaluation

The evaluation content of the measurement system is generally reflected in three aspects:

First, the measurement system must have sufficient stability. It means that the measurement system must have sufficient resolution to measure changes in products or processes sensitively and effectively.

Second, the measurement system must be stable. It means that the measurement system can eliminate the interference caused by special reasons as far as possible.

Third, the error of the measurement system should be consistent within the expected range, and the measurement system is sufficient for all measurement objectives (products or processes).

(2) Variation of measurement system

The variation (fluctuation) of the measuring system is mainly caused by measuring instruments and measuring personnel. The variation (fluctuation) of measurement system mainly includes the following items.

First, the fluctuation of measurement data caused by the measuring instrument can be expressed by the repeatability EV of the measuring instrument, that is, the variation of its repeated measurement value. The smaller the EV, the higher the acceptability of the measuring instrument. Repeatability refers to the fluctuation caused by repeated measurement of the same variable under the premise of the same operator, the same setting, the same parts, the same environmental conditions, and short-term conditions, etc.

Second, the fluctuation of measurement data caused by the factors of the measurement personnel can be expressed by the reproducibility AV of the measurement system, that is, the measurement technology variation of the measurement personnel. The smaller the AV, the higher the acceptability of the measurement personnel. Reproducibility refers to the fluctuation caused by measurement under the premise of different operators, different settings, different test parts, different environmental conditions, and long-term conditions, etc.

In practice, it is not possible to investigate AV only for one operator, but it is necessary to investigate two or more operators for variance estimation. Generally, when the total variance of the measurement system is less than 10%, the measurement system is OK; If more than 30%, it is unreliable.

(3) Influence of measurement system variation

The variation of measurement system will have an impact on product decision and process judgment, including the following items.

First, it may affect the decision on the product. For example, the unqualified measurement system may lead to the wrong reception or wrong scrapping of the product.

Second, it may affect the decision-making of the process, such as identifying the common causes of variation as special causes, or identifying special causes as common causes. Common causes refer to the causes that have always existed objectively and have a certain impact on the process, but are not very obvious. Special causes refer to the causes that occur occasionally and have a great impact on the process. When the process is only affected by common causes, the performance of "process" on the control chart is controlled; When the process is affected by special reasons, abnormal points will appear on the control chart.

Third, it may affect the evaluation of process capability, because the observed process variation is equal to the actual variation plus the measured variation. The process capability judgment error will directly affect whether the production process is controlled.

Fourth, it may lead to errors in the preparation of process operations, such as the original qualified process, which has been incorrectly adjusted due to measurement errors, resulting in unqualified batches.

(4) GR&R evaluation

GR&R(Gauge Repeatability & Reproducibility) is the repeatability and reproducibility of the measurement system, and is one of the important methods to evaluate the measurement system.

The main criterion of GR&R is as follows.

If GR&R is less than 10% of the measured part tolerance, there is no problem with the measurement system.

If GR&R is greater than 10% but less than 20% of the measured part tolerance, the measuring system is acceptable.

If GR&R is greater than 20% but less than 30% of the measured part tolerance, whether the measurement system is acceptable shall be determined according to the importance of the data and the cost of improving the measurement system.

If GR&R is greater than 30% of the measured part tolerance, the measurement system is unacceptable and needs to be improved.

3.8.2 Uncertainty of measurement

The uncertainty of measurement indicates the ability of measurement. The concept of measurement uncertainty was first proposed by Eisenhardt in 1963 when he studied the "precision and accuracy estimation of instrument calibration system", which means that there is no measurement result without error, and the error value obtained according to the rules is the measurement uncertainty. The English expression of uncertainty means indeterminacy and instability. When describing the measurement results, it changes from qualitative nouns to quantitative nouns, representing the quantitative representation of the quality of the measurement results to determine the credibility of the measurement results.

In 1980, the International Metrology Commission requested the Bureau International des Poides et Measures (BIPM) to solicit the opinions of 32 national metrology institutes and 5 international organizations. The BIPM Working Group issued a proposal to recommend the use of measurement uncertainty to evaluate measurement results. The proposal was INC-1 (1980), and recommended the principles for expressing measurement uncertainty to countries. In 1993, on the basis of INC-1 (1980), the Third Working Group (WG3) of the Fourth Technical Advisory Group (TAG4) of the International Organization for Standardization (ISO) drafted and issued the Guide to the Expression of Uncertainty in Measurement (GUM) in the name of seven international organizations. And the second edition of International Vocabulary of Basic and General Terms in Metrology (VIM), laying the foundation for the unified use of uncertainty evaluation and expression of measurement results in the world.

Revised version of GUM issued in 1995. The seven international organizations are BIPM, International Electro Technical Commission(IEC), International Standard Organization(ISO), International Organization for Legal Metrology(OIML), International Union of Pure and Applied Physics (IUPAP), International Union of Pure and Applied Chemistry(IUPAC) and International Federation of Clinical Chemistry(IFCC) respectively. International Laboratory Accreditation Cooperation(ILAC) also acknowledges the GUM. In 1998, China has issued the national metrological technical specification JJF1001-1998 General Metrological Terms and Definitions, in which the contents of the first six chapters completely correspond to the second edition of VIM.

The usual calculation method of uncertainty is the GUM method in the Guide to the Expression of Uncertainty in Measurement. The meaning of coverage factor k is "the digital factor for the combined standard uncertainty to obtain the expanded uncertainty". The value of the coverage factor is equal to the ratio of the expanded uncertainty to the combined standard uncertainty. The coverage factor $k = 2$ is selected in many fields that mainly focus on length measurement. The uncertainty multiplied by the coverage factor k ($k > 1$) is called the expanded uncertainty. When $k = 2$, the measured reliability level reaches 95%, equivalent to $\pm 2\sigma$ of the reliability interval. When $k = 3$, the measured reliability level reaches 99%, euivalent to $\pm 3\sigma$ of the reliability interval. A provider of measurement data indicates the measurement uncertainty in the measurement results, indicating that the provider is responsible for ensuring the quality of the measurement results.

3.9　有试验及试验系统保证

人们为了在有限的条件下对所关心的对象进行研究，于是出现了试验这个方法，并在实践中开发应用了很多试验方法。为了研究一个客观对象（过程）受到多个因素影响的情况，人们开始了试验设计。为了应对复杂的、真实的生产与科研需要，人们运用适当的原理，结合各种设备、仪器、夹具、应用软件等，开发了各种试验系统。一次完美的试验、一个完美的试验设计、一个高质量的试验系统，可以保证所取得的数据是高质量的，是可以达到数据的应用目的的。

3.9.1 试验方法

有关试验方法的内容，可参考第 2 章第 2.5 节中的内容。这里介绍其中的析因试验。

析因试验（Factorial Experimental Design）是一种全面试验法，它能够全面反映各因素对于试验对象指标的影响。在试验时，它是把各因素的全部水平进行排列组合，然后交叉分组开展试验，以考察因素的主效应、因素之间的交互效应；比较各因素不同水平的平均效应、因素间不同水平组合下的平均效应，以找到最佳组合。

析因效应总的试验数是所有因素水平的乘积。比如：有 3 个因素同时进行试验，每个因素取 2 个水平，那么试验组合的总数为 2^3 个，即 8 个。如果是 3 个因素，每个因素水平数是 4 个，则试验总数为 4^3 个，即达到 64 个。所以，通常析因试验设计中，水平不能太多，一般以 2 或 3 个水平为宜，否则试验的工作量巨大。

举例：

某水产养殖场进行鱼饲料配方。现有 4 个鱼塘，假定鱼塘中的鱼，都已经养殖了一年，且大小一致。现在给四个鱼塘分别喂食不同配方的饲料，然后观察三个月后鱼的生长增重情况。饲料分为普通、A 类配方、B 类配方。四个鱼塘喂食的饲料情况如下：

第一个鱼塘，普通饲料；第二个鱼塘，普通饲料加 A 类配方；第三个鱼塘，普通饲料加 B 类配方；第四个鱼塘，普通饲料加 A 类、B 类配方。

分析：

配方有 A 类、B 类两种，且各自都存在"使用"和"不使用"两种水平。因此，这个试验属于 2×2 的析因试验设计。在分析的时候，不仅可以分析 A、B 两种配方对鱼增重的作用，还可以分析 A 类配方与 B 类配方之间是否存在交互作用。

用 A1 表示配方 A 因素 1 水平，A2 表示配方 A 因素 2 水平，B1 表示配方 B 因素 1 水平，B2 表示配方 B 因素 2 水平。则有 2×2 析因试验设计的格式，见表 3-8。

表 3-8　2×2 析因试验设计

配方 A ＼ 配方 B	B1	B2
A1	A1B1	A1B2
A2	A2B1	A2B2

三个月后，检查人员对四个鱼塘的鱼进行 3 次随机抽样，每个鱼塘每次抽样量都相同，得到了四个鱼塘的每条鱼平均增重情况，见表 3-9。

表 3-9　四个鱼塘每条鱼平均增重情况　　　　　　　　　单位：克

项目名称	第一个鱼塘 普通饲料	第二个鱼塘 普通饲料+A	第三个鱼塘 普通饲料+B	第四个鱼塘 普通饲料+A、B	合计
第一次抽样	200	330	230	460	1220
第二次抽样	225	310	270	430	1235
第三次抽样	175	290	250	400	1115
$\sum X$	600	930	750	1290	3570
$\bar{x}$	200	310	250	430	1190
$\sum X^2$	121 250	289 100	188 300	556 500	4 256 850

从表 3-9 可以看出，第四个鱼塘的鱼增重最多，平均每条增加 430 克；第二个鱼塘次之，平均每条增加 310 克；第三个鱼塘平均每条增加 250 克，高于第一个鱼塘平均增加 200 克。

第二个鱼塘与第一个鱼塘的差数为 310-200 = 110 克，这是 A 类配方的作用。

第三个鱼塘与第一个鱼塘的差数为 250-200 = 50 克，这是 B 类配方的作用。

第二个鱼塘、第四个鱼塘都使用了 A 类配方，第四个鱼塘与第二个的差数为 430-310＝120 克，显然这是有 B 配方的作用。

那么，A 类 B 类配方交互作用为："第四个鱼塘与第二个的差数"减去"第三个鱼塘与第一个鱼塘的差数"，等于 120-50 = 70 克。

3.9.2 试验设计

1. 试验设计的含义

试验设计是指在大多数情况下，人们对所研究的问题的机理不十分清楚，或很难用数学公式表达，只能借助试验方法，用尽可能少的试验次数，又经济又快速地求得满意结果。试验设计是事先安排好的试验活动，它依赖于对结果的统计评价，从而在规定的置信水平下得出结论。

试验设计的一般步骤为：分析试验系统，选择试验方案，完成试验方案，分析试验数据，得出试验结论，实现持续改进。

试验设计可以分为单因子试验设计、多因子试验设计、正交试验法等。

2. 单因子试验设计

单因子试验设计是假设在管理活动中只存在一种影响因素，通过设计出相应的试验方法来求得最优解，解决这个影响因素以达到管理目标。一般用数学表达式（即目标函数）来反映出因子和试验结果之间的关系。用 x 表示因素的取值，目标函数为 $y = f(x)$。函数存在的两种情形：第一种情形，函数呈单调上升或单调下降，此时只要在有效取值范围内即可找到最优解。第二种情形，函数呈现出"∪"型或"∩"形，即图形有一个峰值 a 或谷值 b，该峰值 a 或谷值 b 就是最优解。

举例：

现在有一家公司的加工试验室，为了获得最好的产品，基本确定某个添加剂化合物（因子）的值在 12 至 16 克之间，即 [12，16]。那么，具体多少值是最优解呢？

（1）均分法

现在试验室打算做 4 次试验，那么每次试验的间距 =（16−12）/（4+1）= 0.8。

因此，试验人员分别取 X1 = 12.8，X2 = 13.6，X3 = 14.4，X4 = 15.2 进行试验即可。

经过试验，在 X3 = 14.4 时，添加剂的效果最佳。

（2）0.618 法

取 X1 = 12+（16−12）×0.618 = 14.47；

取 X2 = 12+（16−12）×0.382 = 13.53。

现在假设 X1 =14.47 时，试验的效果比 X2 =13.53 好，那么把 13.53 以下的数据去掉，取区间（13.53，16）继续所试验。

则有：X3 = 13.53+（16−13.53）×0.618 = 15.06。

现在假设 X1 =14.47 时，试验的效果比 X3 =15.06 好，则取区间 [13.53，15.06] 继续试验。

则有：X4 = 13.53+（15.06−13.53）×0.618 = 14.48。

试验后发现，X4 的效果为最佳。

（3）数列法

根据试验的取值范围 [12，16]，现在把试验的间隔设为 0.5，则有取值 12，12.5，13，13.5，14，14.5，15，15.5，16 共 9 次，所以试验次数为 9 次。

这样，可以制作试验次序表，见表 3−10。

表 3−10　试验次序表

可能的次序		1	2	3	4	5	6	7	8	9
数列	F0	F1	F2	F3		F4			F5	
	1	2	3	4		5			8	
试验值		12	12.5	13	13.5	14	14.5	15	15.5	16
试验次序				X2		X1				

根据数列，一共有 F5 个数列。因此，第一个试验点 X1，取数列 F（5−1）= F4，试验值为 14。

第二个试验点 X2，取数列 F（5−2）= F3，试验值为 13。

现在假设 X1 的结果优于 X2，则把 X2 以下的部分去掉，那么就有了新的试验次序表，范围从第 4 个可能次序到第 9 个可能次序，见表 3−11。

表 3-11　试验次序表

可能的次序	4	5	6	7	8	9
数列		F4			F5	
		5			8	
试验值	13.5	14	14.5	15	15.5	16
试验次序		X1			X3	

下一个试验点 X3，取 X1 的对称点，则取"可能的次序 5"的对称点为"可能的次序 8"。

所以，X3 的试验值为 15.5。

假设 X1 的试验结果优于 X3，那么，试验值取值 14 为最佳。

3. 双因子试验设计

在管理实践中，由两个因素共同对目标产生影响是常见的。此时，可以通过双因子试验设计来获得对目标的影响。双因素试验设计的效应结果可以简单归纳为四种：一是因子 A 没有影响，因子 B 产生主要影响；二是因子 A 为主要影响，因子 B 没有影响；三是因子 A、因子 B 都产生主要影响，但 A、B 没有交互；四是因子 A 与因子 B 产生交互影响。

当因子 A 没有影响，因子 B 产生主要影响时，不同水平的 B 效应不同。当因子 A 为主要影响，因子 B 没有影响时，不同水平的 A 效应不同。当因子 A、因子 B 都产生主要影响，但 A、B 没有交互时，效应随着 A、B 水平的不同而产生变化。当因子 A 与因子 B 产生交互影响时，不同水平 A、B 的效应产生交错。

双因子（A、B）试验结果用 F 值检验，a 为因子 A 的水平数，b 为因子 B 的水平数，r 为每个试验的重复数，$n = abr$，则相应检验结果见表 3-12。

表 3-12　双因子试验检验结果表

因子 A 主要影响，因子 B 没有影响	因子 A 没有影响，因子 B 有主要影响	因子 A、因子 B 均有影响，但没有交互	因子 A、因子 B 均有影响，而且有交互
原假设 H_0：所有 A 的水平下均无显著差异 备择假设 H_1：至少 A 在两个水平下有显著差异	原假设 H_0：所有 B 的水平下均无显著差异 备择假设 H_1：至少 B 在两个水平下有显著差异	原假设 H_0：A 与 B 的影响无显著差异 备择假设 H_1：至少 A、B 有两次试验呈显著差异	原假设 H_0：A 与 B 的交互没有产生显著影响 备择假设 H_1：A 与 B 的交互产生显著影响
检验值 $F = \dfrac{MS(A)}{MSE}$	检验值 $F = \dfrac{MS(B)}{MSE}$	检验值 $F = \dfrac{MST}{MSE}$	检验值 $F = \dfrac{MS(AB)}{MSE}$
自由度 $\dfrac{a-1}{abr-ab}$	自由度 $\dfrac{b-1}{abr-ab}$	自由度 $\dfrac{ab-1}{abr-ab}$	自由度 $\dfrac{(a-1)(b-1)}{abr-ab}$
拒绝域：$F \geqslant F_\alpha$	拒绝域：$F \geqslant F_\alpha$	拒绝域：$F \geqslant F_\alpha$	拒绝域：$F \geqslant F_\alpha$

4. 正交试验设计法

（1）概念。正交试验设计法是多因子试验设计中的简易方法，它的原理是针对预设的目标情况，使用正交表安排尽量少的试验次数，经济、快捷地求得满意解。正交表是按一定规律排出来的数字表格，它的数字排列具有均衡分散、搭配均匀的平等性，即正交性。

（2）正交表记号。正交表记号用 $L_p(m^n)$ 表示，表示 n 个因素，m 个水平，全面试验 m^n 次，正交试验 P 次。正交表记号所表示的意思见图 3-4。

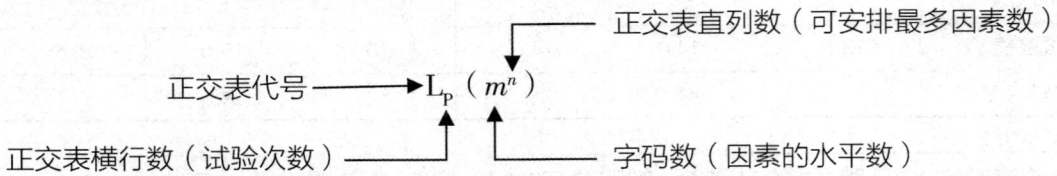

图 3-4　正交表记号

数学可以证明，在运用正交表 $L_p(m^n)$ 做的 P 次试验中的最好组合，虽然不一定是全面试验 m^n 次的最佳组合，但平均说来它较 $\dfrac{P}{P+1} \times (m^n)$ 次为最优。正交表中，任一列中的各水平都出现，而且出现的次数相等；任两列之间各种不同水平的所有可能的组合都出现，而且出现的次数相同。

（3）正交表的特点。一是整齐可比性，每一个因子的各水平间具有可比性。二是均匀分散，用正交表挑选出来的各因子水平的组合，在全部水平组合中的分布是均匀的。

以正交表 $L_8(2^7)$ 为例，来说明正交性，见表 3-13。

表 3-13　正交表 $L_8(2^7)$

列号＼试验号	1	2	3	4	5	6	7
1	1	1	1	1	1	1	1
2	1	1	1	2	2	2	2
3	1	2	2	1	1	2	2
4	1	2	2	2	2	1	1
5	2	1	2	1	2	1	2
6	2	1	2	2	1	2	1
7	2	2	1	1	2	2	1
8	2	2	1	2	1	1	2

正交表记号"$L_8(2^7)$"中的字母、数字的含义如下："L"指正交表代号；"8"是正交表横行数（试验次数）；"2"是字码数（因素的水平数）；"7"是正交表直列数（可安排最多因素数）。正交表中的数字排列具有正交性，即数字排列均衡分散，搭配均匀。比如上表中，其正交性体现在：每个纵列有 4 个"1"和 4 个"2"，即每个因素的每个水平在 8 次试验中各出现 4 次；任两纵列同一横行搭配的数字对（1，1）（1，2）（2，1）（2，2）各出现 2 次，即每两个因素的两个水平间的搭配是均匀的，且每个搭配的出现次数是相等的。

举例：

（1）试验目的：某质量管理小组在"某测量基准站大范围 GPS 测量方法"攻关活动中，面临了"基准站端口参数设置"问题。而对基准站测量有影响的参数主要是"波特率、传输端口、数据格式"三个参数，而且三个参数之间互有影响。为确定三个参数之间的最佳组合，小组决定采用正交分析法开展活动。

（2）评价指标：基准站点的精度高为好。

（3）确定因素：波特率、传输端口、数据格式3个因素。

（4）确定水平：每个因素取3个水平，因素水平表如表3-14。

表3-14　因素水平表

水平 \ 因素	A 波特率	B 传输端口	C 数据格式
水平 1	9600	Port1	GGA
水平 2	19 200	Port2	GGK
水平 3	38 400	Port3	GSA

（5）利用正交表、确定试验方案。$L_9(3^4)$最多能安排4个3个水平的因素。在本例中是3个因素各有3个水平，可以用$L_9(3^4)$来安排试验：按照因素水平表中3个因素的次序，放到$L_9(3^4)$的上首，每列放1个因素；然后根据因素水平表把相应的水平对号入座。得到表3-15。

表3-15　正交试验表

列号 \ 试验号	A 波特率 1	B 传输端口 2	C 数据格式 3	试验结果 精度偏差/mm
1	1（9600）	1（Port1）	3（GSA）	9.6
2	2（19 200）	1（Port1）	1（GGA）	3.9
3	3（38 400）	1（Port1）	2（GGK）	5.4
4	1（9 600）	2（Port2）	2（GGK）	5.5
5	2（19 200）	2（Port2）	3（GSA）	4.7
6	3（38 400）	2（Port2）	1（GGA）	8.8
7	1（9600）	3（Port3）	1（GGA）	8.1
8	2（19 200）	3（Port3）	2（GGK）	2.7
9	3（38 400）	3（Port3）	3（GSA）	3.2
T1	23.1	18.8	20.7	
T2	11.4	18.9	13.5	T =51.9
T3	17.4	14.1	17.4	$\bar{x}$ =51.9/9 =5.77
R	11.7	4.7	7.2	

（6）分析试验结果：

直接看：第8号试验绝对误差最小，所以初定优先水平为：A2B3C2。

算一算：确定较优位级，A：T2＜T3＜T1；B：T3＜T1＜T2；C：T2＜T3＜T1，所以选择A2B3C2。

确定主要因素：A＜C＜B，A为主要因素；C为重要因素；B为次要因素。

有时会碰到有交互作用的正交试验法，则通过交互作用表格，来查明交互作用的存在，并合理安排试验。以 $L_8(2^7)$ 为例来说明试验设计的安排，见表3-16。

表3-16　用 $L_8(2^7)$ 表头设计说明综合考虑主因素和交互作用的试验设计

列号 因素个数	1	2	3	4	5	6	7
3	A	B	A×B	C	A×C	B×C	
4	A	B	A×B C×D	C	A×C B×D	B×C A×D	D
4	A	B C×D	A×B	C B×D	A×C	D B×C	A×D

3.9.3 试验系统

试验系统可以根据需要，形成很多种类，如通用测试系统（拉力试验机）、气动试验系统、压力检测试验系统、热传导试验系统、风洞试验系统、材料试验系统、电力试验系统、通讯试验系统、膜过滤试验系统、汽车试验系统等。各试验系统在各自的领域发挥着重要作用。

1. 通用测试系统

通用测试系统根据不同用途有很多分类。比如：通用机械测试系统可以根据要求，进行拉伸、压缩、弯曲、剥离、撕裂、剪切、摩擦等各种机械试验。通用电子测试系统采用标准化、模块化、通用化的思路，在装配不同的测试软件后可以对不同的电子设备进行测试。比如有的测试系统专门测试无线通信设备，包括手机、芯片、智能家居等。有的可以进行通信网络安全测试、信息风险评估等。

2. 气动试验系统

气动试验系统以压缩空气为介质，组成气动回路来传动和控制机械，以达到试验目的。也可以加注特种气体，如加注氢气、氮气等。气动系统可用于气体填充、管路中气体的置换、检测系统的气密性、检测泄露点、监测温度、检验容器（如氢气瓶）在极端环境下的疲劳寿命等。与生产装置连接在一起的各种气缸，通过压力控制阀、方向控制阀、流量控制阀来控制气动输出力的大小、气缸的运动方向、运动速度。

气动装置结构相对简单、安装维护简便，压力等级低，使用安全。排气处理简单，不影响环境。输出力、输出速度控制方便。

3. 压力检测试验系统

压力检测试验系统应用广泛，可以进行医学上相关压力检测、石油井压力检测、土木工程压力检测（如混凝土抗压强度检测）、化工（管道、压力容器等）压力检测、海洋及深海压力检测、材料压力检测等。

4. 热传导试验系统

热传导试验系统可以进行热传导系数检测，应用在纺织物、金属材料、土壤、工业（炼钢）、机械、冷库等许多行业中。

5. 风洞试验系统

风洞是指按照一定要求设计的、具有动力装置的、用于各种气动力试验的可控气流管道系统。风洞试验系统可以进行各种风速、温度等环境下的测试，应用在矿山、气象、环保、工程、结构、计量、航空、海洋工程、汽车等诸多领域。风洞试验是风工程学科的重要内容，它具有直观性强、近似模拟、试验条件易控制、节约人力物力时间等特点。风洞试验可以进行测力、测压、气动参数测试、风致振动测试、绕流形态观测、风特性与风环境测试、通风排气性能测试等。

6. 材料试验系统

材料试验系统可以对各种材料进行相关性能的检测试验，应用于建筑材料、金属材料、高分析材料、化工材料等行业。比如工程材料试验系统，可以通过压缩、拉伸、动态波形等加载方式，测试沥青混合料、混凝土等建筑材料、金属材料、塑料、包装材料的力学性能。

7. 电力试验系统

电力试验系统可以进行各种与电力有关设备等的检测试验，包括高压耐压试验、蓄电池直流系统监测、电力安全工具器具检定、变压器试验、电机试验、断路器开关监测、继电器保护试验、电缆线路监测、电能及计量监测、接地电阻绝缘电阻测试、油化分析（色谱分析、绝缘油介质损耗测试、颗粒污染度测定、矿物油颗粒度检测、液相锈蚀测定等）、避雷装置绝缘子检测、发电机检测、无功补偿电容检测、电流表测试、高低压 CT 变比测试、金属探测电导率、电力故障检测等。

8. 通讯试验系统

通讯试验系统可以进行 LED 光源特性研究、模拟信号的直接强度调职传输、模拟信号的脉冲频率调制传输、数字光纤通信、数字编码传输、视频信号传输、卫星通信试验、多谐震荡器试验、模拟信号源试验、电磁波传输、接受滤波放大器试验、计算机串口试验、各类编译码系统（PCM、ADPCM、CVSD）试验、FSK（ASK）调制解调试验、数字同步技术试验、数字频率合成试验等。

9. 膜过滤试验系统

膜过滤试验系统可以进行流体分离、纯化开发、料液浓缩、除菌、脱盐、脱除溶剂等，在生物、制药、化工、食品、环保等领域应用广泛。

10. 汽车试验系统

汽车试验系统可以根据汽车的技术特性、可靠性、耐久性、环境适应性等方面，对汽车及其零部件、材料等进行试验，包括汽车有关的静力强度试验、环境试验、发动机试验、发动机台架测试、模型设计试验、紧固件装配设计试验、底盘测试、整车下线检测、机动车排放检测、安全检测、速度表检验、转向角检验等。

3.9   Test and Test System Guarantee

———

In order to study the objects concerned under limited conditions, people have developed and applied many test methods in practice. In order to study the situation that an objective object (process) is affected by multiple factors, people began to design experiments. In order to meet the complex and real needs of production and scientific research, people have developed various test systems by using appropriate principles and combining various equipment, instruments, fixtures, application software, etc. A perfect test, a perfect test design and a high-quality test system can ensure that the data obtained is of high quality and can achieve the purpose of data application.

3.9.1 Test method

For the contents of test methods, refer to Section 5, Chapter 2. Here we introduce the factorial test.

Factorial Experimental Design is a comprehensive test method, which can comprehensively reflect the influence of various factors on the indicators of test objects. In the experiment, it is to arrange and combine all levels of factors, and then conduct the experiment in cross groups to investigate the main effect of factors and the interaction effect between factors; Compare the average effect of different levels of each factor and the average effect of different levels of combinations among factors to find the best combination.

The total number of trials for factorial effects is the product of all factor levels. For example, if three factors are tested at the same time, and each factor takes two levels, then the total number of test combinations is 2^3, that is 8. If there are 3 factors, and each factor takes 4 levels, then the total number of test is 4^3, that is 64. Therefore, generally in the factorial test design, the level should not be too much, and generally 2 or 3 levels are preferred, otherwise the test workload will be huge.

Case:

The fish feed formulation was carried out in an aquaculture farm. There are four fish ponds. It is assumed that the fish in the ponds have been farmed for one year and are of the same size. Now feed the four fish ponds with different formulas, and observe the growth and weight gain of the fish three months later. Feeds are divided into three types: ordinary formula, type A formula and type B formula. The feed of the four fish ponds is as follows:

The first fish pond, common feed; the second fish pond, common feed plus type A formula; the third fish pond, common feed plus type B formula; The fourth fish pond, ordinary feed plus type A and B formula.

Analyze.

There are two kinds of formulas: type A and type B. And there are two levels of "use" and "not use" respectively. Therefore, this test belongs to the factorial experimental design of 2×2. In the analysis, not only we can analyze the effect of formula A and B on fish weight gain, but also can we analyze whether there is interaction between formula A and formula B.

A1 represents formula A factor 1 level, A2 represents formula A factor 2 level, B1 represents formula B factor 1 level, and B2 represents formula B factor 2 level. The format of factorial test design of 2× 2 is shown in Table 3-8.

Table 3-8　Factorial test design of 2×2

Type B / Type A	B1	B2
A1	A1B1	A1B2
A2	A2B1	A2B2

Three months later, the inspectors randomly sampled the fish in the four fish ponds for three times, with the same sampling amount for each fish pond. The average weight gain of each fish in the four fish ponds was obtained, as shown in Table 3-9.

Table 3-9　Average weight gain of each fish in four fish ponds　　　　Unit: g

Project name	The first fish pond ordinary feed	The second fish pond ordinary feed + A	The third fish pond ordinary feed + B	The fourth fish pond ordinary feed +A, B	Total
First sampling	200	330	230	460	1220
Second sampling	225	310	270	430	1235
Third sampling	175	290	250	400	1115
$\sum X$	600	930	750	1290	3570
$\bar{x}$	200	310	250	430	1190
$\sum X^2$	121 250	289 100	188 300	556 500	4 256 850

It can be seen from Table 3-9 that the weight gain of fish in the fourth pond is the largest, with an average increase of 430g per fish; The second fish pond took the second place, with an average increase of 310 grams per fish pond. The average increase of the third fish pond is 250g, which is 200g higher than that of the first fish pond.

The difference between the second pond and the first pond is 310-200 = 110g, which is the affect of type A formula.

The difference between the third pond and the first pond is 250-200 = 50g, which is the affect of type B formula.

The second fish pond and the fourth fish pond both use type A formula. The difference between the fouth pond and the second pond is 430-310 = 120g. Obviously, it has the function of formula B.

Then, the interaction of type A and type B formulas is as follows. The "difference between the fourth fish pond and the second fish pond" minus the "difference between the third fish pond and the first fish pond" equals 120−50 = 70g.

3.9.2 Design of experiment

1. The connotation of the experimental design

Experiment design means that in most cases, people are not very clear about the mechanism of the problem they are studying, or it is difficult to express it in mathematical formulas. They can only use experimental methods to get satisfactory results economically and quickly with as few times as possible. Experimental design is a pre arranged experimental activity, which depends on the statistical evaluation of the results to craw conclusions under the specified confidence level.

The general steps of the experimental design are as follows: analyze the test system, select the test scheme, complete the test scheme, analyze the test data, draw test conclusions, and achieve continuous improvement.

The experimental design can be divided into single factor test design, multi factor test design, orthogonal test method, etc.

2. Single factor test design

Single factor test design assumes that there is only one influencing factor in management activities, and the optimal solution is obtained by designing corresponding experimental methods to solve this influencing factor to achieve management objectives. Generally, mathematical expression (i.e. objective function) is used to reflect the relationship between factors and test results. The value of factor is represented by x, and the objective function is $y = f(x)$. There are two cases of functions: in the first case, the function is monotonically ascending or monotonically descending, and the optimal solution can be found as long as it is within the effective value range. In the second case, the function presents a "∪" or "∩" shape, that is, the graph has a peak value a or valley value b, which is the optimal solution.

Case:

Now there is a processing laboratory of a company. In order to obtain the best product, it is basically determined that the value of an additive compound (factor) is between 12 and 16 grams, that is [12,16]. So, how much is the optimal solution?

(1) Equipartition method

Now the laboratory plans to do four tests, so the spacing of each test = (16−12)/(4+1) = 0.8.

Therefore, the tester can take X1 = 12.8, X2 = 13.6, X3 = 14 4, X4 = 15.2 for test respectively.

Through test, when X3 = 14.4, the effect of additive is the best.

(2) 0.618 method

Take X1 = 12+(16−12)×0.618 = 14.47;

Take X2 = 12+(16−12)×0.382 = 13.53.

Now suppose that when X1 = 14.47, the test effect is better than X2 = 13.53, then remove the data below 13.53, and take the interval [13.53, 16] to continue the test.

Then: X3 = 13.53+(16−13.53)×0.618 = 15.06.

Now suppose that when X1 = 14.47, the test effect is better than X3 = 15.06, then take the interval [13.53,15.06] to continue the test.

Then: X4 = 13.53+(15.06−13.53)×0.618 = 14.48.

After the test, it is found that X4 is the best.

(3) Series method

Now set the test interval is 0.5 according the the test value range [12,16], the values are 12,12.5,13,13.5,14,14.5,15,15.5 and 16 for 9 times, so the number of tests is 9.

In this way, the test sequence table can be made, as shown in Table 3−10.

Table 3−10 Test sequence table Ⅰ

Possible order		1	2	3	4	5	6	7	8	9
Series	F0	F1	F2	F3		F4			F5	
	1	2	3	4		5			8	
Test value			12	12.5	13	13.5	14	14.5	15	15.5	16
Test sequence				X2		X1				

According to the sequence, there are F5 sequences. Therefore, for the first test point X1, take the number sequence F (5−1) = F4, and the test value is 14.

For the second test point X2, take the number sequence F(5−2) = F3, and the test value is 13.

Now suppose that the result of X1 is better than X2, then remove the part below X2, and a new test sequence table will be available, ranging from the fourth possible sequence to the ninth possible sequence, as shown in Table 3−11.

Table 3−11 Test sequence table Ⅱ

Possible order	4	5	6	7	8	9
Series		F4			F5	
		5			8	
Test value	13.5	14	14.5	15	15.5	16
Test sequence		X1			X3	

The next test point is X3. Take the symmetrical point of X1, then take the symmetrical point of "possible order 5" as "possible order 8".

So, the test value of X3 is 15.5.

Assuming that the test result of X1 is better than X3, the test value of 14 is the best.

3. Two factor experimental designs

In management practice, it is common for two factors to influence the objectives together. At this point, the impact on the target can be obtained through a two factor experimental design. The effects of the two factor experimental design can be summarized into four types: first, factor A has no effect, while factor B has the main effect; second, factor A is the main influence, while factor B has no influence; third, both factors A and B have a major impact, but A and B have no interaction; fourth, factor A interacts with factor B.

When factor A has no effect and factor B has the main effect, different levels of B have different effects. When factor A is the main influence and factor B has no influence, the A effect at different levels is different. When both factors A and B have a major impact, but A and B do not interact, the effect changes with the level of A and B. When factor A interacts with factor B, the effects of different levels A and B interact.

The results of the double factor (A, B) test are tested with the F value. a is the level number of factor A. b is the level number of factor B. r is the number of repeats of each test. $n = abr$, and the corresponding test results are shown in Table 3-12.

Table 3-12　Double factor test result table

Factor A has a major impact, while factor B has no impact	Factor A has no impact, while factor B has a major impact	Both factor A and factor B have influence, but no interaction	Both factor A and factor B have influence and interaction
The null hypothesis H_0: No significant difference at all A levels The alternative hypothesis H_1: At least A has significant difference at two levels	The null hypothesis H_0: No significant difference at all B levels The alternative hypothesis H_1: At least B has significant difference at two levels	The null hypothesis H_0: No significant difference between A and B The alternative hypothesis H_1: At least two tests A and B showed significant difference	The null hypothesis H_0: The interaction between A and B has no significant impact The alternative hypothesis H_1: The interaction between A and B has a significant impact
Test value $F = \dfrac{MS(A)}{MSE}$	Test value $F = \dfrac{MS(B)}{MSE}$	Test value $F = \dfrac{MST}{MSE}$	Test value $F = \dfrac{MS(AB)}{MSE}$
Degree of freedom $\dfrac{a-1}{abr-ab}$	Degree of freedom $\dfrac{b-1}{abr-ab}$	Degree of freedom $\dfrac{ab-1}{abr-ab}$	Degree of freedom $\dfrac{(a-1)(b-1)}{abr-ab}$
Rejection region: $F \geqslant F_\alpha$	Rejection region: $F \geqslant F_\alpha$	Rejection region: $F \geqslant F_\alpha$	Rejection region: $F \geqslant F_\alpha$

4. Orthogonal experimental design

（1）Concept. Orthogonal experimental design method is a simple method in multi factor test design. Its principle is to use orthogonal table to arrange as few test times as possible to obtain satisfactory solution economically and quickly according to the preset target situation. Orthogonal table is a numerical table arranged according to a certain rule. Its numerical arrangement has the equality of balanced dispersion and uniform collocation, namely orthogonality.

（2）Orthogonal table mark. The mark of orthogonal table is expressed by $L_p(m^n)$, which represents n factors, m levels, comprehensive tests for m^n times, and orthogonal tests for P times. See Figure 3-4 for the meaning of orthogonal table notation.

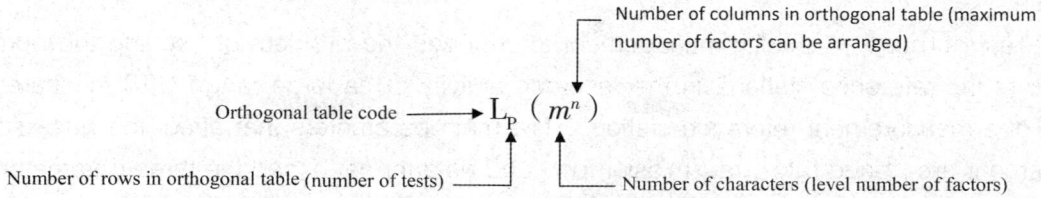

Figure 3-4   Orthogonal table mark

Mathematically, it can be proved that the best combination in the P tests with orthogonal table $L_p(m^n)$ is not necessarily the best combination for the comprehensive test of mn times, but on average, it is better than $\dfrac{P}{P+1} \times (m^n)$ times. In an orthogonal table, all levels in any column appear with equal frequency; All possible combinations of different levels between any two columns occur the same number of times.

(3) The characteristics of orthogonal table. The first is neat comparability, with comparability between levels of each factor. The other is uniform dispersion. The combinations of each factor level selected by the orthogonal table are evenly distributed in all horizontal combinations.

Take orthogonal table $L_8(2^7)$ as an example to illustrate orthogonality. See Table 3-13.

Table 3-13   Orthogonal table $L_8(2^7)$

Column No. \ Test No.	1	2	3	4	5	6	7
1	1	1	1	1	1	1	1
2	1	1	1	2	2	2	2
3	1	2	2	1	1	2	2
4	1	2	2	2	2	1	1
5	2	1	2	1	2	1	2
6	2	1	2	2	1	2	1
7	2	2	1	1	2	2	1
8	2	2	1	2	1	1	2

The meanings of the letters and numbers in the sign "$L_8(2^7)$" of the orthogonal table are as follows: "L" refers to the code of orthogonal table; "8" is the number of rows in the orthogonal table (number of tests); "2" is the number of characters (the level number of factors); "7" is the number of columns in the orthogonal table (the maximum number of factors can be arranged). The number arrangement in the orthogonal table has the orthogonality, that is, the number arrangement is evenly dispersed, and the number arrangement is evenly matched. For example, in the above table, the orthogonality is reflected in: each column has four "1" and four "2", that is, each level of each factor appears four times in eight tests; Number pairs (1, 1) (1, 2) (2, 1) (2, 2) (2, 2) in any two columns of the same horizontal collocation appear twice respectively, that is, the collocation between the two levels of every two factors is uniform, and the number of occurrences of each collocation is equal.

Case：

（1）Test purpose：A quality management team faced the problem of "setting the port parameters of the reference station" in the research activity of "a large range GPS measurement method of a measurement reference station". The main parameters that affect the base station measurement are "baud rate, transmission port and data format", and the three parameters affect each other. In order to determine the best combination of the three parameters, theteam decided to use the orthogonal analysis method to carry out the activity.

（2）Evaluation index：high accuracy of reference station is preferred.

（3）Determined factors：baud rate, transmission port and data format.

（4）Determined levels：Three levels are taken for each factor, and the factor level table is shown in Table 3-14.

Table 3-14　Factor level table

Factor / Level	A Baud rate	B Transmission port	C Data format
Level1	9600	Port1	GGA
Level2	19 200	Port2	GGK
Level 3	38 400	Port3	GSA

（5）Determine the test plan by using orthogonal table. $L_9(3^4)$ can arrange up to 4 factors of 3 levels. In this example, the three factors have three levels, and $L_9(3^4)$ can be used to arrange the test：put the three factors in the factor level table at the top of $L_9(3^4)$, and one factor for each column；Then set the corresponding levels according to the factor level table. Table 3-15 is obtained.

Table 3-15　Orthogonal test table

Column No. / Test No.	A Baud rate 1	B Transmission port 2	C Data format 3	Test result precision deviation/mm
1	1(9600)	1(Port1)	3(GSA)	9.6
2	2(19 200)	1(Port1)	1(GGA)	3.9
3	3(38 400)	1(Port1)	2(GGK)	5.4
4	1(9600)	2(Port2)	2(GGK)	5.5
5	2(19 200)	2(Port2)	3(GSA)	4.7
6	3(38 400)	2(Port2)	1(GGA)	8.8
7	1(9600)	3(Port3)	1(GGA)	8.1
8	2(19 200)	3(Port3)	2(GGK)	2.7
9	3(38 400)	3(Port3)	3(GSA)	3.2

Column No. / Test No.	A Baud rate 1	B Transmission port 2	C Data format 3	Test result precision deviation/mm
T1	23.1	18.8	20.7	
T2	11.4	18.9	13.5	T =51.9
T3	17.4	14.1	17.4	$\bar{x}$ =51.9/9 =5.77
R	11.7	4.7	7.2	

（6）Analyze test results

Direct analysis: the absolute error of No. 8 test is the smallest, so the initial priority level is A2B3C2.

Calculate: determine the preferred bit level, A: T2<T3<T1; B: T3<T1<T2; C: T2<T3<T1, so A2B3C2 is chosen.

Identify key factors: A<C<B, so, A is the key factor; C is an important factor; B is a secondary factor.

Sometimes, the orthogonal test method with interaction is encountered, so the interaction table is used to find out the existence of interaction and arrange the test reasonably. $L_8(2^7)$ is taken as an example to illustrate the arrangement of test design, as shown in Table 3-16.

Table 3-16　$L_8(2^7)$ is Used to explain the experimental design that comprehensively considers the main factors and interactions

Column No. / Number of factors	1	2	3	4	5	6	7
3	A	B	A×B	C	A×C	B×C	
4	A	B	A×B C×D	C	A×C B×D	B×C A×D	D
4	A	B C×D	A×B	C B×D	A×C	D B×C	A×D

3.9.3 Test system

The test system can form many types as required, such as general test system (tensile testing machine), pneumatic test system, pressure detection test system, heat conduction test system, wind tunnel test system, material test system, electric power test system, communication test system, membrane filtration test system, automobile test system, etc. Each test system plays an important role in its own field.

1. General test system

According to different uses, general test systems have many classifications according to different uses. For example, the general mechanical testing system can conduct various mechanical tests such as tensile, compression, bending, peeling, tearing, shearing, friction as required. The general electronic test system adopts the idea of standardization, modularization and generalization, and can test different electronic devices after assembling different test softwares. For example, some testing systems specially test wireless communication equipment, including mobile phones, chips, smart homes. Some can conduct communication network security testing, information risk assessment, etc.

2. Pneumatic test system

The pneumatic test system uses compressed air as the medium to form a pneumatic circuit to drive and control the machinery to achieve the test purpose. Special gases can also be added, such as hydrogen and nitrogen. The pneumatic system can be used for gas filling, replacement of gas in the pipeline, detection of the tightness of the system, detection of the leakage point, monitoring of temperature, inspection of the fatigue life of containers (such as hydrogen cylinders) in extreme environments, etc. Various cylinders connected with the production device control the size of the pneumatic output force, the movement direction and the movement speed of the cylinder through the pressure control valve, direction control valve and flow control valve.

The pneumatic device is relatively simple in structure, simple in installation and maintenance, low in pressure level and safe in use. The exhaust treatment is simple and does not affect the environment. It is convenient to control the output force and the output speed.

3. Pressure detection test system

The pressure detection test system is widely used, which can be used for medical related pressure detection, oil well pressure detection, civil engineering pressure detection (such as concrete compressive strength detection), chemical (pipeline, pressure vessel, etc.) pressure detection, ocean and abysmal sea pressure detection, material pressure detection, etc.

4. Heat conduction test system

The heat conduction test system can be used to detect the heat conduction coefficient, and is applied in many industries such as textiles, metal materials, soil, industry (steel making), machinery, cold storage, etc.

5. Wind tunnel test system

Wind tunnel refers to a controllable air flow pipeline system designed according to certain requirements, with power devices, and used for various aerodynamic tests. The wind tunnel test system can be used for testing under various wind speed, temperature and other environments, and is applied in many fields such as mining, meteorology, environmental protection, engineering, structure, measurement, aviation, ocean engineering, automobile and so on. Wind tunnel test is an important part of wind engineering, which has the characteristics of strong visualization, approximate simulation, easy control of test conditions, saving manpower and material time. Wind tunnel test can be used for force measurement, pressure measurement, aerodynamic parameter test, wind-induced vibration test, flow pattern observation, wind characteristics and wind environment test, ventilation and exhaust performance test, etc.

6. Material test system

The material test system can test the relevant properties of various materials, and is used in building materials, metal materials, high analysis materials, chemical materials and other industries. For example, the engineering material test system can test the mechanical properties of building materials such as asphalt mixture, concrete, metal materials, plastics and packaging materials through compression, tension, dynamic waveform and other loading methods.

7. Electric power test system

The power test system can carry out various detection tests of power related equipment, including high-voltage withstand voltage test, battery DC system monitoring, verification of power safety tools and instruments, transformer test, motor test, circuit breaker switch monitoring, relay protection test, cable line monitoring, electrical energy and metering monitoring, grounding resistance insulation resistance test oil chemical analysis (chromatographic analysis, dielectric loss test of insulating oil, particle contamination test, particle size test of mineral oil, liquid corrosion test, etc.), lightning arrester insulator test, generator test, reactive compensation capacitance test, ammeter test, high and low voltage CT ratio test, metal detection conductivity, power fault detection, etc.

8. Communication test system

The communication test system can conduct LED light source characteristics research, direct strength transfer transmission of analog signals, pulse frequency modulation transmission of analog signals, digital optical fiber communication, digital coding transmission, video signal transmission, satellite communication test, multivibrator test, analog signal source test, electromagnetic wave transmission, receiving filter amplifier test, computer serial port test, various coding and decoding systems (PCM, ADPCM, CVSD) test FSK (ASK) modulation and demodulation test, digital synchronization technology test, digital frequency synthesis test, etc.

9. Membrane filtration test system

The membrane filtration test system can be used for fluid separation, purification and development, feed concentration, sterilization, desalination, solvent removal, etc. It is widely used in biological, pharmaceutical, chemical, alimental, environmental protection and other fields.

10. Automobile test system

The vehicle test system can test the vehicle and its parts and materials according to the technical characteristics, reliability, durability, environmental adaptability and other aspects of the vehicle, including the static strength test, environmental test, engine test, engine bench test, model design test, fastener assembly design test, chassis test, vehicle offline test, vehicle emission test, safety test, speedometer test Steering angle inspection, etc.

3.10 可以高度模拟

高质量的数据，为高质量的模拟提供了保证。关于模拟的原理、过程等相关内容可参考第2章第3.6节的内容。现在模拟主要由计算机来实现，其中建模是模拟的关键工作，建模的质量直接决定了模拟的质量。

3.10.1 建模

1. 数学模型的概念

数学模型是指对客观对象进行的简化和假设后建立起的数学结构，这个结构能够解释客观对象的现实性态，预测对象的未来，提供处理对象的最优决策等。数学模型中包括了概念模型、物理模型、仿真模型（模拟模型）三个内容。

概念模型，是指针对客观对象所进行的文字描述，能形象地反映客观对象的全部内容、构成、有机关系。

物理模型，是指根据客观对象的物理特征所建立起的数学关系，通常可以用相关的函数来表达。

仿真模型（模拟模型），是指用计算方法把数学关系转化为计算机程序。

2. 建模的方法

（1）从程序上区分，建模的方法主要有分析法、测试法、综合法等（第2章第2.6节中有介绍）。把每个方法的关键步骤抓好，是提高建模质量的重要途径。

（2）从机理上区分，建模的方法分为物理机理法、系统辨识法。系统辨识法又可以区分为结构模式识别法、参数估计法。

物理机理法是指根据建模对象的应用场合和模型的使用目的，进行合理的假设，然后根据对象的物理机理（物理变化规律）来建立数学方程，进行自由度分析，然后建立系统模型。

系统辨识法是根据系统的输入输出的时间函数、系统的特性等来确定描述系统的模型。其中的结构模式识别是选择模型类型中的数学模型来表达。参数估计是在知道模型结构后，用参数来估计模型。

3. 建模的依据

建模的依据主要有：

（1）研究的目标，该目标决定了建模的详细程度、精确程度。

（2）先验知识，事先对于客观对象的特点，已经有了相应处理方法的积累，包括函数、曾经的模型等数据。

（3）试验数据，通过试验来获取相关数据，弥补先验知识的不充分。其中又可以分为两种类型，一种是演绎法，先根据还不够充分的理论建立一般规律，然后用先验知识猜测其特殊的规律，再通过试验来确定规律的正确性。另一种是归纳法，直接根据试验来观察到特殊的规律，在设法加入少量信息，来推导出一般的规律。

4. 数学模型的类型

在第 2 章第 2.6 节中介绍了数学模型的各种分类。这里主要介绍连续系统模型、离散系统模型和混合系统模型。

（1）连续系统模型

连续系统模型实质是连续时间的系统模型，它可以用各种相关的传递函数、微分方程、状态空间、S 域结构图等来表示。对于集中参数系统，可以用常微分方程（初始条件）表示；对于分布参数系统，可以用偏微分方程（初始条件加边界）表示。

连续系统的数值积分有欧拉法（矩形法）、梯形法、辛普森法（Simpson 法、抛物线法）、亚当姆斯法、变步长发等，其精度取决于截断误差（算法阶次、步长）、舍入误差（计算机字长）、积累误差（计算时间）。

（2）离散系统模型

离散系统模型可以用差分方程、脉冲传递函数、离散状态空间、Z 域结构图表示。离散可以分为时间连续但空间离散（有限元）、时间离散且空间离散两种。

离散系统中有实体（包括临时和永久）、事件（引起系统的状态发生变化的行为）、活动（实体在两个事件之间保持某一个状态的持续过程）、进程（描述若干事件和活动之间的逻辑关系、时序关系）。

比如高速公路收费处的进程为：车辆到达—排队—落杆收费—收费—抬杆放行。其中的活动为排队和收费；事件为车辆到达、落杆收费、抬杆放行。临时实体为驾驶员，永久实体为收费员。模型的输出为平均等待事件、最大排队长度、收费效率等。

（3）混合系统模型

混合系统模型，又称计算机控制系统，它通过"采样—保持"的模式把连续时间系统 S 近似地表达成离散时间系统 Z，用差分方程来表示。在近似的过程中，"采样—保持"会带来信号重构的误差。

在日常工作中的决策树模型、库存模型等就属于混合系统模型。

（4）离散事件系统模型

离散事件系统模型是指随机系统在事件的驱动下，以一定的概率发生状态转移。它用状态转换图来表示概率模型。

离散系统的仿真语言有：GPSS 语言，用每个语句表示一个动作，临时实体在语句之间传递，形成仿真模型的逻辑结构。SLAM 语言，用事件调度法、进程交换法来描述临时实体的进程，临时实体触发离散时间的发生、状态值变化；状态值超过阈值的时候触发事件。SIMAN 语言，把模拟的过程分为模型处理、试验处理、仿真运行、仿真输出，使建模与模拟试验分离。SIMscriptII.5 语言，它由事件表、时间推进程序来控制模拟运行。

举例：

一个离散事件系统由两个相同的部件 A 和 B，以及一个维修工组成。现在需要掌握该系统是否能够正常工作。为此，进行系统模型的构建。现在有：

A、B 两个部件的寿命统计分布为：$L(t) = 1 - e^{-\lambda t}$，

维修时间的统计分布为：$M(t) = = 1 - e^{-\lambda t}$，

为此，构建系统的网络模型见图 3-5 "部件维修模型图"。

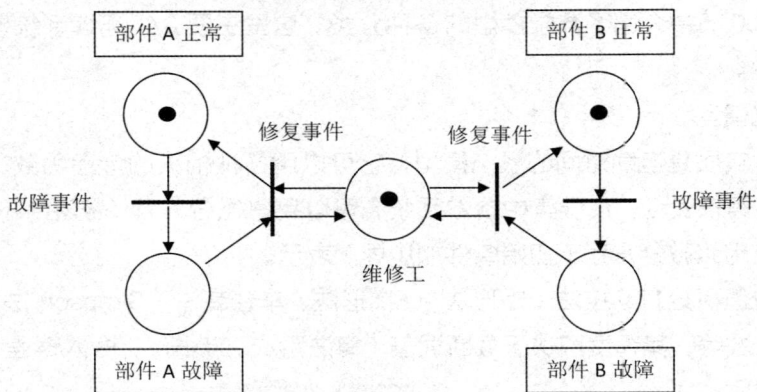

图 3-5　部件维修模型图

现在把图 3-5 的状态命名为 "10110"，即部件 A、部件 B 出现故障，维修后都正常。

还会出现图 3-6 的状态，命名为 "01110"，表示部件 A 维修失败，部件 B 维修后正常。

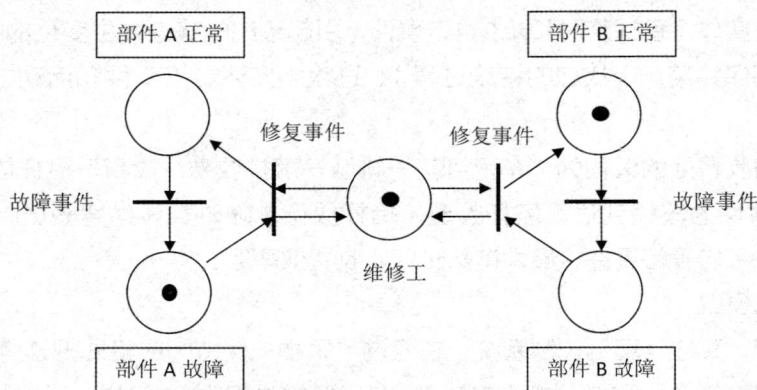

图 3-6　部件维修模型图

接下去还有状态为 "10101"，说明部件 A 维修后正常，部件 B 维修失败，见图 3-7。

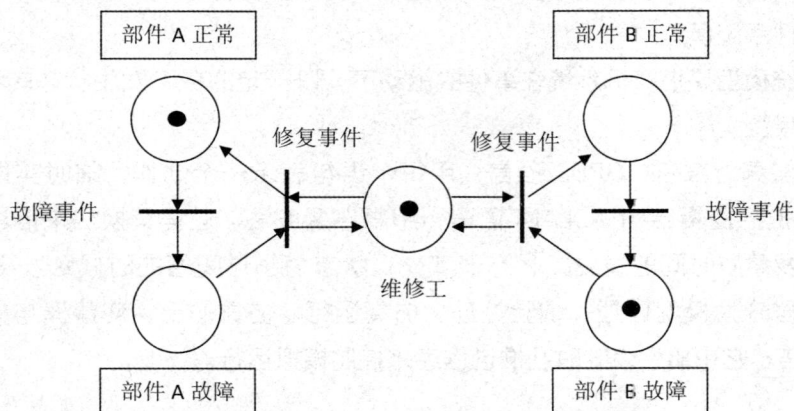

图 3-7　部件维修模型图

另有一个状态为 "01101"，说明部件 A、部件 B 都维修失败，见图 3-8。

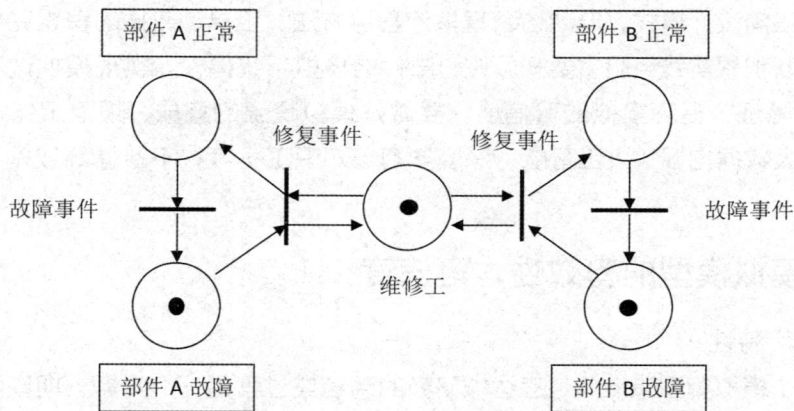

图3-8　部件维修模型图

该系统经过模拟后，得到各状态的稳态概率如表3-17。

表3-17　不同状态下的稳态概率

状态	稳态概率	部件 A	部件 B
10110	0.8635	正常	正常
01110	0.0657	故障	正常
01101	0.0050	故障	故障
10101	0.0657	正常	故障

从表3-13分析，得到：

部件 A、B 都正常的概率最大，为 0.8635；一个部件故障的概率最小，为 $2 \times 0.0657 = 0.1314$；

两个部件都故障的概率为 0.0050。所以，可以认为该系统能正常工作。

5. 面向对象的模拟

这里的对象是指数据结构（属性）加操作方法（成员函数），对象是一个程序模块，是运行的基本实体。面向对象的模拟有几个特征：封闭性，只能被同类对象中的成员函数直接访问；继承性，各子类继承父类的数据结构和操作方法；多态性，接受同一消息时，不同对象产生不同结果；动态联编，可在运行时动态改变数据结构和操作方法。

面向对象的模拟程序语言有三类：一是一般的面向对象的语言，如 C++。二是面向 AI 的面向对象语言。三是专门的面向对象的语言和软件包，如 Smalltalk。

6. 有限元法

有限元法（Finite Element Method），是把一个客观对象（连续体）视为若干个有限大小的、按照一定方式相互连接在一起的单元体的离散化集合，来求解这个客观对象（连续体）的热、力、电磁等方面数值的一种方法。它在建筑、机械、工艺、光学、声学等领域有广泛应用。它可以进行流体模拟、电磁模拟、应力模拟等。有限元的应用软件有 Ansys、Abaque、Fluent 等。

7. 推进基于建模和模拟的模式

为了进一步提高建模和模拟的质量，要开展基于模型的数据库建设，开发"基于模型的系统工程"（MBSE）正向设计模式，推动仿真模拟流程、方法、工具、数据的积累完善。要开展模拟的建模研究，不断积累实践经验。要开展核心部件的模拟系统建设，实现模拟试验与实物试验的相互补充、相互验证，提高模拟的可信度。要做好模拟资源的集成、规范工作，充分利用云计算、量子计算、大数据挖掘、人工智能、深度学习、知识工程等，不断提高效率，减少误差，提高模拟的科学性。

3.10.2 提升模拟模型的有效性、可信度

1. 模型的三个特性

如何建立一个模拟模型来准确地表达被研究的系统或过程是一个难题。期望所建立的模型能够具有三个特性，即有效性（valid）、可信度（credible）、适度详细（appropriately detailed）。

模型的有效性主要包括模型确认（validation）和模型验证（verificaiton）两个方面。模型确认是指所建立的模型与被模拟的对象之间的关系，两者的一致性越接近则有效性越好；模型验证是指所建立的模型与模型计算机实现之间的关系，计算机实现的正确性越高则有效性越好。模型的可信度主要指所建立的模型及参数与真实对象之间的差异，两者的差异越小模型的可信度越高。对于一个尚不太了解的、真实存在的客观对象来说，针对该对象所建立的模型进行可信度判断是有难度的。可信度可以通过做一个实际模型与模拟的结果进行比较完成。可信度的评价可以包括确认、验证、不确定性量化、适用性分析、充分性评价等内容。适度详细是指在建模的过程中，为方便构建模型而遵从化繁为简原则，同时尽量还原客观对象本身。

2. 提升模拟模型有效性、可信度的方法

主要有：收集高质量的关于系统或过程的信息和数据；请教专业人士，观察系统或过程，寻找接近的理论，从相同的模拟研究中寻找可用的信息，寻找建立模型的经验等；建立模拟模型后尽量增加模型的最大似然性，努力增加更多的数据，数据越多，模型越好；运用统计分析、假设检验、Algorithms 等方法；修正模型，处理好缺失值和异常值；修正整个模拟模型的输出等。

3.10 Availability to be Highly Simulated

High quality data guarantees high quality simulation. Refer to Section 6 of Chapter 2 for the principle and process of simulation. Now simulation is mainly realized by computer, in which modeling is the key work of simulation, and the quality of modeling directly determines the quality of simulation.

3.10.1 Modeling

1. Concept of mathematical model

Mathematical model refers to the mathematical structure established after simplifying and assuming the objective object. This structure can explain the reality of the objective object, predict the future of the object, and provide the optimal decision to deal with the object. Mathematical model includes conceptual model, physical model and simulation model (simulation model).

Conceptual model refers to the written description of the objective object, which can vividly reflect all the contents, composition and organic relationship of the objective object.

Physical model refers to the mathematical relationship established according to the physical characteristics of the objective object, which can usually be expressed by relevant functions.

Simulation model refers to the transformation of mathematical relations into computer programs by means of calculation.

2. Modeling method

(1) From the perspective of procedure, modeling methods mainly include analytical method, testing method and comprehensive method (introduced in Section 6 of Chapter II). To grasp the key steps of each method is an important way to improve the quality of modeling.

(2) In terms of mechanism, modeling methods are divided into physical mechanism method and system identification method. The system identification method can also be divided into structural pattern identification method and parameter estimation method.

The physical mechanism method refers to making reasonable assumptions according to the application occasions of the modeling object and the use purpose of the model, then establishing mathematical equations according to the physical mechanism of the object (physical change law), conducting degree of freedom analysis, and then establishing the system model.

The system identification method is to determine the model describing the system according to the time function of the input and output of the system and the characteristics of the system. The structural pattern recognition is expressed by selecting the mathematical model in the model type. Parameter estimation is to estimate the model with parameters after knowing the model structure.

3. Basis of modeling

The basis of modeling mainly includes the following items.

(1) Research objective, which determines the degree of detail and accuracy of modeling.

(2) Priori knowledge. For the characteristics of objective objects in advance, the corresponding processing methods have been accumulated, including functions, previous models and other data.

(3) Test data. Through experiments to obtain relevant data, make up for the lack of prior knowledge. Among them, there are two types. One is deductive method. First, establish general laws based on insufficient theories, then guess their special laws with prior knowledge, and then determine the correctness of laws through experiments. The other is induction, which directly observes special laws based on experiments and tries to add a small amount of information to deduce general laws.

4. Types of mathematical models

In the Section 6 of Chapter 2, various classifications of mathematical models are introduced. This paper mainly introduces continuous system model, discrete system model and hybrid system model.

(1) Continuous system model

Continuous system model is essentially a continuous time system model, which can be represented by various related transfer functions, differential equations, state space, S-domain structure diagram, etc. For lumped parameter system, it can be expressed by ordinary differential equation (initial condition); For distributed parameter system, it can be expressed by partial differential equation (initial condition plus boundary).

The numerical integration of continuous system includes Euler method (rectangle method), trapezoid method, Simpson method (parabola method), Adams method, variable step size method, etc. Its accuracy depends on truncation error (algorithm order, step size), rounding error (computer word length), and accumulation error (calculation time).

(2) Discrete system model

Discrete system model can be represented by difference equation, pulse transfer function, discrete state space and Z-domain structure diagram. Discrete can be divided into two types: time continuous but space discrete (finite element), time discrete and space discrete.

Discrete systems include entities (including temporary and permanent), events (actions that cause changes in the state of the system), activities (continuous processes in which entities maintain a certain state between two events), and processes (describing the logical and temporal relationships between several events and activities).

For example, the process of expressway toll office is: vehicle arrival-queuing-pole dropping and toll collection-toll collection-pole lifting and pass. The activities are queuing and toll collection; The events are vehicle arrival, pole dropping and toll collection, pole lifting and pass. The temporary entity is the driver, and the permanent entity is the toll collector. The outputs of the model are average waiting events, maximum queue length, charging efficiency, etc.

(3) Hybrid system model

The hybrid system model, also known as computer control system, approximates continuous time system S to discrete time system Z through the "sample and hold" mode, which is expressed by difference equation. In the process of approximation, "sample and hold" will bring errors in signal reconstruction.

Decision tree model and inventory model in daily work belong to hybrid system model.

(4) Discrete event system model

The discrete event system model refers to the state transition of a random system driven by events with a certain probability. It uses state transition graph to represent probability model.

The simulation languages of discrete systems include GPSS language include GPSS language, SLAM language and SIMscriptII.5 language. GPSS language uses each statement to represent an action. The temporary entities are transferred between statements to form the logical structure of the simulation model. SLAM language uses event scheduling method and process exchange method to describe the process of temporary entity, which triggers the occurrence of discrete time and the change of state value. The event is triggered when the status value exceeds the threshold value. SIMAN language divides the simulation process into model processing, test processing, simulation operation and simulation output, so as to separate modeling and simulation test. SIMscriptII.5 language is controlled by event table and time advance program.

Case:

A discrete event system consists of two identical components A and B, and a repairman (worker). Now we need to know whether the system can work normally. Therefore, the system model is built. Now here are as follows.

The statistical distribution of the life of components A and B is: $L(t) = 1 - e^{-\lambda t}$,

The statistical distribution of maintenance time is: $M(t) = 1 - e^{-\lambda t}$

Therefore, the network model of the system is built as shown in Figure 3-5 "Component Maintenance Model".

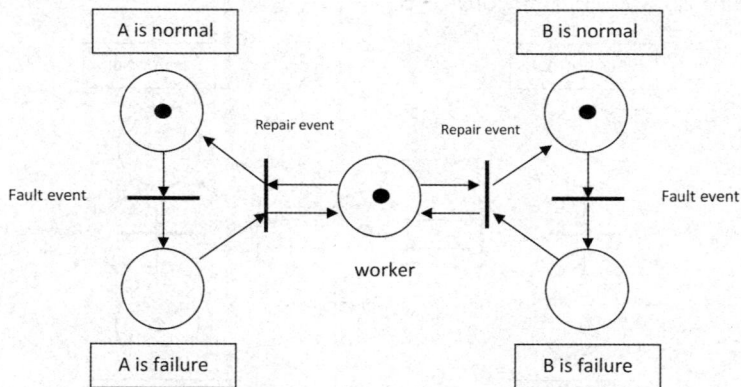

Figure 3-5   Component maintenance model diagram

Now name the status of Figure 3-5 as "10110", that is, component A and component B fail and are normal after maintenance.

The status shown in Figure 3-6 will also appear, named "01110", indicating that component A failed to repair and component B was normal after repair.

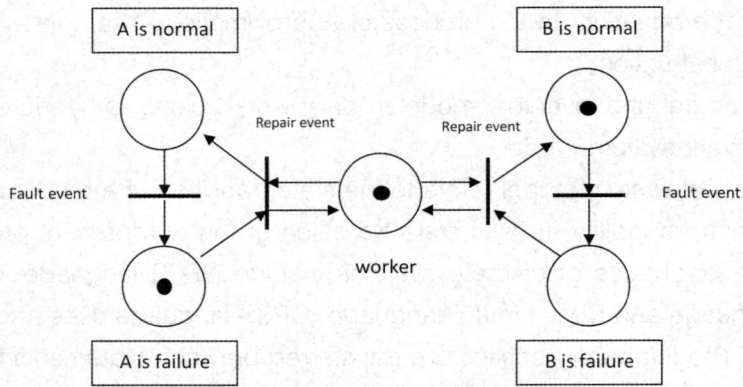

Figure 3-6　Component maintenance model diagram

The next status is "10101", which means that component A is normal after maintenance and component B is failed. See Figure 3-7.

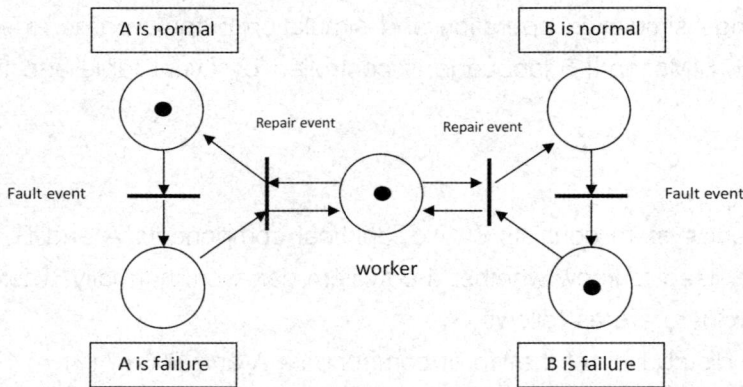

Figure 3-7　Component maintenance model diagram

The other status is "01101", which indicates that both component A and component B have failed to repair, as shown in Figure 3-8.

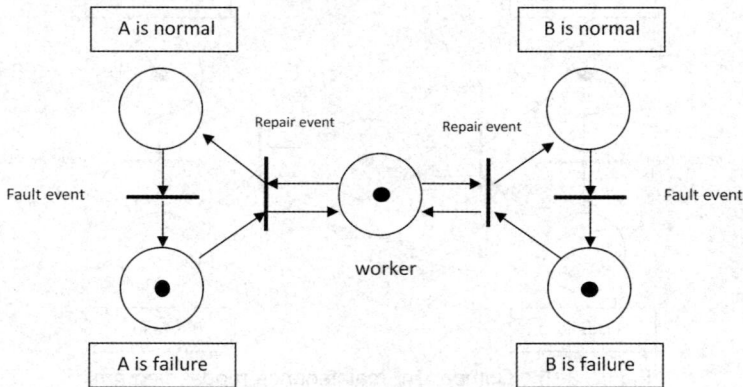

Figure 3-8　Component maintenance model diagram

After the system is simulated, the steady-state probability of each state is shown in Table 3-17.

Table 3-17　Steady-state probability under different states

State	Steady-state probability	Component A	Component B
10110	0.8635	Normal	Normal
01110	0.0657	Fault	Normal
01101	0.0050	Fault	Fault
10101	0.0657	Normal	Fault

According to the analysis in Table 3-13: the maximum probability that components A and B are normal is 0.8635; The minimum probability of one component failure is $2 \times 0.0657 = 0.1314$; The probability of both components failing is 0.0050. Therefore, it can be considered that the system can work normally.

5. Object oriented simulation

The object here refers to data structure (attribute) plus operation method (member function). The object is a program module and the basic entity for operation. Object oriented simulation has several characteristics: closure, which can only be directly accessed by member functions of the same kind of objects; inheritance, each subclass inherits the data structure and operation method of the parent class; polymorphism, when receiving the same message, different objects produce different results; dynamic binding can dynamically change the data structure and operation method at runtime.

There are three types of object-oriented simulation program languages. First, general object-oriented languages, such as C++. The second is an AI oriented object-oriented language. Third, specialized object-oriented languages and software packages, such as Smalltalk.

6. Finite element method

Finite Element Method is a method to solve the thermal, mechanical, electromagnetic and other numerical values of an objective object (continuum) by treating it as a discrete set of several finite size elements connected together in a certain way. It is widely used in architecture, machinery, technology, optics, acoustics and other fields. It can simulate fluid, electromagnetic and stress. The application software of finite element includes Ansys, Abaque, Fluent, etc.

7. Promote modeling and simulation based models

In order to further improve the quality of modeling and simulation, it is necessary to carry out model-based database construction, develop "model-based system engineering" (MBSE) positive design mode, and promote the accumulation and improvement of simulation processes, methods, tools and data. The modeling research of simulation should be carried out to continuously accumulate practical experience. The construction of simulation system for core components shall be carried out to realize mutual complementation and verification between simulation test and physical test, so as to improve the reliability of simulation. It is necessary to integrate and standardize simulation resources, make full use of cloud computing, quantum computing, big data mining, artificial intelligence, deep learning, knowledge engineering, etc., and constantly improve efficiency, reduce errors, and improve the scientificity of simulation.

3.10.2 Improving the effectiveness and credibility of the simulation model

1. Three properties of the model

How to build a simulation model to accurately express the studied system or process is a difficult problem. The model is expected to have three characteristics, namely, validity, credibility, and appropriately detailed.

The effectiveness of the model mainly includes two aspects: model validation and model verification. Model validation refers to the relationship between the established model and the simulated object. The closer their consistency is, the better their effectiveness will be; Model validation refers to the relationship between the established model and the computer implementation of the model. The higher the correctness of the computer implementation, the better the effectiveness. The credibility of the model mainly refers to the difference between the established model and parameters and the real object. The smaller the difference, the higher the credibility of the model. It is difficult to judge the reliability of the model established for an objective object that is not yet known and exists in reality. The credibility can be achieved by comparing the actual model with the simulation results. The evaluation of credibility can include confirmation, verification, uncertainty quantification, applicability analysis, adequacy evaluation, etc. Moderate detail means that in the process of modeling, in order to facilitate the construction of the model, the principle of simplifying complexity is followed, and the objective object itself is restored as much as possible.

2. Methods to improve the effectiveness and reliability of simulation models

It mainly includes the following items: collect high-quality information and data on systems or processes; consult professionals, observe systems or processes, find close theories, find available information from the same simulation research, and find experience in building models; After establishing the simulation model, try to increase the maximum likelihood of the model, and try to add more data. The more data, the better the model; Use statistical analysis, hypothesis test, algorithms and other methods; Correct the model and handle the missing and abnormal values; Correct the output of the whole simulation model, etc.

3.11　形成大数据

———

从数据分析和应用的角度看，大数据的质量无疑是远远超过通常的数据量的。当然，这里并不否定单个数据、少量数据、局部数据的个体特征的优势。如果说望远镜让人们能够感受宇宙，显微镜让人们能够观测微生物，那么大数据将开启重大的时代转型，改变人们的生活和理解世界的方式。

3.11.1　形成大数据思维

1. 大数据思维的起源

2009 年，一种新型的流感病毒甲型 H1N1 在迅速传播开来，全球的公共卫生机构在担心一场致命的流行病即将来袭。当时尚没有对抗这种病毒的疫苗，公共卫生专家只能先减慢它的传播速度，但问题是先要搞清楚这种流感出现在哪儿。由于人们生病后会不会立即就医，多半会先熬上多日，加上疾控中心的数据是每周汇总一次，显然，信息的滞后与病毒的飞速传播让公共卫生机构一时无所适从。而谷歌公司的员工把网上特定的检索词利用了起来，他们把检索词"哪些是治疗咳嗽和发热的药物"的使用频率，与 H1N1 流感在时间和空间上的传播建立起了联系。谷歌公司共处理了 4.5 亿个不同的数学模型，发现有 45 条检索词条的组合用于一个特定的数学模型后，可以判断出流感的起源地。这就为公共卫生机构提供了非常有价值的数据信息。这种以前所未有的方式，通过对海量数据分析来获取有巨大价值的产品、服务，或者深刻洞见，成为了一种新型的能力。这就是大数据的思维。

2. 从量变到质变的思维

就单单一张图片来说，解读该图片就是图片上的信息。在一万七千多年前的法国拉斯科洞穴上，有一幅壁画是一匹标志性的马，从马的图像的角度看，与当代一幅画上的马、一张照片上的马没有质的区别。难怪毕加索在看到拉斯科洞穴的壁画后开玩笑说："自那以后，我们就再也没有创造出什么东西了"。但诺维格尝试在每秒钟播放 24 副不同形态的马的图片，出现了电影，这显然是从量变到质变。

当事物达到分子级别时，其物理性质会改变。铜是导电的物质，但到了纳米级别就不能在磁场中导电。银离子可以抗菌，但以分子形式存在时就无法抗菌。在纳米级别，金属变柔弱，陶土有弹性。同样，当数据量巨量增加时，可以完成很多小数据所无法做到的分析。

3. 从随机抽样到全数据模式

随机抽样是用最少的数据获得最多的信息，它是在有限条件下的统计分析重要方法，而且突出随机性是其核心。而如今，在全部数据可收集、可分析的背景下，全数据成为了最佳选择，可以分析趋势、分析相关、分析异常、分析个别。大数据时代的抽样分析，有点类似于汽车时代的骑马出行。

4. 大数据的多样性

大数据的数据来源广泛，决定了大数据形式的多样性。大数据可以分为结构化数据、非结构化数据、半结构化数据。结构化数据之间因果关系强，比如项目信息管理系统的数据、医疗信息系统的数据等。非结构化数据之间没有因果关系，比如一段音频、若干图片等，各自独立。半结构化数据之间的因果关系弱，比如邮件的记录、网页数据等。

5. 接受大数据的不精确性

从已知的信息来看，只有5%的数据是结构化的，而且能适用于传统数据库。因此，其余的95%的数据基本是非结构化数据、半结构化数据，不精确性是大数据的重要特征。对于小数据而言，最基本、最重要的是减少错误，保证数据质量，所以每一次记录都要尽量精确。比如航天中的时间精度，铯原子钟必须是几百万年误差1秒，尽量保持天地之间的通信联系"绝对"一致、精确，否则会对空间站控制、航天员操作等会产生重大影响。而"不精确"恰恰是大数据中的亮点。如果说大数据的数据出错，那不是大数据的错，而很可能是测量、试验、记录、交流过程中出现的错。大数据的算法与小数据的算法在结果上会存在差异。比如：当数据只有500万的时候，一种简单的算法表现欠佳，准确率只有75%；而当数据达到10亿时，原来的算法其准确率会达到95%以上。在大数据时代，快速获得一个轮廓、发展脉络，其重要性可能会超过精确性。

6. 开发大数据的价值

在现实中，大量的数据是无效或低价值的，因此也称大数据为低价值密度。大数据的最大价值在于通过对各种类型数据的分析，挖掘出大数据的价值，包括预测、相关性分析、机器学习、智能推算等。各种科学理论会贯穿于大数据分析的方方面面，而大数据的分析又会催生出新的分析思路。在数字化时代，数据支持交易的作用会被掩盖，数据只是被交易的对象。而在大数据时代，数据的价值从最初的用途转变为未来的潜在用途，对影响组织评估其拥有数据及访问者的方式，推动各种管理模式的改变。

3.11.2 大数据的优势

大数据时代已经到来，大数据的优势将不断涌现。现举例如下：

（1）对于总体的分析结果肯定超过抽样的分析结果。随机抽样的结果肯定会存在误差，因此，传统的统计中有置信度水平等各种分析。而对于总体的分析结果就是全部的分析，虽然数据中各种混杂的情况都存在。但"混杂"正是人们必须要面对的，而且将会产生新价值的所在。

（2）在应用效率上远远领先小数据。小数据具有样本的局限性，应用基本也受到限制，而且数据的应用更显得单一性。大数据广泛而庞大，信息量充分，在应用时不受数据单一的困扰，分析而得到的结果更科学，故应用效率远胜小数据。

（3）在相关性分析上处于优势地位。大数据在进行不同对象间的相关性分析时，可以不受任何的干涉，两个或多个元素之间都可以构成相关性，这样的优势是小数据无法比拟的。

（4）引领学科的发展。大数据时代的各学科将发生巨大的变化，甚至是本质上的变化与发展，会影响到人们的价值体系、知识体系、工作和生活方式。大数据将成为现代社会基础设施的一部分。

3.11.3 数据创新

（1）数据的再利用。数据不再是一次性使用的对象，而是可以反复使用的资源。

（2）数据的重组。通过对不同数据的重组，产生新的重组数据，对它们的开发会带来新的价值。

（3）数据的扩展。本来是为了某一个目的而获取的数据，但改变一个目的后，会产生新的用途。

（4）数据的估值。数据作为一项重要资产而具备重要价值。

3.11　Forming Big Data

From the perspective of data analysis and application, there is no doubt that the quality of big data is far more than the usual amount of data. Of course, this does not negate the advantages of individual characteristics of single datum, a small amount of data or local data. If telescopes enable people to feel the universe and microscopes enable people to observe microorganisms, then big data will open a major era transformation and change the way people live and understand the world.

3.11.1 Forming big data thinking

1. The origin of big data thinking

In 2009, a new influenza virus, H1N1, was spreading rapidly, and public health institutions around the world were worried about a deadly epidemic. When there is no vaccine against this virus, public health experts can only slow down its spread, but the problem is to find out where the flu appears first. Because people will go to the doctor immediately after they get sick, most of them will stay up for many days. In addition, the data from the CDC is summarized once a week. Obviously, the lag of information and the rapid spread of viruses make public health institutions at a loss. The employees of Google used specific search words on the internet. They linked the frequency of use of the search words "which are medicines for cough and fever" with the spread of H1N1 flu in time and space. Google has processed 450 million different mathematical models, and found that 45 combinations of search terms can be used for a specific mathematical model to determine the origin of the flu. This provides very valuable data information for public health institutions. This unprecedented way, through the analysis of massive data to obtain products and services of great value, or profound insights, has become a new ability. This is the thinking of big data.

2. Thinking from quantitative change to qualitative change

As far as a single picture is concerned, the interpretation of the picture is the information on the picture. On the cave of Lascaux, France, more than 17000 years ago, there was a mural painting of a symbolic horse. From the perspective of the image of the horse, it is not qualitatively different from the horse in a contemporary painting or photograph. No wonder Picasso joked after seeing the murals in Lascaux Cave: "Since then, we have never created anything." However, Novig tried to play 24 pictures of horses in different shapes every second, and the film appeared, which was obviously from quantitative change to qualitative change.

When things reach the molecular level, their physical properties will change. Copper is a conductive material, but it can no longer conduct electricity in the magnetic field at the nanometer level. Silver ions can resist bacteria, but they cannot do so in molecular form. At the nanometer level, the metal becomes weak and the clay is elastic. Similarly, when the amount of data increases dramatically, many small data cannot be analyzed.

3. From random sampling to full data mode

Random sampling is to obtain the most information with the least data. It is an important method of statistical analysis under limited conditions, and highlighting randomness is its core. Now, under the background that all data can be collected and analyzed, full data has become the best choice to analyze trends, correlation, exceptions and individuals. Sampling analysis in the age of big data is somewhat similar to riding in the age of automobiles.

4. Diversity of big data

The wide range of data sources of big data determines the diversity of big data forms. Big data can be divided into structured data, unstructured data and semi-structured data. There is a strong causal relationship between structured data, such as the data of project information management system and medical information system. There is no causal relationship between unstructured data, such as a piece of audio and several pictures, which are independent of each other. The causal relationship between semi-structured data is weak, such as mail records, web page data, etc.

5. Acceptance of the big data inaccuracy

From the known information, only 5% of the data is structured and can be applied to traditional databases. Therefore, the remaining 95% of data are basically unstructured data and semi-structured data, and imprecision is an important feature of big data. For small data, the most basic and important thing is to reduce errors and ensure data quality, so every record should be as accurate as possible. For example, for the time accuracy in aerospace, the cesium atomic clock must have an error of 1 second in millions of years, and try to keep the communication between heaven and earth "absolutely" consistent and accurate, otherwise it will have a significant impact on space station control, astronaut operation, etc. And "imprecision" is just the highlight of big data. If the data of big data is wrong, it is not the fault of big data, but it is probably the fault of measurement, test, recording and communication. There will be differences in the results between big data algorithms and small data algorithms. For example, when the data is only 5 million, a simple algorithm performs poorly, and the accuracy rate is only 75%; When the data reaches 1 billion, the accuracy of the original algorithm will reach more than 95%. In the age of big data, it may be more important than accuracy to quickly obtain a contour profile and development path.

6. Development of the big data value

In reality, a large amount of data is invalid or low value, so it is also called low value density. The greatest value of big data lies in mining the value of big data through the analysis of various types of data, including prediction, correlation analysis, machine learning, intelligent calculation, etc. Various scientific theories will run through all aspects of big data analysis, and the analysis of big data will lead to new analysis ideas. In the digital era, the role of data supporting transactions will be covered up, and data is only the object of transactions. In the age of big data, the value of data has changed from its original use to its potential future use, which affects the way organizations evaluate their own data and visitors, and promotes the change of various management models.

3.11.2 Advantages of big data

The era of big data has come, and the advantages of big data will continue to emerge. Here are some examples:

(1)The analysis results of the population must exceed the analysis results of the sampling. The results of random sampling will certainly have errors. Therefore, traditional statistics include various analyses such as confidence level. The overall analysis result is the whole analysis, although there are all kinds of mixed situations in the data. But "hybrid" is exactly what people have to face and will produce new value.

(2)It is far ahead of small data in application efficiency. Small data has the limitation of samples, and its application is basically limited, and the application of data is more unitary. Big data is extensive and huge, and the amount of information is sufficient. It is not bothered by single data in application. The results obtained from analysis are more scientific, so the application efficiency is far better than small data.

(3)It is in an advantageous position in correlation analysis. Big data can be used to analyze the correlation between different objects without any interference. Two or more elements can form a correlation, which is incomparable to small data.

(4)Lead the development of disciplines. In the age of big data, great changes will take place in various disciplines, even in essence, which will affect people's value system, knowledge system, work and lifestyle. Big data will become a part of modern social infrastructure.

3.11.3 Data innovation

(1)Reuse of data. Data is no longer a one-time object, but a resource that can be used repeatedly.

(2)Reorganization of data. Through the reorganization of different data, new reorganized data will be generated, which will bring new value to their development.

(3)Data expansion. Data originally obtained for a certain purpose will be used for a new purpose after changing a purpose.

(4)Valuation of data. Data has important values as an important asset.

Chapter 4
Data Abnormalities

第 4 章
数据的异常

4.1 观察与记录异常

观察是人们获取事实数据的重要手段，但观察中难免会发生各种错误。举一个例子：1807 年的时候，有一个著名的双缝实验，大致是这样的：一个点上有光源，中间是两块不透光的板，第一个板上有一条竖着的细缝，第二块板上有两条竖着的细缝，另一面放着一张幕布。光从两条细缝中穿过，在幕布上看到了光的干涉现象，证明了光的波的特性。但当时的科学家不愿意就此否定的"光的粒子性质"，用摄像机进行全程摄录，令人惊异的是，在幕布上出现了两条竖线，就此肯定了光的粒子性。这种观察上出现的"错误"，到了"光的波粒二象性"的提出才得以纠正。

4.1.1 观察异常

观察异常（包括观察错误）是指人们在实施观察的过程中，由于一些自觉或不自觉的异常（错误）因素，导致对观察对象做出的与其本来真实情况不相符的描述。

观察异常的发生，如果不算观察者本人的责任性，其发生的概率还是很多的。人们总是相信眼见为实，以为观察到的就是真实的、正确的，但有的时侯"眼见为实"也会出错。下面举例来说明与"眼见为实"出现偏离的结果。

（1）观察结果会与事先的构图有关。图 4-1 是一张模棱两可的图片，如果观察者的脑子里的构图中有酒杯，那么这张图就成为了"酒杯"的图像。如果观察者脑子里的构图有头像，那么这张图就成为"两个对视的头像"。这是一种"先入为主"的观察。

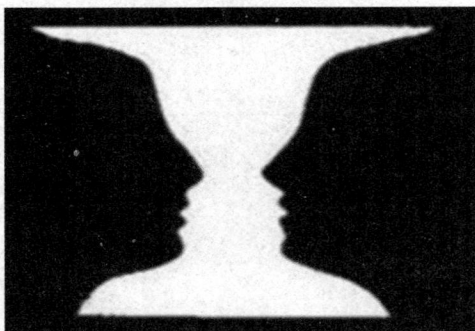

图 4-1　酒杯与头像

（2）观察结果与视角有关。在观察图 4-2 时，如果眼睛盯着中间的圆圈观看，那么会感到这个圆圈从平面上凸显（分离）出来，而事实上这张图片整个就是一张平面图。

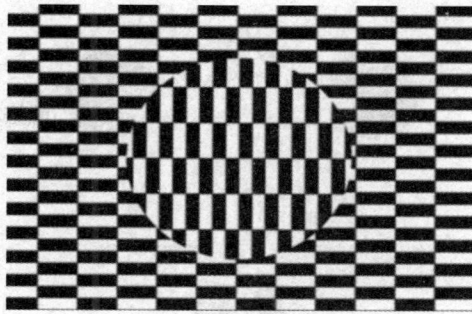
图 4-2　圆圈分离

（3）观察结果受到其他图形的干扰，平行线成为相互有角度的线条。在图 4-3 中，由于各白色方块、黑色方块出现了偏斜，感觉横着的线条是相互间有角度的，而实际上这些横线条是相互平行的线条。

图 4-3　被干扰的平行线

（4）观察结果受到其他图形的干扰，圆圈出现了变形。在图 4-4 中，观察者看到的是一组组向内的螺旋线，而实际上这些圆圈其实是同心圆。

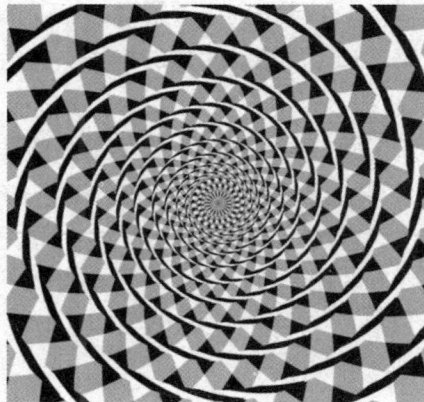
图 4-4　被干扰的同心圆

（5）观察结果受到其他图形的干扰，凭空想象出观察结果。在图 4-5 中，观察者会看到图形中间有一个清晰的白色三角形。而事实上，这个白色三角形是不存在的。

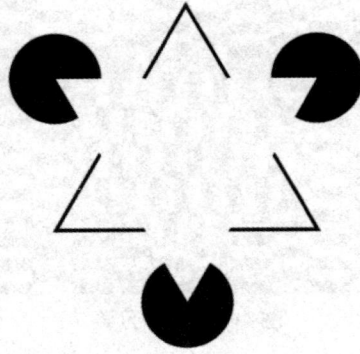

图4-5　想象出的三角形

4.1.2 导致观察异常的因素

导致观察异常的发生有很多原因，主要有：

（1）在实施观察时，观察者受到"先入为主"的概念影响。观察主体（观察者）会按照自己已有的概念、视觉图像等，对所观察到的事物进行描述。先后在耶鲁大学、印第安纳大学任教的美国学者汉森，提出了"感觉中渗透着理论"的观点，即观察者自己的理论观点、文化修养等影响着观察的结果。在汉森的著作《发现的模式》中提到，有两个生物学家观察同一个单细胞动物阿米巴，一个人说看到了单细胞动物，另一个却说看到了一个无细胞动物。

（2）观察者会受自己喜爱或厌恶等情感因素的支配，从而影响观察结果。只要是人在观察，那么情绪肯定会有意无意地支配着观察者的心理。当一个人情绪高昂时，观察者会更加投入到观察活动，所观察到的结果会带有积极的因素。而当一个人情绪低落时，就会蒙上一层阴影，有的人还会失去实事求是的态度。

（3）观察者限于自己的生理能力，不能识别全部因素。"盲人摸象"是一个很好的例子，几个盲人只能从自己所接触到的角度来感知（度量）大象的样子。色盲者在观察时与正常人在色彩上的观察结论也会出现很大的区别。

（4）观察者限于自己当时的知识面，导致观察出现偏差。亚里士多德在《论机械》中提到这么一个不可思议的现象：有一个大轮，半径为 R；大轮上带着一个小轮，半径为 r。大轮的圆周长为 $2\pi R$，小轮的圆周长为 $2\pi r$。显然，$2\pi R > 2\pi r$。两个轮子的圆心相同。

现在大轮滚动一周，则有观察到的现象是：大轮从 A 点滚动一周到达 A' 点；与此同时，小轮从 B 点滚动到达 B' 点。而且，$|AA'| = |BB'|$。见图4 - 6。

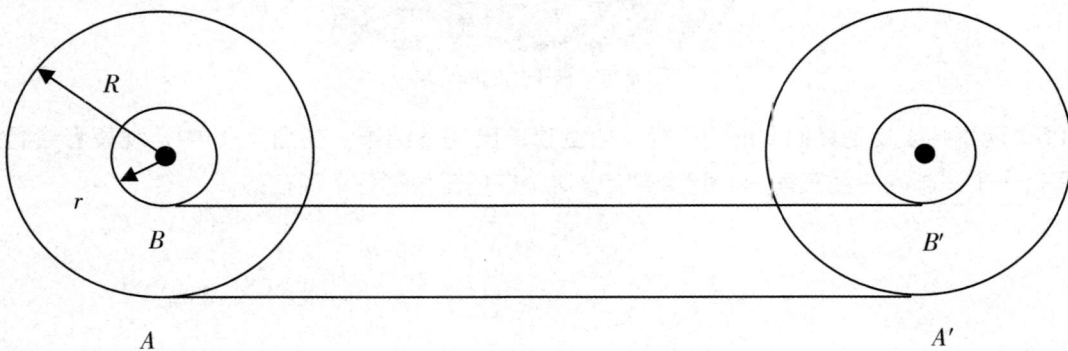

图4-6　带有小轮的大轮滚动一周

由于$|AA'| = 2\pi R$，而$|BB'| = 2\pi r$，而$2\pi R > 2\pi r$，那么结果是$|AA'| \neq |BB'|$。这就成为了悖论。

后来伽利略用六边形来进行滚动。同样，当大六边形滚动一周从A点到达A'点的同时，小六边形从B点滚动到达B'点。而且，$|AA'| = |BB'|$。见图4－7。

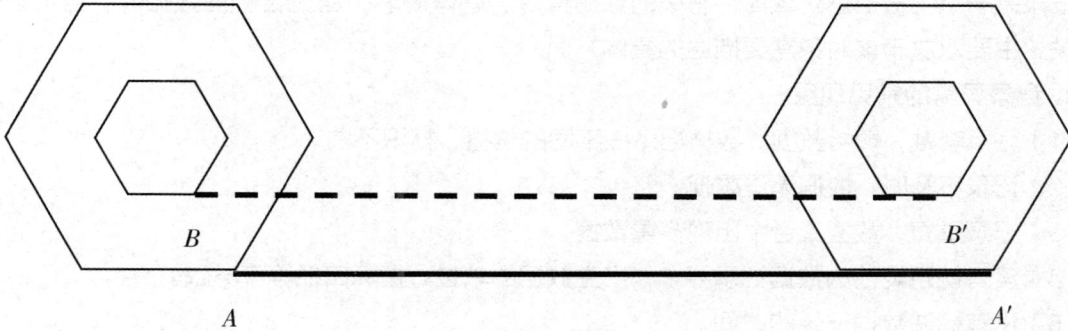

图4-7 带有小六边形的大六边形滚动一周

但伽利略发现，线段AA'是实线，而线段BB'是虚线。他认为，把六边形改为12边形，则虚线中的空白就更小。而圆是由无穷个边组成的，当出现n边形且n趋向无穷时，则图形变成了圆，虚线中的空白就消失了。事实上，这样的认知还是错误的。

导致以上观察错误的原因是，分析者受到了知识面有限的影响。当时的人们没有认识到小轮在滚动的同时，还存在着滑行的距离。也就是说，$|AA'| = 2\pi R = |BB'| = 2\pi r +$小轮的滑行距离。

（5）观察者受到了周围环境的影响，使观察结果不正确。比如观察者受权威或现成结论的影响，不敢超越已有的结论。哥白尼提出日心说的时候，对当时的地心说构成了巨大挑战。比如周边的媒体通过信息加工等方法，为某个客观事物（事件）构建起一个"媒体中的事物"，往往带有媒体的倾向性，会直接影响观察结果。另外，环境的嘈杂、气温、光线的亮度、不速之客的到来、小动物的干扰等都会构成影响。

（6）观察的方法出错。观察者在执行观察任务时，没有按照观察计划执行，观察的方法不恰当，使观察失败。观察者的观察角度、观察维度（平面与立体）、观察路径、没有使用恰当的观察仪器等，都会影响观察结果。

（7）观察客体出错。观察者应当确定观察对象，但如果客体发生了变化，那么观察结果就会发生迥异。

4.1.3 避免观察异常发生的做法

（1）观察者有全面的知识基础。完成一次观察，不只是简单的观察行为，更重要的是拥有足够的知识面，以确保观察正确、全面完成。

（2）制定详细的观察计划，做好充分的准备工作，并认真执行。

（3）观察者要有稳定的心理状态，优化自身素质，消除主体局限性对观察活动造成的负面影响。

（4）观察者避免先入为主的概念，观察做到客观、真实，勤于记录，不遗漏细节。

（5）观察时要集中注意力。要调动所有的身体感官，把眼睛、耳朵、手、脑、肢体等全部调动起来。

（6）借助适当的观察仪器辅助观察，突破观察者的生理限制，真实记录观察结果。

4.1.4 记录异常

记录是一组与被描述对象属性有关的数据集合。具体记录中应包含哪些数据项，与记录的目的有关，主要取决于该对象需要描述的具体方面。

1. 记录异常的常见现象

（1）张冠李戴，填错数据，数据记录在不同的位置，标识不清。

（2）记录不及时，数据无法重现。

（3）记录重复，数据集合中出现多重数据。

（4）没有使用规范的数据记录要求，产生数据的单位、指标口径等不一致。

（5）没有标注数据记录的时间。

（6）小数点标错，小数点前移或后移，形成错误数据。

（7）科学计数法出错，$a \times 10^n$ 的表述中，n 出错，数据无效。

（8）记录人员自身业务不熟练，没有具备相关记录的技能。

（9）数据没有得到确认，数据缺少编辑整理。

（10）数据的存储介质（硬盘、内存条、U 盘等）损坏。

（11）计算机设备导致的记录异常，如 CPU 产生异常、软件模拟产生异常等。

（12）计算机程序本身的错误导致数据记录异常。程序错误（Bug）是程序设计中的本来就存在的功能不正常、不齐全而导致的软件运行不正常，导致记录错误。

2. 避免记录异常的措施

（1）提高相关业务能力，具备必要的记录技能，增强责任心。

（2）严格按照记录的规范要求执行记录。

（3）记录的数据有专人进行复核、确认，避免不必要的错乱。

（4）做好计算机的维护工作，保证程序正常运行，避免存储介质损坏等。

（5）一旦发生记录异常后及时补救，比如：相对做到数据记录过程的再现。

4.1　Abnormality of Observation and Record

Observation is an important means for people to obtain factual data, but it is inevitable that various errors will occur in observation. Take an example: in 1807, there was a famous double slit experiment, which was roughly as follows: there was a light source on one point, two opaque plates in the middle, a vertical slit on the first plate, two vertical slits on the second plate, and a curtain on the other side. The light passes through two slits, and the interference phenomenon of light is seen on the screen, which proves the characteristics of light waves. However, the scientists at that time were unwilling to deny the "particle property of light". They used cameras to record the whole process. Surprisingly, two vertical lines appeared on the screen, confirming the particle property of light. This "error" in observation was corrected only when "wave particle duality of light" was proposed.

4.1.1 Observation abnormality

Observation anomaly (including observation error) refers to the description made by people to the observation object that is inconsistent with its original reality due to some conscious or unconscious abnormal (error) factors during the implementation of observation.

If the responsibility of the observer is not included in the observation of the occurrence of anomalies, the probability of occurrence is still large. People always believe that seeing is believing, and they think that what they observe is true and correct, but sometimes "seeing is believing" can also make mistakes. The following is an example to illustrate the result of deviation from "seeing is believing".

(1)The observation results will be related to the previous composition. Figure 4 − 1 is an ambiguous picture. If there is a wine glass in the composition of the observer's mind, this picture becomes an image of "wine glass". If the composition in the observer's mind has a head portrait, this picture will become "two heads looking at each other". This is a "preconceived" observation.

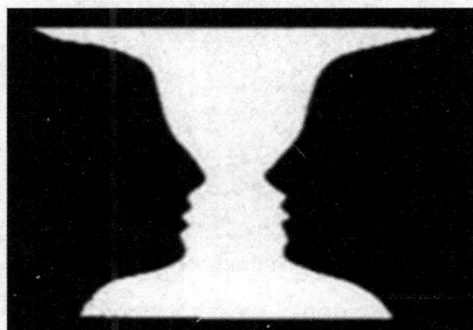

Figure 4−1　Wineglass and head portrait

(2) The observation results are related to the perspective. When looking at Figure 4-2, if you look at the circle in the middle, you will feel that the circle protrudes (separates) from the plane. In fact, this picture is a plane view.

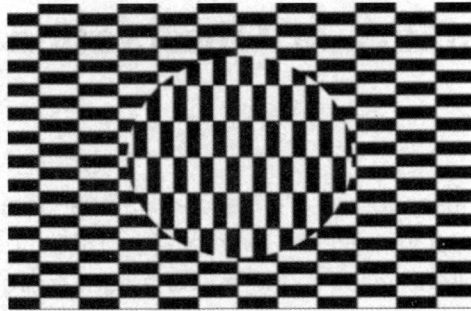

Figure 4-2　Separated circle

(3) The observation results are interfered by other figures, and the parallel lines become mutually angled lines. In Figure 4-3, due to the deviation of each white square and black square, it is felt that the horizontal lines are at an angle to each other, but actually these horizontal lines are parallel to each other.

Figure 4-3　Disturbed parallel lines

(4) The observation results are interfered by other figures, and the circle is deformed. In Figure 4-4, what the observer sees is a group of inward spirals, while actually these circles are concentric circles.

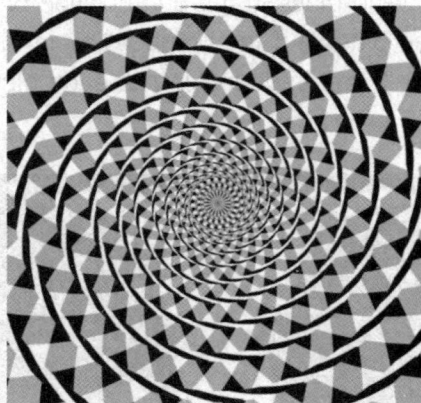

Figure 4-4　Disturbed concentric circles

(5) The observation results are interfered by other figures, so you can imagine the observation results out of thin air. In Figure 4-5, the observer will see a clear white triangle in the middle of the figure. In fact, this white triangle does not exist.

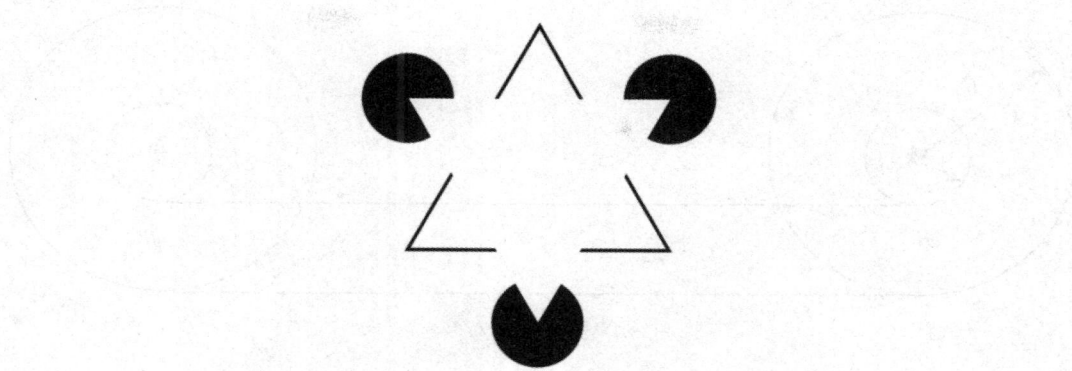

Figure 4-5　Imagined triangle

4.1.2 Factors leading to abnormal observation

There are many reasons for abnormal observation, mainly include the following items.

（1）When observing, the observer is influenced by the concept of "preconception". The observing subject (observer) will describe the observed things according to his existing concepts, visual images, etc. Norwood Russell Hansen, an American scholar who has successively taught at Yale University and Indiana University, put forward the view that "feeling permeates theory", that is, the observers' own theoretical views, cultural accomplishments, etc. affect the observation results. In Hansen's book *The Finding Pattern*, it is mentioned that two biologists observed the same single celled animal amoeba. One said he saw a single celled animal, while the other said he saw a cellless animal.

（2）Observers will be dominated by their own likes or dislikes and other emotional factors, thus affecting the observation results. As long as people are observing, emotions will surely dominate the observer's psychology, either intentionally or unintentionally. When a person is in high mood, the observer will be more involved in observation activities, and the observed results will have positive factors. When a person is in low mood, it will cast a shadow, and some people will lose the attitude of pursuing truth.

（3）The observer is limited to his/her own physiological ability and cannot identify all factors. "The blind man feels the elephant" is a good example. Several blind people can only perceive (measure) the appearance of the elephant from the perspective they are exposed to. The observation conclusion of color blindness and normal people in color will also be very different.

（4）The observer is limited to his/her knowledge at that time, resulting in deviation in observation. Aristotle mentioned such an incredible phenomenon in *On Machinery*: there is a big wheel with a radius of R; The big wheel is accompanied by a small wheel with a radius of r. The circumference of the big wheel is $2\pi R$, and the circumference of the small wheel is $2\pi r$. Obviously, $2\pi R > 2\pi r$. The centers of two wheels are the same.

Now, when the big wheel rolls one circle, it is observed that: the big wheel rolls one circle from point A to point A'; At the same time, the small wheel rolls from point B to point B'. In addition, $|AA'| = |BB'|$. See Figure 4-6.

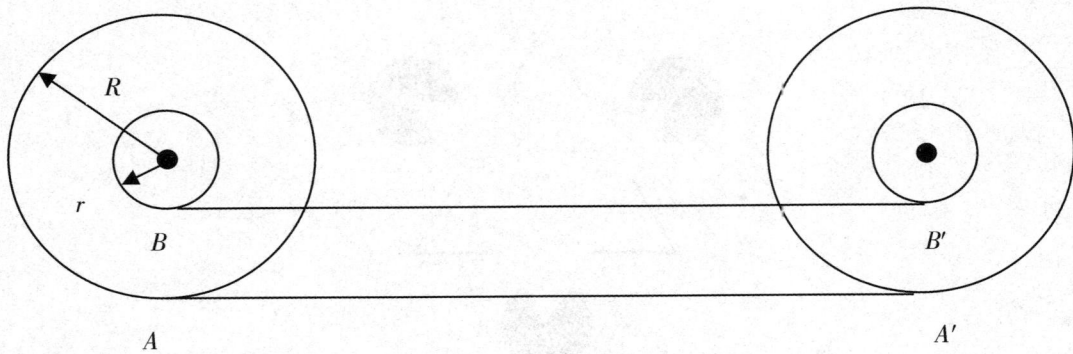

Figure 4-6   A large circle with a small circle rolls one turn

Because $|AA'| = 2\pi R$, $|BB'| = 2\pi r$, and $2\pi R > 2\pi r$, then the result is $|AA'| \neq |BB'|$. This becomes a paradox.

Later Galileo used hexagon to roll. Similarly, when the big hexagon rolls a circle from point A to point A', the small wheel rolls from point B to point B'. In addition, $|AA'| = |BB'|$. See Figure 4-7.

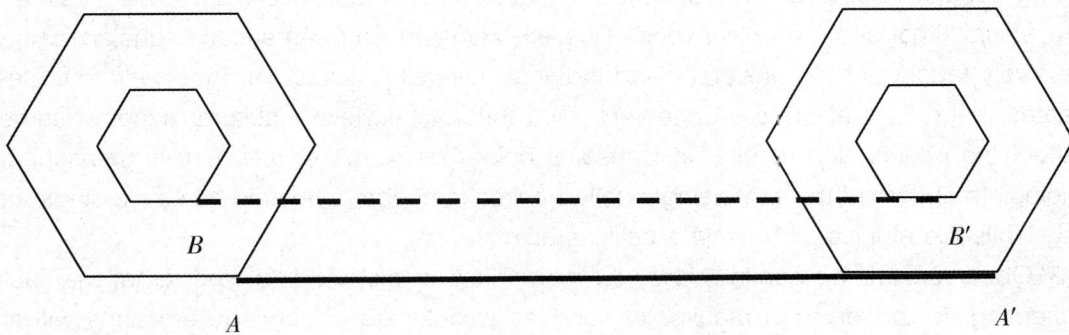

Figure 4-7   A large hexagon with a small hexagon rolls one turn

However, Galileo found that the segment AA' is a solid line, while the segment BB' is a dotted line. He believed that if the hexagon was changed to a 12 sided shape, the blank space in the dotted line would be smaller. The circle is composed of infinite edges. When n edges appear and n tends to infinity, the figure becomes a circle, and the blank space in the dotted line disappears. In fact, such cognition is still wrong.

The reason for the above observation errors is that the analyst is affected by limited knowledge. At that time, people did not realize that there was a sliding distance when the small wheel rolled. In other words, $|AA'| = 2\pi R = |BB'| = 2\pi r +$ sliding distance of small wheel.

(5) The observer is affected by the surrounding environment, which makes the observation result incorrect. For example, observers are influenced by authority or ready-made conclusions and dare not go beyond existing conclusions. When Copernicus put forward the heliocentric theory, it posed a great challenge to the geocentric theory at that time. For example, the surrounding media, through information processing and other methods, build a "thing in the media" for an objective thing (event), which often has a media bias and will directly affect the observation results. In addition, the noise of the environment, the temperature, the brightness of the light, the arrival of unexpected guests, and the interference of small animals will all have an impact.

(6) Error in observed method. The observer failed to observe in accordance with the observation plan, and the observation method was inappropriate. The observation results will be affected by the observer's observation angle, observation dimension (plane and three-dimensional), observation path, and the absence of comfortable observation instruments.

(7) Observation object error. The observer should determine the object of observation, but if the object changes, the observation results will be very different.

4.1.3 Methods to avoid abnormal observation

(1) The observer has a comprehensive knowledge base. Completing an observation is not just a simple observation, but more importantly, having enough knowledge to ensure that the observation is correct and complete.

(2) Formulate a detailed observation plan and make full preparations and implement it carefully.

(3) Observers should have a stable psychological state, optimize their own quality, and eliminate the negative impact of the limitations of the subject on observation activities.

(4) Observers should avoid preconceived concepts, observe objectively and truthfully, be diligent in recording and not omit details.

(5) Pay attention when observing. It is necessary to mobilize all the body senses, eyes, ears, hands, brain, limbs, etc.

(6) With the aid of appropriate observation instruments, we can break through the physiological limitations of observers and record the observation results truly.

4.1.4 Abnormal record

A record is a collection of data related to the attributes of the object being described. What data items should be included in a specific record is related to the purpose of the record and mainly depends on the specific aspects of the object to be described.

1. Common phenomena of abnormal record

(1) Attribute data to the wrong thing, wrong data, data recorded in different positions, unclear identification.

(2) The record is not timely and the data cannot be reproduced.

(3) The record is duplicate, and multiple data appears in the data set.

(4) There are no standardized data recording requirements, the units and indicators that generate data are inconsistent.

(5) The time of the data record is not marked.

(6) The decimal point is marked wrong, and the decimal point is moved forward or backward to form error data.

(7) Error in Scientific notation, in the expression of $a \times 10^n$, n error, invalid data.

(8) The recording personnel are not skilled in their own business and do not have relevant recording skills.

(9) The data was not confirmed and lacked editing and sorting.

(10) The data storage medium (hard disk, memory module, USB flash disk, etc.) is damaged.

(11) Abnormal record caused by computer equipment, such as CPU abnormality, abnormal

software simulation, etc.

(12)The error of the computer program itself leads to abnormal data recording. Program error (bug) is the abnormal operation of software caused by the original abnormal and incomplete functions in the program design, resulting in recording errors.

2. Measures to avoid abnormal record

(1)Improve relevant business capabilities, have the necessary recording skills, and enhance the sense of responsibility.

(2)The record shall be executed in strict accordance with the specification requirements of the record.

(3)The recorded data shall be rechecked and confirmed by special personnel to avoid unnecessary confusion.

(4)Maintain the computer operation, ensure the normal operation of programs, and avoid damage to storage media.

(5)In case of any abnormal recording, it shall be remedied in a timely manner, such as relatively reproducing the data recording process.

4.2　测量异常

引起测量异常的原因各有不同，主要可以分为测量仪器异常、测量方法异常、测量者导致异常三个方面。

4.2.1 测量仪器异常

1. 测量仪器的涵义

测量仪器又称为计量器具，是所有用来测量并能得到被测量对象量值的技术工具或装置的统称。严格来说，计量器具是测量仪器和测量工具的统称，本文中，测量仪器涵盖了测量工具。测量仪器是认识和改造物质社会的工具。在计量学中，测量既是核心的概念，又是研究的对象。测量工作通过测量仪器来实现，因此，测量仪器是测量的基础，测量仪器的质量是保证测量结果质量的重要条件。

测量仪器在各级组织中的生产过程中，承担着重要角色，是重要的劳动工具。在一般的机械类型的工厂，万能量具的数量是金属切削机械的 6 倍左右，在用的量具是加工人员数量的 1.5 倍左右，各种长度、热、力、电、理化计量器具在一个企业固定资产的占比很大。

2. 测量仪器的种类

测量仪器有很多种类，可分为几何量测量仪器、力学测量仪器、热学测量仪器、化学测量仪器、时间频率测量仪器、光学测量仪器、电磁测量仪器、无线电测量仪器等，简要介绍如下：

（1）几何量测量仪器，它可以进行角度测量、粗糙度测量、平面度测量、直线度测量等。几何量测量仪器的品种非常多，比如：量块、千分尺、通用卡尺、直尺、塞尺、卷尺、量规、针规、指示类量具、线纹类器具、各种水平仪、测厚仪、各种显微镜、光学计、投影仪、干涉仪、试验筛、焊接检验尺、线缆计米器、跳动检查仪、测高度仪、气动测量仪、大量程百分表、各种测距仪、各种测径仪、试模、水准仪、倾角仪、测试模板、测试圆杆、经纬仪、角度块、轴角编码器、对中仪、三维扫描仪、引伸计、测斜仪、投线仪、咬力测试仪、点焊分析仪、刻线尺、纤维细度分析仪、锥空端面直径测量表、多面棱体、齿轮测量仪、垂准仪、水位计、条码检测仪等。

（2）力学测量仪器，可以进行质量、力值、能量、容量、转速、硬度、压力、振动、密度、流量等的测量。力学测量仪器主要有：砝码、各种指示秤、各种天平、扭矩扳子、各种压力表、压力变送器、测力仪、拉力压力试验机、各种硬度计（金属洛氏、布氏、维氏、里氏、巴氏、肖氏）、转速表、定负荷硬度计、压力计、冲击试验台、冲击试验机、碰撞试验台、扭力天平、各种振动试验台、比重天平、测功装置、各种万能试验机、微压计、冷媒检漏仪、差压检漏仪、加速度计、测振仪、张力计、试验箱、压力头、液位计、脆度碎度检查仪、耐磨试验仪、气压表、移液器、桩基动测仪、比重瓶、模拟运输试验台、速度传感器、多分量力传感器、物理性能（耐久、疲劳、寿命）性能测试仪、装料衡器、各种流量计（超声、电磁、液体浮子、液体质量、液体容积式、涡轮、皂膜、湿气体、热气体）、气压计、冲击测量仪、界面张力仪、气体密度控制器、动平衡测量分析仪、电梯限速器测试仪、噪声检测仪、机车速度测试表、轴重仪、扭矩仪、扭转试验机、角速度试验机、离心机、密度计、噪声振动计、电离真空计、非连续累计自动衡器、配料秤、摩擦试验机、扭矩倍增器、线速度测量仪、容重器、杆秤、油气回收检测仪、质量比较仪、螺栓检测仪、稀释配标仪、瓶口分液器、转速传感器等。

（3）热学测量仪器，可以进行温度、湿度、热流等的测量。热学测量仪器主要有：温度调节仪、各种热电偶、各种温度计（标准水银、玻璃液体、工业热电阻）、热处理炉、温度指示仪、光照培养仪、自然通风老化试验箱、工业过程测量记录仪、双金属温度计、压力式温度计、机械式温度计、辐射温度计、温度变送器、温度巡回检测仪、热像仪、恒温槽、环境试验设备、表面温度计、盐务试验设备、铠装热电偶、温度校准仪、电阻炉、淋雨试验设备、温度采集仪、温度校准器、臭氧老化试验箱、可焊性测试仪、微波消解仪、真空干燥箱、冻干机、磁力搅拌器、湿度传感器、通风干湿表、过程校验仪、露点仪、制冷器具、钢轨测温仪、温度开关、紫外老化箱、热敏电阻测试仪、软化击穿试验仪、日晒气候色牢度试验仪、高温动态老化系统、腐蚀气体试验设备、沥青延伸度仪、沙尘试验设备、数字温度湿度表、红外温度计、辐射源测试仪、太阳辐射试验箱、湿度发生器等。

（4）化学测量仪器，可以进行光谱分析、色谱分析、质谱分析、水质分析、气体分析、生化分析、热化学分析、元素分析等。化学测量仪器主要有：紫外可见近红外分光光度计、红外分光光度计、傅里叶变换红外光谱仪、发射光谱仪、荧光光度计、火焰光度计、原子吸收分光光度计、旋光仪、手持折射仪、多晶X射线衍射仪、测汞仪、光谱仪、农药残留检测仪、微量分光光度计、气相色谱仪、凝胶色谱仪、离子色谱仪、液相色谱仪、毛细管电泳仪、薄层色谱扫描仪、（气相、液相）色谱质谱联用仪、等离子质谱仪、飞行时间质谱仪、水中油分浓度分析仪、浊度计、溶解氧测定仪、BOD测定仪、COD测定仪、总有机碳分析仪、氨氮检测仪、硝酸盐检测仪、硅酸根分析仪、磷酸根分析仪、余氯测定仪、重金属分析仪、高锰酸盐检测仪、游离氯分析仪、水质硬度仪、烷基汞分析仪、微量总有机碳分析仪、尿素检测仪、细菌浊度分析仪、水质分析仪、石油产品水分测定工艺、烘干法水分测定工艺、木材含水率测量仪、各种粘度计、各种酸度计、熔体流动速率仪、pH计、实验室离子计、电位滴定仪、电阻率仪、电导率仪、各种盐度计、电泳仪、溶解性固体总量测定仪、示波极谱仪、滴定仪、臭氧分析仪、一氧化碳检测仪、二氧化碳分析器、二氧化硫气体检测仪、硫化氢气体检测仪、烟尘采样器、甲烷测定器、氨气检测仪、大气采样器、尘埃粒子计数器、液体颗粒计数器、悬浮颗粒物采样器、空气微生物测验仪、微粒检测仪、粒度分析仪、凝胶成像系统、细菌内毒素分析仪、抗生素测定仪、聚合酶链反应分析仪、菌落计数器、开口闭口闪点测定仪、热量分析仪、熔点测定仪、酸值测定仪、凝点测定工艺、苯胺点测定仪、测硫仪、溶出度仪、氮元素分析仪、冰点测定仪、渗透仪、氧气透过率测试仪等。

（5）时间频率测量仪器，可以进行与时间、周期运动有关的物理量的测量。时间频率计量仪器主要有：频率计、通用计数器、频率分析仪、石英晶体频率标准、计时器、频标比对器、时间继电器、钟表分析仪、铷原子频率标准、全球卫星导航系统接收机、汽车行驶记录仪、振弦式频率读数仪、时间间隔测量仪、时间检定仪、标准数字时钟、同步时钟系统、GNSS时间同步系统、时间间隔发生器、时间合成器、时钟测试仪、测速仪、脉冲计数器、谐振式波长计等。

（6）光学测量仪器，可以研究光的行为和性质，进行光度、色度、辐射度、激光参数、城乡光学、光源、光功率、光衰减、激光能量等的测量。光学测量仪器主要有：光功率计、光衰减器、光谱分析仪、光波长计、光时域反射计、光纤熔接机、光万用表、光回波损耗仪、光纤偏振模色散测试仪、阿贝折射仪、光泽度仪、亮度机、光照度机、测色色差计、标准光源箱、白度计、橙明度检测仪、漫透射视觉密度计、紫外辐射照度计、光色电综合测试仪、光通量灯、雾度计、彩色分析仪、色温标、发光强度测试仪、紫外分析仪、分布光度计、LED 光强分布测试仪、反射率测定工艺、反射式密度计、紫外分析仪、光生物安全测试仪、紫外曝辐射量表、光学透过率测定仪、石油产品颜色分析仪及比色板、比色计、标准色板、干涉滤光片、水质色度仪、光谱光度计标准滤光器、激光衰减器、黑白密度片、激光能量计、光纤识别仪、偏振依赖损耗测试仪、光纤折射率分布和几何参数测量仪、无源光网络功率计、角膜曲率计、光纤损耗和模场直径测量仪等。

（7）电磁测量仪器，可以测量与电磁现象有关的物理量，包括测试电压、电流、电阻、电容、电感、磁感应强度、磁通、磁矩等。电磁测量仪器主要有：静电测试仪、电阻测试仪、高绝缘电阻测量仪、直流高压高值电阻器、直流低电表、微欧计、直流数字电桥、接地电阻表、耐电压测试仪、泄露电流测试仪、交流电参数测量仪、功率表、多功能校准源、交流电参数测量仪、交直流电表校验仪、高压静电电压表、电流表、直流电阻器、直流电桥、绝缘电阻表、电压互感器、直流电流源、互感器负荷箱、数字高压表、交流电能表、接入式电能标、电量变送器、工频单相相位表、磁通计、磁强计、高压谐振试验装置、变压器空负载损耗测试仪、变压器短路阻抗测试仪、三倍频试验变压器装置、输电线路工频参数测试仪、互感器综合特性测试仪、直流高压分压器、工频高压分压器、绝缘油介电强度测试仪、雷击计数器校验仪、调压器、变压器、综合测试仪点检装置、电子镇流器性能分析仪、氧化锌避雷器测试仪、保护回路矢量分析仪、大型地网接地电阻测试仪、电量记录分析仪、高压电容电桥、分流器、电流标、电压标、功率标、电阻标、回路电阻测试仪、直阻仪、工频磁场发生器、电阻应变仪、功率指示器量程校准器、变压器绕阻变形测试仪、电流电压传感器、电能质量分析仪、绝缘油介质损耗及体积电阻率测试仪、电压检测仪检定装置、电缆故障定位电源、验电器、直流电压互感器、磁粉探伤机、交流充电设备、电压检测仪、弱磁场交变磁场计、试验变压器操作箱、超低频高压发生器、氧化锌避雷器直流参数测试仪、模拟交直流标准电阻器、磁轭式磁粉探伤机、低压验电笔、涡流探伤仪、万用表、钳形漏电流表、磁力式磁强计、互感器校验仪、电压互感器、变压器铁芯接地电流测试仪、容性设备检测装置、耐电压测试校验仪、交流电能表检定装置、绕组线击穿电压试验仪、乳化沥青微粒离子电荷试验仪、故障滤波分析装置、网络充电设施接入节点计电能量与计时装置、直流电变送器、恒定磁场线圈、高电压耐电压测试仪、接触电流测试仪、直流高压试验装置、机电式交流电能表、涡流电导率仪、电雷管测试仪、交流峰值电压表、数据采集仪、绝缘电阻表多功能试验箱、功率分析仪等。

（8）无线电测量仪器，可以进行电磁波、高频和微波功率、5G 设备等的试验检测。无线电计量仪器有：LCR 数字电桥、交流电桥、标准电容器、标准电感器、半导体管特性图示仪、电容分选仪、电容器容量损耗分选仪、半导体特性图示仪校准仪、电感箱、交流电阻箱、模拟示波器、示波器校准仪、电子电压表、射频电压表、数字示波器、网络分析仪和校准件、射频阻抗材料分析仪、低频信号发生器、脉冲信号发生器、射频通信测试仪、频谱分析仪、函数任意波信号发生器、调制度测量仪、噪声系数分析仪、误码测试仪、规程分析仪、频谱分析仪、矢量信号分析仪、中功率计、电视场强电平检测仪、失真度测量仪、电话分析仪、电视信号发生器、音频分析仪、衰减器、轻型仿真线、数字传输分析仪、功率指示器、数字移动通信综合测试仪、动态信号分析仪、电磁骚扰测量接受机、话路特性分析仪、天馈线分析仪、抖晃仪、蓝牙测试仪、高速串行误码仪、频率特性测试仪、选频电平表、电平振荡器、噪声发生器、断续干扰分析仪、网络线路分析仪、逻辑分析仪、射频微波开关、矢量示波器、数据网路性能测试仪、电场测量仪、场强探头、电磁辐射仪、失真度校准器、时序噪声分析仪、尖峰信号发生器、脉冲磁场发生器、电磁发射和敏感度脉冲信号发生器、光示波器、射频识别测试仪、无线信道模拟器、光继电保护测试仪、耦合去耦网络、阻抗稳定网络、人工电源网络、取样示波器、电磁干扰发生器、振荡波发生器、空气线、高清视频信号发生器、频率响应分析仪、调制域分析仪等。

3. 测量仪器异常的原因

导致测量仪器异常的原因很多，主要有：

（1）由于测量仪器的结构原因所产生的异常。由于发生摩擦、测量压力等的变化，没有及时调整好各部分的结构而导致异常。

（2）测量条件发生变化所产生异常。比如温度、湿度、照度、震动等发生了变化，引起测量仪器异常。

（3）测量部件、电路等的连接线松动导致异常；测量仪器的防尘性能、密封性能不足，导致精确度变差。

（4）测量仪器长时间缺少保养，稳定性差。

（5）测量的配比不正确。有些测量涉及一些溶液配比等，由于配比的异常导致测量仪器结果异常。

（6）测量设备明显超出其允许范围导致测量结果异常。

4. 测量仪器异常的检查方法

判断测量仪器是否出现异常的方法主要有：

（1）直接观察。用人的感觉器官来直接找出异常的发生部位。比如：测量仪器发生断线、接触不良、元器件过热、冒烟、机械传动部件缺油、磨损、部件间隙过大、异常响声等。

（2）电路参数测量。用万用表对电路的电压、电流、电阻值等进行测量，与正常值进行对比，来确定异常的部位。

（3）测试仪器检查。使用专门的测试仪器，对测量仪器进行测试，找到异常点。

（4）缩小范围检查。对于存在疑问的部件、插件，使用正常的同类部件、插件进行替换，可以缩小异常的范围，在替换的过程中找到异常部件、插件。也可以把有疑问的电路从测量仪器中切除，通过观察电流的变化等来缩小异常范围。

（5）讯号查找。对于有疑问的测量仪器，选用不同的讯号发生器输出的讯号，逐级观测讯号在电路中的传输情况，以找到异常点。

（6）波形查找。用示波器来观测测量仪器的电路和部件的波形，与正常波形进行比较，来判断是否异常。

（7）短路查找。针对干扰、自激等异常，把电路中某两个点之间暂时短路。如果短路后异常消失，说明异常在短路点之前；如果短路后还是有异常，则在短路点之后继续查找。

（8）用检验模式查找。把计量设备中出现的数据，绘制控制图，用检验模式来判断是否异常，异常通常反映为测量点超出控制界限、测量点的分布不呈随机状态。在国际标准 ISO8258：1991、国家标准 GB/T 4091—2001 两个文件中，对常见的测量过程异常规定了 8 种分布模式（可参阅《质量管理统计应用》）。如果平均值控制图出现异常，说明测量过程受到了不受控的系统效应的影响；如果标准偏差控制图出现异常，说明测量过程受到了不受控的随机效应的影响。

4.2.2 测量仪器异常的应对措施

1. 测量仪器的校准

（1）校准的作用

仪器校准是指在规定的条件下，用参考测量标准给予包括实物量具在内的测量仪器的特性赋值，并确定它的示值误差，把测量仪器所代表的量值按照比较链或者校准链，将其朔源到测量标准所复现的量值上。校准的依据是校准规范和校准方法，通常有统一规定，特殊情况下可以自行规定。校准的结果记录在校准报告、校准证书中，也可以用校准因数、校准曲线来表示。对于同一台测量设备的同一参数、指标，通过校准、核查发现超差，要及时进行调整、修理、降级使用，直至报废处理。

（2）校验装置与仪器

一是仪电校验设备，有检验台，比如热电偶热电阻检定装置、热工检定系统、气动仪表校验台、微机热电阻检定装置、电动变送器校验台、直流信号校准器、称量校准器、温度校验台等。有仪表回路校验仪，比如过程信号校验仿真仪、热工校验仪、热工宝典等。有信号发生器，比如频率发生器、数显信号发生器、校验信号发生器、仿真器、全功能校验仪等。有压力类校验器，比如精密数字压力计、压力校验仪、压力表校验仪、过程信号校验仪、低真空测试仪。有温度类校验器，比如制冷恒温槽、水槽、热电偶校验仿真仪、微机热电阻检定装置等。

二是几何量的校验设备，比如扭簧式比较仪、测厚仪、静力触探仪、三等金属线纹尺、标准钢卷尺、步矩规等。

三是力学校验设备，如活塞压力计、质量比较仪、能量发生器百分比精密天平、微差压检定装置、振动冲击分析系统、流量校准装置等。

四是热学校验设备，如温度湿度采集系统、标准恒温槽、湿度检定箱、温度校验仪等。

五是理化校验设备，如酸度计校准装置、电导率校准装置、紫外可见分光光度计校准装置、标准色板、标准光泽度板、标准旋光管、标准灯等。

六是时间校验设备，如铯原子频标、频标比对器、相位噪声测试系统等。

七是无线电校准设备，如信号分析仪、网路分析仪、信号源等。

（3）校准报告的内容

校准报告主要包括：计量校准实验室名称、地址、校准地点、证书或报告的唯一性标识，每页及总页数标识，仪器检测、仪器维修、仪器校正、计量校验，电磁流量计检测、隔膜压力变送器维修、封开计量检测、送校单位名称和地址，被校对象的描述和标识，仪器校验日期，校准结果有效性，对校准所依据的技术规范名称及代号，本次校准所用计量检测测量标准的朔源性及有效性说明，校准环境的描述，仪器校准结果及测量不确定度的说明，校准证书或校准报告签发人的签名、职务或等效标识，签发日期，校准结果仅对被校对象有效的声明，未经实验室书面批准不得部分复制证书或报告的声明等。

2. 停止使用

（1）对于偶尔出现的计量性能异常的情况，要立即暂停使用，进行标识，先查清楚问题所在，然后继续使用。

（2）在测量过程中偶尔发生的一些异常，如读错、记错、仪器突然波动等，对于异常值可以立即删除。

（3）如果不能确定哪个测量值发生异常，可以采用统计方法来进行判别。针对一个具体的测量仪器，找到一个量值比较稳定的核查标准并进行连续定期观测。根据定期观测结果，计算数据的平均值、标准差、极差等变化情况，以此推断出测量过程是否处于统计控制状态。对于在测量过程中，通过测量数据的状态趋势分析、控制图分析等，显示某些参数正在超出预定控制限的测量设备，要立即停止使用，进行标识，查找问题，纠正后进行校准或检定，然后继续使用。

3. 保持适当环境

（1）对某些测量仪器要保持适当的环境，包括温度、湿度、灰尘粒子等，进行防锈、防尘、防潮等。室内环境要清洗、通风、散热等，要做好防护工作。

（2）做好保养工作。对于不常用的测量仪器要定期进行部位检测、性能检测，电子仪器要进行通电，保证良好的工作状态。机械设备要进行清洁、润滑、防腐等工作。

4.2.3 测量方法异常

测量方法的异常，属于系统误差。一旦测量方法出现异常，那么无论测量是如何进行、测量多少的次数，其结果都是异常的。

1. 测量方法异常的原因

（1）测量方法本身不完善。测量的操作规程（方法）没有得到确认和批准；没有根据新的测量要求、条件等进行及时更新。

（2）测量标志（主要指临时性的）设置错误。

（3）众多的偶然误差合并所导致。由于各种不确定的细小因素的客观存在，在一定的条件下它们相互作用，从而形成较大的异常。

（4）测量仪器与测量内容不相符。不同规格、型号的测量仪器，其测量方法、测量范围、精度等各不相同，但具体测量时，没有与相应的测量要求匹配。

（5）在执行测量时，对于测量仪器上的注意事项没有引起重视，导致测量出现异常。

2. 测量方法异常的应对措施

（1）测量前要确保测量的操作规程（方法）的正确性，要经过确认和批准。测量要求、条件等发生变化要及时调整更新测量操作规程（方法）。

（2）保持测量标志的正确性，永久性的测量标志要保护好，临时性的测量标志也要确保正确。

（3）确保测量过程中的每一个步骤执行完整、正确，不发生每一个步骤的细小误差，避免积累后形成较大异常。

（4）要选择与测量要求、测量目的相匹配的测量仪器，型号和规格要适当。

（5）关注测量仪器上的注意事项，遵照执行。

（6）在测量过程中，要小心使用测量仪器，仪器避免掉落、挤压、震动、冲击、磨损。

（7）保持操作的清洁，及时清除测量仪器上及测量仪器周围的各种垃圾、灰尘。

（8）在测量仪器适宜的环境温度下进行测量。

（9）定期校准测量仪器。

4.2.4 测量者导致异常

1. 测量者导致异常的原因

（1）测量人员能力不足，没有接受正确的相关培训。有的测量人员缺乏操作经验，测量动作不正确。

（2）测量人员面对新的测量仪器，不熟悉测量仪器的操作要求。

（3）因测量人员的性格、个人倾向性等造成测量异常，是属于测量感觉上的差异所造成。

（4）测量过程中，测量人员误读数值、记录错误、计算错误等。比如对于模拟式的仪表、带有刀形指针的仪表，眼睛的视线应当经指示器尖端与仪表度盘垂直，否则会因视线不垂直而读数错误；带有镜面标度尺的仪表，眼睛的视线应该经指示器的尖端与镜面反射像重合，否则因视觉角度误差而误读。

（5）测量人员个体情绪的波动引起的测量误差。或者在不适的环境下精力不集中等造成测量误差。

（6）相关测量人员没有按照操作规程和要求来进行测量。比如一些测量仪器有恒温恒湿的时间要求，但具体测量时没有达到规定的时间。

（7）测量人员的身体体征对于测量结果的干扰。比如一些精密天平的使用，由于人体的呼吸而造成数值波动。有时环境温度达标，但手的温度过高或过冷使测量物体发生温差，如手握量块会导致温度敏感变化（金属材料有线膨胀系数），给千分尺检定带来零位误差。

（8）长时间测量工作造成身体疲倦，影响测量结果。

（9）两个人及多人进行测量时，相互之间缺少默契配合。由于各自对于操作规程理解的差异、熟练程度的差异、反应速度的差异等，造成测量结果异常。

2. 测量者异常的应对措施

（1）测量人员必须接受测量相关培训，掌握测量的知识。

（2）面对新的测量仪器，要尽快熟悉相关测量操作要求，不能盲目测量。

（3）在执行测量时，测量人员要保持平和的情绪，避免周围环境的干扰。

（4）在执行测量时，测量人员要有客观、公正的心态，避免个人倾向。

（5）测量过程中，要有正确的测量姿势，要防止眼睛位置不同、读取刻度时的视差等引起误差。

（6）测量时，必须严格按照测量的操作规程和要求进行测量。

（7）取得测量数据后，不做任何的修饰。

（8）时间较长的测量，要注意测量人员的休息等。

（9）多人测量时，要加强沟通、磨合，通过比对试验找到最佳的测量人员组合。

4.2　Abnormal Measurement

———

The causes of measurement anomalies are different, which can be mainly divided into three aspects: abnormal measurement instrument, abnormal measurement method, and abnormal surveyors.

4.2.1 Abnormal measuring instrument

1. Connotation of the measuring instrument

Measuring instrument is also called measuring devices, which is the general name of all technical tools or devices used to measure and obtain the quantity value of the measured object. Strictly speaking, measuring instruments are collectively referred to as measuring instruments and measuring tools. In this paper, measuring instruments cover measuring tools. Measuring instruments are tools to understand and transform the material society. In metrology, measurement is not only the core concept, but also the object of study. Measurement is realized by measuring instruments. Therefore, measuring instruments are the basis of measurement, and the quality of measuring instruments is an important condition to ensure the quality of measurement results.

Measuring instruments play an important role in the production process of organizations at all levels and are important labor tools. In general machinery type factories, the number of universal measuring tools is about 6 times that of metal cutting machines, and the number of measuring tools in use is about 1.5 times that of processing personnel. Various length, heat, force, electricity, physical and chemical measuring instruments account for a large proportion of fixed assets in an enterprise.

2. Types of measuring instrument

There are many kinds of measuring instruments, including geometric quantity measuring instruments, mechanical measuring instruments, thermal measuring instruments, chemical measuring instruments, time and frequency measuring instruments, optical measuring instruments, electromagnetic measuring instruments, radio measuring instruments, etc. A brief introduction is as follows.

(1) Geometric quantity measuring instrument, which can be used for angle measurement, roughness measurement, flatness measurement, straightness measurement, etc. There are many kinds of geometric measuring instruments, such as measuring block, micrometer, universal caliper, ruler, feeler gauge, tape measure, gauge, needle gauge, indicating measuring tool, linear instrument, various levelers, thickness gauges, various microscopes, optical meters, projectors, interferometers, test sieves, welding inspection ruler, cable meter, runout tester, altimeter, pneumatic measuring instrument, large range dial indicator, various distance

measuring instruments, various calipers, test molds, levels, inclinometers, test templates, test round rods, theodolites, angle blocks, shaft angle encoders, centering instruments, three-dimensional scanners, extensometers, inclinometers, line projectors, bite force testers, spot welding analyzers, scribers, fiber fineness analyzers, cone and hollow end diameter meters, polyhedrons, gear measuring instruments, verticals, water level meters, bar code detectors, etc.

（2）Mechanical measuring instruments can measure mass, force value, energy, capacity, speed, hardness, pressure, vibration, density, flow, etc. Mechanical measuring instruments mainly include: weights, various indicating scales, various scales, torque wrenches, various pressure gauges, pressure transmitters, force measuring instruments, tensile pressure testing machines, various hardness meters (metal Rockwell, Brinell, Vickers, Leeb, Babbitt, Shaw), tachometers, constant load hardness meters, pressure gauges, impact testing machines, impact testing machines, impact testing platforms, torque scales, various vibration testing platforms, specific gravity scales Dynamometer, various universal testing machines, micro manometers, refrigerant leak detectors, differential pressure leak detectors, accelerometers, vibration meters, tensiometers, test chambers, pressure heads, level gauges, brittleness and fragility tester, abrasion resistance tester, barometer, pipette, pile foundation dynamic tester, pycnometer, simulated transportation test bench, speed sensor, multi-component force sensor, physical performance (durability, fatigue, life) performance tester, loading scale, various flow meters (ultrasonic, electromagnetic, liquid float, liquid mass, liquid volumetric, turbine, soap film, wet gas, hot gas), barometer, impact gauge Interface tensiometer, gas density controller, dynamic balance measurement analyzer, elevator speed limiter tester, noise detector, locomotive speed tester, axle load meter, torque meter, torsion tester, angular speed tester, centrifuge, densitometer, noise vibrometer, ionization vacuum meter, discontinuous accumulation automatic weighing instrument, batching scale, friction testing machine, torque multiplier, linear speed measuring instrument, weight container, steelyard, oil and gas recovery detector, mass comparator, bolt detector, dilution and matching instrument, bottle mouth liquid separator, speed sensor, etc.

（3）Thermal measuring instrument can measure temperature, humidity, heat flow, etc. Thermal measuring instruments mainly include: temperature regulators, various thermocouples, various thermometers (standard mercury, glass liquid, industrial thermal resistance), heat treatment furnaces, temperature indicators, light incubators, natural ventilation aging test chambers, industrial process measurement recorders, bimetal thermometers, pressure thermometers, mechanical thermometers, radiation thermometers, temperature transmitters, temperature itinerant detectors, thermal imagers, thermostatic baths, environmental testing equipment, surface thermometers, salt test equipment, armored thermocouple, temperature calibrator, resistance furnace, rain test equipment, temperature acquisition instrument, temperature calibrator, ozone aging test box, weldability tester, microwave digestion instrument, vacuum drying box, freeze-drying machine, magnetic stirrer, humidity sensor, ventilation psychrometer, process calibrator, dew point meter, refrigeration appliance, rail thermometer, temperature switch, ultraviolet aging box, thermistor tester, softening breakdown tester, sunshine climate color fastness tester, high temperature dynamic aging system, corrosion gas test equipment, asphalt elongation tester, sand and dust test equipment, digital temperature and humidity meter, infrared thermometer, radiation source tester, solar radiation test box, humidity generator, etc.

(4) Chemical measuring instruments can be used for spectral analysis, chromatographic analysis, mass spectrometry analysis, water quality analysis, gas analysis, biochemical analysis, thermochemical analysis, element analysis, etc. Chemical measuring instruments mainly include ultraviolet visible near-infrared spectrophotometer, infrared spectrophotometer, Fourier transform infrared spectrometer, emission spectrometer, fluorescence photometer, flame photometer, atomic absorption spectrophotometer, polarimeter, handheld refractometer, polycrystalline x-ray diffractometer, mercury detector, spectrometer, pesticide residue detector, micro spectrophotometer, gas chromatograph, gel chromatograph, ion chromatograph, liquid chromatograph, capillary electrophoresis apparatus, thin-layer chromatography scanner, (gas phase, liquid phase) chromatography-mass spectrometer, plasma mass spectrometer, time-of-flight mass spectrometer, oil concentration analyzer in water, turbidity meter, dissolved oxygen analyzer, BOD analyzer, COD analyzer, total organic carbon analyzer, ammonia nitrogen detector, nitrate detector, silicate analyzer, phosphate analyzer, residual chlorine analyzer, heavy metal analyzer, permanganate detector, free chlorine analyzer, water hardness meter, alkyl mercury analyzer, micro total organic carbon analyzer, urea detector, bacterial turbidity analyzer, water quality analyzer, petroleum product moisture determination process, drying method moisture determination process, wood moisture content meter, various viscometers, various acidity meters, melt flow rate meter, pH meter, laboratory ion meter, potential titrator, resistivity meter, conductivity meter, various salinity meters, electrophoresis apparatus, total dissolved solids detector, oscillopolarograph, titrator, ozone analyzer, carbon monoxide detector, carbon dioxide analyzer, sulfur dioxide gas detector, hydrogen sulfide gas detector, smoke sampler, methane detector, ammonia detector, gas sampler, dust particle counter, liquid particle counter, suspended particle sampler, air microbe tester, particle detector, particle size analyzer, gel imaging system, bacterial endotoxin analyzer, antibiotic analyzer, polymerase chain reaction analyzer, colony counter, open and closed flash point tester, heat analyzer, melting point tester, acid value tester, condensation point determination process, aniline point tester, sulfur tester, dissolution tester, nitrogen analyzer, freezing point tester, permeability tester, oxygen permeability tester, etc.

(5) Time frequency measuring instrumentcan measure physical quantities related to time and periodic movement. Time frequency measuring instruments mainly include: frequency meter, universal counter, frequency analyzer, quartz crystal frequency standard, timer, frequency standard comparator, time relay, clock analyzer, rubidium atomic frequency standard, global satellite navigation system receiver, vehicle traveling recorder, vibrating wire frequency reader, time interval meter, time verification instrument, standard digital clock, synchronous clock system, GNSS time synchronization system, time interval generator , time synthesizer, clock tester, velocimeter, pulse counter, resonant wavelength meter, etc.

(6) Optical measuring instrument can study the behavior and property of light, and measure the luminosity, chromaticity, radiance, laser parameters, urban and rural optics, light source, light power, light attenuation, laser energy, etc. Optical measuring instruments mainly

include: optical power meter, optical attenuator, spectral analyzer, optical wavelength meter, optical time domain reflectometer, optical fiber fusion splicer, optical multimeter, optical return loss meter, optical fiber polarization mode dispersion tester, Abbe refractometer, glossmeter Haze meter, color analyzer, color temperature scale, luminous intensity tester, ultraviolet analyzer, distribution photometer, LED light intensity distribution tester, reflectivity measurement process, reflective densitometer, ultraviolet analyzer, photobiosafety tester, ultraviolet exposure meter, optical transmittance tester, petroleum product color analyzer and color comparison plate, colorimeter, standard color plate, interference filter, water quality colorimeter, spectrophotometer standard filter, laser attenuator, black and white density sheet, laser energy meter, optical fiber identifier, polarization dependent loss tester, optical fiber refractive index distribution and geometric parameter tester, passive optical network power meter, corneal curvature meter, optical fiber loss and mode field diameter tester, etc.

（7）Electromagnetic measuring instruments can measure physical quantities related to electromagnetic phenomena, including testing voltage, current, resistance, capacitance, inductance, magnetic induction intensity, magnetic flux, magnetic moment, etc. Electromagnetic measuring instruments mainly include: electrostatic tester, resistance tester, high insulation resistance tester, DC high-voltage high value resistor, DC low ammeter, micro ohmmeter, DC digital bridge, grounding resistance meter, withstand voltage tester, leakage current tester, AC parameter tester, power meter, multi-function calibration source, AC parameter tester, AC/DC meter calibrator, high-voltage static voltmeter, ammeter, DC resistor, DC bridge Insulation resistance meter, voltage transformer, DC current source, transformer load box, digital high-voltage meter, AC electric energy meter, access type electric energy indicator, electric quantity transmitter, power frequency single-phase indicator, fluxmeter, magnetometer, high-voltage resonance test device, transformer no-load loss tester, transformer short-circuit impedance tester, triple frequency test transformer device, power frequency parameter tester of transmission line, transformer comprehensive characteristic tester, DC high voltage divider, power frequency high voltage divider, dielectric strength tester of insulating oil, lightning strike counter calibrator, voltage regulator, transformer, spot check device of comprehensive tester, electronic ballast performance analyzer, zinc oxide arrester tester, protection circuit vector analyzer, large grounding grid grounding resistance tester, electricity recording analyzer, high-voltage capacitance bridge, shunt, current scale, voltage scale, power scale, resistance scale, loop resistance tester, direct resistance meter, power frequency magnetic field generator, resistance strain gauge, power indicator range calibrator, transformer winding resistance deformation tester, current voltage sensor, power quality analyzer, dielectric loss and volume resistivity tester of insulating oil, voltage detector verification device, cable fault locating power supply, electroscope, DC voltage transformer, magnetic particle flaw detector, AC charging equipment, voltage detector, weak magnetic field alternating magnetometer, test transformer operation box, ultra-low frequency and high voltage generator, zinc oxide arrester, DC parameter tester, analog AC and DC standard resistor, magnetic yoke type magnetic particle flaw detector, low-voltage electroprobe, eddy current flaw detector, multimeter, clamp type leakage current meter, magnetic magnetometer, transformer calibrator, voltage transformer, transformer core grounding current tester, capacitive equipment detection device, withstand voltage test calibrator, AC electric energy meter verification device, winding line breakdown voltage tester, emulsified asphalt particle ion charge tester, fault filter analysis device, network charging facility access node electrical energy and timing device, DC transmitter, constant magnetic field coil, high voltage withstand voltage tester, contact current tester, DC high-voltage test device, electromechanical AC electric energy meter, eddy current conductivity meter, electric detonator tester, AC peak voltmeter, data acquisition instrument, insulation resistance meter multi-function test box, power analyzer, etc.

(8) Radio measuring instruments can be used for testing electromagnetic wave, high-frequency and microwave power, 5G equipment, etc. Radio measuring instruments include LCR digital bridge, AC bridge, standard capacitor, standard inductor, semiconductor tube characteristic plotter, capacitance sorter, capacitor capacity loss sorter, semiconductor characteristic plotter calibrator, inductance box, AC resistance box, analog oscilloscope, oscilloscope calibrator, electronic voltmeter, RF voltmeter, digital oscilloscope, network analyzer and calibrator, RF impedance material analyzer, low-frequency signal generator, pulse signal generator, RF communication tester, spectrum analyzer, function arbitrary wave signal generator, modulation meter, noise figure analyzer, code error tester, procedure analyzer, spectrum analyzer, vector signal analyzer, medium power meter, TV field strength level detector, distortion meter, telephone analyzer, TV signal generator, audio analyzer, attenuator, light analog cable, digital transmission analyzer, power indicator, digital mobile communication integrated tester, dynamic signal analyzer, electromagnetic interference measuring receiver, session characteristic analyzer, antenna feeder analyzer, shake tester, bluetooth tester, high-speed serial bit error tester, frequency characteristic tester, frequency selective level meter, level oscillator, noise generator, intermittent interference analyzer, network line analyzer, logic analyzer, RF microwave switch, vector oscilloscope, data network performance tester, electric field tester, field strength probe, electromagnetic radiator, distortion calibrator, timing noise analyzer, spike signal generator, pulse magnetic field generator, electromagnetic emission and sensitivity pulse signal generator, optical oscilloscope, radio frequency identification tester, wireless channel simulator, optical relay protection tester, coupling decoupling network, impedance stabilization network, artificial power network, sampling oscilloscope, electromagnetic interference generator, oscillating wave generator, air line, high-definition video signal generator, frequency response analyzer, modulation domain analyzer, etc.

3. Reasons for abnormal measuring instruments

There are many reasons for the abnormality of measuring instruments, mainly include:

(1) Abnormalities due to the structure of the measuring instrument. Due to changes in friction, measuring pressure, etc., the structure of each part is not adjusted in time, resulting in abnormalities.

(2) Abnormalities caused by changes in measurement conditions. For example, changes in temperature, humidity, illumination, vibration, etc. cause abnormalities of measuring instruments.

(3) The connection wires of measuring components and circuits are loose, causing abnormalities; Inadequate dustproof and sealing performance of measuring instrument lead to poor accuracy.

(4)The measuring instrument lacks maintenance for a long time, and its stability is poor.

(5)The measured ratio is incorrect. Some measurements involve some solution proportions, and the results of measuring instruments are abnormal due to the abnormal proportions.

(6)Measuring equipment significantly exceeds its allowable range, resulting in abnormal measurement results.

4. Inspection method for measuring instrument abnormality

The main methods to judge whether the measuring instrument is abnormal are as follows.

(1)Direct observation. The human sensory organs are used to directly find out the place where the abnormality occurs. For example, the measuring instrument has broken wires, poor contact, overheating and smoking of components, lack of oil in mechanical transmission components, wear, excessive clearance between components, abnormal noise, etc.

(2)Circuit parameter measurement. Measure the voltage, current and resistance of the circuit with a multimeter and compare them with the normal values to determine the abnormal parts.

(3)Inspection of test instruments. Use special test instruments to test the measuring instruments and find abnormal points.

(4)Narrow the scope for inspection. For parts and plug-ins in question, replace them with normal similar parts and plug-ins to narrow the scope of exceptions and find abnormal parts and plug-ins in the process of replacement. The circuit in question can also be removed from the measuring instrument, and the abnormal range can be reduced by observing the change of current.

(5)Signal search. For questionable measuring instruments, select signals output by different signal generators and observe the transmission of signals in the circuit step by step to find abnormal points.

(6)Waveform search. Use an oscilloscope to observe the waveforms of the circuits and components of the measuring instrument and compare them with normal waveforms to determine whether they are abnormal.

(7)Short circuit search. For interference, self excitation and other abnormalities, short circuit between two points in the circuit temporarily. If the abnormality disappears after short circuit, it indicates that the abnormality is before the short circuit point; If there is still something abnormal after the short circuit, continue to search after the short circuit point.

(8)Find with inspection mode. Draw the control chart for the data in the measuring equipment, and use the inspection mode to judge whether it is abnormal. The abnormality is usually reflected in that the measuring points exceed the control limit, and the distribution of measuring points is not random. In the international standard ISO 8258:1991 and the national standard GB/T 4091-2001, eight distribution modes are specified for common measurement process anomalies (refer to the Statistical Application of Quality Management). If the average value control chart is abnormal, it indicates that the measurement process is affected by uncontrolled systematic effects; If the standard deviation control chart is abnormal, it indicates that the measurement process is affected by uncontrolled random effects.

4.2.2 Countermeasures for abnormal measuring instruments

1. Calibration of measuring instruments

(1) The role of calibration. Instrument calibration refers to assigning values to the characteristics of measuring instruments, including physical measuring tools, with reference measurement standards under specified conditions, and determining their indication errors. The value represented by the measuring instrument is then derived from the value reproduced by the measurement standard according to the comparison chain or calibration chain. The calibration is based on the calibration specifications and methods, which are generally uniformly specified, and can be specified under special circumstances. Calibration results are recorded in calibration reports and certificates, and can also be represented by calibration factors and calibration curves. For the same parameters and indicators of the same measuring equipment, if they are found out to be out of tolerance through calibration and verification, they shall be adjusted, repaired, degraded and used in a timely manner until they are scrapped.

(2) Calibration devices and instruments. The first is the instrument and electrical calibration equipment. There are inspection platforms, such as thermocouple thermal resistance verification device, thermal verification system, pneumatic instrument verification platform, microcomputer thermal resistance verification device, electric transmitter verification platform, DC signal calibrator, weighing calibrator, temperature verification platform, etc. There are instrument loop calibrators, such as process signal calibration simulator, thermal calibrator, thermal dictionary, etc. There are signal generators, such as frequency generator, digital display signal generator, calibration signal generator, emulator, full function calibrator, etc. There are pressure calibrators, such as precision digital pressure gauge, pressure calibrator, pressure gauge calibrator, process signal calibrator, and low vacuum tester. There are temperature calibrators, such as refrigeration thermostatic bath, water tank, thermocouple calibration simulator, microcomputer thermal resistance calibration device, etc.

The second is the calibration equipment of geometric quantities, such as torsion spring comparator, thickness gauge, static penetrometer, third class metal line ruler, standard steel tape, step gauge, etc.

The third is mechanical calibration equipment, such as piston pressure gauge, mass comparator, percentage precision balance of energy generator, micro differential pressure verification device, vibration impact analysis system, flow calibration device, etc.

The fourth is thermal calibration equipment, such as temperature and humidity acquisition system, standard thermostatic bath, humidity verification box, temperature calibrator, etc.

The fifth is physical and chemical calibration equipment, such as acidity meter calibration device, conductivity calibration device, UV visible spectrophotometer calibration device, standard color plate, standard gloss plate, standard polarimeter, standard lamp, etc.

The sixth is time verification equipment, such as cesium atomic frequency standard, frequency standard comparator, phase noise testing system, etc.

The seventh is radio calibration equipment, such as signal analyzer, network analyzer, signal source, etc.

(3) Contents of calibration report. The calibration report mainly includes: the name, ad-

dress, calibration location, unique identification of certificate or report, identification of each page and total pages, instrument testing, instrument maintenance, instrument calibration, metrological verification, electromagnetic flowmeter testing, diaphragm pressure transmitter maintenance, sealing measurement testing, name and address of the sending unit, description and identification of the calibrated object, instrument calibration date, validity of calibration results. The name and code of the technical specification on which the calibration is based, the description of the source and validity of the measurement, detection and measurement standards used in this calibration, the description of the calibration environment, the description of the instrument calibration results and measurement uncertainty, the signature, job title or equivalent identification of the issuer of the calibration certificate or calibration report, the issuing date, and the statement that the calibration results are only valid for the object being calibrated. The statement of certificate or report shall not be partially copied without the written approval of the laboratory.

2. Stop using

(1) For the occasional abnormal metering performance, it is necessary to immediately suspend the use, identify the problem, find out the problem, and then continue to use.

(2) For some anomalies that occasionally occur during the measurement process, such as reading errors, memory errors, sudden fluctuations of the instrument, the abnormal values can be deleted immediately.

(3) If we cannot determine which measurement value is abnormal, you can use statistical methods to distinguish. For a specific measuring instrument, find a relatively stable verification standard and conduct continuous regular observation. According to the regular observation results, calculate the average value, standard deviation, range and other changes of the data to infer whether the measurement process is under statistical control. During the measurement process, the measuring equipment that shows that some parameters are exceeding the predetermined control limit through the state trend analysis and control chart analysis of the measurement data shall be immediately stopped, identified, find out the problem, calibrated or verified after correction, and then continue to use.

3. Maintain an appropriate environment

(1) Some measuring instruments shall be kept in an appropriate environment, including temperature, humidity, dust particles, etc., and protected from rust, dust, moisture, etc. The indoor environment shall be cleaned, ventilated and cooled, and the fire prevention work shall be done well.

(2) Maintain the instrument. For the measuring instruments that are not commonly used, the position inspection and performance inspection shall be carried out regularly, and the electronic instruments shall be powered on to ensure a good working condition. Mechanical equipment shall be cleaned, lubricated and protected against corrosion.

4.2.3 Abnormal measurement method

The abnormality of measurement method belongs to systematic error. Once the measurement method is abnormal, no matter how the measurement is carried out or how many times it is measured, the results will be abnormal.

1. Reasons for abnormal measurement methods

(1) The measurement method itself is not perfect. The operating procedures (methods) for measurement have not been confirmed and approved; It was not updated in time according to new measurement requirements and conditions.

(2) The measurement mark (mainly temporary) is set incorrectly.

(3) Many accidental errors are combined. Due to the objective existence of various uncertain small factors, they interact under certain conditions, thus forming a larger anomaly.

(4) The measuring instrument is inconsistent with the measurement content. The measuring method, measuring range and accuracy of measuring instruments of different specifications and models are different, but the specific measurement does not match the corresponding measurement requirements.

(5) During the measurement, attention was not paid to the precautions on the measuring instrument, which caused the measurement to be abnormal.

2. Countermeasures for abnormal measurement methods

(1) Before measurement, ensure the correctness of the measurement operation procedures (methods), which shall be confirmed and approved. In case of changes in measurement requirements and conditions, timely adjust and update the measurement operation procedures (methods).

(2) The correctness of survey marks shall be maintained. Permanent survey marks shall be well protected, and temporary survey marks shall also be correct.

(3) Ensure that each step in the measurement process is carried out completely and correctly, without small errors in each step, and avoid large abnormalities after accumulation.

(4) The measuring instrument matching the measurement requirements and purposes shall be selected, and the model and specification shall be appropriate.

(5) Pay attention to the precautions on measuring instruments and follow them.

(6) During the measurement, the measuring instrument shall be used carefully to avoid falling, squeezing, vibration, impact and abrasion.

(7) Keep the operation clean, and timely remove all kinds of garbage and dust on and around the measuring instrument.

(8) The measurement shall be conducted at the appropriate ambient temperature of the measuring instrument.

(9) Calibrate measuring instruments regularly.

4.2.4 Abnormal caused by the measurer

1. The reasons of the abnormality caused by the measurer

(1) The measurers are not competent enough and did not receive the correct training. Some researchers lack operation experience and measure incorrectly.

(2) The measurers are not familiar with the operation requirements of new measuring instruments.

(3) The measurement abnormality is caused by the personality and personal tendency of the measuring personnel, which is caused by the difference in measurement sense.

(4) During the measurement, the measurers misread the values, recorded errors, calculated errors, etc. For example, for analog instruments and instruments with knife shaped pointers, the

line of sight of the eyes should be vertical to the instrument dial through the tip of the indicator, otherwise the reading will be wrong because the line of sight is not vertical; For an instrument with a mirror scale, the line of sight of the eye should coincide with the mirror reflection image through the tip of the indicator, otherwise it will be misread due to visual angle error.

(5) The measurement error caused by the fluctuation of the individual emotion of the measuring personnel. Or the measuring personnel lack concentration in uncomfortable environments, etc.

(6) The relevant surveyors did not measure according to the operating procedures and requirements. For example, some measuring instruments have time requirements for constant temperature and humidity, but the specific measurement time does not reach the specified standards.

(7) The physical signs of the measurers interfere with the measurement results. For example, the use of some precision scales may cause numerical fluctuations due to human respiration. Sometimes the ambient temperature reaches the standard, but the temperature of the hand is too high or too low, which causes the temperature difference of the measured object. For example, holding the measuring block in the hand will cause temperature sensitive changes (linear expansion coefficient of metal materials), which will bring zero error to the micrometer verification.

(8) Long time measurement work causes physical fatigue and affects measurement results.

(9) There is a lack of tacit cooperation between two or more people when measuring. The measurement results are abnormal due to the different understanding of the operating procedures, proficiency and speed of reflection.

2. Countermeasures for measurer's abnormality

(1) Measurers must receive training on measurement and master the knowledge of measurement.

(2) When dealing with the new measuring instruments, it is necessary to get familiar with relevant measuring operation requirements as soon as possible, and blind measurement is not allowed.

(3) During the measurement, the measurers shall keep calm and avoid the interference of the surrounding environment.

(4) When carrying out the measurement, the measuring personnel should have an objective and fair attitude and avoid personal inclination.

(5) During measurement, correct measurement posture shall be provided to prevent errors caused by different eye positions and parallax when reading scales.

(6) The measurement must be carried out in strict accordance with the operating procedures and requirements for measurement.

(7) After obtaining the measurement data, do not make any modification.

(8) For long time measurement, pay attention to the rest of the measuring personnel.

(9) When measuring with more than one person, communication and collaboration shall be strengthened to find the best combination of measuring personnel through comparison test.

4.3　试验异常

———

试验异常是试验中经常遇到的问题，通常试验异常可以从试验设计异常、试验过程异常、试验人员异常等方面进行分析。

4.3.1 试验设计异常

试验设计（Design of Experiment）起源于农业试验，由英国生物统计学家费舍尔（Ronald Aylmer Fisher）在 20 世纪 20 年代创立。1935 年，费舍尔出版《试验设计》，正式标志试验设计的诞生。20 世纪 50 年代，日本统计学家田口玄一（G. Taguchi）创立了正交试验设计。我国数学家华罗庚提出了优选法。1978 年我国数学家王元、方开泰提出了均匀设计。

试验设计的异常由很多原因引起，分析时可以从"试验的因子、受试的对象、试验的效应"三个方面入手，而出现的问题通常在"样本（因子）是否随机、是否遵从重复原则、对照组是否真的具备对照、均衡原则是否得到执行"等几个方面展开。应该说，试验设计与统计紧密相连。试验设计异常的情形也有很多，下面予以简单介绍。

1. 用单因子试验设计替代了多因子试验设计

单因子试验设计是常见的试验设计方法之一，由于它设计简单，只有一个因子在发生变化，因此研究分析比较容易。而多因子在客观世界中真实存在，对于多因子的试验设计比单因子试验设计复杂很多，研究分析的过程也更难。由于单因子分析的优势，人们在进行试验设计和分析时，会有意或无意地用单因子试验设计代替了多因子试验设计，造成得出的试验结论不正确。

举例：

某水产养殖场饲养食用螺，饲养人员希望提高食用螺的增长情况。于是，他们打算先摸索出一种饲料添加配方，再探索出最佳的底泥、水体 pH 值、水温等环境，以促进食用螺的快速增长。为此，他们开展了相关试验，有关配方组合的试验数据见表 4-1。

表 4-1　配方组合试验数据表

配方	配方组别			
	配方一	配方二	配方三	配方四
玉米/%	10	×	×	×
豆饼/%	5	10	×	×
杂鱼饼/%	10	10	10	×
麦麸/%	15	10	20	30
肉末/%	10	20	20	20
螺增长/g	19	16	18	16

饲养人员分析：在配方一中，螺的增长最快，但配方所需的品种为五种，供货有点难度；配方二中，用到了豆饼、杂鱼饼、麦麸、肉末等四个品种，螺的增长一般；配方三中，用到了杂鱼饼、麦麸、肉末，螺的增长较好，且与配方一的结果接近；配方四中只用到了麦麸和肉末，螺的增长居中。因此，饲养人员选用配方三作为一种新的配方。

随后，饲养人员继续进行试验，首先就相同的底泥厚度下饲养螺，得到最佳的水体 pH 值为 7~8。接着，饲养人员在固定了水体 pH 值后，确定最佳的水温环境为 28~29 ℃。然后，饲养人员在固定了 pH 值、水温后，又确定了最佳的底泥厚度为 12~14 cm。

分析：如果单单从表面看饲养人员的整个试验，似乎无懈可击。饲养人员的思路是通过多次试验，逐一排摸出最佳饲料配方、最佳水体 pH 值、最佳水温、最佳底泥厚度。但实际上，饲养人员所采用的方法都是单因子试验设计。在研究配方时，饲养人员先锁定某个成分不用，看螺的增长情况。在后续的试验中，饲养人员还是先锁定其他因子的条件不变，然后得到所谓的最佳水体 pH 值等。本例中正确的试验方法应该是多因子试验设计，饲养人员可以通过正交试验设计来完成整个试验。另外，饲养人员也可以先进行均匀试验设计，然后采用正交试验设计来得出最佳组合。

2. 对照的两组之间违反了均衡原则

均衡原则要求在试验设计时，对照的两组之间除了要考察的试验因子取不同的水平之外，还要确保其他所有的可能干扰因子对各组产生的效应处于相等的水平。

举例：

某公司进行应急训练。选择了本部的 30 名青年员工为试验组、某片区项目上 30 名青年员工为对照组开展培训。在培训 2 天后，进行应急的相关考试，见表 4-2。

表 4-2　试验组与对照组应急考试结果　　　　　　单位：分

组别	培训前应急事故处理能力考核	培训后应急事故处理能力考核
试验组	52 ±8	86 ±5
对照组	61 ±6	87 ±4

因此，该公司得出结论，本次的应急培训取得了很好效果。

分析：该公司进行应急培训采用了对照试验的方法。但是，组织者对于对照组的选择不合理，没有遵循均衡原则。很明显的，如果本部的青年员工与某片区项目上的青年员工在应急的基本技能上是存在区别的，那么就无法进行相互对照比较。组织者应当把随机抽取本部的员工、随机抽取项目上的员工统一归拢起来，然后进行打乱，再随机各取一半人数，分别纳入试验组和对照组，这样才能保证试验的结果是可信的。

3. 对照过剩

有的试验设计时采用了对照方法，但对照过剩，使试验过多，增加数据分析工作量。而且多余的对照没有实际意义。

举例：

某制药公司需要进行 A、B 两种药物疗效的比较，设计了相应的对照试验。第一组为空白对照组，第二组只使用 A 药，第三组同时使用 A 药和 B 药，第四组为第三组的空白对照组，见表 4-3。

表 4-3　对照试验内容

试验组别	试验内容
第一组	空白对照
第二组	只使用 A 药
第三组	同时使用 A 药和 B 药
第四组	第三组的空白对照组

分析：这是属于对照过剩的设计，因为第一组和第四组的试验效应是相同的，试验者只需要采纳第一组的试验结果就可以了。实际上，还有更好的对照方法，试验者可以把第四组的试验内容改为"只使用 B 药"。这样就形成了 A 药、B 药两个因子，A 药、B 药各自都有"用与不用"两个水平，共 4 个试验条件，这个试验设计就成为了"两因子的析因设计"。

4. 因素和水平之间的概念混淆

在试验中，因子是被选中进行试验的因素，或者是它们的组合。水平是因子的某一个值、某一个级别、或者是某一种状态。比如某个试验中的一个因子是温度，那么，具体的 25 ℃、26 ℃、27 ℃就是因子温度的三个水平。

举例：

某公司要进行金属材料的黏结，但总是发生黏结不够牢固的情况。为此，技术人员想找到一种粘力较强的配方。胶水有 A、B、C 三种型号，黏结温度有 100 ℃、120 ℃、150 ℃三种情况，黏结时间有 10 s、30 s、60 s 三种情形。

技术人员取了 60 根金属材料进行试验，共分六组，每组 10 根材料。第一组是胶水 A，温度 100 ℃、黏结 10 s。第二组胶水 A，温度 120 ℃、黏结 10 s。第三组是胶水 B，温度 120 ℃，黏结 30 s。第四组是胶水 B，温度 100 ℃，黏结 30 s。第五组是胶水 C，温度 150 ℃，黏结 30 s。第六组是胶水 C，温度 150 ℃，黏结 60 s。

在试验过程中，技术人员还分别就胶水在每个金属材料的单位面积上的数量进行了比较，然后看胶水的黏结力（黏结力 1 MPa = 10.197 kg/cm²）。试验数据见表 4-4。

表 4-4　金属黏结试验数据

组别	涂胶水数量/（g·cm⁻²）	黏结力/MPa
第一组胶水 A，温度 100 ℃，黏结 10 s	5	2590
第二组胶水 A，温度 120 ℃，黏结 10 s	6	3060
第三组胶水 B，温度 120 ℃，黏结 30 s	6	3160

组别	涂胶水数量/（g·cm⁻²）	黏结力/MPa
第四组胶水 B，温度 100 ℃，黏结 30 s	6	2810
第五组胶水 C，温度 150 ℃，黏结 30 s	6	2680
第六组胶水 C，温度 150 ℃，黏结 60 s	7	3150

技术人员根据表 4-4 数据，得出结论为第三组的胶水 B、温度 120 ℃、黏结 30 s，胶水 B 的使用数量为 6 g/cm²，此时黏结力达到最好，为 3160 MPa。

分析：

技术人员混淆了试验中因子和水平的概念。比如：第一组与第二组中，胶水 A 是两种水平（5 和 6），温度是两种水平（100 ℃和 120 ℃）；第二组与第三组中，胶水 A 与胶水 B 是两个因子，但都只有一个水平；第四组与第五组中，胶水 B 和胶水 C 是两个因子，且都只有一个水平，温度是两个水平（100 ℃和 150 ℃）。这些因子和水平之间的安排相当混乱，有些是同一个因子有两个水平；有些是不同的因子，但都各有一个水平；有些是不同的因子，不同的水平等。因此，这样的分析结果其可信度是打折扣的。

5. "重复试验" 没有按照重复原则设计

重复试验中的"重复"，是指在试验过程中应当遵循重复原则。重复原则是指在相同的试验条件下，做两次或两次以上的独立试验。这里的独立是指用不同的样品做试验。比如重复取样，就是从同一个总体中多次取样；重复测量，就是对某个客观对象，随着时间的推移进行多次测量。

举例：

某公司进口一台断层扫描仪 A（价格昂贵），为了检验其断层扫描的精度和质量，技术人员用公司内原有的断层扫描仪 B（价格相对便宜）进行比较。扫描的对象是某公司生产的一块钢材。

第一次试验，技术人员使用断层扫描仪 A 和 B，分别对钢材各进行了 2 次扫描试验，并检测相应的精度、质量指标。

过了两个星期，技术人员使用断层扫描仪 A 和 B，第二次分别对原来的钢材继续各进行了 2 次扫描，并检测相应的精度、质量指标。

根据两次扫描试验的结果，技术人员进行了统计分析，发现两次的精度、质量指标都没有显著性差异。因此，技术人员认为，今后采购断层扫描时，使用型号 B 即可，可以节约大笔开支。

分析：

在本例中，技术人员试验使用的是断层扫描仪 A 和 B，试验的样品对象保持不变，都是同一块钢材。技术人员所得出的结论属于从"个别"推算出"一般"的结论，没有遵循重复原则进行试验。正确的试验做法是，技术人员应当在同一天时间内，对钢材进行两次重复取样，然后使用不同的钢材样本来开展试验。

6. 其他试验异常的情形

（1）受试对象选用不当。试验者没有结合试验的具体情况来确定"受试对象"的标准，受试对象的纳入和剔除显得随意，增加了非试验因子的干扰。

（2）观测指标选用不当。观测指标未能很好地反映试验的效应。观测指标是用来反映试验因子的作用是否强弱的指标，要结合试验因子的性质、特点等，综合考虑仪器、技术水平等条件，找出特异性强、灵敏度高、准确、可靠的观测指标。观测指标应当强调客观。

（3）试验的成分配比异常。试验用的因子配比没有统一按照要求执行，导致试验缺少可比性。

（4）正交试验没有按照正交表执行。有时候还发生有的试验人员把因子和水平颠倒的情形。

4.3.2 试验过程异常

1. 试验过程异常的情形

试验过程中发生异常的情况有很多情形，比如：

（1）突然停电。

（2）试验用的仪器、仪表等发生故障。比如电路接触不良、电压表无指示等。

（3）试验用的机械、设备等发生故障。

（4）试验的样品异常，如样品不符合要求、样品包装破损、样品丢失、试剂纯度不够等。

（5）试验中产生异味，且不明异味的来源和原因。有些可能是有毒气味等。

（6）试验人员自身发生身体不适情形，不能正常开展试验工作。有时是试验人员受到了外界的干扰，使试验结果可能失真。

（7）试验的环境不符合相关要求，包括温度、湿度、照度、粉尘等。

（8）试验中发生火苗，有可能导致火灾。有时会发生爆炸等。

（9）试验的周期上发生了变化，比如时间上突然压缩等。

2. 试验过程异常的应对方法

（1）试验过程发生异常，如停电、仪表故障等，应立即终止试验，切断电源，查明原因，及时向相关负责人报告，原有的试验结果作废。但试验的全部数据应当予以保存。

（2）试验之间对于相关机械设备进行检查。一旦发生故障，应立即停止试验，及时向相关负责人报告，原有试验结果作废，保存好所有的数据。

（3）样品送达后，在试验之前应进行检查。使用的试剂品质等要确保合格、有效。发生样品异常，要及时告知送样人员，用合格样品来进行试验。

（4）试验过程中，要坚持有人值守。特别是有加热现象的，防止烫伤、火灾。试验人员应掌握灭火技能和灭火器使用方法，定期检查灭火器材。

（5）试验中要防止有毒物质的排放、泄露，不对环境构成污染。试验过程中发现有异味产生，务必查明原因。要开窗通风，疏散人员，防止中毒。要佩戴防毒用具清理现场。

（6）试验人员发生身体不适，要停止试验。以防发生意外事故。试验人员要有责任心和独立判断能力，不能受外界的干扰。

（7）保持试验室良好的环境，温度、湿度、照度、粉尘等指标必须符合试验要求。

（8）试验过程坚持用电安全。电线接头等用绝缘胶布包好，定期检查。防止触电事故。

（9）防止机械伤害、防止触电发生等。一旦有发生意外，立即救治。

（10）完善试验室的相关操作规定，仪器设备妥善保管，仪器器皿等要清洁干净，无任何残留。器皿仪器保存良好，必要的有高温消毒，有清洗程序，无二次污染，有复核检查。保持试验环境良好。

（11）严格试验记录的规定，把试验过程完整记录下来。

（12）发生异常后，应当在排除异常原因后，重新开展试验。

4.3.3 试验结果的处理

试验结果的处理，主要有试验原始记录的管理、试验数据的处理、试验报告的编写等。

1. 试验原始记录的管理

（1）原始记录由专职人员进行如实记录，不做任何修饰、删减、增加。

（2）原始记录应记录规范，要注明日期、试验地点、取样内容、样本情况、试验人员、复核人员等信息。

（3）原始记录必须用碳素笔填写，内容完整，妥善保管。电脑记录的数据要及时存档。

（4）原始记录中涉及到有计算的内容，必须有人校核。

（5）原始记录应妥善保管，未经批准，原始记录不能外借。

2. 试验数据的处理

（1）试验人员要明确数据的管理原则。

（2）试验人员要清楚数字的修约规则。

（3）对于试验中的异常值要有明确的处理方法，要区分可以剔除的异常数值、不可剔除的异常数值。

（4）数据要科学、规范记录。数据的有效位数与准确度要求匹配，不足部分以"0"补齐，以便数据位数相等。

（5）产生异常数据后，需要对试验方法等进行重复性试验来判断。

3. 试验报告

（1）试验报告要填写规范、准确，表述清晰，各种信息齐备，填写要求清楚。

（2）试验报告严格使用法定计量单位。

（3）试验报告中的项目齐全、手工记录的字迹清晰、结论明确。记录不涂改，使用碳素笔填写。

（4）所有试验人员在报告上签字，注明日期。

（5）试验报告有相关负责人的批准。

（6）试验报告的存放要符合要求。

4.3.4 试验结果异常

严格来说，试验设计符合要求、试验过程正常、试验人员满足要求，那么试验结果是真实的、有效的。这里所指的试验结果异常，指试验结果的指标没有达到试验者期望的状态。导致试验结果异常的原因，还是要从试验设计、试验过程、试验人员、试验环境等方面进行分析。包括试验设计是否正确、试验是否受到干扰、试验的条件是否符合要求、试验用仪器设备的状态如何、试验人员的技能是否满足要求、试验校准是否正确等。

1. 试验结果异常的处理

（1）明确相关人员进行调查，包括出现异常时的试验人员、负责人等。

（2）对试验结果进行复核、确认，对可能的原因进行客观的评估，不冒然推论原因。

（3）对试验人员的经验、正确执行试验方法的能力进行确认和评估。

（4）检查试验用的材料、仪器、试剂、溶液等有无异常，是否满足管理要求。

（5）检查仪器设备的性能、校验情况、以往的使用情况，确认是否满足要求。

（6）确定是否需要进行试验室调查，如有必要，指定专人按照要求对试验室进行调查分析。

（7）查明发生异常的原因，确定差错的来源，采取纠正和预防措施，保证以后的试验成功率。必要时，对于相关人员进行培训。

（8）试验过程中出现的明显错误，如突然停电、停机、玻璃仪器破裂等，应停止试验，做好记录，该次试验无效。后续重新试验以获得有效数据。

2. 试验室调查

一是初步调查：

（1）对试验过程中的各因素进行检查，包括对试验相关的人员、样品、仪器、设备、试剂、标准、分析方法、计算方法、环境等全部进行检查。

（2）进行偶然误差或系统误差的判断，以便后续采取措施。

（3）实验室负责人与试验人员要及时讨论分析整个试验的全部内容，确认有无操作、理解上的问题，检查原始记录，查找异常和可疑信息，检查仪器及操作过程是否正确、完整。

（4）由于某种误差而中断的试验，要记录中断的原因，在调查分析后对结果无影响的前提下，才可以继续试验。

二是深入调查：

如果试验室初步调查不能识别、确认原因，可进行深入调查，以识别、查找可能的原因。

（1）通过具体的调查测试方案，尝试操作测试系统，以便再现、得到与原始超过标准、异常结果出现时相同类型的问题。

（2）调查测试方案一般采用原样品复验、重新取样复验等方法。如果发现存在非取样原因的实验室偏差、或者不能排除存在试验室偏差可能性时，采用原样复验。当调查发现初检样品有误或者样品本身不具有代表性时，要重新取样进行复验。

（3）复验后发现的确属于试验室偏差后，要安排原试验人员在排除偏差后自行复验。

（4）如果未发现确切的偏差原因，且不能排除存在实验室偏差可能性时，安排原试验人员和专职复检人员共同进行原样复检，进行平行测试。要核对试剂是否异常，是否在有效期，仪器量具等是否经过校正；操作是否规范。复检合格，则判为试验结果合格；如果复检不合格，按首次检验不合格处理。

（5）发现超标原因并在复检合格后，用复检结果取代原结果。原不合格的记录保存归档。

4.3 Abnormal Experiment

———

Experiment abnormality is often encountered in the test. Generally, experiment abnormality can be analyzed from the aspects of experiment design abnormality, experiment process abnormality, and experiment personnel abnormality.

4.3.1 Abnormal test design

Design of Experiment originated from agricultural experiments and was founded by British biostatistician Ronald Aylmer Fisher in the 1920s. In 1935, Fisher published *The Design of Experiments*, officially marking the birth of experimental design. In the 1950s, Japanese statistician G. Taguchi founded the orthogonal experimental design. Chinese mathematician Hua Luogeng proposed the optimization method. In 1978, Chinese mathematicians Wang Yuan and Fang Kaitai proposed uniform design.

The abnormality of the experimental design is caused by many reasons. The analysis can start from three aspects: "factors of the experiment, subjects, and effects of the experiment". The problems that arise are usually "whether the samples (factors) are random, whether the principle of repetition is followed, whether the control group is really prepared for comparison, and whether the principle of balance is implemented". It should be said that test design is closely related to statistics. There are also many cases of abnormal test design, which are briefly introduced below.

1. Single factor experimental design replaces multifactor experimental design

Single factor experimental design is one of the common experimental design methods. Because of its simple design, only one factor is changing, so research and analysis is relatively easy. However, multiple factors really exist in the objective world. The experimental design for multiple factors is much more complex than the single factor experimental design, and the process of research and analysis is also more difficult. Because of the advantages of single factor analysis, people will intentionally or unintentionally replace the single factor test design with the multi factor test design when conducting test design and analysis, resulting in incorrect test conclusions.

Case:

An aquaculture farm raises edible snails, and the keepers hope to improve the growth of edible snails. Therefore, they plan to find out a feed addition formula, and then explore the best sediment, water pH, water temperature and other environments to promote the rapid growth of edible snails. To this end, they carried out relevant tests, and the test data of the formula combination are shown in Table 4-1.

Table 4-1　Test data of formula combination

Formula composition	Formula group			
	Formula I	Formula II	Formula III	Formula IV
Corn/%	10	×	×	×
Soybean cake/%	5	10	×	×
Miscellaneous fish cake/%	10	10	10	×
Wheat bran/%	15	10	20	30
Minced meat/%	10	20	20	20
Snail growth/g	19	16	18	16

Analysis of breeders: in Formula 1, the growth of snails is the fastest, but the formula requires five varieties, which makes it difficult to supply; In Formula 2, four varieties including Soybean cake, miscellaneous fish cake, wheat bran and minced meat are used, and the growth of snails is average; In Formula 3, miscellaneous fish cake, wheat bran and minced meat are used. The growth of snails is good, and the result is close to that of Formula 1; Formula 4 only uses wheat bran and minced meat, with the growth of snails in the middle. Therefore, the breeders choose Formula 3 as a new formula.

Then, the breeders continued the experiment. First, they raised snails under the same sediment thickness, and obtained the best pH value of the water body as 7-8. After fixing the pH value of the water body, the breeders determined that the best water temperature environment was 28-29 ℃. Then, after fixing the pH value and water temperature, the breeders determined that the optimal sediment thickness was 12-14 cm.

Analysis: If we only look at the whole experiment of the breeders from the surface, it seems impeccable. The idea of the breeders is to find out the best feed formula, the best water pH value, the best water temperature and the best sediment thickness one by one through many tests. But in fact, the methods adopted by breeders are all single factor experimental designs. When studying the formula, the breeders should first focus on a certain ingredient to see the growth of snails. In the subsequent experiment, the breeders first locked the conditions of other factors unchanged, and then obtained the so-called optimal pH value of the water body. The correct test method in this example should be multifactor experimental design, and the breeders can complete the whole experiment through orthogonal experimental design. In addition, breeders can also carry out uniform experimental design first, and then use orthogonal experimental design to obtain the best combination.

2. The balance principle was violated between the two control groups

The principle of equilibrium requires that in the design of the experiment, in addition to the different levels of test factors to be investigated between the two control groups, it is also necessary to ensure that the effects of all other possible interference factors on each group are at the same level.

Case:

A company conducts emergency training. 30 young employees from the headquarters were selected as the experimental group, and 30 young employees from a regional project were selected as the control group for training. After 2 days of training, emergency related examinations shall be conducted, as shown in Table 4-2.

Table 4-2 Emergency examination results of test group and control group   Unit: score

Group	Assessment of emergency handling ability before training	Assessment of emergency handling ability after training
Test group	52 ±8	86 ±5
Control group	61 ±6	87 ±4

Therefore, the company concluded that the emergency training had achieved good results.

Analysis: The company adopted the control test method for emergency training. However, the choice of the control group by the organizer was unreasonable. It did not follow the principle of balance. Obviously, if there is a difference in the basic emergency skills between the youth employees of the department and the youth employees of a project in a certain area, it is impossible to compare them with each other. The organizer should gather the employees of the department and the projects randomly selected together, then make a mess, and take half of them at random to be included in the test group and the control group respectively, so as to ensure that the test results are reliable.

3. Control surplus

Some of the experiments are designed with the control method, but the control is surplus, which increased the workload of data analysis. The redundant contrast has no practical significance.

Case:

A pharmaceutical company needs to compare the efficacy of A and B drugs, and has designed a corresponding control trial. The first group is the blank control group while the second group uses only drug A. The third group uses both drug A and drug B and the fourth group is the blank control group of the third group. See Table 4-3.

Table 4-3 Control test contents

Test group	Test contents
Group I	Blank control
Group II	Only using drug A
Group III	Simultaneous use of drug A and drug B
Group IV	Blank control group of Group III

Analysis: This is a control surplus design, because the experimental effects of the first group and the fourth group are the same, and the experimenter only needs to adopt the experimental results of the first group. In fact, there is a better control method. The experimenter can change the test content of the fourth group to "only use drug B". In this way, two factors, drug A and drug B, are formed. Drug A and drug B each have two levels of "use or not". There are four test conditions in total. This test design becomes the "factorial design of two factors".

4. Conceptual confusion between factors and levels

In an experiment, factors are selected for the experiment, or a combination of them. A level is a value, a rank, or a state of a factor. For example, one factor in a test is temperature, and the specific 25 ℃, 26 ℃ and 27 ℃ are the three levels of factor temperature.

Case:

A company wants to bond metal materials, but the bonding is always not firm enough. To this end, technicians want to find a formula with strong viscosity. There are three types of glue: A, B and C; the bonding temperature is 100 ℃, 120 ℃ and 150 ℃; and the bonding time is 10 seconds, 30 seconds and 60 seconds.

The technicians took 60 pieces of metal materials for test, which were divided into six groups, with 10 pieces of materials for each group. The first group is glue A at 100 ℃ for 10 seconds. The second group is glue A, temperature 120 ℃, bonding for 10 seconds. The third group is glue B at 120 ℃ for 30 seconds. The fourth group is glue B at 100 ℃ for 30 seconds. See Table 4-3 for the data after the test. The fifth group is glue C, temperature 150 ℃, bonding for 30 seconds. The sixth group is glue C at 150 ℃ for 60 seconds.

During the test, technicians also compared the quantity of glue on the unit area of each metal material, and then looked at the adhesive force of glue(1 MPa = 10.197 kg / cm^2). See Table 4-4 for test data.

Table 4-4　Metal bond test data

Group	Quantity of glue/g · cm^{-2}	Cohesiveness/MPa
Group I: glue A, temperature 100 ℃, bonding for 10 seconds	5	2590
Group II: glue A, temperature 120 ℃, bonding for 10 seconds	6	3060
Group III: glue B, temperature 120 ℃, bonding for 30 seconds	6	3160
Group IV: glue B, temperature 100 ℃, bonding for 30 seconds	6	2810
Group V: glue C, temperature 150 ℃, bonding for 30 seconds	6	2680
Group VI: glue C, temperature 150 ℃, bonding for 60 seconds	7	3150

According to the data in Table 4-3, the technicians concluded that the Group III glue B, with a temperature of 120 ℃ and a bonding time of 30 seconds, used 6 g/cm^2. At this time, the bonding force reached 3160 MPa, which was the best.

Analysis: The technician confused the concept of factor and level in the test. For example, in the first group and the second group, glue A is at two levels (5 and 6), and the temperature is at two levels (100 ℃ and 120 ℃); In the second and third groups, glue A and glue B are two factors, but both have only one level; In the fourth and fifth groups, glue B and glue C are two factors with only one level, and the temperature is at two levels (100 ℃ and 150 ℃). The arrangement between these factors and levels is quite chaotic, some of which have two levels for the same factor; Some are different factors, but each has a level; Some are different factors, different levels, etc. Therefore, the reliability of such analysis results is compromised.

5. Not designing "repeated experiments" according to the principle of repetition

The "repetition" in repeated refers to the principle of repetition that should be followed during the experimental process. The principle of repetition refers to two or more independent tests under the same test conditions. Independence here refers to testing with different samples. For example, repeated sampling refers to multiple sampling from the same population; Repeated measurement refers to multiple observations of an object over time.

Case:

A company imported an expensive tomographic scanner A (expensive). In order to test the accuracy and quality of its tomographic scanning, technicians compared it with the original tomographic scanner B (relatively cheap) in the company. The scanning object is a piece of steel produced by a company.

For the first test, the technicians used the fault scanners A and B to conduct two scanning tests on the steel respectively, and detect the corresponding accuracy and quality indicators.

After two weeks, the technicians used tomographic scanners A and B to scan the original steel twice respectively for the second time, and detect the corresponding accuracy and quality indicators.

According to the results of the two scanning tests, the technicians carried out statistical analysis and found that there was no significant difference in the accuracy and quality indicators between the two tests. Therefore, technicians believed that model B can be used when purchasing tomography in the future, which can save a lot of money.

Analysis:

In this case, the technician used the tomographic scanners A and B, and the test sample objects remained unchanged, all of which were the same piece of steel. The conclusions obtained by the technicians belong to the "general" conclusions derived from "individual", and the test did not follow the principle of repetition. The correct test method is that technicians should take two repeated samples of steel on the same day, and then use different steel samples to carry out the test.

6. Other abnormal test conditions

(1) Improper selection of subjects. The experimenter did not determine the standard of "subject" in combination with the specific conditions of the experiment, so the inclusion and elimination of subjects seemed random, which increased the interference of non-experimental factors.

(2) Improper selection of observation indicators. The observation index can not reflect the effect of the experiment well. Observation indicators are used to reflect whether the effect of test factors is strong or weak. It is necessary to find out the observation indicators with strong specificity, high sensitivity, accuracy and reliability by combining the nature and characteristics of test factors, taking into account of the conditions such as instrument and technical level. The observation indicators should emphasize objectivity.

(3) The composition ratio of the test is abnormal. The factor ratio used in the test was not u-

niformly implemented according to the requirements, resulting in the lack of comparability of the test.

(4) The orthogonal test was not performed according to the orthogonal table. Sometimes the experimenter reverses the factor and level.

4.3.2 Abnormal test process

1. Abnormal situations during the test

There are many abnormal situations during the test as follows.

(1) Sudden power failure.

(2) The instruments and meters used for the test have faults, such as poor circuit contact, no indication of voltmeter, etc.

(3) Failure of machinery and equipment for test.

(4) Abnormal test samples, such as unqualified samples, damaged sample packages, lost samples, insufficient reagent purity, etc.

(5) Peculiar smell is produced during the test, and the source and cause of the odor are unknown. Some may be toxic odor.

(6) The tester cannot carry out the test normally due to physical discomfort. Sometimes the test personnel are interfered by the outside world, which may distort the test results.

(7) The test environment does not meet the relevant requirements, including temperature, humidity, illumination, dust, etc.

(8) Fire may occur during the test, which may lead to fire. Sometimes there will be explosions, etc.

(9) The test cycle changes, such as suddenly shortening the test time.

2. Handling methods for abnormal test process

(1) In case of any abnormality during the test, such as power failure and instrument failure, the test shall be terminated immediately and the power supply shall be cut off. The cause shall be found out, and the relevant person in charge shall be reported in time. The original test results shall be invalidated. However, all test data shall be saved.

(2) Relevant mechanical equipment shall be checked between tests. In case of any fault, the test shall be stopped immediately and reported to the relevant person in charge in a timely manner. The original test results shall be invalidated and all data shall be saved.

(3) After the sample is delivered, it shall be checked before the test. The quality of reagents used shall be qualified and effective. In case of any abnormal sample, the sample delivery personnel shall be informed in time and qualified samples shall be used for test.

(4) During the test, personnel shall be on duty. Especially in case of heating, scald and fire shall be prevented. The test personnel shall master the fire fighting skills and use methods of fire extinguishers, and regularly check the fire extinguishers.

(5) During the test, the discharge and leakage of toxic substances shall be prevented so as not to pollute the environment. If peculiar smell is found during the test, be sure to identify the cause. Open windows for ventilation. Evacuate people and prevent poisoning. Wear anti-poison equipment to clean up the site.

(6) If the tester is unwell, the test shall be stopped. In case of accidents, the tester shall

have a sense of responsibility and independent judgment ability, and shall not be interfered by the outside world.

(7) Keep a good environment for the laboratory, and the temperature, humidity, illumination, dust and other indicators must meet the test requirements.

(8) During the test, electrical safety shall be adhered to. Wire joints shall be wrapped with insulating tape and inspected regularly. Prevent electric shock accidents.

(9) Prevent mechanical injury and electric shock. In case of accident, treat the injured immediately.

(10) Improve the relevant operating regulations of the laboratory. Keep the instruments and equipment properly, and clean the instruments and vessels without any residue. Utensils and instruments shall be kept well, and if necessary, they shall be subject to high-temperature disinfection, cleaning procedures, no secondary pollution, and recheck inspection. Keep the test environment in good condition.

(11) Strictly regulate the test record and completely record the test process.

(12) In case of abnormality, the test shall be carried out again after the cause of abnormality is eliminated.

4.3.3 Treatment of test results

The processing of test results mainly includes the management of original test records, the processing of test data, and the preparation of test reports.

1. Management of original test records

(1) Original records shall be truthfully recorded by full-time personnel without any modification, deletion or addition.

(2) The original record shall be recorded in a standardized manner, indicating the date, test location, sampling content, sample conditions, testers, reviewers and other information.

(3) The original record must be filled in with a carbon pen, complete and properly kept. Data recorded by computer shall be archived in time.

(4) The original record involves the content with calculation, which must be checked by someone.

(5) The original records shall be kept properly and cannot be lent without approval.

2. The processing of test data

(1) The tester shall be explicit about the data management principles.

(2) The tester shall be clear about the rounding off rules of figures.

(3) There should be a clear treatment method for the abnormal values in the test, and distinguish between the abnormal values that can be eliminated and the abnormal values that cannot be eliminated.

(4) The data shall be recorded scientifically and normatively. The significant digits of the data should match the accuracy requirements, and the insufficient parts should be filled with "0", so that the data digits are equal.

(5) After generating abnormal data, it is necessary to conduct repeated tests on test methods to judge.

3. Test reports

（1）The test report shall be filled in normatively and accurately, with clear description, complete information and clear filling requirements.

（2）The test report shall strictly use the legal metrological unit.

（3）The items in the test report are complete. The manual records are clear, and the conclusions are clear. The record shall not be altered and shall be filled in with a carbon pen.

（4）All test personnel shall sign and date the report.

（5）The test report is approved by the relevant person in charge.

（6）The storage of test report shall be rechecked.

4.3.4 Abnormal test results

Strictly speaking, if the test design meets the requirements, the test process is normal, and the test personnel meet the requirements, then the test results are true and effective. The abnormal test results here refer to the fact that the indicators of the test results do not meet the expectations of the experimenter. The causes of abnormal test results should be analyzed from the aspects of test design, test process, test personnel, test environment, etc. It includes whether the test design is correct, whether the test is interfered, whether the test conditions meet the requirements, the status of the test instruments and equipment, whether the skills of the test personnel meet the requirements, and whether the test calibration is correct.

1. Handling of abnormal test results

（1）Specify relevant personnel for investigation, including test personnel and person in charge in case of abnormality.

（2）Recheck and confirm the test results. Objectively evaluate the possible causes, and do not presume to infer the causes.

（3）Confirm and evaluate the experience of test personnel and their ability to correctly implement test methods.

（4）Check whether the materials, instruments, reagents, solutions, etc. used for the test are abnormal and meet the management requirements.

（5）Check the performance, calibration and previous use of instruments and equipment to confirm whether they meet the requirements.

（6）Determine whether laboratory investigation is required, and if necessary, assign special personnel to conduct investigation and analysis on the laboratory as required.

（7）Find out the cause of abnormality. Determine the source of error, and take corrective and preventive measures to ensure the success rate of future tests. If necessary, relevant personnel shall be trained.

（8）In case of obvious errors during the test, such as sudden power failure, shutdown, glass instrument breakage, etc., the test shall be stopped and recorded. The test is invalid. Subsequently retest to obtain valid data.

2. Laboratory investigation

First, preliminary investigation.

（1）Check all factors during the test, including personnel, samples, instruments, equipment, reagents, standards, analytical methods, calculation methods, and environment related to the test.

(2) Judgment of accidental error or systematic error for subsequent measures.

(3) The person in charge of the laboratory and the test personnel shall timely discuss and analyze all contents of the whole test, confirm whether there are problems in operation and understanding, check the original records, find abnormal and suspicious information, and check whether the instruments and operation process are correct and complete.

(4) If the test is interrupted due to some error, the reason for the interruption shall be recorded, and the test can be continued only after the investigation and analysis have no impact on the results.

Second, in-depth investigation.

If the laboratory preliminary investigation cannot identify and confirm the causes, in-depth investigation can be carried out to identify and find possible causes.

(1) Through the specific investigation and test plan, try to operate the test system, so as to reproduce and get the same type of problems as when the original exceeding standard and abnormal results occur.

(2) The survey and test plan generally adopts the methods of original sample reinspection, resampling reinspection, etc. If the laboratory deviation not caused by sampling is found, or the possibility of laboratory deviation cannot be ruled out, the original sample shall be used for reinspection. When the investigation finds that the initial sample is wrong or the sample itself is unrepresentative, it is necessary to take a new sample for reinspection.

(3) After reinspection, if it is found that the deviation does belong to the laboratory, the original tester shall be arranged to reinspect by himself after the deviation is eliminated.

(4) If the exact cause of the deviation is not found and the possibility of laboratory deviation cannot be ruled out, the original test personnel and full-time reinspection personnel shall be arranged to conduct the original reinspection and parallel test. It is necessary to check whether the reagent is abnormal, whether it is within the validity period, and whether the instruments and measuring tools have been calibrated; Whether the operation is standardized. If the reinspection is qualified, the test result is qualified; If the reinspection is unqualified, it shall be treated as unqualified in the first inspection.

(5) If the reason for exceeding the standard is found and the reinspection is qualified, the reinspection results shall replace the original results. The original unqualified records shall be kept and archived.

4.4　抽样异常

抽样发生异常是数据收集中经常遇到的问题。抽样异常主要分为抽样方法异常、抽样人员异常、抽样器具异常、抽样过程异常、抽样条件异常等几种情形。

4.4.1 抽样方法异常

抽样方法有很多种，比如有简单抽样、分层抽样、系统抽样等方法。抽样方法发生异常，会直接导致分析结论的异常。

1. 片面抽样

片面抽样主要是调研人员没有认真分析样本的分布情况，对于样本构成的复杂性、全面性分析不够，在抽样的时候只对部分样本进行抽样，导致对于总体的判断失真。

比如在美西战争时期，有的美国人从参战海军士兵中调查，得出海军士兵的死亡率为 0.9%。而同一时期，纽约市的市民平均死亡率为 1.6%。于是得出结论：在战争期间，在海军服役的军人比普通居民还要安全。这是片面抽样的典型。因为海军士兵的年龄都是处于 20 岁上下，身体健康，服役前都有体检合格证明。而纽约市民的构成则复杂很多，其中有老年人、婴幼儿，还有各种病人，而当时的老年人的寿命普遍不算很高，当时的医疗条件也导致了婴幼儿容易夭折，各种病人的死亡率相对也高，再加上各种事故构成的样本，其死亡率自然偏高。而要得到准确的结论，研究人员应当采用分层分析的方法，即把纽约市的青年人（年龄要集中在 20 岁左右）作为样本来统计，这样才具有很好的可信度。

再比如某著名学校开设了一个升学复习班。抽样调查表明，该复习班的升学率几乎达到 99%，在当地处于高水准的状态。实质上，该复习班在招收学员时是有条件的，必须具备相当好的基本功，而且进入复习班是要经过严格的考试筛选。其实学校已经预先把拔尖的学生遴选出来组成了升学复习班，再加上师资力量的雄厚，其升学率固然达到很高的水准。因此，这样的抽样调查结论，是不能与其他地方的复习班进行平行比较的。

2. 简单抽样理解为随意抽样

简单抽样因为其方法简单而广受欢迎。但有的抽样人员把简单抽样理解为随意抽样、任性抽样，从而导致抽样失败。其中的原因在于抽样的一个重要原则就是随机原则，如果缺少了随机，那么样本就缺少代表性，用于分析的结论会发生很大偏差。因此，"简单抽样"的全称是"简单随机抽样"，在抽样样品的过程中，重点在于随机，以确保总体中的每个个体被抽取出来的概率相等。随机主要体现在抽样样本时不受任何主观因素、其他系统性因素的影响。而随意、任性则回避了科学的样本分布，带有明显的个人倾向。

比如某电影院刚新上映一部大片，为了获取该影片受欢迎的程度数据，院方安排了调研人员进行调查。在电影刚上映的第一天，调研人员等候在影院门口，等电影散场后，马上以自认为随机的方式抽取了 50 名观众进行调研，结果表明，50 受访人中 39 人认为该电影拍得特别好，属于国际公认的大片；9 人认为该电影拍得相当好，水平明显高于其他电影；还有 2 人持基本肯定的态度，认为值得一看。于是调研人员得出结论，该电影是难得的佳片，受广大市民的欢迎率几乎达到了 100%，并为此广为宣传。其实这样的抽样就是随意抽样。因为调研人员在抽取样本时，总体就是首场观众，而通常首场观众可能出于各种渠道已经得知了该影片的相关消息，几乎就是冲着这部电影来的，得出的结论自然好。客观的抽样方法是，抽样人员在一定的片区内，随机抽取行人进行调查，这样，喜欢这部电影的、不喜欢这部电影的、没有看过这部电影的人等都会被抽到，得出的"受市民欢迎的百分比"结论就可靠得多。

3. 等距抽样中抽样间隔与系统性异常的周期相同

等距抽样每隔一定的抽样间隔来抽取样本，因此，一旦抽样间隔与系统性异常的周期相同，则抽样异常的结果在所难免。

比如：某加工机械加工产品，由于不可知的系统性原因，导致每次加工完 9 个产品后，第 10 个产品会出现质量问题。现在质量检验人员到生产线上进行抽样检查，采用等距抽样法，间隔是 10 个产品。这样，如果第一个次品正好位于间隔点之外，那么抽样的结果就是产品全部合格。显然，这样的结论与实际发生了很大差异。而如果次品正好位于间隔点上，即第一个抽样样品正好是次品，则其后所有的抽样样品也正好都是次品，那么抽样的结果是产品全部不合格。显然这样的结论也是不正确的。

4. 随意确定样本容量

有的抽样人员在抽样中，没有根据恰当的方法来确定样本容量。有的人简单地根据总体的数量来确定抽样量。如果抽样量过少，则使抽样结果的分析不准确。如果抽样量过大，会增加大量的人力、物力、财力，是不经济的行为。

在第 3 章第 3.2 节中，介绍了简单随机抽样中给定均值来估计抽样量，以及给定比例来估计抽样了。在进行均值估计样本容量时，抽样人员需要先设定标准误差的置信水平、总体标准差、可接受的抽样误差范围等，然后可以计算得出样本容量。在根据比例来估计样本容量时，抽样人员需要事先设定标准误差的置信水平、抽样中的样本比例（通常 $P \leqslant 5\%$）、可接受的抽样误差范围，然后通过计算得出样本容量。

对于给定均值的样本容量确定，可以按照一些经验值来进行参考，见表 4-5 "给定均值的样本容量"，抽样人员可以根据不同的可接受抽样误差来找到对应的样本容量。

表 4-5　给定均值的样本容量

可接受的抽样误差/%	样本容量	可接受的抽样误差/%	样本容量
1.0	10 000	6.0	277
1.5	4500	6.5	237
2.0	2500	7.0	204
2.5	1600	7.5	178
3.0	1100	8.0	156
3.5	816	8.5	138
4.0	625	9.0	123

可接受的抽样误差/%	样本容量	可接受的抽样误差/%	样本容量
4.5	494	9.5	110
5.0	400	10.0	100
5.5	320		

从表 4-5 可以看出，可接受的抽样误差越小，样本容量越大。其实，根据以往计算样本容量所积累的经验，也可以得到一些值得借鉴的做法：当总体越小的时候，如果要有较高的概率得到与总体统计分析相同结果的样本，也就是说样本的质量很高、具有很好的推论精确性，那么抽样的比例必须越大。而如果抽样的比例小，那么就需要有较大的总体来获得好的统计分析结果。

经验表明，如果总体在 1000 以下，那么抽样的比例达到 30% 时会有较高的统计精确性，那么样本量要达到 300。如果总体在 10 000，那么抽样的比例达到 10% 时会有同样较高的统计精确性，那么样本量要达到 1000。如果总体在大于 15 000，那么抽样的比例达到 1% 时，也会有同样较高的统计精确性，那么样本量要达到 1500。如果总体非常大，达到了 1000 万，那么抽样的比例达到 0.025% 时，其统计精确性也非常好，那么样本量只需达到 2500。显然可以看出，随着总体数量级别的大大提高，抽样的比例很小也不会对统计分析的精确性构成很大的影响了。假设总体的数量达到 1 亿以上，那么，样本容量达到 2500 时，其统计分析的精确性与 1000 万中抽取 2500 个样本进行分析的精确性是相同的。

5. 样本轮换周期过长

样本轮换是指调研人员需要对于样本进行持续的跟踪调查，由于样本处于不断地变化之中，因此每隔一段时间需要对于样本进行部分或者全部地轮换，以获取新的样本信息，保持调研结果的准确性。通常样本轮换适用于调查周期比较长的对象。比如进行人口调查时，如果进行长时间（比如连续几年）的调查，那么样本（人口）本身会随着时间的推移发生重大的变化，当若干年后的样本（人口）出来时，早期的样本（人口）中，有的学历已经提升，有的户籍已经变更等。再加上有些人员因工作、学习等原因，对于长期调查产生了不合作的态度等等，因此必须采用样本轮换的方法。

但是，如果样本轮换周期过长，由于客观对象发生比较大的变化，那么抽样的结果与实际会发生不相符。发生比较大的变化有很多情形，比如企业生产线的技术进步，生产效率的提升，前后样本之间发生很大变化，尤其是信息产业，几乎每年都会有技术上的新突破。再如社会不断进步、人口的变化等，尤其是一些原来比较落后的地区，因为当地建设、经济极大改善等，早先的样本就会不适用于现在的情形。

当然，样本轮换周期过长时，早期的样本也可以进行利用，比如进行回归分析时，可以把早期的样本资料作为辅助信息，可以提高样本估计的精度。

6. 点估计与区间估计混淆

点估计是指用样本的统计量来估计总体参数，形象地说，把样本统计量视为数轴上的某一个点值，其估计的结果也是数轴上的一个点值。区间估计是指用一个区间范围来估计总体参数，反映在数轴上，就是由样本统计量加减估计误差，从而得到相应的估计区间。

虽然点估计和区间估计都是属于参数估计的范畴，但两者还是存在明显的区别，主要反映在两个方面。一是结果不同。点估计是用样本数据来估计总体分布所含未知参数的一个具体的数值，它的精确程度可以用置信区间来表示。区间估计是根据样本数据来估计总体的未知参数最可能所在的一个区间，它的精确程度是用概率来表示，也就是置信度。二是是否考虑抽样误差。点估计是直接用抽取的样本指标来推断总体的指标，在抽样推断中不考虑抽样误差。而事实上用抽样的指标来代替总体的指标，肯定会存在误差。区间估计是根据样本的指标和抽样误差来估计总体指标的可能范围，它是用一定的概率来说明误差不超出一个给定的区间。

有的调研人员用点估计来代替区间估计，会造成结论不准确。比如在某城市进行当地居民家庭日常开支的调查中，调研人员用抽样调查的数据，计算后得到该市居民家庭平均每月正常开支，且这为 2000 元。于是调研人员得出结论：该市居民家庭每月正常开支为 2000 元。在结论中，没有对抽样误差和置信度进行描述，这就是用点估计代替了对总体的每个家庭每月正常开支的区间估计。

正确的方法是，调研人员采用区间估计的方法。

调研人员可以用简单随机抽样的方法（重复抽样），共取得 300 户家庭的每月正常开支,且这些家庭的开支属于非正态分布。经过计算，得到样本的平均值为 2000 元，样本的标准差 s 为 150 元。

虽然样本的总体分布未知，但因为 n =300，可以用正态分布来近似。用样本标准差 s 代替总体标准差 σ。现设定置信度为 95%，则 α =0.05。查正态分布表，得到 $Z_{0.05/2}$ = 1.96。

则有，总体均值 μ 的置信区间为

$$\bar{X} \pm Z_{\alpha/2} \frac{s}{\sqrt{n}}$$

$$= 2000 \pm 1.96 \frac{150}{\sqrt{300}}$$

$$= （1983.03，2016.97）$$

这样，调研人员得出结论：在置信度 95% 下，该市居民家庭每月正常开支的区间为 1983.03 元到 2016.97 元。

一般地，对于简单随机抽样、等距抽样，总体标准差 σ 已知，总体均值的区间估计所用的公式见表 4-6。

表 4-6 总体均值 μ 的区间估计

（置信度为 1-α、总体标准差 σ 已知）

总体分布	样本容量	重复抽样	不重复抽样
正态分布	小样本（$n<30$）	$\bar{X} \pm Z_{\alpha/2} \dfrac{\sigma}{\sqrt{n}}$	$\bar{X} \pm Z_{\alpha/2} \dfrac{\sigma}{\sqrt{n}} \sqrt{\dfrac{N-n}{N-1}}$
	大样本（$n \geq 30$）		
非正态分布	小样本（$n<30$）	—	—
	大样本（$n \geq 30$）	$\bar{X} \pm Z_{\alpha/2} \dfrac{\sigma}{\sqrt{n}}$	$\bar{X} \pm Z_{\alpha/2} \dfrac{\sigma}{\sqrt{n}} \sqrt{\dfrac{N-n}{N-1}}$

对于简单随机抽样、等距抽样，总体标准差 σ 未知，总体均值的区间估计所用的公式见表 4-7。

表4-7　总体均值 μ 的区间估计

（置信度为 $1-\alpha$ 、总体标准差 σ 未知）

总体分布	样本容量	重复抽样	不重复抽样
正态分布	小样本（ $n<30$ ）	$\bar{X} \pm t_{(\frac{\alpha}{2},n-1)} \frac{s}{\sqrt{n}}$	$\bar{X} \pm t_{(\frac{\alpha}{2},n-1)} \frac{S}{\sqrt{n}} \sqrt{\frac{N-n}{N-1}}$
	大样本（ $n \geqslant 30$ ）	$\bar{X} \pm Z_{\alpha/2} \frac{s}{\sqrt{n}}$	$\bar{X} \pm Z_{\alpha/2} \frac{S}{\sqrt{n}} \sqrt{\frac{N-n}{N-1}}$
非正态分布	小样本（ $n<30$ ）	—	—
	大样本（ $n \geqslant 30$ ）	$\bar{X} \pm Z_{\alpha/2} \frac{s}{\sqrt{n}}$	$\bar{X} \pm Z_{\alpha/2} \frac{S}{\sqrt{n}} \sqrt{\frac{N-n}{N-1}}$

7. 不重视抽样框的设计

有的调研人员在执行抽样时，对于抽样框的设计不重视，有的甚至没有抽样框。抽样框也称为抽样结构或抽样框架，是指对可以选择作为样本的总体列出名册或者进行排序编号，以确定总体的抽样范围和抽样结构。抽样范围、单位和结构是根据调研的目的和内容来确定的。确定完成后，就可以进行随机抽样。如果没有抽样框，就不能计算样本单位的概率，也谈不上进行概率选样。

比如某加工厂加工工程用支架，检查人员只是简单地到流水线上抽取 100 个支架作为样本来进行分析，那就是缺少抽样框的设计。

正确的方法是：假设该工厂有 5 台支架加工机械，每台加工机械每天加工 1000 个支架，那么每天共有 5000 个支架加工完成。这个"5000 个支架"就构成了抽样框。

抽样框可以理解为备选作为样本的全部抽样单位（即总体单位）的顺序或编排形式。在编排之后，有关总体的信息应当齐全，包括总体的单位名称、编号、流水号、地址等。抽样框可以用多种形式呈现，有的抽样框是具体的，如目录结构、区域结构、目录区域复合结构等。在本例中，抽样框是具体的，即抽样单位可以列成表册。有些抽样框的形式是抽象的，它是呈开放的形态，凡是符合调研条件的就构成抽样框中的元素，比如在大街上、商场内进行消费者的随机调查等。

由于抽样框的设计问题而导致抽样异常的有：

一是目标总体单位涵盖不足，即抽样框没有覆盖全部目标总体单位，使一些目标单位没有机会入选样本，它会导致对总体的估计偏低。

二是目标总体单位涵盖过分，即抽样框中包含了一些非目标总体单位，如果不能及时发现并剔除，会导致对总体的估计偏高。

三是在目标总体单位没有全部涵盖的同时，又包含了非目标总体单位。也及时说，有些目标单位没有被入选样本，而非目标单位则被选中。这种情况下，如果缺失的目标单位与非目标单位的特征有着显著差异，那么对总体的估计也会出现大的偏差。

四是抽样单元与目标总体单元没有完全一一对应，存在着一对多、多对一、多对多的情形。包含所有抽样单元的总体为抽样框，而构成抽样框的单元为抽样单元。比如某公司对材料供应商进行抽样调查，构成总体的是该公司的所有材料供应商。抽样的时候，抽样人员选择的抽样框是近5年来与该公司签订采购的合同。在这个抽样框中，签订合同比较多的供应商被抽中的概率大，那么就构成了多对一的情形，也就是抽样框单元多于目标总体单元。这样的总体估计也会随着供应商特征指标的不同而出现偏差。

五是抽样框过时，即随着时间的推移，抽样总体与目标总体存在大的差异。比如在企业的材料中，随着科技的发展，许多新材料已经在项目上应用，而抽样框还是老的材料品种，那么会导致估计的精度出现异常。

8. 没有区分系统性误差与偶然误差

系统性误差是一种非随机性的误差。偶然误差是随机误差，它是由于客观对象中的有关因素出现微小的随机波动而形成的误差，而且这些误差之间可以相互抵偿。有时候，调研人员没有认真区分系统性误差还是偶然误差，使调查结论异常。比如一个粮食加工厂加工大米，标准为每袋5 kg。现在检查人员随机在流水线上抽样10袋，得到的数据为：5.02 kg、5.01 kg、5.05 kg、5.12 kg、5.02 kg、5.10 kg、5.06 kg、5.03 kg、5.01 kg、5.07 kg。公司要求误差控制在2.5%以内，因此，检查人员认为加工的大米全部符合要求。但是复检人员发现，这10袋大米的重量都大于5 kg，没有一袋是低于5 kg的，而正常加工的大米重量应该在5 kg上下波动。故认为该加工流水线存在着系统性误差。

产生系统性误差的原因有很多，比如：加工流水线本身有问题；测量仪器本身的误差，如零点没有校正；试验时的条件没有达到理论所规定的要求；个人的偏好；环境温度偏热或偏冷；从一个存在严重缺陷的抽样框中抽样等。

9. 没有区分不等概率抽样与等概率抽样

等概率抽样是指抽样时总体中的个体被抽中的概率相等，它适用于总体内的个体之间大小差异不是很明显。不等概率抽样是指抽样时，总体中的个体被抽中的概率是不相等的，它适用于总体内的个体之间存在着大小差异明显，而且所要调查的标志与个体呈强正相关关系的情形。采用不同概率抽样是为了减少对于总体估计的误差。

在一个总体内，个体之间绝对相等是几乎不可能的，所以不等概率的抽样通常适用于两种情况，一种是总体内的个体有比较完整的标志，能够根据个体的大小不同来确定被抽中的概率。另一种是调查的标志与辅助标志之间存在着清晰的关系，比如两者呈正相关关系且又是等比例关系。

有时候，调研人员没有区分不等概率抽样与等概率抽样，会造成结论异常。比如某调研人员调研四家企业的员工人数与工资总额。得到的数据见表4-8。

表4-8　四家公司的员工人数与工资总额

企业名称	A	B	C	D	合计
企业人数/人	30	50	80	120	280
企业的工资总额/万元	180	240	450	900	1770

现在假设样本的容量为2，则样本的构成有 $4^2 = 16$。

从表4-8可以看出，四个企业被抽中的概率是不同的，A公司被抽中的概率为30/280 = 10.7%，B公司为17.8%，C公司为28.6%，D公司为42.9%。那意味着，四个公司被抽中的概率是不一样的。如果是等概率的话，那么每家公司被抽中的概率都是 25%。因此，如果按照等概率来计算工资总额的估计量方差会明显偏大，而不等概率计算工资总额估计量的方差会小很多。

10. 缺少抽样控制程序

有些调研人员在执行抽样任务时，缺少规范的抽样控制程序作为指导，抽样活动的质量不符合相关标准要求。

在质量管理体系中，一个完整的抽样程序包含的主要内容有：制定抽样程序的目的；程序适用范围；相关人员的职责；程序的主要内容，包括抽样工作的准备、对抽样人员的要求、抽样方法、抽样时间、抽样过程中需要控制的因素；对于偏离抽样程序的检测规定；正常、加严、放宽检验的规定；抽样检验的控制要求；抽样检验的标准；抽样记录和要求等。

4.4.2 抽样人员异常

由于抽样人员的异常而导致抽样异常的情形主要有：

（1）抽样人员缺少抽样的知识培训，缺少对于抽样相关理论、方法的掌握。

（2）抽样人员缺少责任心，对待抽样缺少必要的认识和重视。

（3）抽样人员缺少对于应急情况的处理能力，比如对于样本的代表性、抽样误差的处理等不知所云，一旦发生问题，无所适从。

（4）抽样人员不熟悉抽样对象的相关要求，使抽样不规范。

（5）抽样单填写不规范，缺少必要的信息，如产品名称、规格型号、等级、执行标准、抽样依据、抽样数量、检验性质、抽样地点、验证情况、批量等。

（6）人员的职责分配不清。

4.4.3 抽样器具设备等异常

由于抽样器具设备异常而导致抽样异常的情形主要有：

（1）缺少专用的抽样器具。有些抽样必须有专用的取样器具和存储器皿。

（2）对抽样器具缺少有效的管理。抽样的器具很多，如样品的容器、抽样用设备、样品标签、样品记录表、电子秤、安全防护用品等，必须有严格的管理措施，否则会导致器具的准确性下降，也会造成物品缺失等，使抽样工作无法正常开展、效率低下等。

（3）抽样设备不能正常工作。比如：接触不良、管路破损、堵塞、流量不稳定等。

（4）抽样设备缺少安全保护措施、缺少定期检查，存在安全风险等。

（5）使用废旧抽样器具设备进行抽样。

（6）抽样器具设备之间存在相互干扰。

4.4.4 抽样过程异常

抽样过程的异常情形主要有：

（1）没有规范的抽样流程，抽样随意性强。

（2）没有按照抽样的方案执行，没有根据抽样方案规定的样本数量进行抽样。

（3）抽样方案有具体的要求，但具体执行时没有随机抽样，抽样缺少代表性。

（4）抽取的样品没有及时做标识或者缺少标识。

（5）抽样中遇到一些意外时，缺少相应的替代方案。比如：人手不够时用熟人来帮忙；抽取样本有困难时，轻易放弃被抽中的样本；临时更换样本、增减样本等。

（6）抽样中不计算抽样误差的存在。

（7）抽样过程中缺少监督机制。

4.4.5 抽样条件异常

抽样条件发生异常的情形主要有：

（1）对于总体的界定不清晰，使抽样的样本脱离总体的范围。

（2）总体的分布过于分散，使抽取的样本也过于分散，导致后续分析困难。

（3）抽取的样本容量过小，导致样本的代表性不够。

（4）客观对象的某种特征已经清晰掌握，且该特征会影响研究结果，但抽样还是按照简单随机抽样方法进行，使样本代表性不足。

（5）抽取的样本量发生异常，缺少相应的修正方案。

（6）抽样的环境发生变化。比如样品保存不当，发生变质、污染等情况；有些见光分解的样品缺少避光容器等。对于抽样的环境缺少相应管理方法。

（7）存在抽样的仪器设备混用、交叉污染等情况，导致样本研究的准确性下降。

（8）抽样的安全存在风险。抽取的样品没有有效隔离。样品缺少监控，存在样品损毁风险。

（9）抽样失败后没有弥补方案和措施。

4.4.6 抽样异常的应对方法

对于抽样异常的情况发生，应当有相应的纠正和预防措施，主要的应对方法有：

（1）完善抽查制度，明确相关责任，提高责任心。

（2）加强业务培训，正确开展抽样方案设计、确定样品等工作，抽样过程开展质量控制，执行数据处理规范。

（3）抽样前学习相关标准、规定，掌握抽样方法，熟悉抽样的环境条件、技术要求。涉及到施工生产、工艺流程等情形，要了解施工生产、工艺流程的相关内容。

（4）严格制定抽样方案，选择适当的抽样方法、确定抽样内容和相关要求，严格过程操作。对于客观对象具有明确特征的抽样，应当选择配额抽样方法，以提高准确率。

（5）明确抽样的总体，清晰界定总体的边界，不发生模糊现象。

（6）抽样要保证足够的样本容量。样本量发生异常时，要有相应的修正方案。

（7）保证抽样的环境稳定、安全。对样本有检查、保护制度并正确执行。要确保设备设施仪器等的状况符合要求，保证输出产品的稳定性。

（8）出于抽样的需要，应当添置必备的、专业的抽样器具。

（9）对于抽样误差和各种原因导致的非抽样误差进行有效控制。要采取必要措施，减少疏忽带来的风险，避免抽样误差。

（10）抽样时要严格执行抽样的规范要求。抽样的车辆、装备、人员等满足工作要求。

（11）对于易碎物品、危险化学品、有特殊储存条件的样品，要采取正确措施，保证样品在运输过程中不发生变化。

（12）对于样品有封装要求的，样品封装应符合规范要求。要采取防拆封措施，保证样品真实性。

（13）抽样人员应不少于 2 人，要确保抽样公平公正。

（14）抽样记录使用规范的抽样文书，详细记录抽样信息。

4.4.7 抽样方案的设计

一个抽样方案的设计步骤，通常包括：

（1）明确抽样调查的目的。

（2）明确抽样调查的范围和内容，包括确定总体与总体的目标量，确定抽样框，确定调查的具体内容。

（3）确定抽样的方法，结合总体的具体情况进行确定。

（4）确定本次抽样的精度要求。

（5）计算样本容量。如果分阶段抽样，还要确定各阶段样本量的配置。

（6）确定总体目标量的估计方法和估计效果。

（7）确定开展抽样的时间、地点，准备必要的抽样工具、装备。

（8）明确抽样的组织形式，确定开展抽样工作的相关人员。

（9）制定实施抽样方案的方法和步骤。

（10）样本的审核与管理。

（11）发生异常情况的修正办法。

（12）其他事项，如经费保障等。

4.4.8 零缺陷的抽样方案

（1）零缺陷的来历。零缺陷的概念最早由质量管理大师菲利普·克劳士比（Philip B. Crosby）在 20 世纪 60 年代提出。美国国防部于 1950 年发布了 MIL−STD−105A 规范。美国一位大学教授尼古拉斯·斯托格利亚（Nicholas L. Squeglia）在 1965 年发表 C = 0 的抽样方案。后来的 C = 0 抽样方案是依据 MIL−STD105 标准而制定，其接纳准则为"0 收 1 退"，即抽样全部合格才接受。1994 年美国三大汽车生产商（通用、福特、克莱斯勒）发布了 QS9000 质量体系规范，其中明确了零缺陷的接纳准则。2020 年，美国发布 ANSI/ASQ/Z1.4—2020 规范。

（2）零缺陷方案的特点。主要是检验数量相对少，经济；易于管理和运用；主要适用于计数型抽样。

（3）采用零缺陷方案时，也需要根据不同客观对象类别的质量特性，规定接收质量限（AQL）。

4.4 Sampling Anomaly

——

Abnormal sampling is a common problem in data collection. Sampling exceptions are mainly divided into sampling method exceptions, sampling personnel exceptions, sampling equipment exceptions, sampling process exceptions, and sampling conditions exceptions.

4.4.1 Abnormal sampling method

There are many sampling methods, such as simple sampling, stratified sampling, systematic sampling. Abnormal sampling methods will directly lead to abnormal analysis conclusions.

1. Unilateral sampling

The unilateral sampling is mainly due to the fact that the investigators did not carefully analyze the distribution of the samples, did not fully analyze the complexity and comprehensiveness of the sample composition. When sampling, only a part of the personnel are sampled, resulting in distortion of the population judgment.

For example, during the Spanish American War, some Americans surveyed navy soldiers who participated in the war and found that the death rate of navy soldiers was 0.9%. In the same period, the average death rate of New York City residents was 1.6%. It was concluded that during the war, soldiers serving in the navy were safer than ordinary residents. This is typical of one-sided sampling. Because naval soldiers are all around 20 years old, they are healthy, and have medical certificates before serving. The composition of New York citizens is much more complex, including the elderly, infants and children, and various patients. At that time, the life span of the elderly was generally not very high. The medical conditions at that time also led to infants and children dying prematurely, and the mortality of various patients was relatively high. In addition, the death rate of various accidents was naturally high. To get accurate conclusions, researchers should adopt the method of hierarchical analysis, that is, take the young people in New York City (the age should be about 20 years old) as a sample for statistics, so that they can have good reliability.

Another example is that a famous school has opened a review class for further study. The sampling survey shows that the enrollment rate of the class reaches 99%, which is at a high level in the local area. In fact, the review class is conditional when recruiting students, and must have fairly good basic skills. In addition, entering the review class is subject to strict examination screening. In fact, the school has selected the top students in advance to form a review class for further study. With the strong faculty, the rate of further study has reached a very high level. Therefore, such a sampling survey conclusion cannot be compared with the review classes in other places in parallel.

2. Simple sampling is understood as random sampling

Simple sampling is popular because of its simplicity. However, some sampling personnel understand simple sampling as random sampling and capricious sampling, which leads to sampling failure. The reason is that one of the important principles of sampling is the random principle. If there is no random, the sample will lack representativeness, and the conclusions used for analysis will be greatly biased. Therefore, the full name of "simple sampling" is "simple random sampling". In the process of sampling, the emphasis is on randomization to ensure that the probability of each individual in the population being sampled is equal. Randomization is mainly reflected in that the sampling is not affected by any subjective factors or other systematic factors. However, randomness and capriciousness avoid the scientific sample distribution, with obvious personal tendency.

For example, a movie theater has just released a blockbuster film. In order to obtain the popularity data of the film, the movie theatre has arranged researchers to investigate. On the first day of the film's release, the researchers waited at the gate of the cinema. After the film ended, they immediately selected 50 audiences in a random way. The results showed that 39 of the 50 interviewees believed that the film was shot very well and belonged to an internationally recognized blockbuster; 9 people thought that the film was made very well, and the level was obviously higher than other films; 2 held a basically positive attitude and thought it was worth seeing. So the researchers concluded that the film is a rare good film, and its popularity among the general public has almost reached 100%, which has been widely publicized. In fact, such sampling is random sampling. Because when researchers draw samples, they are generally the first audience, and usually the first audience may have learned about the film through various channels, almost for the film, so the conclusion is naturally good. The objective sampling method is that the sampling personnel randomly select pedestrians in a certain area for investigation. In this way, people who like the film, don't like the film, and haven't seen the film will be selected, and the conclusion of "the percentage of people who are popular with the public" will be much more reliable.

3. The sampling interval in the sampling is the same as the period of systematic anomaly

Equidistance sampling samples are taken at regular sampling intervals. Therefore, once the sampling interval is the same as the period of systematic anomaly, abnormal sampling results are inevitable.

For example, due to unknown systematic reasons, the 10th product will have quality problems after 9 products are processed each time. Now the quality inspectors go to the production line for sampling inspection, using the equal interval sampling method, with an interval of 10 products. In this way, if the first defective product is just outside the interval point, the sampling result is that all products are qualified. Obviously, such a conclusion is quite different from the reality. If the defective products are located at the interval, that is, the first sampling is defective products, then all the subsequent samples are also defective products, and the sampling result is that all the products are unqualified. Obviously, this conclusion is also incorrect.

4. Randomly determine the sample size

Some sampling personnel did not determine the sample size according to appropriate methods during sampling. Some people simply determine the sample size according to the total number. If the sampling quantity is too small, the analysis of the sampling results will be inaccurate. If the sample size is too large, it will increase a lot of human, material and financial resources, which is uneconomic.

In the Section 2 of the Chapter 3, we introduce the simple random sampling in which the sampling quantity is estimated by a given mean value and the sampling is estimated by a given proportion. When estimating the sample size by means, the sampler needs to first set the confidence level of standard error, the overall standard deviation, the acceptable sampling error range, etc., and then calculate the sample size. When estimating the sample size according to the proportion, the sampling personnel need to set the confidence level of the standard error, the sample proportion in the sampling (usually $P \leqslant 5\%$), and the acceptable sampling error range in advance, and then calculate the sample size.

For the determination of the sample size of a given mean, some empirical values can be used for reference, as shown in Table 4-5 "Sample size of a given mean". Samplers can find the corresponding sample size according to different acceptable sampling errors.

Table 4-5 Sample size for a given mean

Acceptable sampling error/%	Sample size	Acceptable sampling error/%	Sample size
1.0	10 000	6.0	277
1.5	4500	6.5	237
2.0	2500	7.0	204
2.5	1600	7.5	178
3.0	1100	8.0	156
3.5	816	8.5	138
4.0	625	9.0	123
4.5	494	9.5	110
5.0	400	10.0	100
5.5	320		

It can be seen from Table 4-5 that the smaller the acceptable sampling error is, the larger the sample size is. In fact, according to the experience accumulated in calculating the sample size in the past, we can also get some practices that are worth learning from: when the population is smaller, if we want to have a higher probability to get samples with the same results as the population statistical analysis, that is to say, the sample quality is high and the inference accuracy is good, then the sampling rate must be larger. If the sampling rate is small, then a larger population is needed to obtain good statistical analysis results.

Experience shows that if the population is below 1000, the sampling proportion will reach 30% with high statistical accuracy, and the sample size will reach 300. If the population is 10 000, the sampling proportion reaches 10%, which will have the same high statistical accuracy, and the sample size should reach 1000. If the population is greater than 15 000, the sampling proportion reaches 1%, which will also have the same high statistical accuracy, and the sample size should reach 1500. If the population is very large, reaching 10 million, the statistical accuracy is also very good when the sampling proportion reaches 0.025%, and the sample size only needs to reach 2500. Obviously, with the great improvement of the overall quantity level, the small sampling proportion will not greatly affect the accuracy of statistical analysis. Assuming that the total number reaches more than 100 million, the accuracy of statistical analysis is the same as that of 2500 samples from 10 million when the sample size reaches 2500.

5. The sample rotation cycle is too long

Sample rotation means that investigators need to conduct continuous tracking investigation on the samples. Because the samples are constantly changing, they need to rotate the samples partially or completely at regular intervals to obtain new sample information and maintain the accuracy of the research results. Usually, the sample rotation is applicable to the objects with a long investigation period. For example, if the population survey is conducted for a long time (such as for several consecutive years), the sample (population) itself will change significantly over time. When the sample (population) comes out several years later, some of the early samples (population) have improved their academic records and some registered residence has changed. In addition, due to work, study and other reasons, some personnel have a non cooperative attitude towards long-term investigation, so the method of sample rotation must be adopted.

However, if the sample rotation cycle is too long and the objective object has relatively large changes, the sampling results will be inconsistent with the actual situation. There are many situations where great changes have taken place, such as the technological progress of the enterprise's production line, the improvement of production efficiency, and the great changes between the samples before and after. Especially in the information industry, there are new breakthroughs in technology almost every year. Moreover, with the continuous social progress and demographic changes, especially in some formerly backward areas, the previous sample will not be applicable to the current situation because of the great improvement of local construction and economy.

Of course, when the sample rotation cycle is too long, early samples can also be used. For example, early sample data can be used as auxiliary information in regression analysis, which can improve the accuracy of sample estimation.

6. Confusion between point estimation and interval estimation

Point estimation refers to estimating the overall parameters with the sample statistics. Figuratively, the sample statistics is regarded as a point value on the number axis, and the estimated result is also a point value on the number axis. Interval estimation refers to using an interval range to estimate the overall parameters, which is reflected on the number axis. It is to add or subtract the estimation error from the sample statistics to obtain the corresponding estimation interval.

Although both point estimation and interval estimation belong to the category of parameter estimation, there are still obvious differences between them, mainly reflected in two aspects. First, the results are different. Point estimation is to use sample data to estimate a specific value of unknown parameters contained in the population distribution, and its accuracy can be expressed by confidence interval. Interval estimation is to estimate the most probable interval of the unknown parameters of the population based on sample data. Its accuracy is expressed by probability, that is, confidence. Second, whether to consider sampling error. Point estimation is to infer population indicators directly from sampled sample indicators, and sampling error is not considered in sampling inference. In fact, there will be errors if the sampled indicators are used to replace the overall indicators. Interval estimation is to estimate the possible range of the overall index according to the sample index and sampling error. It uses a certain probability to indicate that the error does not exceed a given interval.

Some investigators use point estimation instead of interval estimation, which will lead to inaccurate conclusions. For example, in a survey of household daily expenditure of local residents in a city, the researchers calculated that the average monthly normal expenditure of households in the city was 2000 yuan based on the sampling survey data. So the investigators concluded that the normal monthly expenditure of the city's households was 2000 yuan. In the conclusion, the sampling error and confidence are not described, that is, the point estimation is used to replace the interval estimation of the monthly normal expenditure of each family in the whole.

The correct method is to use interval estimation.

Investigators can use a simple random sampling method (repeated sampling) to obtain a total of 300 households' monthly normal expenditure. Some households' expenditure belongs to a non normal distribution. After calculation, the average value of the sample is 2000 yuan, and the standard deviation s of the sample is 150 yuan.

Although the overall distribution of the sample is unknown, the normal distribution can be used to approximate it because $n = 300$. Substitute sample standard deviation s for population standard deviation σ. Now the confidence level is set as 95%, then $\alpha = 0.05$. Check the normal distribution table to get $Z_{0.05/2} = 1.96$.

Then the population mean μ The confidence interval of is

$$\bar{X} \pm Z_{\alpha/2} \frac{s}{\sqrt{n}}$$

$$= 2000 \pm 1.96 \frac{150}{\sqrt{300}}$$

$$= (1983.03, 2016.97)$$

In this way, the investigators concluded that under the confidence level of 95%, the range of normal monthly expenditure of households in the city was 1983.03 yuan to 2016.97 yuan.

Generally, for simple random sampling and equidistant sampling, the population standard deviation σ is known, then the formula used for interval estimation of population mean is shown in Table 4-6.

Table 4-6　Interval estimation of population value μ

(Confidence is $1- \alpha$ and population standard deviation σ is known)

Population distribution	Sample size	Repeated sampling	Non repeated sampling
Normal distribution	Small sample size ($n < 30$)	$\bar{X} \pm Z_{\alpha/2} \frac{\sigma}{\sqrt{n}}$	$\bar{X} \pm Z_{\alpha/2} \frac{\sigma}{\sqrt{n}} \sqrt{\frac{N-n}{N-1}}$
	Large sample size ($n \geqslant 30$)		

Population distribution	Sample size	Repeated sampling	Non repeated sampling
Non normal distribution	Small sample size ($n < 30$)	—	—
	Large sample size ($n \geqslant 30$)	$\bar{X} \pm Z_{\alpha/2} \dfrac{\sigma}{\sqrt{n}}$	$\bar{X} \pm Z_{\alpha/2} \dfrac{\sigma}{\sqrt{n}} \sqrt{\dfrac{N-n}{N-1}}$

For simple random sampling and equidistant sampling, the population standard deviation σ is unknown, then the formula used for interval estimation of population mean is shown in Table 4-7.

Table 4-7　Interval estimation of population value μ

(Confidence is $1-\alpha$, population standard deviation σ is unknown)

Population distribution	Sample size	Repeated sampling	Non repeated sampling
Normal distribution	Small sample size ($n < 30$)	$\bar{X} \pm t_{(\frac{\alpha}{2},n-1)} \dfrac{s}{\sqrt{n}}$	$\bar{X} \pm t_{(\frac{\alpha}{2},n-1)} \dfrac{S}{\sqrt{n}} \sqrt{\dfrac{N-n}{N-1}}$
	Large sample size ($n \geqslant 30$)	$\bar{X} \pm Z_{\alpha/2} \dfrac{s}{\sqrt{n}}$	$\bar{X} \pm Z_{\alpha/2} \dfrac{S}{\sqrt{n}} \sqrt{\dfrac{N-n}{N-1}}$
Non normal distribution	Small sample size ($n < 30$)	—	—
	Large sample size ($n \geqslant 30$)	$\bar{X} \pm Z_{\alpha/2} \dfrac{s}{\sqrt{n}}$	$\bar{X} \pm Z_{\alpha/2} \dfrac{S}{\sqrt{n}} \sqrt{\dfrac{N-n}{N-1}}$

7. Pay no attention to the design of sampling frame

Some investigators do not pay attention to the design of the sampling frame when carrying out sampling, and some even do not have a sampling frame. A sampling frame is also called a sampling structure or a sampling frame. It refers to a list or sorting number of populations that can be selected as samples to determine the sampling range and structure of the population. The sampling scope, unit and structure are determined according to the purpose and content of the survey. After confirmation, random sampling can be carried out. If there is no sampling frame, the probability of the sample unit cannot be calculated, nor is it possible to select samples by probability.

For example, the inspector simply took 100 supports from the assembly line for analysis, which is the lack of sampling frame design.

The correct method is as follows. Suppose that the factory has 5 support processing machines, and each processing machine processes 1000 supports every day, then 5000 supports are processed every day. The "5000 supports" constitute the sampling frame.

The sampling frame can be understood as the sequence or arrangement form of all sampling units (i.e. population units) that can be selected as samples. After the arrangement, the overall information shall be complete, including the overall unit name, number, serial number, address, etc. Sampling frames can be presented in many forms. Some sampling frames are specific, such as directory structure, region structure, directory region composite structure, etc. In this example, the sampling box is specific, that is, the sampling units can be listed in a table book. Some sampling frames are abstract in form, and they are open in shape. All the elements that meet the survey conditions constitute the elements of the sampling frame, such as random surveys of consumers on the street and in the mall.

The sampling exceptions caused by the design of the sampling frame are：

First, the target population units are not covered enough, that is, the sampling frame does not cover all the target population units, so that some target units have no chance to be included in the sample, which will lead to the low estimation of the population.

Second, the target population units are over covered, that is, some non target population units are included in the sampling frame. If they cannot be found and eliminated in time, the estimation of the population will be too high.

Third, while the overall target units are not fully covered, they also include non target overall units. It is also timely to say that some target companies are not selected in the sample, while non target companies are selected. In this case, if there is a significant difference between the characteristics of the missing target units and those of the non target units, the overall estimation will also have a large deviation.

Fourth, the sampling unit and the target population unit are not completely one-to-one corresponding, and there are one to many, many to one, and many to many situations. The population containing all sampling units is a sampling frame, and the units constituting the sampling frame are sampling units. For example, a company conducts a sampling survey on material suppliers, and the overall composition is all the material suppliers of the company. When sampling, the sampling frame selected by the sampling personnel is the procurement contract signed with the company in recent five years. In this sampling box, suppliers with more contracts are more likely to be selected, which constitutes a many to one situation, that is, there are more units in the sampling box than in the target population. Such overall estimation will also be biased with different supplier characteristic indicators.

Fifth, the sampling frame is out of date, that is, there is a big difference between the sampling population and the target population over time. For example, in enterprise materials, with the development of science and technology, many new materials have been used in projects, but the sampling frame is still an old material variety, which will lead to abnormal estimation accuracy.

8. No distinction between systematic error and accidental error

Systematic error is a non random error. Accidental error is random error, which is caused by small random fluctuations of relevant factors in the objective object, and these errors can compensate each other. Sometimes, investigators fail to carefully distinguish between systematic error and accidental error, which makes the investigation conclusion abnormal. For example, a grain processing plant processes rice with a standard of 5kg per bag. Now the inspectors randomly sample 10 bags on the assembly line, and the data obtained are：5.02 kg, 5.01 kg, 5.05 kg, 5.12 kg, 5.02 kg, 5.10 kg, 5.06 kg, 5.03 kg, 5.01 kg and 5.07 kg. The company requires that the error be controlled within 2.5%. Therefore, the inspectors believe that all the processed rice meets the requirements. However, the reinspection personnel found that the weight of these 10 bags of rice was more than 5 kg, and none of the bags was less than 5 kg, while the weight of normal processed rice should fluctuate around 5 kg. Therefore, there are systematic errors in the processing line.

There are many reasons for systematic errors. For example, the processing line itself has problems; The error of the measuring instrument itself, such as the zero point is not corrected; The conditions during the test do not meet the requirements specified in the theory; Personal preferences exist; The ambient temperature is too high or too low; Sampling from a sampling frame has serious defects, etc.

9. No distinction between unequal probability sampling and equal probability sampling

Equal probability sampling refers to that the probability of individuals being sampled in the population is equal when sampling, and it is applicable to that the size difference between individuals in the population is not very obvious. Unequal probability sampling refers to that when sampling, the probability of individuals being selected in the population is not equal. It is applicable to situations where there are significant differences in size between individuals in the population, and the signs to be investigated are strongly positively correlated with individuals. Different probability sampling is used to reduce the error of population estimation.

In a population, it is almost impossible for individuals to be absolutely equal, so sampling with unequal probability is usually applicable to two situations. One is that individuals in the population have relatively complete signs, and the probability of being selected can be determined according to the size of individuals. The other is that there is a clear relationship between the survey signs and auxiliary signs, such as a positive correlation and an equal proportion between them.

Sometimes, investigators fail to distinguish between unequal probability sampling and equal probability sampling, which will lead to abnormal conclusions. For example, a researcher investigated the number of employees and total wages of four enterprises. The data obtained are shown in Table 4-8.

Table 4-8   Number of employees and total wages of the four companies

Enterprise name	A	B	C	D	Total
Number of employees/person	30	50	80	120	280
Total wages of the enterprise/10 thousand	180	240	450	900	1770

Now suppose the sample size is 2, then the sample composition is $4^2 = 16$.

It can be seen from Table 4-8 that the probability of the four enterprises being selected is different. The probability of company A being selected is $30/280 = 10.7\%$, that of company B is 17.8%, that of company C is 28.6%, and that of company D is 42.9%. That means that the probability of four companies being selected is different. If the probability is equal, then the probability of each company being selected is 25%. Therefore, if we calculate the variance of the payroll estimator with equal probability, it will be significantly larger, while the variance of the payroll estimator with unequal probability will be much smaller.

10. Lack of sampling control procedure

Some investigators lack standardized sampling control procedures as guidance when performing sampling tasks, and the quality of sampling activities does not meet the requirements of relevant standards.

In the quality management system, a complete sampling procedure includes the following main contents: the purpose of formulating the sampling procedure; Scope of application of the procedure; Responsibilities of relevant personnel; The main contents of the procedure include the preparation of sampling work, requirements for sampling personnel, sampling method, sampling time, and factors to be controlled in the sampling process; Inspection regulations for deviation from sampling procedures; Normal, tightened and relaxed inspection regulations; Control requirements for sampling inspection; Standards for sampling inspection; Sampling records and requirements, etc.

4.4.2 Abnormal sampling personnel

The abnormal sampling caused by the abnormality of the sampling personnel mainly include the following items.

(1) The sampling personnel lack the training of sampling knowledge and the mastery of sampling related theories and methods.

(2) The sampling personnel lack the sense of responsibility and the necessary understanding and attention to sampling.

(3) The sampling personnel lack the ability to deal with emergency situations, such as the representativeness of samples, the handling of sampling errors, etc.

(4) The sampling personnel are not familiar with the relevant requirements of the sampling object, so that the sampling is not standardized.

(5) The sampling form was filled out irregularly and lacked necessary information, such as product name, specification, grade, executive standard, sampling basis, sampling quantity, inspection nature, sampling location, verification status, batch size, etc.

(6) Unclear assignment of responsibilities of personnel.

4.4.3 Abnormal sampling equipment

The abnormal sampling caused by the abnormal sampling equipment mainly include the following items.

(1) Lack of special sampling equipment. Some samples must have special sampling instruments and storage dishes.

(2) Lack of effective management of sampling equipment. There are many sampling appliances, such as sampling containers, sample equipment, sample labels, sample record forms, electronic scales, safety protection articles, etc., which must have strict management measures, otherwise, the accuracy of the appliances will be reduced, and items will be missing, making the sampling work impossible to carry out normally and inefficient.

(3) The sampling equipment does not work properly. For example, poor contact, pipeline damage, blockage, unstable flow, etc. exist.

(4) The sampling equipment lacks safety protection measures and regular inspection so that safety risks exist.

(5) Use waste sampling equipment for sampling.

(6) Mutual interference between sampling equipment.

4.4.4 Abnormal sampling process

The abnormal sampling process mainly include the following items.

(1) There is no standardized sampling process and sampling is random.

(2) The sampling plan was not implemented and the sampling was not conducted according to the sample quantity specified in the sampling plan.

(3) The sampling plan has specific requirements, but there is no random sampling in the specific implementation, and the sampling is not representative.

(4) The sampled samples are not identified in time or are lack of identification.

(5) When encountering some accidents in sampling, there is a lack of corresponding alternatives. For example, when there are not enough people, use acquaintances to help; When it is difficult to take samples, give up the samples easily; Make temporary replacement of samples and increase or decrease of samples, etc.

(6) The existence of sampling error not calculated in sampling.

(7) Lack of supervision mechanism in the sampling process.

4.4.5 Abnormal sampling conditions

Abnormal sampling conditions mainly include the following characteristics.

(1) The definition of the population is not clear, which makes the sampled samples out of the range of the population.

(2) The population distribution is too scattered, so that the samples are also too scattered, resulting in difficulties in subsequent analysis.

(3) The sample size is too small, resulting in insufficient representativeness of the sample.

(4) A certain feature of the objective object has been clearly understood, and this feature will affect the research results, but the sampling is still carried out according to the simple random sampling method, which makes the sample representative insufficient.

(5) The sample size is abnormal, and there is no corresponding correction scheme.

(6) The sampling environment changes. For example, the samples are improperly stored, deteriorated, polluted, etc; Some light decomposing samples lack light proof containers. The sampling environment lacks corresponding management methods.

(7) The accuracy of sample research is reduced due to the mixed use and cross contamination of sampled instruments and equipment.

(8) The safety of sampling is at risk. The sampled samples were not effectively isolated. The sample is lack of monitoring, and there is a risk of sample damage.

(9) There are no remedial plans and measures after sampling failure.

4.4.6 Countermeasures for sampling anomaly

There should be corresponding corrective and preventive measures for the occurrence of sampling abnormalities. The main countermeasures are as follows.

(1) Improve the spot check system, clarify relevant responsibilities and improve the sense of responsibility.

(2) Strengthen business training, correctly carry out sampling plan design, determine sam-

ples, etc., carry out quality control in the sampling process, and implement data processing specifications.

(3) Before sampling, learn relevant standards and regulations, master sampling methods, and be familiar with the environmental conditions and technical requirements of sampling. When it comes to construction production, process flow and other situations, it is necessary to understand the relevant contents of construction production and process flow.

(4) The sampling plan shall be strictly formulated. The appropriate sampling method shall be selected. The sampling content and relevant requirements shall be determined, and the process operation shall be strictly carried out. For the sampling of objective objects with clear characteristics, the quota sampling method should be selected to improve the accuracy.

(5) The sampled population shall be clearly defined, and the boundary of the population shall be clearly defined without ambiguity.

(6) Sufficient sample size shall be ensured for sampling. If the sample size is abnormal, there should be a corresponding correction plan.

(7) Ensure the stability and safety of the sampling environment. There are inspection and protection systems for samples and they are implemented correctly. Ensure that the conditions of equipment, facilities and instruments meet the requirements and ensure the stability of output products.

(8) Necessary and professional sampling instruments shall be added for sampling.

(9) The sampling error and non-sampling error caused by various reasons shall be effectively controlled. Necessary measures shall be taken to reduce the risk caused by negligence and avoid sampling errors.

(10) The sampling specification shall be strictly implemented. The sampled vehicles, equipment and personnel meet the work requirements.

(11) For fragile articles, hazardous chemicals and samples with special storage conditions, correct measures should be taken to ensure that the samples do not change during transportation.

(12) If there are packaging requirements for the sample, the sample packaging shall meet the specification requirements. Anti-unpacking measures shall be taken to ensure the authenticity of samples.

(13) The number of sampling personnel shall not be less than 2 to ensure the fairness of sampling.

(14) The sampling records shall use standardized sampling documents to record the sampling information in detail.

4.4.7 Design of sampling plan

The design steps of a sampling plan usually include:

(1) Clarify the purpose of sampling survey.

(2) Define the scope and content of the sampling survey, including determining the population and the target quantity of the population, determining the sampling frame, and determining the specific content of the survey.

(3) Determine the sampling method according to the specific situation of the population.

（4）Determine the precision requirements of this sampling.

（5）Calculate the sample size. If sampling is conducted in stages, the configuration of sample size for each stage shall also be determined.

（6）Determine the estimation method and effect of the population target quantity.

（7）Determine the time and place of sampling, and prepare necessary sampling tools and equipment.

（8）Specify the organization form of sampling and determine the relevant personnel for sampling.

（9）Formulate the methods and steps for implementing the sampling plan.

（10）Review and management of samples.

（11）Corrective measures for abnormal conditions.

（12）Other matters, such as sufficient sufficient fund.

4.4.8 Zero defect sampling plan

（1）The origin of zero defect. The concept of zero defect was first proposed by Philip B. Crosby, a master of quality management, in the 1960s. The US Department of Defense issued MIL-STD-105A specification in 1950. Nicholas L. Squeglia, an American university professor, published a sampling plan with C=0 in 1965. The subsequent C=0 sampling plan was formulated according to MIL-STD105 standard, and its acceptance criterion was "0 acceptance and 1 withdrawal", that is, the sampling was accepted only after all samples were qualified. In 1994, the three major American automobile manufacturers (General Motors, Ford, Chrysler) issued the QS9000 quality system specification, which defined the acceptance criteria of zero defects. In 2020, the United States issued ANSI/ASQ/Z1.4-2020 specifications.

（2）Characteristics of the zero defect scheme. Mainly because the inspection quantity is relatively small and economic, easy to manage and use; It is mainly applicable to sampling by attributes.

（3）When adopting the zero defect scheme, it is also necessary to specify the acceptance quality limit (AQL) according to the quality characteristics of different objective object categories.

4.5 模拟异常

模拟异常是数据发生异常的主要问题之一。常见的模拟异常主要有：建模异常、随机数错误、分布曲线引用不当、模拟的条件异常等。

4.5.1 建模异常

建模异常的情形主要有数学模型错误、几何模型的简化处理、分析中的相关设置异常等。

1. 数学模型错误

（1）非线性问题简化为线性问题

在建模中，最容易理解的就是在散点图中求取回归直线，这个直线的函数是对于模型条件的要求为最少，其基本原理是 OLS（Ordinary Least Squares），意思是普通最小二乘法。图 4-8 就是运用 OLS，在各散点中求得一条回归直线。

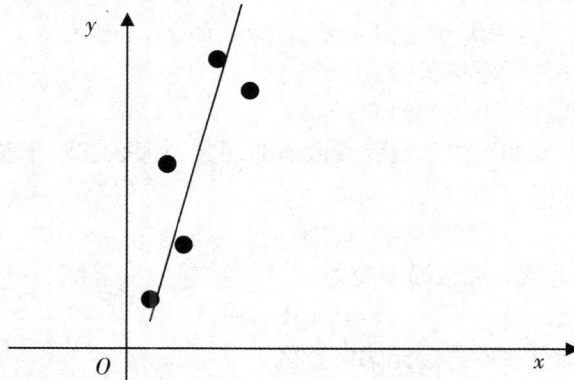

图 4-8 运用 OLS 得到回归直线

人们为了分析方便，用回归直线来建立最简单的模型。但在实践中，单纯的回归直线是不能满足建模需求的。因此，如果把非线性问题简化为线性问题，那么模型就发生异常。如果把图 4-8 中的散点继续收集下去，那么就得到了图 4-9 所显示的散点。

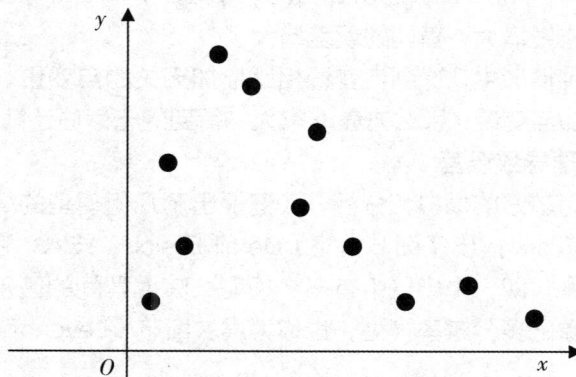

图 4-9 更多的散点图

针对图4-9的散点分布情形，应用OLS来求取函数，于是找到一条最接近于原来数据点的曲线（见图4-10），这就是一条符合实际的模型曲线。

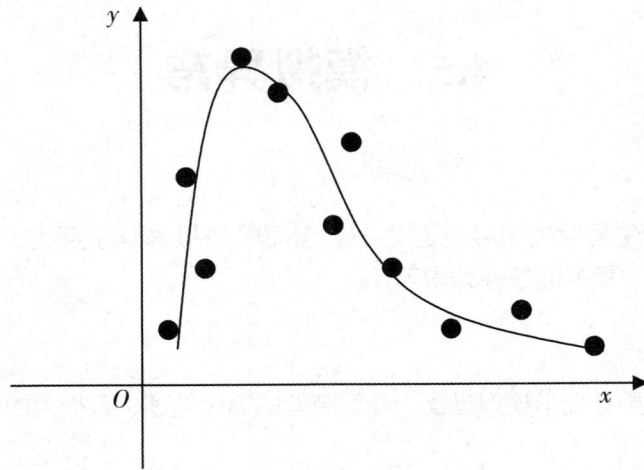

图4-10　OLS获得的模型曲线

利用各散点的历史数据，可以计算出这条曲线的函数表达式，由此而得到了模型参数。用这样的一个模型来近似地表达原来的数据点，则在模型中测出的值与实际值之间会存在偏差（残差）。由此，得到了一个多元线性回归模型：

$$y_i = \beta_0 + \beta_1 x_{i1} + \cdots + \beta_m x_i + \varepsilon_{im}, i = 1, 2, \cdots, n$$

这样才完成了一个非线性问题的建模。

（2）自变量有遗漏或增加了无关的自变量

在一个正常的模型中，假如发生了自变量遗漏的现象，那么它对于整个模型的正确性产生很大偏差。比如，一个正常的模型为

$$y = \beta_0 + \beta_1 x_1 + \beta_2 x_2 + \varepsilon$$

假如模型中的 x_2 发生遗漏，那么模型变为

$$y = \beta_0 + \beta_1 x_1 + \varepsilon$$

则整个模型的参数估计量将不再是无偏估计量。

另一种情形，假如在一个正常的模型中，增加了无关的自变量，那么，原来的参数估计量仍然是无偏，但会增大估计量的方差。比如，一个正常的模型为

$$y = \beta_0 + \beta_1 x_1 + \beta_2 x_2 + \varepsilon$$

假如现在增加了自变量 x_3，那么模型变为

$$y = \beta_0 + \beta_1 x_1 + \beta_2 x_2 + \beta_3 x_3 + \varepsilon$$

显然，模型的计算工作量增大，模型的误差增大了。

客观地说，在上述两种情形中，遗漏自变量相比增加无关的自变量，其对于模型的影响程度更大。当然，如果随意增加自变量，那么方差非常大，得到的无偏估计量也就没有多大意义。

2. 几何模型的简化处理导致误差

为了对客观对象实体有较好的模拟和分析，人们提出了几何模型的简化，以此来生成客观对象的简化模型。1976年，Clark提出了细节省略（Detail Elision）技术，它是指用具有多层次结构的物体几何来描述一个场景，即一个场景中有多个模型，其模型间的区别在于细节的描述程度。这样就引申出构造客观对象的多分辨率模型，也称为层次细节（Level of Detail）模型，简称LOD技术。

在简化过程中，会用到各种简化的算法。比如"顶点删除算法"，它是一种应用广泛的模型

简化算法，其思路是：在三角网格中，如果一个顶点与它周围的三角面片可以被认为是共面的时候，而且删除该点不会带来拓扑结构的概念，那么就可以把这一点删除。删除该点以后，所有与该点相连的面也同时被删除，然后重新对其邻域进行三角化。顶点删除法示意见图4-11。

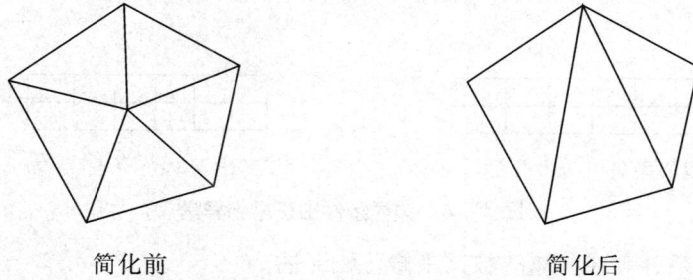

简化前　　　　　　　　　　　　　　　简化后

图4-11　顶点删除算法示意

几何模型的简化处理固然好，但会导致一定的误差发生。假设现在有简化处理后的模型，见图4-12。

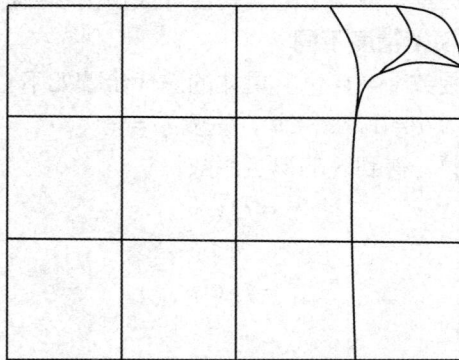

图4-12　几何模型简化处理图

在图4-12中可以看到，几何模型简化处理后出现了一些不规则图形，见图的右上角。一定程度上，它反映了一些误差的产生。那么这些图形是否可以删除，或者是用其他信息代替，形成了模型的异常。

3. 建模过程中的相关参数设置异常

在建模过程中，如果相关的参数设置异常，那么直接导致模拟的结果异常。图4-13简单地表述了由圆的参数代替了菱形参数的设置，图形发生很大不同。

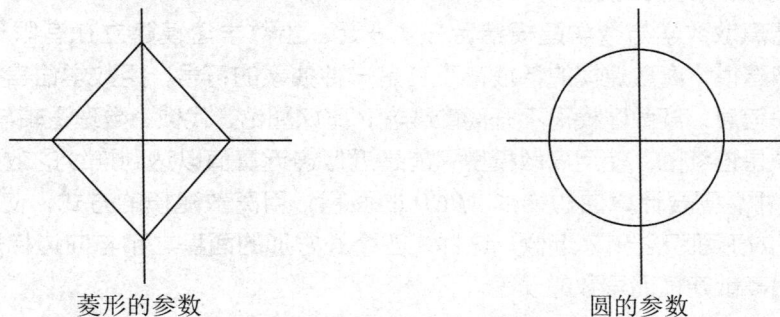

菱形的参数　　　　　　　　　　　　　　圆的参数

图4-13　参数的设置不同

实际建模中的参数很多，比如有失效准则、连接关系、材料参数、等效参数等，需要通过模型来修正。

4. 模型的边界条件使用异常

模型的边界条件是指模型在求解区域边界上的变量随时间和地点的变化而变化的规律。如果边界条件被正确使用，那么边界条件可以恰当地反映简化模型的复杂性。图4-14表述了边界条件由正常到异常的结果。

边界条件正常　　　　　　　　　　　边界条件异常

图4-14　边界条件由正常到异常

在建模中，边界条件需要大量的数据来修正和评估。

4.5.2 算法上的误差导致建模异常

数学建模有很多算法，比如蒙特卡洛法、数据拟合法、参数估计法、线性规划法、二分图、神经网络法、网格算法、穷举法等。算法上的误差会导致建模和模拟的异常，以下简单说明。

1. 数值的迭代路径不同使模型精度不同

第一种迭代法，使用拟合函数 $y = f(x)$，具体的迭代步骤如下：

设有数据集 $S = (x_i, y_i) \mid i = 1, 2, \cdots, n$，先设 $n = 1$，

第一步，用最小二乘法计算，得到 S 的拟合函数：

$$y = f_n(x)$$

第二步，用该函数来预测

$$y_{n+1} = f_n(x_{n+1})$$

第三步，更新数据集，得到

$$S = S \cup \{(x_{n+1}, y_{n+1})\}$$

继续重复第一步，不断循环往下，直到函数 $y = f_n(x)$ 收敛，输出收敛函数 $y = f(x)$。

第二种迭代法，使用函数

$$f_n(x) = \sum_{i=1}^{n} a_i\, h_{(b_i, c_i)}(x)$$

这是一种越阶函数，通过这样的一种迭代法，精度会逼近任何闭区间上的连续函数或者开区间上的绝对连续函数。这种方法是目前深度学习、神经网络中的算法。

2. 用离散的方法建模导致误差

离散建模是把离散数学与数学建模结合在了一起。当数学建模建立在有限集或可列集之上时，就称之为离散建模。离散建模的特征在于它是一种抽象的符号，用数学语言来表示，建模时可以屏蔽大量无关信息。离散建模离不开离散数学，比如图论、代数、数理逻辑与关系等。

由于客观对象是连续的，使用离散模型只能是近似逼近真值以达到连续的效果，所以产生了误差。在图4-15中，需要计算曲边梯形 $ABCD$ 的面积，用离散建模的方式，使用了编号为1、2、3、4的四个长方形面积之和来近似。这样，四个长方形的面积之和与曲边梯形面积之间存在着误差。这就是由离散建模而导致的误差。

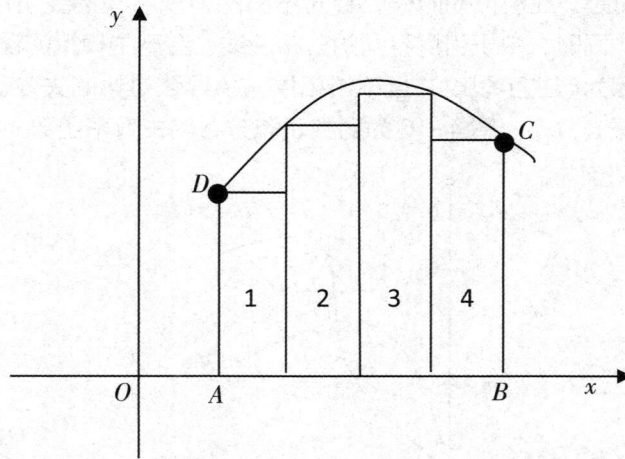

图 4-15 离散模型逼近曲线

离散化误差是存在较多的一种误差，比如在有限元分析中，其自由度是有限的，而关于自由度的数字模型有无数个，这就导致了误差的存在。

3. 由算法的收敛性而导致误差

以蒙特卡洛法为例，蒙特卡洛法是以概率统计理论为基础的一种方法，它按照概率模型所描述的过程，来解决一些数值方法难以解决的问题。运用蒙特卡洛法所得到的近似值与真值之间存在着误差。

根据中心极限定理，设 X_1，X_2，$\cdots$，X_n 为独立且同分布的变量序列，则它们的方差（误差）为

$$\sigma^2 = \int (x - E(x))^2 f(x) \, \mathrm{d}x$$

$f(x)$ 是 X 的分布密度函数。则有

$$\lim_{n \to \infty} P\left(\frac{\sqrt{n}}{\sigma} \mid \bar{x}_n - E(X) \mid < x \right) = \frac{1}{\sqrt{2\pi}} \int_{x}^{x} \mathrm{e}^{\frac{t^2}{2}} \mathrm{d}t$$

当 n 充分大的时候，不等式 $\mid \bar{x}_n - E(X) \mid$ 近似地收敛于概率（$1 - \alpha$）。

运用蒙特卡洛法，其误差为

$$\varepsilon = \frac{Z\sigma}{\sqrt{n}}$$

上式中，Z 为置信水平，可以通过查证态分布表而得到。Z 值与 α 一一对应，常用的三个对应值见表 4-9。

表 4-9　Z 值与 α 对应表

α	0.5	0.05	0.003
Z 值	0.6745	1.96	3

在工程建模中，收敛性问题是其误差的重要来源。

4.5.3 引用分布曲线的不同形成建模异常

在模拟中，经常用到各种不同的函数曲线来作为分布曲线。常用的分布曲线有：均匀分布、指数分布、伽马分布、贝塔分布、正态分布、威布尔分布、三角分布、对数正态分布、二项分

布、几何分布、泊松分布等。在模拟的时候，针对不同的客观事物需要引用不同的分布曲线，这样才有更好的模拟效果。同时，由于引用不同的分布曲线，会产生模拟的误差。

比如，有某种材料因为产生了裂纹并且不断拓展，会导致材料的失效。在模拟的时候，正确的分布曲线应当是对数正态分布。图 4-16 为对数正态分布的失效率函数 $\lambda(t)$ 的函数曲线。

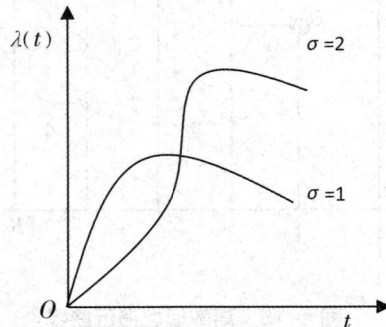

图 4-16　对数正态分布的失效率函数曲线

但如果在模拟的时候，所引用的分布曲线为威布尔曲线，则模拟的结果会有很大的差别。图 4-17 为威布尔分布的失效率函数 $\lambda(t)$ 的函数曲线，图中 m 为形状参数。

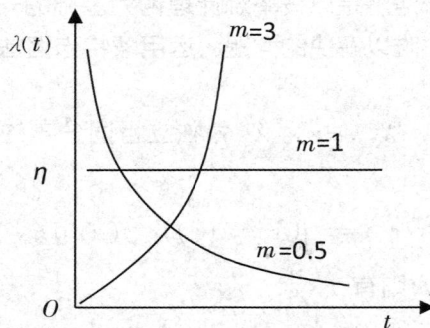

图 4-17　威布尔分布的失效率函数曲线

显然，由于分布曲线的引用不同，模拟的结果肯定存在着较大的误差。

举例：

某质量小组研究某种型号的温度传感器的可靠性，小组随机抽取了 15 个温度传感器进行寿命试验的长期跟踪，测得的寿命数据为（单位：日）：90，170，280，360，550，730，920，1100，1300，1600，2000，2400，2900，3400，4000+。现在小组需要找到最适合这组数据的参数分布模型。

具体分析：从采集的 15 个数据看，前 14 个数据为精确寿命数据，第 15 个数据属于右删失数据。现在使用 Minitab 软件进行不同分布的 ID 确认分析。在分析过程中，使用的分布有威布尔分布、对数正态分布、指数分布和对数分布四种。不同分布的寿命概率分析结果见图 4-18。

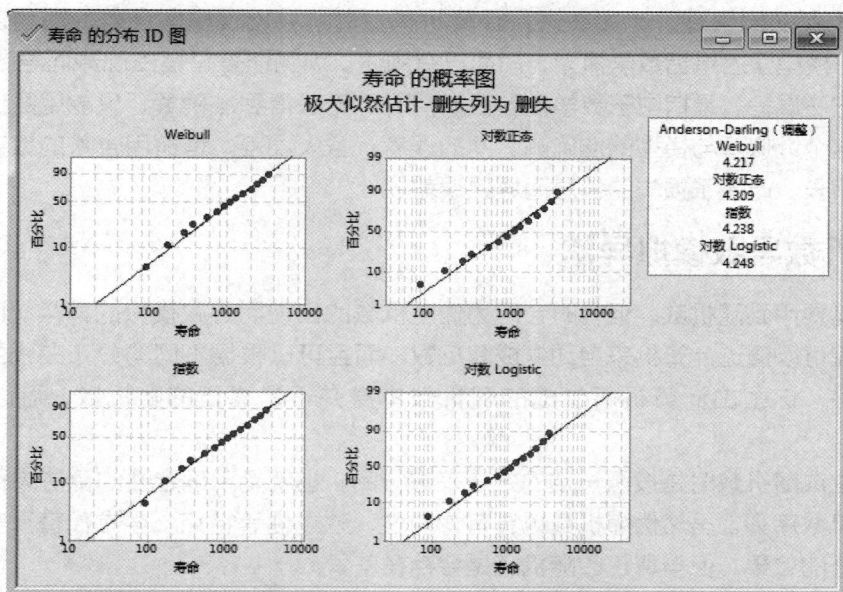

图 4-18　不同分布的寿命概率图

输出的拟合优度值见图 4-19。

图 4-19　不同分布的拟合优度值

从图 4-19 中可以看出，Weibull 分布的拟合优度值 4.217 为最小，说明拟合最好。那也说明如果使用其他分布曲线，则误差更大。

4.5.4 条件的局限性导致模拟异常

模拟的过程需要相应的条件来支持。由于相关条件的局限性，会导致模拟的异常产生。主要有：

（1）环境条件的局限性。比如一些材料的老化试验模拟，在不同的温度、速度、高度（大气压）、湿度、照度等情况下的结果会有很大的不同，但由于环境条件所限，构成了模拟的误差。

（2）思维的局限性。每个人对于建模有着不同的思路，而思维的局限性会限制建模的过程和结果。有的人能够在复杂现象的背后看到普遍规律，而有的人则不能；有的人能够使用不同的算法，而有的人只局限在有限的算法中。要突破思维的局限性，需要在两个方面努力：一是横向拓展，即针对模拟的对象，把它与其他相关的知识、概念进行对比，发现有哪些相似之处、哪些不同之处等，实现知识的链接。二是纵向拓展，即针对模拟的对象，对其纵向所有的知识进行串联，然后实现知识的突破，构建创新的模型。

（3）使用工具的局限性。比如在进行计算机模拟时，会出现模型误差、观测误差、截断误差、舍入误差。模型误差是指数学模型，前面已有阐述。观测误差是指由观测而导致的误差。截断误差，也称方法误差，是指由于数学模型很多情况下无法得到准确解，只能用数值方法求得模型的近似解，这个近似解与模型的准确解之间的误差。舍人误差，是指用舍入的数字来适应一个有限的计算机字长，产生了损失的信息而导致误差。

4.5.5 伪随机数导致模拟异常

很多模拟需要用到随机数，但有时会因为伪随机数的出现而导致模拟的异常。比如蒙特卡洛法的应用，现在可以通过计算机来自动生成随机数，而且可以根据不同的分布曲线生成相应的随机数。而事实上，这些由计算机所生成的随机数本身并不是真正的随机数，可以称之为伪随机数。

计算机在生成随机数时是按照一定的方法来进行的，也就是用确定性的算法来计算出［0,1］均匀分布的随机数序列。虽然伪随机数也具有均匀性、独立性等特点，通常代替真实的随机数进行使用也有很好的效果，但与真正的随机还是会存在差异。

4.5　Abnormal Simulation

Abnormal Simulation is one of the main problems of data anomaly. Common simulation anomalies mainly include: modeling anomaly, random number error, improper reference of distribution curve, simulated condition anomaly, etc.

4.5.1 Abnormal modeling

Abnormal modeling mainly includes errors in mathematical models, simplification of geometric models, and exceptions in relevant settings in analysis.

1. Mathematical model error

(1) The nonlinear problem is reduced to a linear problem

In modeling, it is easiest to understand that the regression line is obtained in the scatter plot. The function of this line is to minimize the requirements for model conditions. Its basic principle is OLS, which means ordinary least squares. Figure 4-8 uses OLS to find a regression straight line among scattered points.

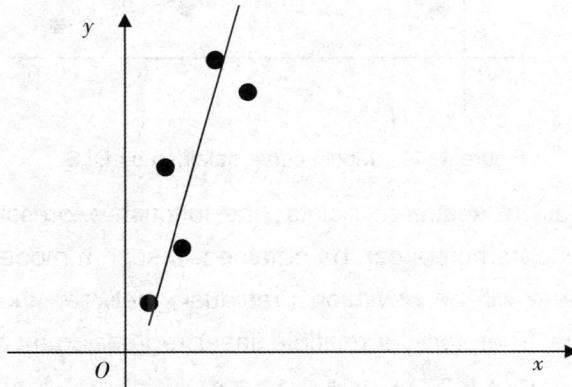

Figure 4-8　Using OLS to get regression straight line

For the convenience of analysis, people use regression lines to build the simplest model. But in practice, the simple regression line cannot meet the modeling requirements. Therefore, if the nonlinear problem is simplified to a linear problem, the model will be abnormal. If we continue to collect the scatter points in Figure 4-8, we will get the scatter points shown in Figure 4-9.

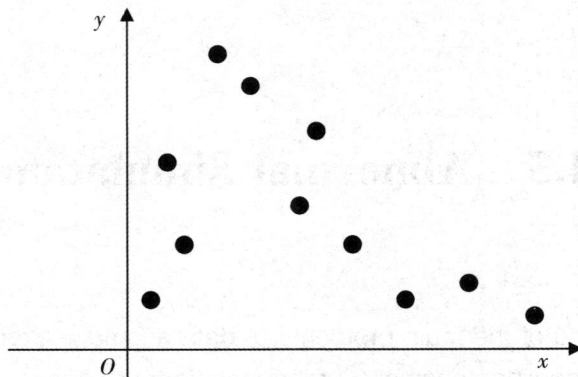

Figure 4-9　More scatter points

For the distribution of scattered points in Figure 4-9, OLS is applied to obtain the function, and then a curve that is closest to the original data points (see Figure 4-10) is found. It is a model curve in line with the actual situation.

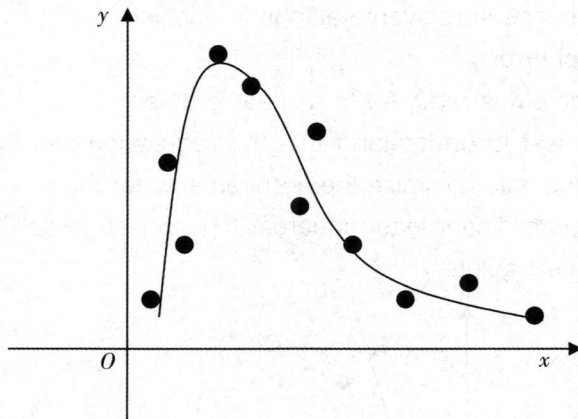

Figure 4-10　Model curve obtained by OLS

Using the historical data of scattered points, the function expression of this curve can be calculated, and the model parameters can be obtained. If such a model is used to approximate the original data point, there will be deviation (residual) between the measured value in the model and the actual value. Therefore, a multiple linear regression model is obtained:

$$y_i = \beta_0 + \beta_1 x_{i1} + \cdots + \beta_m x_i + \varepsilon_{im}, i = 1, 2, \cdots, n$$

In this way, the modeling of a nonlinear problem is completed.

(2) The independent variable is omitted or irrelevant independent variable is added.

In a normal model, if the omission of independent variables occurs, it will cause great deviation to the correctness of the whole model. For example, a normal model is

$$y = \beta_0 + \beta_1 x_1 + \beta_2 x_2 + \varepsilon$$

If x_2 in the model is omitted, the model becomes:

$$y = \beta_0 + \beta_1 x_1 + \varepsilon$$

Then the parameter estimator of the whole model will no longer be unbiased.

Another kind, if an independent variable is added to a normal model, the original parameter estimator is still unbiased, but the variance of the estimator will increase. For example, a normal model is

$$y = \beta_0 + \beta_1 x_1 + \beta_2 x_2 + \varepsilon$$

If the independent variable x_3 is now added, the model becomes:

$$y = \beta_0 + \beta_1 x_1 + \beta_2 x_2 + + \beta_3 x_3 + \varepsilon$$

Obviously, the calculation workload of the model increases, and the error of the model increases.

Objectively speaking, in the above two cases, the omission of independent variables has a greater impact on the model than the addition of independent variables. Of course, if the independent variable is added at will, the variance is very large, and the unbiased estimator obtained is meaningless.

2. Simplification of geometric models leads to errors

In order to have a better simulation and analysis of the objective object entity, people put forward the simplification of the geometric model to generate the simplified model of the objective object. In 1976, Clark proposed the Detail Elision technology, which refers to the use of object geometry with a multi-level structure to describe a scene, that is, there are multiple models in a scene. The difference between models is the degree of detail description. This leads to a multi-resolution model for constructing objective objects, also known as the Level of Detail model, or LOD technology for short.

In the process of simplification, various simplified algorithms will be used. For example, the "vertex deletion algorithm" is a widely used model simplification algorithm. Its idea is: in a triangular mesh, if a vertex and its surrounding triangular patches can be considered coplanar, and deleting the point does not bring about the concept of topology, then this point can be deleted. After the point is deleted, all the faces connected to the point are deleted at the same time, and then its neighborhood is triangulated again. See Figure 4-11 for vertex deletion method.

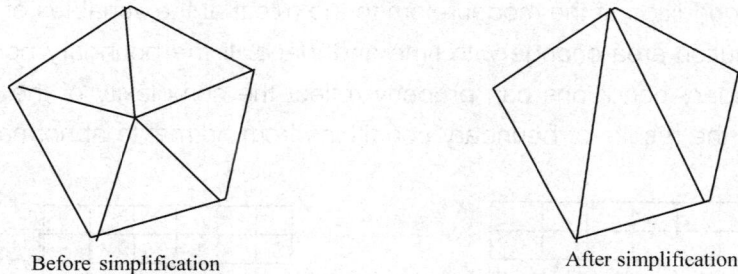

Before simplification   After simplification

Figure 4-11   Schematic algorithm of vertex deletion

Although the simplification of geometric model is good, it will lead to some errors. Suppose there is a simplified model, as shown in Figure 4-12.

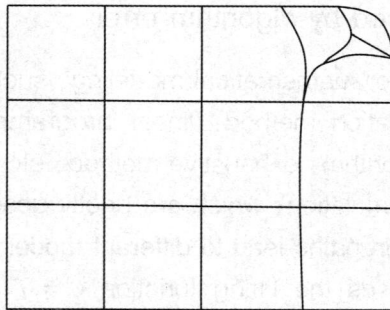

Figure 4-12   Simplification of geometric model

As can be seen in Figure 4-12, some irregular figures appear after the geometric model is simplified, as shown in the upper right corner of the figure. To some extent, it reflects the generation of some errors. Then whether these figures can be deleted or replaced with other information, forming the model exception.

3. Abnormal parameter setting during modeling

During the modeling process, if the relevant parameter settings are abnormal, the simulation results will be abnormal directly. Figure 4-13 simply shows that the setting of diamond parameters is replaced by circle parameters, and the figures are very different.

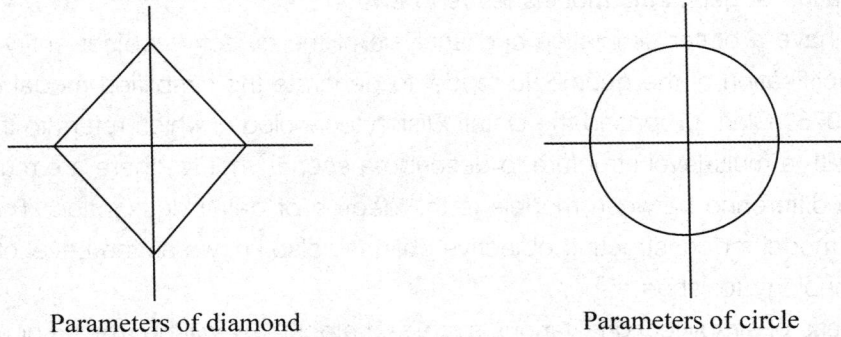

Parameters of diamond   Parameters of circle

Figure 4-13　Different parameter settings

There are many parameters in the actual modeling, such as failure criteria, connection relations, material parameters, equivalent parameters, etc., which need to be modified through the model.

4. Abnormal use of boundary conditions of the model

The boundary condition of the model refers to the rule that the variables of the model on the boundary of the solution area change with time and place. If the boundary conditions are used correctly, the boundary conditions can properly reflect the complexity of the simplified model. Figure 4-14 shows the results of boundary conditions from normal to abnormal.

Normal boundary conditions   Abnormal boundary conditions

Figure 4-14　Boundary conditions from normal to abnormal

In modeling, boundary conditions need a lot of data to correct and evaluate.

4.5.2 Modeling anomaly caused by algorithm error

There are many algorithms for mathematical modeling, such as Monte Carlo method, data fitting method, parameter estimation method, linear programming method, bipartite graph, neural network method, grid algorithm, exhaustive method, etc. Errors in the algorithm will lead to exceptions in modeling and simulation, which are briefly described below.

1. Different numerical iteration paths lead to different model accuracy

The first iteration method uses the fitting function $y = f(x)$, and the specific iteration steps are:

Set: with data set $S = (x_i, y_i) \mid i = 1, 2, \cdots, n$,

Let $n = 1$ first,

The first step is to calculate with the least square method to obtain the fitting function of S:

$$y = f_n(x)$$

The second step, use this function to predict

$$y_{n+1} = f_n(x_{n+1})$$

The third step, update the dataset to get

$$S = S \cup \{(x_{n+1}, y_{n+1})\}$$

Continue to repeat the first step and loop down until the function $y = f_n(x)$ converges, and output convergence function $y = f(x)$.

The second iterative method is to use functions

$$f_n(x) = \sum_{i=1}^{n} a_i h_{(b_i, c_i)}(x)$$

This is a kind of over order function. Through such an iterative method, the accuracy will approximate any continuous function on a closed interval or an absolute continuous function on an open interval. This method is the algorithm of deep learning and neural network.

2. Modeling with discrete methods leads to errors

Discrete modeling is the combination of discrete mathematics and mathematical modeling. When mathematical modeling is based on finite sets or countable sets, it is called discrete modeling. The characteristic of discrete modeling is that it is an abstract symbol, expressed in mathematical language, and can shield a lot of irrelevant information during modeling. Discrete modeling is inseparable from discrete mathematics, such as graph theory, algebra, mathematical logic and relations.

Because the objective object is continuous, the discrete model can only approximate the true value to achieve continuous effect, so errors are generated. In Figure 4-15, it is necessary to calculate the area of curved trapezoid $ABCD$, which is approximated by the sum of the areas of four rectangles numbered 1, 2, 3, and 4 in a discrete modeling manner. In this way, there is an error between the sum of the areas of the four rectangles and the area of the curved trapezoid. This is the error caused by discrete modeling.

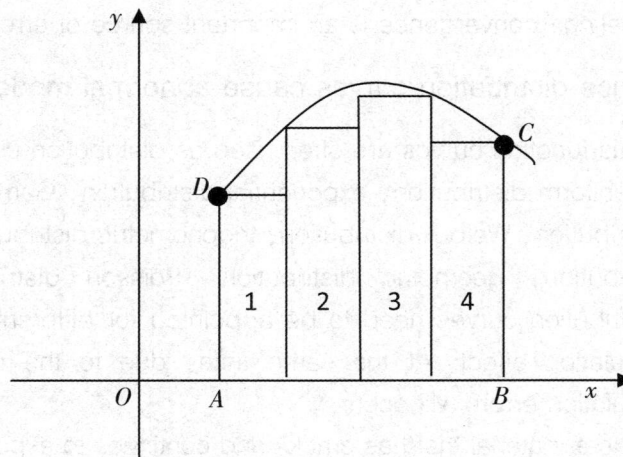

Figure 4-15   Approximation curve of discrete model

Discretization error is a kind of error. For example, in finite element analysis, the degree of freedom is limited, and there are countless digital models about the degree of freedom, which leads to the existence of error.

3. Error caused by convergence of algorithm

Take the Monte Carlo method as an example. The Monte Carlo method is a method based on the theory of probability and statistics. It solves some problems that are difficult to solve by numerical methods according to the process described by the probability model. There is an error between the approximate value obtained by Monte Carlo method and the true value.

According to the central limit theorem, let X_1, X_2, $\cdots$, X_n be independent and identically distributed variable sequences, then their variance (error) is

$$\sigma^2 = \int (x - E(x))^2 f(x) \, \mathrm{d}x$$

$f(x)$ is the distribution density function of X. Then there is

$$\lim_{n \to \infty} P\left(\frac{\sqrt{n}}{\sigma} \mid \bar{x}_n - E(X) \mid < x\right) = \frac{1}{\sqrt{2\pi}} \int_{-x}^{x} e^{-\frac{t^2}{2}} \mathrm{d}t$$

When n is sufficiently large, the inequality $\mid \bar{x}_n - E(X) \mid$ converges approximately to probability $(1 - \alpha)$.

The error of Monte Carlo method is

$$\varepsilon = \frac{Z\sigma}{\sqrt{n}}$$

In the formula, Z is the confidence level, which can be obtained by looking up the normal distribution table.

It is one-to-one correspondence between Z value and α. See Table 4-9 for three commonly used correspondence values.

Table 4-9　Z value vs α Corresponding table

α	0.5	0.05	0.003
Z value	0.6745	1.96	3

In engineering modeling, convergence is an important source of error.

4.5.3 Different reference distribution curves cause abnormal modeling

In simulation, various function curves are often used as distribution curves. Common distribution curves include: uniform distribution, exponential distribution, Gamma distribution, Beta distribution, normal distribution, Weibull distribution, trigonometric distribution, lognormal distribution, binomial distribution, geometric distribution, Poisson distribution, etc. During simulation, different distribution curves need to be appointed for different objective things, so as to have better simulation effect. At the same time, due to the reference of different distribution curves, simulation errors will occur.

For example, there is a material that has cracks and continues to expand, which will lead to material failure. During simulation, the correct distribution curve should be lognormal distribution. Figure 4-16 shows the function curve of the failure rate function $\lambda (t)$ of lognormal distribution.

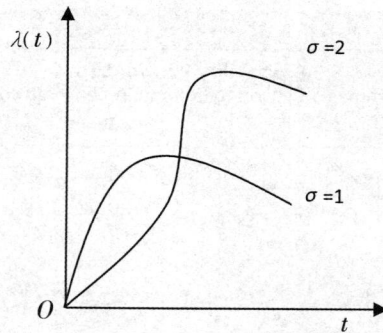

Figure 4-16   Failure rate function curve of lognormal distribution

However, if the referenced distribution curve is Weibull curve during simulation, the simulation results will be very different. Figure 4-17 shows the function curve of the failure rate function $\lambda(t)$ of Weibull distribution, and m is the shape parameter.

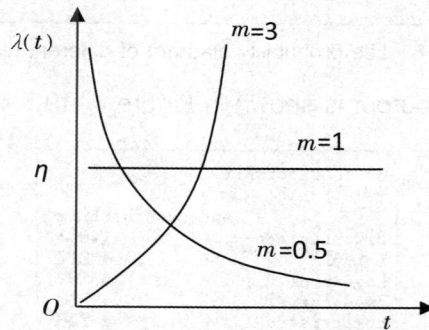

Figure 4-17   Failure rate function curve of Weibull distribution

Obviously, due to different references of distribution curves, there must be large errors in the simulation results.

Case:

A quality management team studied the reliability of a certain type of temperature sensor. The team randomly selected 15 temperature sensors for long-term tracking of life test. The measured life data is (unit: day): 90, 170, 280, 360, 550, 730, 920, 1100, 1300, 1600, 2000, 2400, 2900, 3400, 4000+. Now the team needs to find the best parameter distribution model for this group of data.

Analysis as follows: from the 15 collected data, the first 14 data are accurate life data, and the 15th data is the right censored data. Now use Minitab software for ID confirmation analysis of different distributions. Select "Reliability/Survival" in "Statistics", choose "Distribution ID Plot" in "Distribution Analysis (Right Censoring)". Click "Censor", select "Censored column" into "Use censoring columns".

In the process of analysis, there are four distributions used: Weibull distribution, lognormal distribution, exponential distribution and log-logistic distribution. See Figure 4-18 for the life probability analysis results of different distributions.

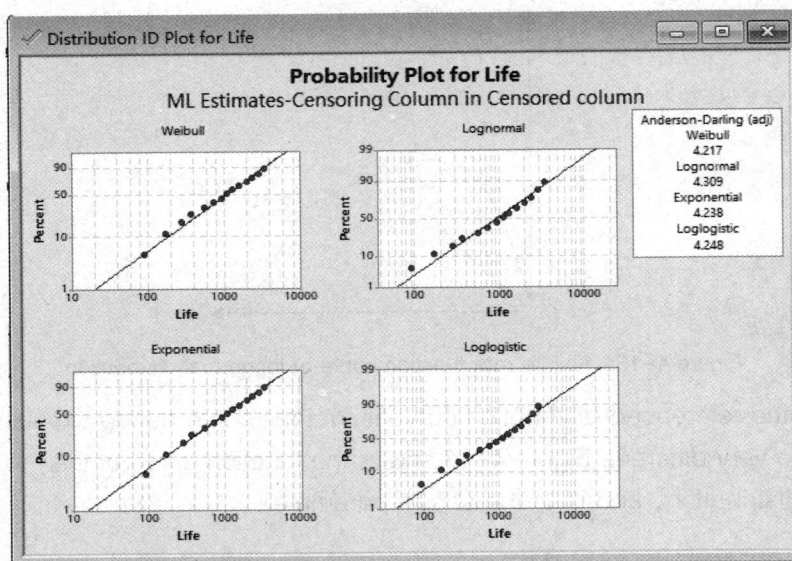

Figure 4-18   Life probability diagram of different distributions

The goodness of fit value output is shown in Figure 4-19.

Figure 4-19   Goodness of fit values for different distributions

It can be seen from Figure 4-19 that the goodness of fit value 4.217 of Weibull distribution is the smallest, indicating the best fit. It also shows that if other distribution curves are used, the error will be greater.

4.5.4 The limitation of conditions leads to simulation anomaly

The simulation process needs corresponding conditions to support. Due to the limitations of the relevant conditions, simulation anomalies will occur. It mainly includes the following items.

(1) Limitations of environmental conditions. For example, the aging test simulation of some materials will have very different results under different temperature, speed, height (atmospheric pressure), humidity, illumination and other conditions, but due to environmental conditions, the simulation error is formed.

(2) Limitations of thinking. Everyone has different ideas about modeling, and the limitations of thinking will limit the process and results of modeling. Some people can see universal laws behind complex phenomena, while others can't; Some people can use different algorithms, while others are limited to a limited number of algorithms. In order to break through the limitations of thinking, efforts should be made in two aspects: first, horizontal expansion, that is, comparing the simulated object with other relevant knowledge and concepts, finding out what is similar and what is different, and realizing the link of knowledge. The second is vertical expansion, that is, for the simulated object, all its vertical knowledge is connected in series, and then

knowledge breakthrough is achieved to build an innovative model.

(3)Limitations of using tools. For example, model error, observation error, truncation error and rounding error will occur during computer simulation. Model error refers to the mathematical model, which has been described previously. Observation error refers to the error caused by observation. Truncation error, also known as method error, refers to the error between the approximate solution of the model and the accurate solution of the model, which can only be obtained by numerical method because the mathematical model can not be accurately solved in many cases. Rounding error refers to the error caused by the loss of information when rounding numbers are used to adapt to a limited computer word length.

4.5.5 Analog anomaly caused by pseudo random number

Many simulations need to use random numbers, but sometimes the emergence of pseudo random numbers will lead to simulation anomalies. For example, the application of Monte Carlo method can now automatically generate random numbers through computers, and corresponding random numbers can be generated according to different distribution curves. In fact, these random numbers generated by computers themselves are not real random numbers, which can be called pseudo random numbers.

The computer generates random numbers in a certain way, that is, it uses a deterministic algorithm to calculate the sequence of random numbers with uniform distribution of [0,1]. Although pseudo random numbers also have the characteristics of uniformity and independence, and usually replace real random numbers for use with good results, there are differences between them and real random numbers.

4.6 统计分析及计算过程异常

当数据收集整理好以后，接下来的工作通常是进行统计分析及相关的计算。在开展统计分析和计算的时候，也存在着各种异常的情形，使数据出现异常。

4.6.1 统计分析异常

这里的统计分析异常主要指描述性统计方法在应用过程中所出现的问题。统计分析异常的情形非常多，只能把一些常见的问题进行说明。有关统计推断异常的内容放在本章第 7 节予以介绍。

1. 应用统计方法不适宜

统计方法有很多，在应用到具体的场合时，关键在于其适宜性。也就是说，在应用统计方法时，并不是在于应用统计方法的种类多，不在于统计分析的难度有多么大，而在于所应用的方法能够恰当地解剖、分析所面对的客观对象，能够得到正确的分析结论。在实践中，统计方法的应用不适宜情形还是存在的。

举例：

某质量管理小组的活动课题是提高铝模卫生间反坎成型合格率。在现状调查中，小组找到症结为"反坎出现裂缝"和"反坎缺棱掉角"。小组通过原因分析得到的一条末端原因是"拆模不规范"。

随后，小组进行了主要原因的确认程序。在确认过程中，小组成员调查发现在施工现场存在木工暴力拆模的情况，影响了铝模卫生间反坎成型施工质量。小组把工人是否拆模用力过猛的情况统计成"调查分析前后反坎不合格点数统计表"（见表 4–10）和"反坎合格率统计表"（见表 4–11）。

表 4–10　调查分析前后反坎不合格点数统计表

类别	第一天调查结果	第二天调查结果	平均
调查前	不合格点数 4 个	不合格点数 4 个	不合格点数 4 个
调查后	不合格点数 3 个	不合格点数 3 个	不合格点数 3 个

表 4–11　反坎合格率统计表

事件	检测点数/个	合格点数/个	不合格点数/个	合格率/%	合格率差值/%
调查前	35	31	4	88.57	2.85
调查后	35	32	3	91.42	

根据以上数据，小组认为调查前，铝模卫生间反坎成型一次合格率为88.57%，调查后，铝模卫生间反坎成型一次合格率为91.42%，差值为2.85%。因此工人暴力拆模情况对铝模卫生间反坎成型一次合格率影响程度判定为小。

在这个案例中，小组虽然采用了前后对比的方法，但仅仅以调查前后的数据来作为判断的依据是不适宜的。正确的做法是：小组应当选择两组不同的工人，一组施工时拆模不规范，另一组拆模时比较规范，这样才能形成真正的对照，以发现两组工人拆模时各自对于症结的影响程度。

2. 应用统计方法的前置条件不符合要求

有些统计方法在应用时是有前置条件的，比如：绘制直方图时，数据量不能少于50个；绘制散点图的时候，成对的数据应当在30对以上；绘制排列图时，数据量建议大于50个等。否则，所绘制的图形有可能反映数据的规律不够清晰，影响判断。

举例：

有一个调研小组要寻找"圆弧骨架偏差"与"龙骨弯曲加工误差"之间的相关关系。该小组收集了15组数据，并绘制成散点图，见图4-20。

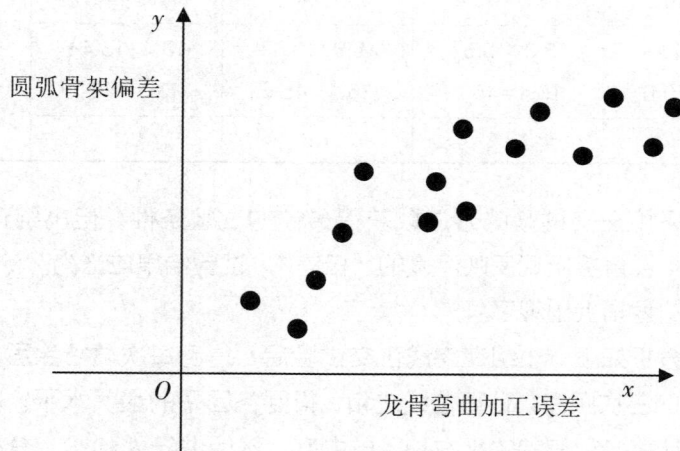

图4-20　散点图

从图4-20可以看出，由于该调研小组收集成对的数据只有15组，数量偏少，因此，难以从图中判断出"圆弧骨架偏差"与"龙骨弯曲加工误差"之间是什么样的关系。如果小组收集的数据量在30组以上乃至更多，那么两个变量之间的关系就显得清晰。

3. 正交试验出现的结果不正确

正交试验方法是针对不同因子、因子的不同水平，求得相对最佳组合的一种试验方法。正交试验的结果应该为得率，即不同的因子在不同的水平下共同作用所得到的结果。但是，也有出现异常的情况。

举例：

某质量管理小组在喷涂活动中，确定了四个因素，即喷涂遍数、喷涂时间、喷涂面积、喷涂距离。每项因素按照标准范围确定三个水平进行正交试验。试验后得到结果，见表4-12。

表 4-12　正交试验结果

因素试验号	喷涂遍数/遍 A	喷涂时间/min B	喷涂面积/m² C	喷涂距离/cm D	实验结果厚度极差值/μm
1	1（1）	1（50）	3（180）	2（30）	15.8
2	2（2）	1（50）	1（120）	1（20）	17.4
3	3（3）	1（50）	2（150）	3（40）	19.4
4	1（1）	2（60）	2（150）	1（20）	12.2 *
5	2（2）	2（60）	3（180）	3（40）	13.4
6	3（3）	2（60）	1（120）	2（30）	14.8
7	1（1）	3（70）	1（120）	3（40）	13.3
8	2（2）	3（70）	2（150）	2（30）	15.5
9	3（3）	3（70）	3（180）	1（20）	16.4
位级 1 之和	15.8+12.2+13.3=41.3	15.8+17.4+19.4=52.6	17.4+14.8+13.3=45.5	17.4+12.2+16.4=46	
位级 2 之和	17.4+13.4+15.5=46.3	12.2+13.4+14.8=40.4	19.4+12.2+15.5=47.1	15.8+14.8+15.5=46.1	因素重要程度次序：B→A→C→D
位级 3 之和	19.4+14.8+16.4=50.6	13.3+15.5+16.4=45.2	15.8+13.4+16.4=45.6	19.4+13.4+13.3=46.1	
极差（R）	9.3	12.2	1.6	0.1	

在本例中，小组采用正交试验的方法以获得最佳的试验条件，但小组正交试验的结果不正确。正交试验的结果不会直接出现反映厚度的"极差"，试验结果应当为"涂层的厚度"。说明小组在试验实施、分析的逻辑上出现了异常。

正确的试验结果分析如下：小组在完成正交试验后，得到每次试验涂层厚度的结果，然后计算极差 R，排定因子的主次顺序；计算位级之和，得到各因子的最优水平。另外小组可以进行试验结果的方差分析，计算各列偏差的平方和、自由度；然后进行 F 检验，分析检验结果后得出相应结论。

4. 针对复杂现象的分层分析异常

分层法是统计中的一个重要方法，通过分层分析，可以把复杂的现象予以简化，能够更清晰地进行分析判断。但是，分层不清或混淆的情况或导致分析数据异常。

举例：

某质量管理小组在现状调查中采用了分层分析，具体如下。

调查一：为了了解墙面砖施工现状情况，小组成员对 1#楼、4#楼已完成的 6～11 层外墙墙面砖进行检查。经抽样检查后统计结果见表 4-13。

表 4-13　1#楼、4#楼 6~11 层的外墙墙面砖合格率检查统计表

楼层	1#楼			4#楼			合计
	6~7 层	8~9 层	10~11 层	6~7 层	8~9 层	10~11 层	
外墙面砖检查点数	55	55	55	55	55	55	330
合格点数/个	33	36	31	42	46	44	232
不合格点数/个	22	19	24	13	9	11	98
合格率/%	60	65.5	56.4	76.4	83.6	80	70.3

调查二：小组对检查的 330 个点按不同的施工队组进行分层分析。1#楼、4#楼外墙墙面砖分别由 A、B 两个施工班组进行施工，施工队组人员技能参差不齐。小组将调查一中检查的 330 个点按 A、B 两个施工班组进行划分，继续统计合格点数及不合格点数，统计结果如表 4-14。

表 4-14　不合格点数按不同施工班组统计

施工队组	A 班组（1#楼）	B 班组（4#楼）
检查外墙面砖点数/个	165	165
合格外墙面砖点数/个	100	132
不合格外墙面砖点数/个	65	33
合格率/%	60.6	80
不合格外墙面砖总点数/个	98	
外墙面砖平均合格率/%	70.3	

调查三：小组再次到现场对外墙墙面砖按具体质量问题进一步调查。检查的内容主要是阴阳角方正差、接缝直线度差、接缝高低差、表面平整度差、空鼓、色差等。调查统计结果见表 4-15。

表 4-15　外墙墙面砖质量问题调查统计表

序	项目	调查数量/点	合格/点	不合格/点	合格率/%
1	阴阳角方正差	100	22	78	22
2	接缝直线度差	100	27	73	27
3	接缝高低差	100	80	20	80
4	表面平整度差	100	90	10	90
5	空鼓	100	95	5	95
6	色差	100	97	3	97
7	其他	100	98	2	98
	合计	700	509	191	72.7

小组在现状调查中进行了分层分析，但分层分析时的数据基数不同，分层之间缺少严密的逻辑性。小组的第一层调查是在 1#、4#楼的 6~11 层，共分析了 330 个点；第二层分析是在同样的地方分析了 330 个点；而第三层则是直接分析了阴阳角方正差、接缝直线度差等 7 类质量问题

700 个点，这 700 个点与前面的 330 个点的关系不清晰，这样的分析结果会造成偏差。

5. 图形绘制异常导致对数据特征的判断不准确

在统计分析中，根据数据来绘制相应的图形是常用的方法。在实践中，由于一些图形绘制异常，会对数据的判断出现不准确。

举例：

某小组开发开槽机，在效果检查阶段，随机抽检 50 个点进行路缘石误差尺寸检测，具体检测数据见表 4-16。

表 4-16　现场实测数据统计表

组别	随机抽检 50 个点进行路缘石尺寸误差/cm										合格点数/个
第一组	0	0	1.2	1	0.5	0	0	1.2	1.3	−1.5	10
第二组	0	−1.5	−1.8	0	−1.2	0	0	0	−1.2	−1.2	10
第三组	0.5	−0.6	0.5	−0.8	1		−0.8	−0.6	−0.9	−1.1	10
第四组	0.6	0.3	−0.5	1.2	−1.3	1.8	0	−0.6	0.8	2.7	9
第五组	0.6	2.6	0.5	−0.9	0.5	0		0.5	0.5		9

小组根据表 4-16 的数据，绘制了路缘石尺寸误差直方图，见图 4-21。

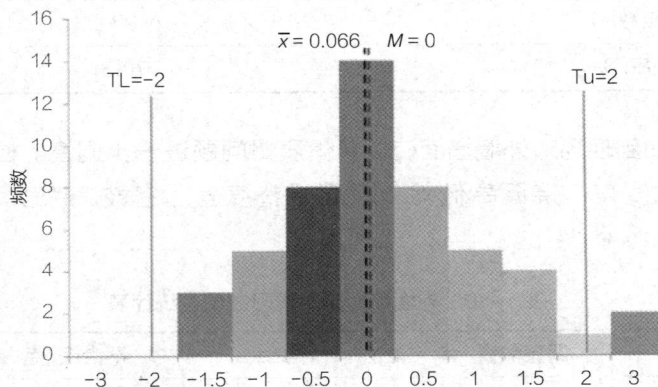

图 4-21　路缘石尺寸误差直方图

小组在效果检查中使用了直方图，并以此来判断数据的分散程度。但小组在取"路缘石尺寸误差实测数据"后绘制直方图，中间跳过了几个重要的步骤，比如缺少组距、缺少对于上下界限的界定、没有编制频数分布表等。

当小组收集完数据后，正确绘制直方图的步骤是：一是确定数据的极差，本例中极差 $R =$ 2.7−（−1.8）= 4.5。二是确定组数。本例 50 个数据，根据常用的"组数 k 选用表"，k =10。三是确定组距 h。$h = R/k =$ 4.5/10 = 0.45。四是确定各组的界限值。组的界限值单位取最小测量单位的 1/2，本例中，界限值取 0.05。第一组下限值为−1.8−0.05 = −1.85，上限值为−1.85+0.45 = −1.4；第二组下限值为−1.4，上限值为−1.4+0.45 = −0.95。后面各组的上下界限值以此类推。五是编制频数分布表。把各组的上下界限值分别填入频数分布表内，把表 4-16 中的数据"对号入座"列入相应的组中，然后统计各组的频数。可以得到：第一组频数至第十组频数分别为：1，

3，8，4，14，10，6，1，1，2。六是画直方图，横坐标为数据值，纵坐标为频数。得到直方图，见图4-22。

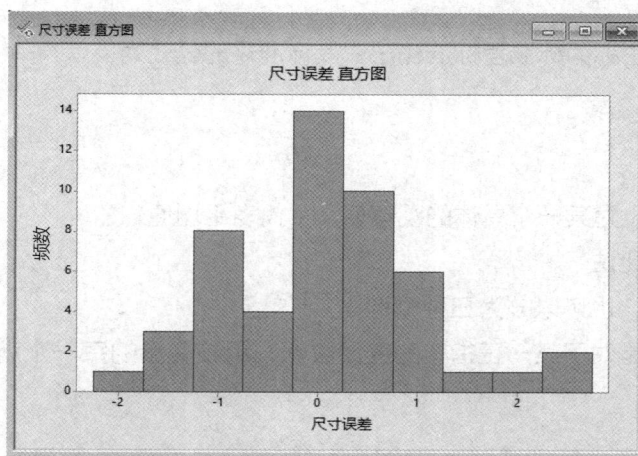

图4-22　尺寸误差直方图

通过图4-21与图4-22的比对，两者的图形是有区别的：从图4-21判断，除了尺寸误差为"2.6""2.7"的数据显得突兀，其余数据落点基本完美，说明开槽机的性能基本正常。但根据图4-22来判断，则该开槽机的性能不稳定，因为尺寸误差在"-0.5、-0.6"处有4个点，使直方图显得有上下交错现象。

6. 虚假相关

当两个变量之间相关的唯一原因是它们受到同一因素的影响时，出现了虚假相关。这需要统计分析人员能够辨识其中的真正原因。

比如，某调研小组对某个地区作社会调查，根据统计数据显示，该地区存在这么一个有关联的现象，即拥有家庭汽车与家中老人长寿之间存在相关关系。但仔细分析后发现，"拥有家庭汽车"与当地"家庭收入比较富有"呈明显相关关系，同时"家中老人长寿"这个现象也与"家庭收入比较富有"呈明显相关关系。显然"拥有家庭汽车"与"家中老人长寿"都受到家庭经济富有的影响。因此可以认为，"拥有家庭汽车"与"家中老人长寿"之间存在虚假的相关关系。

7. 不同属性的总体间进行比较时没有进行标准化处理

在现实的数据采集过程中，会有各种不同性质的总体混合在一起。对于这样的不同属性的总体进行比较时，应当先进行标准化处理。这里的属性包括不同的科目、类别、数量级别、数据单位等。通过使用统一的标准化转化方式后，使这些不同属性的总体能够处于同一个量级，然后可以进行比较分析。否则，数据之间就存在牛头不对马嘴的现象，造成数据分析异常。

数据的标准化方法主要有以下三种。

（1）极差标准化法

极差标准化法是把不同量级的数据通过标准化后，统一到相同量级平台进行评价的方法。设标准化以后的数据为 x' 其标准化的公式为

$$x' = \frac{x_i - x_{min}}{x_{max} - x_{min}}$$

公式的具体涵义：先计算对应指标的最大值和最小值，获得两者的极差 R；然后计算每一个

观察值 x_i 减去最小值，得到 R_1；R_1 与 R 之比即为标准化的值。通过极差标准化以后，所得到的数值变动范围在 0 和 1 之间。

（2） Z -score 标准化法

Z -score 标准化法也称为标准差标准化法，其标准化的公式为

$$x' = \frac{x_i - \bar{x}}{SD}$$

其中，SD 为标准差，x_i 为某一个具体的观察值，$\bar{x}$ 代表平均值。

（3）线性比例标准化法

线性比例标准化法，也可以分为三种情形。

一是极大化法。它是对于某一正指标的变量数据，用该变量的每一个观察值除以变量中的最大值。公式为

$$x' = \frac{x_i}{x_{max}}$$

二是极小化法。它是对于某一逆指标的变量数据，用该变量的每一个观察值除以变量中的最小值。公式为

$$x' = \frac{x_i}{x_{min}}$$

三是 log 函数标准化法。它是先对变量的每一个观察值，取值为以 10 为底的 log 值，然后用该值除以变量中最大值的 log 值。公式为

$$x' = \frac{\log_{10} x}{\log_{10} x_{max}}$$

其中，x 应当 ≥ 1。

4.6.2 计算过程异常

在数据收集以后，计算的过程基本是按照分析所需要的公式、内在的逻辑等进行未知量的推算、运算。计算过程异常的情形主要有以下四种。

1. 应用公式错误

适用于计算的公式非常之多，针对不同的场合需要应用相应的计算公式。有时会发生计算公式应用错误的情况。

比如在进行总体标准差的计算时，应用的公式为

$$\sigma = \sqrt{\frac{1}{N} \sum_{i=1}^{N} (x_i - \mu)^2}$$

当计算人员把这个公式应用在样本标准差计算时，就发生了异常，因为样本标准差的计算公式为

$$\sigma = \sqrt{\frac{\sum_{i=1}^{n} (x_i - \bar{x})^2}{n-1}}$$

两个公式的内涵是有区别的，计算标准差的时候，自由度必须为 $n-1$。

2. 数据修约造成

对数据进行修约是计算过程中常用的方法，目的是对某个已知数根据保留位数的要求，根据一定的规则，将多余位数的数字进行取舍。对数据进行修约，要把握两个方面：一是对数值的修约，它是省略原数值的最后若干位数字，调整所保留的末位数字，使修约后得到的值最接近原数值。经修约后的数值称为修约值。二是修约间隔，也称为修约区间、化整间隔。它是确定修约保留位数的一种方式，是指修约值的最小数值单位，也就是说修约到哪一个数位。

修约间隔通常以 10^n 的形式来表示，n 的取值可以为正整数，也可以为负整数。修约间隔的数值一旦确定，修约值即为该数值的整数倍。比如：修约间隔为 0.1，则修约值应当在 1 的整数倍中选取，相当于把数值修约到一位小数；修约间隔为 10，则修约值应当在 10 的整数倍中选取，相当于把数值修约到"十"的数位。

由于数据修约造成数据异常的情形主要有以下两种。

（1）随意修约

有的人在进行数据计算的时候随意修约。

一是修约的间隔不一致。有时候修约到某一个位数，有时候修约到另一个位数，使前后的数据位数不统一。比如 12.6365，有的修约为 12.64，有的修约为 12.637，造成计算异常。

二是修约规则不一致。有时候使用"四舍五入法"，有时候使用"四舍六入五单双法"。比如 12.6365，修约间隔为 0.001，按照"四舍五入法"，修约为 12.637；按照"四舍六入五单双法"，修约为 12.636。

三是不了解修约规则的具体涵义。

"四舍五入法"比较容易理解，具体使用时，在需要保留数字的位次后一位，逢 5（含大于 5）就进 1，逢 4（含小于 4）就舍。

"四舍六入五单双法"，是指在需要保留数字的位次后一位，逢 4（含小于 4）就舍；逢大于 5 则进 1。

逢 5 的时候，如果 5 的后面有非 0 的数字，则进 1，比如 12.636 51，修约间隔为 0.001，则修约为 12.637。

逢 5 的时候（5 的后面全为 0，或者 5 的后面没有其他数字），如果 5 的前面为奇数，则进 1，比如 12.6375，修约间隔为 0.001，则修约为 12.638。

逢 5 的时候（5 的后面全为 0，或者 5 的后面没有其他数字），如果 5 的前面为偶数，则舍去，比如 12.6365，修约间隔为 0.001，则修约为 12.636。

四是连续修约。连续修约本身就是一个错误的做法。比如：12.4745，先修约至 12.475，再修约至 12.48，进而修约至 12.5，乃至修约至 13，整个步骤是错误的。

（2）修约本身造成误差

由于数据的修约过程是按照修约规则，对于数据保留位数上数字的取舍，有的数字进一，有的数字舍去，其本身就构成了数据的差异，使数据发生一定的异常。

3. 随意舍弃数据

在数据采集或分析的过程中，有的人自认为某些数据不妥，存在过大或过小的问题，或者直接认为某些数据不派用处而随意舍去。这种情况下，直接导致了数据异常。

而重视所有的数据会给一些人带来丰厚的回报。获得 1996 年诺贝尔化学奖的美国莱斯大学教授居尔（Robert F. Curl）长期从事化学研究。在居尔教授发现足球状分子之前，已经有其他实验者发现了一个奇怪的质谱线，但他们都没有引起足够的重视，认为这个质谱线没有什么用处而轻易舍弃。相反，居尔教授则认为该质谱线的数据相当重要，应该存在某种物质，于是他发现了C_{60}，从此让人们发现了一个全新的化学世界，从平面的、低对称性的分子到全对称的球形分子，从平面的芳香性到球面的芳香性，从简单的分子到超原子分子，从一维的超导性到三维的超导性、碳纳米管的电子特性等，极大丰富和提高了科学理论，也彰显出巨大的应用价值。

4. 数据收集完毕后对于数据的含义定义不清

有些人在数据收集完成后，认为工作已经结束。等到整理数据时，却无法对于各数据的具体含义进行界定，使计算过程发生异常。表 4-17 是某公司对不同型号的加工机械，其加工质量的对照试验。

表 4-17　不同型号加工机械加工产品质量一览表

组别	发生次数	机械型号	机械功率/kW	加工质量合格率/%
对照一组	1	A	500	99.5
	2	B	500	98.8
	3	C	600	96.9
	4	D	800	97.4
对照二组	1	A	1000	95.6
	2	B	1000	92.1
	3	C	1200	93.5
	4	D	1500	92.7

在表 4-17 中，存在数据含义定义不清的问题。其中，"发生次数"的表述有误，根据具体的内容，应当改为"序号"。如果按照发生次数来计算，则出现很大的差异。

4.6 Abnormal Statistical Analysis and Calculation Process

——

After the data is collected and sorted out, the next work is usually statistical analysis and related calculations. When statistical analysis and calculation are carried out, there are also various abnormal situations, which make the data abnormal.

4.6.1 Abnormities of statistical analysis

The abnormities of statistical analysis here mainly refer to the problems in the application of descriptive statistical methods. There are many abnormal situations in statistical analysis, and only some common problems can be explained. The content of statistical inference anomaly is introduced in Section 7 of this chapter.

1. The application of statistical methods is not appropriate

There are many statistical methods. When applied to specific occasions, the key lies in their suitability. That is to say, when applying statistical methods, it is not because of the variety of statistical methods and the difficulty of statistical analysis, but because the applied methods can properly dissect and analyze the objective objects they face and get the correct analysis conclusions. In practice, the inappropriate application of statistical methods still exists.

Case:

The activity topic of a quality management team is to improve the forming qualification rate of aluminum mold toilet flip bucket. In the status survey period, the team found that the crux was "cracks appeared on the flip bucket" and "the flip bucket lacked edges and corners". One end cause obtained by the team through cause analysis is "non-standard formwork removal".

Subsequently, the team conducted a procedure to identify the main causes. During the confirmation process, the team members found that there was forcible formwork removal by carpenters at the construction site, which affected the forming quality of the aluminum formwork toilet flip bucket. The team counted whether the workers used too much force to remove the formwork into the "Statistical table of unqualified points before and after investigation and analysis" (see Table 4-10) and the "Statistical table of unqualified rate of flip bucket" (see Table 4-11).

Table 4-10　Statistical table of unqualified points before and after investigation and analysis

Category	Survey results of the first day	Survey results of the next day	Average
Before investigation	4 unqualified points	4 unqualified points	4 unqualified points
After investigation	3 unqualified points	3 unqualified points	3 unqualified points

Table 4-11 Statistical table of unqualified rate of flip bucket

Event	Number of detection points	Number of qualified points	Number of unqualified points	Qualification rate/%	Qualification rate difference/%
Before investigation	35	31	4	88.57	2.85
After investigation	35	32	3	91.42	

Based on the above data, the team believes that before the investigation, the one-time qualification rate of the aluminum mold toilet flip bucket is 88.57%, and after the investigation, the one-time qualification rate of the aluminum mold toilet flip bucket is 91.42%, the difference is 2.85%. Therefore, it is determined that the impact of forcible formwork removal by workers on the one-time qualification rate of aluminum mold toilet flip bucket forming is small.

In this case, although the team adopted the method of comparison before and after the investigation, it is not appropriate to use the data before and after the investigation as the basis for judgment. The correct approach is: the team should choose two different groups of workers. One group is not standard in formwork removal during construction, and the other group is relatively standard in formwork removal, so as to form a real comparison to find out the influence of the two groups of workers on the crux when they remove formwork.

2. Preconditions for applying statistical methods do not meet the requirements

Some statistical methods have preconditions when they are applied. For example, when drawing histograms, the amount of data cannot be less than 50; When drawing a scatter chart, the paired data should be more than 30 pairs; It is recommended that the quantity of the arrangement diagram should be more than 50. Otherwise, the drawn graph may not reflect the law of the data clearly enough, affecting the judgment.

Case:

A research team is looking for the correlation between "arc skeleton deviation" and "keel bending processing error". The team collected 15 sets of data and drew a scatter plot, as shown in Figure 4-20.

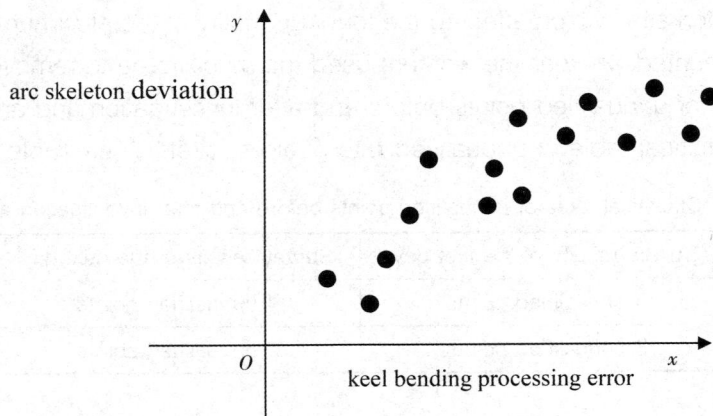

Figure 4-20 Scatter diagram

It can be seen from Figure 4-20 that, because only 15 pairs of data were collected by the research team, the number was small, so it was difficult to judge the relationship between "arc skeleton deviation" and "keel bending processing error" from the figure. If the amount of data collected by the team is more than 30 groups or more, the relationship between the two variables will be clear.

3. The result of orthogonal test is incorrect

Orthogonal test method is a test method to find the relatively optimal combination according to different factors and different levels of factors. The result of orthogonal test should be the yield, that is, the result obtained by different factors acting together at different levels. However, there are incorrect results.

Case：

During the activity, a quality management team determined 4 factors, namely, the number of spraying times, spraying time, spraying area and spraying distance. Three levels of each factor are determined according to the standard range for orthogonal test. The results obtained after the test are shown in Table 4-12.

Table 4-12　Orthogonal test results

Factor test No.	Spraying times /times A	Spraying time /min B	Spraying area /m² C	Spraying distance /cm D	Experimental results thickness range / μm
1	1（1）	1（50）	3（180）	2（30）	15.8
2	2（2）	1（50）	1（120）	1（20）	17.4
3	3（3）	1（50）	2（150）	3（40）	19.4
4	1（1）	2（60）	2（150）	1（20）	12.2 *
5	2（2）	2（60）	3（180）	3（40）	13.4
6	3（3）	2（60）	1（120）	2（30）	14.8
7	1（1）	3（70）	1（120）	3（40）	13.3
8	2（2）	3（70）	2（150）	2（30）	15.5
9	3（3）	3（70）	3（180）	1（20）	16.4
Sum of level 1	15.8+12.2+ 13.3＝41.3	15.8+17.4+ 19.4＝52.6	17.4+14.8+ 13.3＝45.5	17.4+12.2+ 16.4＝46	
Sum of level 2	17.4+13.4+ 15.5＝46.3	12.2+13.4+ 14.8＝40.4	19.4+12.2+ 15.5＝47.1	15.8+14.8+ 15.5＝46.1	Importance order of factors： B→A→C→D
Sum of level 3	19.4+14.8+ 16.4＝50.6	13.3+15.5+ 16.4＝45.2	15.8+13.4+ 16.4＝45.6	19.4+13.4+ 13.3＝46.1	
Range（R）	9.3	12.2	1.6	0.1	

In this case, the group adopted the method of orthogonal test to obtain the best test conditions, but the results of the group orthogonal test were incorrect. The result of orthogonal test will not directly reflect the "range" of thickness, and the test result should be "coating thickness". It indicates that the team has abnormal logic in test implementation and analysis.

The correct test results are analyzed as follows: after completing the orthogonal test, the team obtains the results of coating thickness for each test, calculates the range R, and arranges the primary and secondary order of factors; Calculate the sum of levels to obtain the optimal level of each factor. In addition, the team can perform variance analysis of the test results to calculate the sum of squares and degrees of freedom of the deviations of each column; Then carry out F test, analyze the test results and draw corresponding conclusions.

4. Hierarchical analysis anomaly for complex phenomena

Hierarchical method is an important method in statistics. Through hierarchical analysis, complex phenomena can be simplified and analysis and judgment can be made more clearly. However, unclear layering or confusion may lead to abnormal analysis data.

Case:

A quality management team adopted hierarchical analysis method in the status survey period, as shown below.

Survey I: In order to understand the current situation of wall tile construction, the team members checked the completed wall tiles from the sixth floor to the eleventh floor of Building 1# and Building 4#. See Table 4-13 for the statistical results after sampling inspection.

Table 4-13 Statistical Table for acceptance rate inspection of exterior wall tiles from the 6th to 11th floors of Building 1# and Building 4#

Floor	Building 1#			Building 4#			Total
	F6~F7	F8~F9	F10~F11	F6~F7	F8~F9	F10~F11	
Inspection points of exterior wall tiles	55	55	55	55	55	55	330
Qualified points	33	36	31	42	46	44	232
Unqualified points	22	19	24	13	9	11	98
Qualified rate/%	60	65.5	56.4	76.4	83.6	80	70.3

Survey II: The team analyzed 330 points inspected by different construction teams in layers. The external wall tiles of Building 1# and Building 4# are constructed by two construction teams, A and B. The skills of the construction team members are uneven. The team divided the 330 points inspected in Survey I into two construction teams, A and B, and continued to count the qualified points and unqualified points. The statistical results are shown in Table 4-14.

Table 4-14 Unqualified points counted by different construction teams

Construction team	Team A (Building 1#)	Team B (Building 4#)
Total number of exterior wall tiles inspected	165	165
Number of qualified exterior wall tiles	100	132
Number of unqualified exterior wall tiles	65	33
Qualified rate/%	60.6	80
Total number of unqualified exterior wall tiles	98	
Average qualification rate of exterior wall tiles/%	70.3	

Survey Ⅲ: The team went to the site again to further investigate the external wall tiles according to the specific quality problems. The inspection contents mainly include square difference of internal and external corners, poor straightness of joints, high and low difference of joints, poor surface flatness, hollowing, color difference, etc. See Table 4-15 for the survey results.

Table 4-15　The quality problems of the external wall tiles

No.	Item	Survey points	Qualified points	Unqualified points	Qualified rate/%
1	Square difference of internal and external corners	100	22	78	22
2	Poor straightness of joints	100	27	73	27
3	High and low difference of joints	100	80	20	80
4	Poor surface flatness	100	90	10	90
5	Hollowing	100	95	5	95
6	Color difference	100	97	3	97
7	Other problems	100	98	2	98
	Total	700	509	191	72.7

The team conducted hierarchical analysis method in the status quo investigation, but the data base of hierarchical analysis was different, and there was a lack of strict logic between the layers. The team's first floor survey was conducted from the sixth to eleventh floors of 1# and 4# buildings, and 330 points were analyzed; The second analysis is to analyze 330 points in the same place; The third layer directly analyzed 700 points of 7 types of quality problems, such as the square difference of internal and external corners, and the poor straightness of joints. The relationship between these 700 points and the previous 330 points is not clear, and such analysis results will cause deviation.

5. Abnormal drawing results in inaccurate judgment of data characteristics

In statistical analysis, it is a common method to draw corresponding graphs according to data. In practice, due to some abnormal graphics drawing, the judgment of data will be inaccurate.

Case:

A team developed the slotting machine. In the effect inspection stage, 50 points were randomly selected for curb error size inspection. See Table 4-16 for specific inspection data.

Table 4-16　Statistics of field measured data

Group	Randomly spot check 50 points to investigate the size error of curb/cm										Qualified points
Group Ⅰ	0	0	1.2	1	0.5	0	0	1.2	1.3	-1.5	10
Group Ⅱ	0	-1.5	-1.8	0	-1.2	0	0	0	-1.2	-1.2	10
Group Ⅲ	0.5	-0.6	0.5	-0.8	1	0	-0.8	-0.6	-0.9	-1.1	10
Group Ⅳ	0.6	0.3	-0.5	1.2	-1.3	1.8	0	-0.6	0.8	2.7	9
Group Ⅴ	0.6	2.6	0.5	-0.9	0.5	0	0	0	0.5	0.5	9

According to the data in Table 4-16, the team has drawn a histogram of curb size error, as shown in Figure 4-21.

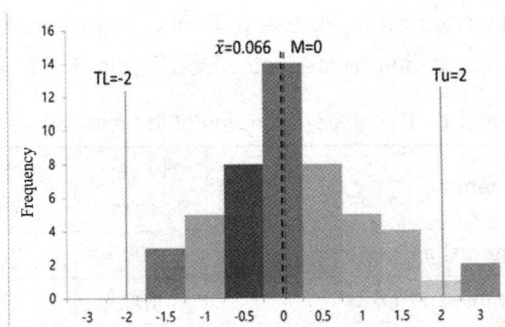

Figure 4-21   Histogram of curb size error

The team used histogram in the effect inspection to judge the degree of data dispersion. However, after taking the "measured data of curb size error", the team drew a histogram, skipping several important steps, such as the lack of group spacing, the lack of definition of upper and lower boundaries, the lack of preparation of frequency distribution table, etc.

After the team has collected the data, the steps to correctly draw the histogram are as follows.

The first, determine the range of the data. In this example, the range $R = 2.7 - (-1.8) = 4.5$.

The second is to determine the number of groups. This case has 50 data, according to the commonly used "group number k selection table", $k = 10$.

The third, determine the group spacing h. $h = R/k = 4.5/10 = 0.45$.

The fourth is to determine the limit value of each group. The limit value unit of the group is 1/2 of the minimum measurement unit. In this case, the limit value is 0.05. The lower limit of the first group is $-1.8-0.05 = -1.85$, and the upper limit is $-1.85+0.45 = -1.4$; The lower limit of the second group is -1.4, and the upper limit is $-1.4+0.45 = -0.95$. The upper and lower limit values of the following groups can be deduced in this way.

Fifth, prepare frequency distribution table. Fill the upper and lower limit values of each group into the frequency distribution table respectively, and list the data in Table 4-16 into the corresponding groups accordingly, and then count the frequency of each group. It can be obtained that the frequencies from the first group to the tenth group are: 1, 3, 8, 4, 14, 10, 6, 1, 1, 2.

The sixth is to draw a histogram. The abscissa is the data value and the ordinate is the frequency. The histogram is obtained, as shown in Figure 4-22.

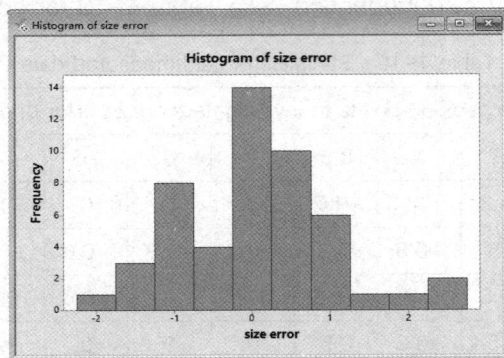

Figure 4-22   Histogram of size error

According to the comparison between Figure 4-21 and Figure 4-22, there is a difference between the two figures: judging from Figure 4-21, except for the data with size errors of "2.6" and "2.7", the rest of the data are basically perfect, indicating that the performance of the slotting machine is basically normal. However, judging from Figure 4-22, the performance of the notching machine is unstable, because there are four points at the "-0.5, -0.6" position of the size error, which makes the histogram appear to be staggered up and down.

6. False correlation

When the only reason for correlation between two variables is that they are affected by the same factor, false correlation occurs. This requires statistical analysts to be able to identify the real reason.

For example, a research group conducted a social survey on a certain area. According to the statistics, there is a correlation between the ownership of family cars and the longevity of the elderly in the area. However, after careful analysis, it was found that "owning a family car" was significantly related to the local "relatively rich family income", and "the elderly in the family live longer" was also significantly related to "relatively rich family income". Obviously, "owning a family car" and "the elderly in the family live a long life" are both affected by the wealth of the family economy. Therefore, it can be considered that there is a false correlation between "owning a family car" and "the elderly in the family live longer".

7. The comparison between populations with different attributes is not standardized

In the actual data collection process, there will be a variety of different natures of the populations mixed together. When comparing such different attribute populations, standardization should be carried out first. The attributes here include different accounts, categories, quantity levels, data units, etc. By using a unified standardized transformation method, the populations of these different attributes can be at the same level, and then they can be compared and analyzed. Otherwise, there will be a phenomenon that the data are different from each other, resulting in abnormal data analysis.

The data standardization methods are as follows.

(1) Range standardization method

The range standardization method is a method that unifies the data of different magnitudes to the platform of the same magnitudes for evaluation after standardization. Set the standardized data as and the standardized formula is

$$x' = \frac{x_i - x_{min}}{x_{max} - x_{min}}$$

The specific meaning of the formula is: first calculate the maximum and minimum values of the corresponding indicators to obtain their range R; Then each observation value subtract the minimum value to obtain R_1; The ratio of R_1 to R is the standardized value. After the range is standardized, the range of the values obtained is between 0 and 1.

(2) Z-score standardization method

Z-score standardization method is also known as standard deviation standardization method, the standardized formula is

$$x' = \frac{x_i - \bar{x}}{SD}$$

In the formula, SD is the standard deviation, x_i is a specific observation value, $\bar{x}$ represents average.

(3) Linear proportional standardization method

Linear proportional standardization method can be divided into three cases.

The first is the maximization method. For the variable data of a positive indicator, divide each observed value of the variable by the maximum value in the variable. The formula is

$$x' = \frac{x_i}{x_{max}}$$

The second is the minimization method. For the variable data of an inverse indicator, divide each observed value of the variable by the minimum value of the variable. The formula is

$$x' = \frac{x_i}{x_{min}}$$

The third is the log function standardization method. For each observed value of a variable, it takes the log value at the bottom of 10, and then divides this value by the log value of the maximum value in the variable. The formula is

$$x' = \frac{\log_{10} x}{\log_{10} x_{max}}$$

In the formula, x should be $\geqslant 1$.

4.6.2 Abnormal calculation process

After data collection, the calculation process is basically to calculate and operate the unknown quantity according to the formula and internal logic required for analysis. Abnormal calculation process mainly include the following four categories.

1. Incorrect formula applied

There are many formulas applicable to calculation, the corresponding calculation formulas are required for different occasions. Sometimes the calculation formula is applied incorrectly.

For example, when calculating the population standard deviation, the applied formula is

$$\sigma = \sqrt{\frac{1}{N} \sum_{i=1}^{N} (x_i - \mu)^2}$$

When the calculator applies this formula to the calculation of sample standard deviation, an abnomity occurs because the calculation formula of sample standard deviation is

$$\sigma = \sqrt{\frac{\sum_{i=1}^{n} (x_i - \bar{x})^2}{n-1}}$$

The connotation of the two formulas is different. When calculating the standard deviation, the degree of freedom must be $n-1$.

2. Abnormal data caused by data rounding

Rounding off the data is a common method in the calculation process. The purpose is to round off the number of redundant digits according to certain rules for a known number according to the requirements of reserved digits. To round off data, we should grasp two aspects: one is to round off the value. It is to omit the last digits of the original value and adjust the last digits retained so that the value obtained after rounding off is closest to the original value. The rounded value is called the rounding value. The second is rounding interval, also known as rounding interval and rounding interval. It is a way to determine the number of digits reserved for rounding off. It refers to the minimum numerical unit of the rounding off value, that is, to which digit.

The rounding interval is usually expressed in the form of 10^n. The value of n can be a positive integer or a negative integer. Once the value of the rounding interval is determined, the rounding value is an integral multiple of the value. For example, if the rounding interval is 0.1, the rounding value should be selected from the integer multiple of 1, which is equivalent to rounding the value to one decimal place; If the rounding interval is 10, the rounding value should be selected from the integer multiple of 10, which is equivalent to rounding the value to the digits of "10".

Abnormal data caused by data rounding mainly include the following two items.

(1) Rounding off data at will

Some people round off at will when doing data calculation.

First, the rounding interval is inconsistent. Sometimes round to a certain number of digits, sometimes round to another number of digits, so that the data digits before and after are not uniform. For example, 12.6365, some are rounded off to 12.64, and some are rounded off to 12.637, causing abnormal calculation.

Second, the rules of revision are inconsistent. Sometimes the "the end is rounded off by 4; in case of 5, enter one digit method" is used, and sometimes the "the end is rounded off by 4; the end is 6, enter one digit; in case of 5, judge according to the position of single or double method" is used. For example, 12.6365, the rounding interval is 0.001, and according to the "the end is rounded off by 4; in case of 5, enter one digit method", the rounding is 12.637; rounded to 12.636 in accordance with the "the end is rounded off by 4; the end is 6, enter one digit; in case of 5, judge according to the position of single or double method".

Third, the specific meaning of the amendment rules are not understood. The method of "the end is rounded off by 4; in case of 5, enter one digit" is easy to understand. For specific use, in the last digit of the digit to be reserved, every 5 (including more than 5) will be rounded to 1, and every 4 (including less than 4) will be rounded off.

The meaning of the method of "the end is rounded off by 4; the end is 6, enter one digit; in case of 5, judge according to the position of single or double method" is as follows: for the last digit of the digit to be reserved, if it is 4 (including less than 4), it will be rounded off; if it is greater than 5, the digit number will be added by 1.

When the digit number is 5, if there is a number other than 0 after the number 5, the number is increased by 1. For example, for 12.636 51, the rounding interval is 0.001, and the rounding is 12.637.

When the digit number is 5, and the digit number 5 is all followed by 0, or there is no other number after 5, if the front of 5 is an odd number, the number is increased by 1. For example, for 12.6375, if the rounding interval is 0.001, the rounding is 12.638.

When the digit number is 5, and all the digits after 5 are 0, or there are no other digits after 5, if the front of 5 is an even number, it is rounded down. For example, for 12.6365, if the rounding interval is 0.001, it is rounded down to 12.636.

Fourth, continuous rounding. Continuous rounding is a wrong practice in itself. For example, for 12.4745, first round to 12.475, then round to 12.48, then round to 12.5, and even round to 13. The whole step is wrong.

（2）Rounding off itself causes errors

Because the rounding off process of data is based on the rounding off rules, the rounding off of the numbers on the reserved digits of data, some of which are rounded to one, and some of which are rounded off, constitutes a difference in the data itself, causing certain anomalies in the data.

3. Discard data at will

In the process of data collection or analysis, some people think that some data is inappropriate—too large or too small of data, or they directly think that some data is not useful and discard it at will. In this case, abnormal data is directly caused.

And paying attention to all the data will bring generous returns to some people. Robert F. Curl, a professor at Rice University who won the Nobel Prize in chemistry in 1996, has been engaged in chemical research for a long time. Before Professor Curl discovered the football shape molecule, other experimenters had found a strange mass spectrum line, but they did not pay enough attention to it and thought it was useless and easily abandoned. On the contrary, Professor Curl believed that the data of the mass spectrum line was very important, and there should be some substance, so he discovered C_{60}, which led to the discovery of a new chemical world, from planar and low-symmetry molecules to fully symmetrical spherical molecules, from planar aromaticity to spherical aromaticity, from simple molecules to superatomic molecules, from one-dimensional superconductivity to three-dimensional superconductivity, and the electronic properties of carbon nanotubes, etc. It has greatly enriched and improved scientific theory, and also demonstrated great application value.

4. The meaning of data is unclear after data collection

Some people think that the work is over after the data collection is completed. When the data is sorted out, the specific meaning of each data cannot be defined, which makes the calculation process abnormal. Table 4-17 is a comparison test of the processing quality of different types of processing machinery by a company.

Table 4-17　Quality of processed products of different models

Group	Number of occurrences	Machine model	Mechanical power/kW	Qualified rate of the processed products/%
Control group 1	1	A	500	99.5
	2	B	500	98.8
	3	C	600	96.9
	4	D	800	97.4
Control group 2	1	A	1000	95.6
	2	B	1000	92.1
	3	C	1200	93.5
	4	D	1500	92.7

In Table 4-17, the definition of data meaning is unclear. Among them, the expression of "number of occurrences" is incorrect and should be changed to "serial number" according to the specific content. If calculated according to the number of occurrences, there will be a big difference.

4.7　统计推断异常

——

统计推断是指在一定的置信度下，根据样本的特征，对总体的特征做出估计和预测的一种方法。通常的统计推断包括了参数估计和假设检验两大内容。从广义的角度，把相关分析、回归分析、t 检验、F 检验、x^2 检验等也纳入统计推断。统计推断中有很多异常的情形，以下进行简要阐述。

4.7.1 参数估计的异常

参数估计是在一定的置信区间的前提下所进行的推论，这就存在着真实数据的发生不在置信区间的情形。

举例：

A 娱乐节目在 B 省有电视播送。为了更好地制定营销策略，需要估计当前在 B 省的收视情况，就委托某市场调研公司做市场预测。该调研公司把 A 节目最近一个季度的收视情况做调查，并随机抽取了 50 个收视记录作为样本，得到收视情况表，见表 4-18。

表 4-18　A 节目收视情况　　　　　　　　　　　　　　单位：万人次

收视人数									
50	53	37	48	86	24	45	65	90	18
32	40	23	59	68	36	52	50	72	47
74	19	64	47	115	28	40	61	36	31
35	24	44	32	46	98	69	40	87	28
41	32	25	40	54	31	30	105	41	60

现在可以分析：$n = 50$，假设总体的标准差 $\sigma = 18$，置信度为 95%，计算样本的均值：

$$\bar{x} = \sum_{i=1}^{50} x_i / n = 2472/50 = 49.44$$

查正态分布概率表，$z_{\alpha/2} = Z_{0.025} = 1.96$

因此，在 95% 置信水平下，置信区间为

$$\bar{x} \pm z_{\alpha/2} \frac{\sigma}{\sqrt{n}} = 49.44 \pm \frac{18}{50} = 49.44 \pm 2.55$$

即置信区间为（46.89，51.99），

说明在 95% 的置信水平下，收视人数的置信区间在 46.89 万人次至 51.99 万人次之间。

解读这个数据时，置信区间在 95% 的概率下得到。其实从表 4-18 就可以看到，不在这个区

间的数据还是很多的。因此，这个答案并不是百分百的正确，只是一个概率下的推断。

4.7.2 相关分析的异常

相关分析是研究两个变量直接的相关关系的一种统计方法。开展相关分析时，可以收集相关变量的成对数据后绘制散点图，从散点图的图形来进行判断存在怎样的相关关系。更好的方法是求取两个变量之间的相关系数 r。如果 r 的绝对值越接近 1，则两个变量的相关程度越强；如果 r 的绝对值越接近 0，则两个变量的相关程度越弱。

在实际应用中，有些人没有按照上述方法来判断相关性，而是凭自己的感觉来进行相关分析，出现相关分析的异常。

举例：

有一个质量管理小组开展活动，小组在原因分析中找到的一条末端原因为"初支成环距离长"。小组在现状调查中找到的症结为"钢架扭曲"。在开展主要原因确认过程中，小组成员采用研究"初支成环距离"与"钢架扭曲发生率"之间的相关关系的方法。小组共随机收集了 30 组数据，形成"初支成环距离"与"钢架扭曲发生率"统计表，见表 4-19。

表 4-19　"初支成环距离"与"钢架扭曲发生率"统计表

初支成环距离/m	钢架扭曲发生率/%	初支成环距离/m	钢架扭曲发生率/%	初支成环距离/m	钢架扭曲发生率/%	初支成环距离/m	钢架扭曲发生率/%	初支成环距离/m	钢架扭曲发生率/%
34.1	18.5	33.4	15.0	34.6	18.9	33.1	16.3	33.5	14.9
31.2	12.3	34.6	16.9	30.9	10.3	34.2	18.8	32.0	13.6
32.6	13.6	31.7	11.9	33.7	14.5	32.4	13.3	32.7	13.2
30.0	10.1	33.6	17.0	30.6	11.0	30.0	10.5	32.4	12.8
34.9	16.6	33.7	17.5	30.6	10.7	32.0	14.4	30.1	10.0
33.5	15.3	31.4	12.3	31.4	11.2	33.0	14.0	31.1	11.6

根据表 4-19 的数据，小组成员绘制了"初支成环距离与钢架扭曲发生率散布图"，见图 4-23。

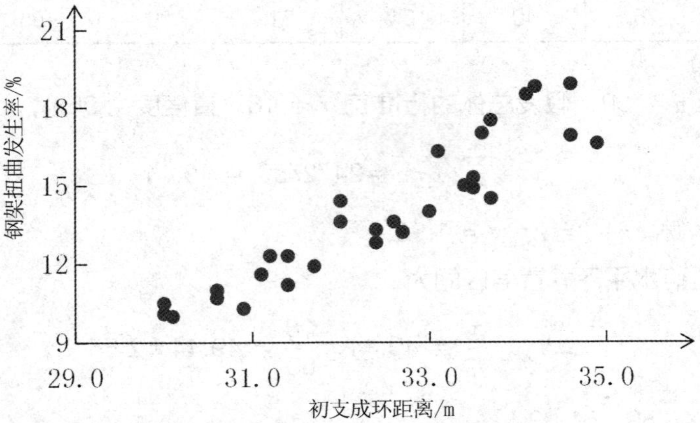

图 4-23　初支成环距离与钢架扭曲发生率散布图

小组成员根据图 4-23 散布图进行分析，认为"钢架扭曲发生率"随"初支成环距离"增加而增大，二者呈强正相关关系，钢架扭曲发生率最大与最小之间相差 8.9%，属于末端原因对于症结的影响大。因此，小组认为末端原因"初支成环距离长"为影响症结的主要原因。

在本例中，小组为了进行主要原因的确认，运用了相关分析来判定"初支成环距离长"与"钢架扭曲"的相关关系。为此，小组在收集 30 组数据后绘制了"初支成环距离与钢架扭曲发生率"散布图。小组的判定依据"二者呈强正相关关系"是目测的结果，而且依据"钢架扭曲发生率最大与最小之间相差 8.9%"来进行最终判定末端原因"初支成环距离长"为症结，属于主观臆测，是缺少说服力的。正确的做法应当研究"初支成环距离长"与"钢架扭曲"的相关系数，通过相关系数进行确认是比较有说服力的。

4.7.3 回归分析的异常

在回归分析中出现异常的情形比较多，主要有以下五种。

（1）回归分析与相关分析混淆

回归分析与相关分析容易混淆，是因为回归分析与相关分析关系很密切。其实两者之间还是存在着区别的。

回归分析是研究两个变量之间相互依赖程度的一种统计方法，两个变量之间是不对等的关系，一个变量是可以控制的自变量，而另一个变量是随机的因变量。它只能证明两个变量之间存在某种关系，确定两个变量之间这种关系的具体形式，可以用数学模型来表示两个变量之间的具体关系。通过回归分析，可以用一个变量来预测另一个变量，也就是说，当一个变量发生变化的时候，可以预测另一个变量的变化数据。

相关分析是研究两个变量之间相关关系的一种统计方法，相关分析的两个变量都是随机的，也可以是一个变量随机，另一个非随机。它可以分析相关的方向、相关的程度，但不区别其中哪个是自变量、哪个是因变量，两个变量之间是对等关系，是成对出现的。相关关系的确定不等于确认两个变量之间存在因果关系。当一个变量的发生变化的时候，是不能够确定另外一个变化会发生怎样的变化。

相关分析是回归分析的基础。如果两个变量之间不进行相关分析，直接作回归分析，则所建立的回归方程没有实际意义。回归分析是相关分析的深入，比如两个变量之间的确存在密切的相关关系，但无法确定两个变量之间是否有确定的依存关系，这是就需要回归分析来确定。

（2）线性回归关系与非线性回归关系混淆

在进行回归分析时，往往会把线性回归作为第一选择，因为线性回归分析相对简单。在实际应用中，非线性回归的线性比比皆是。因此，有的应用者会把两者混淆。尽管可以通过一定的转换方式把非线性回归转化为线性回归，以便于分析，但两者的区别还是显著的。

一元线性回归分析是回归分析中最基本的形式，其表达式为：$y = a + bx + \varepsilon$，其中，x 为自变量，y 为因变量，ε 为误差，a 为常数，b 为斜率。

在回归方程中，因变量是自变量的一次以上函数形式，在图形上回归线呈现为各种形式的曲线，则为非线性回归。非线性回归是回归函数关于未知回归系数具有非线性结构的回归。其表达式有很多，比如：$y = a x^2$。

非线性回归方程的转换：

有的非线性回归表达式可以通过变量的转换，成为线性回归表达式，见表 4-20。

表 4-20　非线性回归方程转换为线性方程

非线性回归方程	转换公式	转换后的线性回归方程
$\dfrac{1}{y} = a + \dfrac{b}{x}$	$X = \dfrac{1}{x}$，$Y = \dfrac{1}{y}$	$Y = a + bX$
$y = a x^b$	$X = \ln x$，$Y = \ln y$	$Y = a' + bX$（$a' = \ln x$）
$y = a + b\ln x$	$X = \ln x$，$Y = y$	$Y = a + bX$
$y = a e^{bx}$	$X = x$，$Y = \ln y$	$Y = a' + bX$（$a' = \ln x$）
$y = a e^{b/x}$	$X = \dfrac{1}{x}$，$Y = \ln y$	$Y = a' + bX$（$a' = \ln x$）

非线性回归方程的不可转换：

有的非线性回归表达式并不能转换为线性回归表达式。对于这种情形，可以使用单纯形法（simplex method），这是一种对无约束极值问题的数学解法，它同样基于回归分析使用的最小二乘法，然后求出误差平方和最小的极值。

单纯形法由美国数学家丹齐格（George Dantzig）于 1947 年提出，它的基本思路是：从一个基本可行解出发，判断其是否为最优。如果是最优，则停止迭代。如果不是最优解，则换一种可行解。通过这样的方式，求得使目标函数值下降的另一个基本可行解。

单纯形法的基本步骤如下：

确定初始的基本可行解。假设在标准型线性规划中，系数矩阵 A 中的前 m 个系数列向量正好构成一个可行基，可行基确定，则可行解也确定。那么有

$$A = (B\ N)$$

其中，$B = (P_1,\ P_2,\ \cdots,\ P_m)$ 为基变量 X_1，X_2，$\cdots$，X_m 的系数列向量构成的可行基；$N = (P_{m+1},\ P_{m+2},\ \cdots,\ P_n)$ 为非基变量 X_{m+1}，X_{m+2}，$\cdots$，X_n 的系数列向量构成的矩阵。

因此，约束方程 $AX = B$ 可以表示为

$$AX = (B\ N)\begin{pmatrix} X_B \\ X_N \end{pmatrix} = B X_B + N X_N = b$$

用可行基 B 的逆阵 B^{-1} 左乘等式两端，移项后可得

$$X_B = B^{-1}b - B^{-1}N X_n$$

令左右非基变量 $X_N = 0$，则有基变量 $X_B = B^{-1}b$

得到初始的基本可行解 $X = \begin{pmatrix} B^{-1}b \\ 0 \end{pmatrix}$

归纳一下，由 $AX = b$，得到：$B X_B + N X_N = b$，再推出 $X_B = B^{-1}b - B^{-1}N X_n$，最后得到 $X_N = 0$，$X_B = B^{-1}b$。

（3）指数函数和对数函数的混淆

指数函数与对数函数在模型的图示上有相像之处，都有呈单调上升或单调下降的形式。在回归分析运用时，容易引起混淆。两者的区别如下：

指数函数的表达式为：$y = a^x$，（$a > 0$，且 $a \neq 1$）

指数函数的图形见图 4-24。

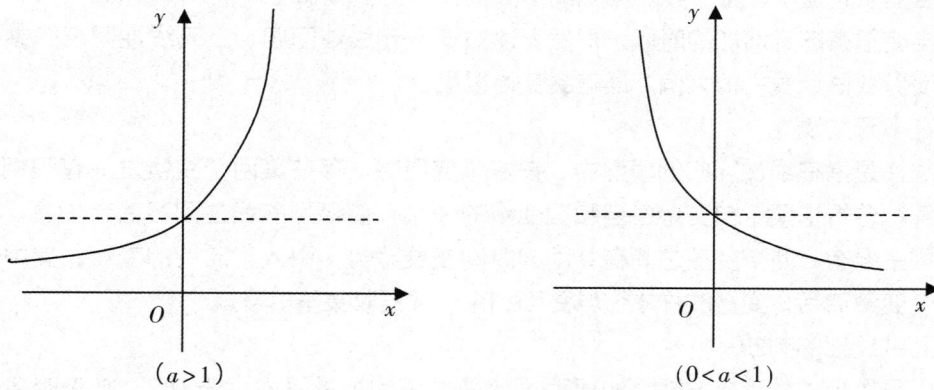

（a>1）　　　　　　　　　　　　　　　　　　　（0<a<1）

图 4-24　指数函数

对数函数的表达式为：$y = \log_a x$，（$a > 0$，且 $a \neq 1$）

对数函数的图形见图 4-25。

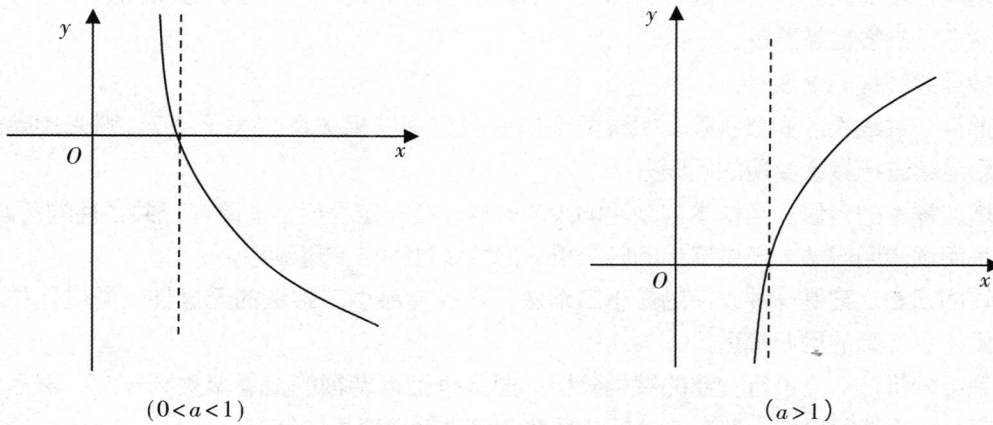

（0<a<1）　　　　　　　　　　　　　　　　　　　（a>1）

图 4-25　对数函数

（4）重要变量遗漏或变量随意增加

在回归分析中，自变量的遗漏或自变量太多，将会使数据发生很大的偏差。比如，一个二元线性回归的方程式为为

$$y = \beta_0 + \beta_1 x_1 + \beta_2 x_2 + \varepsilon$$

其中，x 为自变量，y 为因变量，$\beta_0, \beta_1, \beta_2$ 为待估参数，ε 为随机误差。

对于二元线性回归的参数估计是

$$\Sigma y = n\beta_0 + \beta_1 \Sigma x_1 + \beta_2 \Sigma x_2$$

$$\Sigma x_1 y = \beta_0 \Sigma x_1 + \beta_1 \Sigma x_1^2 + \beta_2 \Sigma x_1 x_2$$

$$\Sigma x_2 y = \beta_0 \Sigma x_2 + \beta_2 \Sigma x_2^2 + \beta_1 \Sigma x_1 x_2$$

如果变量发生遗漏，则原来的方程式变为一元线性回归方程：

$$y = \beta_0 + \beta_1 x + \varepsilon$$

对于一元线性回归的参数估计为

$$\Sigma y = n\beta_0 + \beta_1 \Sigma x$$

$$\Sigma xy = \beta_0 \Sigma x + \beta_1 \Sigma x^2$$

显然，当变量有遗漏，整个线性回归的参数估计发生很大偏差。

同样，当变量有随意增加的时候，根据上述对于一元线性回归、二元线性回归的参数估计的分析，显然使参数估计发生很大偏，误差会变得很大。

（5）存在多元共线性

多元共线性是指在线性回归的模型中，在多个变量中，存在某两个变量之间有强相关关系，那就无法单纯地分析这两个自变量各自和因变量的关系，使模型的线性回归估计失真。导致多元共线性的原因主要有：两个变量之间有共同的时间发展趋势；引入了滞后的变量，其中一个变量总是比另一个变量滞后；变量的样本资料受到约束，数据收集不够多。

对于多元共线性的判断：

一是可以采用相关性分析方法，如果两个变量之间的相关系数大于 0.7，则表明多元共线性存在；但反过来，如果两个变量之间的相关系数小于 0.7，并不能说明多元共线性不存在。

二是采用 VIF 检验。VIF（Variance Inflation Factor）是方差扩大因子值。通常当 VIF 大于 5 时，可认为多元共线性存在；VIF 越大，则多元共线性越严重。

也可以用容差值判断。容差值 = 1/VIF，当容差值小于 0.2 时，可认为多元共线性存在；容差值越小，多元共线性越严重。

对于多元共线性的处理：

一是把存在共线性的变量剔除。如果两个自变量之间的相关系数大于 0.7，则先剔除一个自变量，然后继续进行其余变量的回归分析。

二是增加样本的容量。当样本容量足够大的时候，会一定程度上消除多元共线性的问题。

三是采用逐步回归法。通过逐步回归分析，把共线性的自变量剔除。

四是岭回归法。它是一种改良的最小二乘法，它放弃最小二乘法的无偏性，通过损失部分信息来寻找更符合实际的回归模型。

五是聚类分析法。先进行变量的聚类分析，把具有相近特征的变量聚类到一起，然后进行回归分析。聚类分析法可以结合 VIF 法、相关系数法等来筛选变量的特征。

4.7.4 脱离数据进行推断

在进行统计推断的时候，分析的对象是某一类的样本数据，而推断的对象是另外一类的总体。这样，来自样本特征的数据分析结果，显然与具有另外样本特征的数据是不一致的。比如一个项目上进行作业工人的劳动工效的统计分析。先是得到了一组具有 15 年工作经验工人的工效数据。现在用这样的工效数据去预测另一组只有 3 年工作经验的工作的工效，显然推论结果是不正确的。

4.7.5 应用统计检验异常

统计检验异常是指检验的应用条件不符合要求，主要有以下六种情形。

1. t 检验应用不符合要求

t 检验是一种常见的统计方法，使用方便，应用广泛。但是 t 检验应用条件是：样本来自于正态分布（或近似正态分布）的总体，两个总体的方差相等。如果样本不符合以上条件，则不适用 t 检验。

2. 用参数检验代替非参数检验

参数检验有 t 检验、μ 检验、方差分析等。参数检验的应用条件是资料呈正态分布，各组的方差相同。对于偏态或方差不齐的资料要进行数据转换，符合正态分布、方差相同的条件后，才能进行参数检验。在应用时，直接对偏态、方差不齐的数据进行参数检验是不正确的；若转换以后不符合条件，也不能开展参数检验，只能进行非参数检验。

非参数检验有卡方检验、中位数等。非参数检验是直接对数据的分布进行检验，不需要假设总体的分布形式。其中的卡方检验是检验两个或多个样本比率之间是否存在显著性，以说明两种属性（现象）之间是否存在相关关系。

当测量两个定量变量之间的相关程度时，参数检验的方法是皮尔逊（Pearson）相关系数法，非参数检验的方法是斯皮尔曼（Spearman）秩相关法。

3. 应用最小二乘法不符合要求

应用最小二乘法有相应的条件：随机误差项必须服从均值为 0 的正态分布；随机误差项的方差不受自变量的影响，而且随机误差项的方差是固定值；自变量之间不存在完全共线性；自变量是确定变量，不是随机变量；随机误差项与自变量之间不相关。

4. 最大似然估计与最小二乘法混淆

最大似然估计是利用已知的样本结果，反推最大概率上导致这样结果的参数值，它是基于概率意义上出现的概率为最大，需要知道分布曲线的概率密度函数。

最小二乘法是通过最小化误差的平方和，来寻找数据的最佳函数匹配，它是基于变量在几何意义上的距离最小，对于数据的分布没有要求。

5. 双尾检验与单尾检验混淆

双尾检验，又称双侧检验，它只强调差异，不强调方向性（如大小、多少）。双尾检验在假设时，原假设 $H_0 : \mu_1 = \mu_0$；备择假设 $H_1 : \mu_1 \neq \mu_0$。双尾检验是检验两个参数之间是否有差异，比如：检验样本和总体均值有无差异；检验样本数之间有无差异等。双尾检验适用的情形是不清楚后测的数据是否高于前测的数据，想判断前测、后测数据的均值是否不同。

单尾检验，又称单侧检验，它强调某一方向。单尾检验在假设时，原假设 $H_0 : \mu_1 = \mu_0$；备择假设 $H_1 : \mu_1 > \mu_0$（或 $H_1 : \mu_1 < \mu_0$）。单尾检验是进行一个参数和另一个参数方向性的比较，方向性体现为"大于""小于""好于""差于"等。比如要检验样本所取自总体的参数值大于或小于某个特定值等。单尾检验适用的情形是已知后测数据不可能低于前测数据，准备判断后测数据是不是高于前测数据。

6. t 检验与 μ 检验混淆

t 检验和 μ 检验有相同之处：两者的样本都来自正态分布的总体，样本的总体方差相等。但两者存在区别。

t 检验是参数检验。采用 t 检验时，要求样本容量比较小，通常 $n < 30$，总体的标准差 σ 未知。

μ 检验属于非参数检验。采用 μ 检验时，样本的容量足够大。它需要把两个独立样本的数据合并在一起，转化为顺序数据，然后来检验两本样本的平均值是否有显著差异。

4.7.6 中位数与平均数应用不当

使用中位数和平均数都可以衡量一组数据的集中趋势。但两者存在着很大的区别，即平均数对于异常值很敏感，会很大程度上改变平均数的数值。而中位数对于异常值不敏感，不会受到极值的影响。

4.7 Statistical Inference Abnormalities

Statistical inference is a method to estimate and predict the characteristics of the population according to the characteristics of the sample under a certain degree of confidence. The usual statistical inference includes parameter estimation and hypothesis test. In a broad sense, correlation analysis, regression analysis, t- test, F- test, x^2 test are also included in statistical inference. There are many abnormal situations in statistical inference, which are briefly described below.

4.7.1 Abnormal parameter estimation

Parameter estimation is a deduction under the premise of a certain confidence interval, which means that the occurrence of real data is not within the confidence interval.

Case：

Entertainment program A is broadcast on TV in B province. In order to better develop marketing strategies, it is necessary to estimate the current ratings in B Province and entrust a market research company to make market forecasts. The research company surveyed the ratings of program A in the last quarter, and randomly selected 50 ratings records as a sample to get the ratings table, as shown in Table 4-18.

Table 4-18　The viewing rating of Program A　　　unit：10 000 person-times

Number of viewers									
50	53	37	48	86	24	45	65	90	18
32	40	23	59	68	36	52	50	72	47
74	19	64	47	115	28	40	61	36	31
35	24	44	32	46	98	69	40	87	28
41	32	25	40	54	31	30	105	41	60

Analyze as follow：$n =50$, assuming the standard deviation of the population $\sigma = 18$. The confidence level is 95%. Calculate the mean value of the sample：

$$\bar{x} = \sum_{i=1}^{50} x_i/n = 2472/50 = 49.44$$

Look up the normal distribution probability table, $z_{\alpha/2} = Z_{0.025} = 1.96$

Therefore, at 95% confidence level, the confidence interval is

$$\bar{x} \pm z_{\alpha/2} \frac{\sigma}{\sqrt{n}} = 49.44 \pm \frac{18}{50} = 49.44 \pm 2.55$$

That is, the confidence interval is(46.89,51.99).

It shows that at the 95% confidence level, the confidence interval of the number of viewers is between 468 900 and 519 900.

When interpreting this data, the confidence interval is obtained under 95% probability. In fact, we can see from Table 4-18 that there are still many data out of this range. Therefore, this answer is not 100% correct, but an inference under probability.

4.7.2 Abnormality of correlation analysis

Correlation analysis is a statistical method to study the direct correlation between two variables. When carrying out correlation analysis, we can collect the paired data of relevant variables and draw a scatter chart to judge what kind of correlation exists from the graph of the scatter chart. A better method is to calculate the correlation coefficient r between two variables. The closer the absolute value of r is to 1, the stronger the correlation between the two variables; If the absolute value of r is closer to 0, the correlation between the two variables is weaker.

In practical applications, some people do not judge the correlation according to the above methods, but carry out correlation analysis based on their own feelings, and there are abnormalities in correlation analysis.

Case：

A quality management team carried out activities, and the team found one end cause in the cause analysis as "long distance of temporary support ring". The crux found by the team in the status survey is "distortion of steel frame". In the process of identifying the main causes, the team members adopted the method of studying the correlation between the "long distance of temporary support ring" and the "incidence of steel frame distortion". The team collected 30 groups of data at random to form a statistical table of "temporary support ring distance" and "steel frame distortion rate". see Table 4-19.

Table 4-19   "Temporary support ring distance" and "steel frame distortion rate"

Temporary support ring distance/m	Steel frame distortion rate/%	Temporary support ring distance/m	Steel frame distortion rate/%	Temporary support ring distance/m	Steel frame distortion rate/%	Temporary support ring distance/m	Steel frame distortion rate/%	Temporary support ring distance/m	Steel frame distortion rate/%
34.1	18.5	33.4	15.0	34.6	18.9	33.1	16.3	33.5	14.9
31.2	12.3	34.6	16.9	30.9	10.3	34.2	18.8	32.0	13.6
32.6	13.6	31.7	11.9	33.7	14.5	32.4	13.3	32.7	13.2
30.0	10.1	33.6	17.0	30.6	11.0	30.0	10.5	32.4	12.8
34.9	16.6	33.7	17.5	30.6	10.7	32.0	14.4	30.1	10.0
33.5	15.3	31.4	12.3	31.4	11.2	33.0	14.0	31.1	11.6

According to the data in Table 4-19, the team members drew a "scatter diagram of the temporary support ring distance and the steel frame distortion rate", as shown in Figure 4-23.

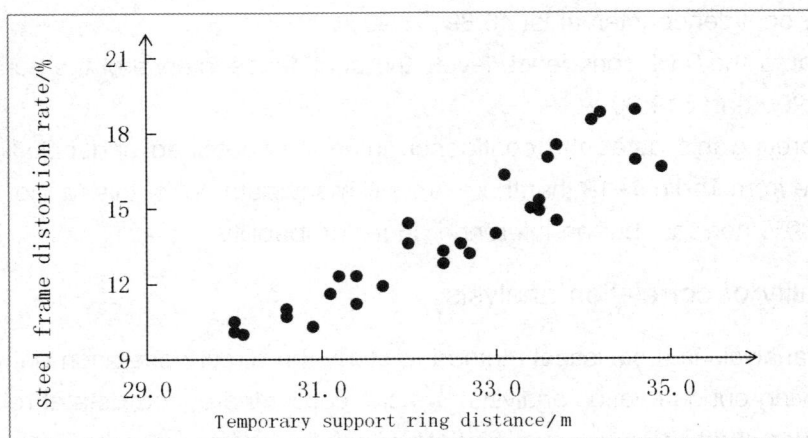

Figure 4-23 Diagram of the "Temporary support ring distance" and the "Steel frame distortion rate"

According to the scatter diagram in Figure 4-23, the team members believed that the "occurrence rate of steel frame distortion" increased with the increase of the "temporary support ring distance", and the two were in a strong positive correlation. The difference between the maximum and minimum steel frame distortion occurrence rate was 8.9%, so the end cause had a great impact on the crux. Therefore, the team confirmed that the end cause "the long temporary support ring distance" was the main cause that affected the crux.

In this case, in order to confirm the main causes, the team used correlation analysis to determine the correlation between "the long distance from temporary support ring" and "the distortion of steel frame". To this end, after collecting 30 sets of data, the team drew a scatter map of "temporary support ring distance and steel frame distortion rate". The judgment of the team is based on the "strong positive correlation between the two", which is the result of visual inspection, and the final judgment of the end cause "the long distance between the temporary support ring" is based on the "difference between the maximum and the minimum steel frame distortion rate of 8.9%", which is subjective speculation and is not convincing. The correct way to do this is to study the correlation coefficient between "long distance of temporary support ring" and "steel frame distortion". It is more convincing to confirm through the correlation coefficient.

4.7.3 Abnormality of regression analysis

There are many abnormal situations in regression analysis, mainly including the following five items.

（1）Confusion between regression analysis and correlation analysis

Regression analysis and correlation analysis are easily confused because they are closely related. In fact, there are differences between the two.

Regression analysis is a statistical method to study the degree of interdependence between two variables. There is an unequal relationship between the two variables. One variable is a controllable independent variable, and the other is a random dependent variable. It can only prove that there is a certain relationship between the two variables, and determine the specific form of this relationship between the two variables, which can be expressed by mathematical models. Through regression analysis, one variable can be used to predict another variable, that is, when one variable changes, the change data of another variable can be predicted.

Correlation analysis is a statistical method to study the correlation between two variables. The two variables of correlation analysis are both random, or one of them is random variable and the other is non-random variable. It can analyze the direction and degree of correlation, but it does not distinguish which is the independent variable and which is the dependent variable. The two variables are equivalent and appear in pairs. The determination of the correlation does not mean the confirmation of the causal relationship between the two variables. When one variable changes, it is impossible to determine how another variable will change.

Correlation analysis is the basis of regression analysis. If the correlation analysis is not conducted between the two variables and the regression analysis is conducted directly, the established regression equation has no practical significance. Regression analysis is the deepening of correlation analysis. For example, there is indeed a close correlation between the two variables, but it is impossible to determine whether there is a certain dependency between the two variables, which requires regression analysis to determine.

(2) Confusion between linear regression relationship and nonlinear regression relationship

When performing regression analysis, linear regression is often the first choice, because linear regression analysis is relatively simple. In practical applications, the linearity of nonlinear regression is everywhere. Therefore, some users will confuse the two. Although nonlinear regression can be transformed into linear regression in a certain way to facilitate analysis, the difference between the two is still significant.

Univariate linear regression analysis is the most basic form of regression analysis, and its expression is: $y = a + bx + \varepsilon$, where x is the independent variable and y is the dependent variable, ε is error, a is constant, and b is slope.

In the regression equation, the dependent variable is the functional form of the equation with more than one degree of independent variable, and the regression line on the graph shows various forms of curves, which is non-linear regression. Nonlinear regression is a regression function with nonlinear structure about unknown regression coefficients. There are many expressions, such as: $y = a x^2$.

Transformation of nonlinear regression equation:

Some nonlinear regression expressions can be converted into linear regression expressions through variable transformation, as shown in Table 4-20.

Table 4-20 Conversion of nonlinear regression equation to linear equation

Nonlinear regression equation	Conversion formula	Converted linear regression equation
$\dfrac{1}{y} = a + \dfrac{b}{x}$	$X = \dfrac{1}{x}, \ Y = \dfrac{1}{y}$	$Y = a + bX$
$y = a x^b$	$X = \ln x, \ Y = \ln y$	$Y = a' + bX \ (a' = \ln x)$
$y = a + b\ln x$	$X = \ln x, \ Y = y$	$Y = a + bX$
$y = a e^{bx}$	$X = x, \ Y = \ln y$	$Y = a' + bX \ (a' = \ln x)$
$y = a e^{b/x}$	$X = \dfrac{1}{x}, \ Y = \ln y$	$Y = a' + bX \ (a' = \ln x)$

Non-conversion of nonlinear regression equation:

Some nonlinear regression expressions cannot be converted into linear regression expressions. In this case, we can use the simplex method. This is a mathematical solution to the unconstrained extreme value problem. It is also based on the least square method used in regression analysis, and then the extreme value with the minimum sum of squares of errors is obtained.

The simplex method was proposed by the American mathematician George Dantzig in 1947. Its basic idea is to judge whether it is optimal from a basic feasible solution. If it is optimal, stop the iteration. If it is not the optimal solution, replace it with a feasible solution. In this way, another basic feasible solution that reduces the value of the objective function is obtained.

The basic steps of simplex method are as follows:

Determine the initial basic feasible solution. Suppose that in the standard linear programming, the first m series vectors in the coefficient matrix A exactly constitute a feasible basis, and if the feasible basis is determined, the feasible solution is also determined. So there are

$$A = (B \ N)$$

Wherein $B = (P_1, P_2, \cdots, P_m)$ is a feasible basis composed of series vectors of basic variables $X_1, X_2, \cdots, X_m$;

$N = (P_{m+1}, P_{m+2}, \cdots, P_n)$ is a matrix composed of series vectors of non-basic variables X_{m+1}, $X_{m+2}, \cdots, X_n$

Therefore, the constraint equation AX=B can be expressed as:

$$AX = (B \ N)\binom{X_B}{X_N} = B X_B + N X_N = b$$

Multiply the two ends of the equation with the inverse matrix B^{-1} of the feasible base B, after item transfer, we get:

$$X_B = B^{-1}b - B^{-1}N X_n$$

Set left and right non-basic variables $X_N = 0$, then there is a basic variable $X_B = B^{-1}b$

Get the initial basic feasible solution $X = \binom{B^{-1}b}{0}$

To sum up, from $AX = b$, we get: $B X_B + N X_N = b$, then deduce $X_B = B^{-1}b - B^{-1}N X_n$. Finally we get $X_N = 0$, $X_B = B^{-1}b$.

(3) Confusion between exponential function and logarithmic function

There are similarities between exponential function and logarithmic function in the diagram of the model, both of which are in the form of monotonic rise or monotonic decline. It is easy to cause confusion when applying regression analysis. The differences between the two are as follows:

The expression of the exponential function is: $y = a^x$, ($a > 0$, 且 $a \neq 1$)

The graph of exponential function is shown in Figure 4-24.

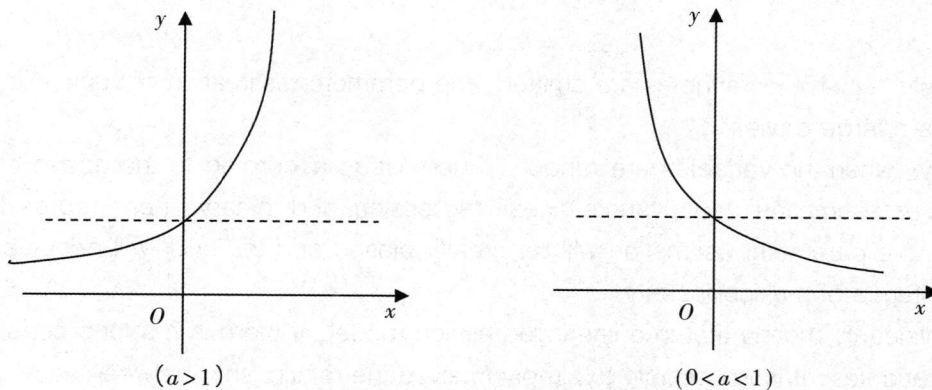

$(a>1)$　　　　　　　$(0<a<1)$

Figure 4-24　Exponential function

The expression of logarithmic function is: $y = \log_a x, (a > 0, \text{且 } a \neq 1)$

The graph of logarithmic function is shown in Figure 4-25.

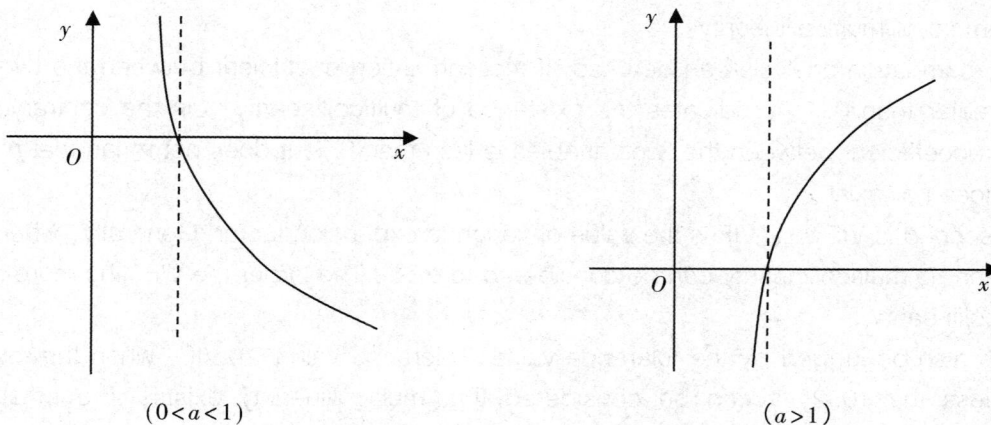

$(0<a<1)$　　　　　　　$(a>1)$

Figure 4-25　Logarithmic function

(4) Important variables are omitted or added randomly

In the regression analysis, the omission of independent variables or too many independent variables will cause great deviation in the data. For example, the equation of a binary linear regression is

$$y = \beta_0 + \beta_1 x_1 + \beta_2 x_2 + \varepsilon$$

In the formula, x is the independent variable, y is the dependent variable, β_0, β_1, β_2 are the parameters to be estimated, and ε is the random error.

The parameter estimation of binary linear regression is

$$\Sigma y = n\beta_0 + \beta_1 \Sigma x_1 + \beta_2 \sum x_2$$

$$\sum x_1 y = \beta_0 \Sigma x_1 + \beta_1 \Sigma x_1^2 + \beta_2 \Sigma x_1 x_2$$

$$\sum x_2 y = \beta_0 \Sigma x_2 + \beta_2 \Sigma x_2^2 + \beta_1 \Sigma x_1 x_2$$

If the variables are missed out, the original equation becomes an univariate linear regression equation, which is

$$y = \beta_0 + \beta_1 x + \varepsilon$$

For univariate linear regression, the parameter estimation is

$$\Sigma y = n\beta_0 + \beta_1 \Sigma x$$
$$\Sigma xy = \beta_0 \Sigma x + \beta_1 \Sigma x^2$$

Obviously, when the variables are omitted, the parameter estimation of the whole linear regression has a large deviation.

Similarly, when the variables are randomly increased, according to the above analysis of the parameter estimation of univariate linear regression and binary linear regression, it is obvious that the parameter estimation will be greatly biased and the error will become large.

(5) Existence of multicollinearity

Multicollinearity means that in a linear regression model, if there is a strong correlation between two variables, it is impossible to simply analyze the relationship between each of the two independent variables and the dependent variables, thus distorting the linear regression estimation of the model. The main reasons for multicollinearity are: there is a common time trend between the two variables; A lagging variable is introduced, one of which is always lagging behind the other; The sample data of variables are constrained, and data collection is not enough.

Judgment of multicollinearity:

First, correlation analysis can be used. If the correlation coefficient between the two variables is greater than 0.7, it indicates the existence of multicollinearity; On the contrary, if the correlation coefficient between the two variables is less than 0.7, it does not mean that multicollinearity does not exist.

The second is VIF test. VIF is the value of variance expansion factor. Generally, when VIF is greater than 5, multicollinearity can be considered to exist; The larger the VIF, the more serious the multicollinearity.

It can also be judged by the tolerance value. Tolerance value = 1/VIF, when the tolerance value is less than 0.2, it can be considered that multicollinearity exists; The smaller the tolerance value, the more serious the multicollinearity.

Treatment of multicollinearity:

The first is to eliminate the variables with collinearity. If the correlation coefficient between the two independent variables is greater than 0.7, first remove one independent variable, and then continue the regression analysis of the other variables.

The second is to increase the sample size. When the sample size is large enough, the problem of multicollinearity will be eliminated to some extent.

The third is the stepwise regression method. Through stepwise regression analysis, the independent variables of collinearity are eliminated.

The fourth is ridge regression method. It is an improved least squares method. It abandons the unbiased nature of the least squares method and seeks a more practical regression model by losing some information.

The fifth is cluster analysis. First, cluster the variables with similar characteristics, and then conduct regression analysis. Cluster analysis can combine VIF method and correlation coefficient method to screen the characteristics of variables.

4.7.4 Infer without correct data

When making statistical inference, the object of analysis is the sample data of one type, while the object of inference is the population of another type. In this way, the data analysis results from the sample characteristics are obviously inconsistent with the data with other sample characteristics. For example, a statistical analysis of the labor efficiency of workers is carried out on a project. First, we got a group of work efficiency data of workers with 15 years of work experience. Now using such ergonomic data to predict the ergonomics of another group of workers with only 3 years of work experience is obviously incorrect.

4.7.5 Abnormality of applying statistical test

Abnormal statistical test means that the application conditions of the test do not meet the requirements, mainly in the following six cases.

1. The application of t-test does not meet the requirements

t-test is a common statistical method, which is convenient to use and widely used. However, the application condition of t-test is that the sample is from a population with normal distribution (or approximate normal distribution), and the variance of the two populations is equal. If the sample does not meet the above conditions, the t-test is not applicable.

2. Replacing nonparameter test with parameter test

Parameter test includes t-test μ-Test, analysis of variance, etc. The application condition of parameter test is that the data is normally distributed and the variance of each group is the same. For data with skewness or uneven variance, data conversion shall be carried out, and parameter test can only be carried out after meeting the conditions of normal distribution and the same variance. In application, it is incorrect to directly test parameters of skewed and uneven data; If the conditions are not met after conversion, parameter inspection cannot be performed, and only non-parametric inspection can be performed.

Nonparameter test includes chi-square test, median, etc. Nonparameter test is to test the distribution of data directly, without assuming the distribution form of the population. The chi-square test is to test whether there is a significant relationship between two or more sample ratios to show whether there is a correlation between the two attributes (phenomena).

When measuring the degree of correlation between two quantitative variables, Pearson correlation coefficient method is used for parameter test, and Spearman rank correlation method is used for non-parameter test.

3. The application of least square method does not meet the requirements

There are corresponding conditions for applying the least square method: the random error term must obey the normal distribution with the mean value of 0; The variance of the random error term is not affected by the independent variable, and the variance of the random error term is a fixed value; There is no complete collinearity between independent variables; The independent variable is a definite variable, not a random variable; There is no correlation between random error terms and independent variables.

4. Confusion between maximum likelihood estimation and least squares method

The maximum likelihood estimation is to use the known sample results to deduce the parameter values that lead to such results in the maximum probability. It is based on the maximum probability of occurrence in the sense of probability. It is necessary to know the probability density function of the distribution curve.

The least square method is to find the best function matching of data by minimizing the sum of squares of errors. It is based on the minimum distance of variables in the geometric sense, and has no requirements for the distribution of data.

5. Confusion between two-tailed test and one-tailed test

The two-tailed test, also known as the two-sided test, only emphasizes the difference, but not the directionality (such as size and quantity). When the two-tailed test is assumed, the null hypothesis is H_0: $\mu_1 = \mu_0$; the alternative hypothesis is H_1: $\mu_1 \neq \mu_0$. The two-tailed test is used to test whether there are differences between the two parameters, such as whether there are differences between the sample and the population mean; Check whether there is any difference between the number of samples. The two-tailed test is applicable when it is not clear whether the post-test data is higher than the pre-test data, and it is ready to judge whether the mean value of the pre-test and post-test data is different.

The one-tailed test, also known as the one-sided test, emphasizes a certain direction. When the one-tailed test is assumed, the null hypothesis is H_0: $\mu_1 = \mu_0$; the alternative hypothesis H_1: $\mu_1 > \mu_0$ (or H_1: $\mu_1 < \mu_0$). One-tailed test is to compare the directionality of one parameter with that of another parameter. The directionality is reflected as "greater than" "less than" "better than" "worse than", etc. For example, the parameter value of the sample taken from the population is greater than or less than a specific value. The one-tailed test is applicable when the post-test data cannot be lower than the pre-test data, and it is ready to judge whether the post-test data is higher than the pre-test data.

6. Confusion between the t -test and the μ -test

There are similarities in the t -test and the μ -test: both samples are from the population with normal distribution, and the total variance of the samples is equal. But there are differences between the two.

t -test is the parameter test. When using t -test, the sample size is required to be small, usually $n < 30$, and the standard deviation σ of the population is unknown.

μ -test is the non parameter test. When t -test is applied, the sample size must be large enough. It needs to combine the data of two independent samples, convert them into sequential data, and then test whether there is significant difference between the average values of the two samples.

4.7.6 Improper application of median and average

Both median and average can be used to measure the concentration trend of a group of data. However, there is a big difference between the two, that is, the average is very sensitive to the abnormal value and will change the value of the average to a large extent. The median is not sensitive to outliers and will not be affected by extreme values.

Chapter 5
Data Missing

第 5 章
数据的缺失

数据缺失主要指出于有意或无意的情形，在本应完整收集、记录的全部数据集合中，发生了部分数据或全部数据的丢失。数据缺失的后果是比较严重的，它使数据集不能正确、完整地反映客观对象的特性，使数据产生一系列的问题：

一是数据的完整程度受损。数据发生缺失的情形，可能是全部数据记录缺失，也可能是部分数据记录缺失，比如数据属性不完整、数据条目不完整、数据过程不完整等。不完整的数据其使用价值会大大降低。

二是数据的有效性受损。完整的数据可以对客观对象进行有效的描述，但缺失的数据很可能导致数据集合的失效，比如缺少时间标示的数据，在分析时失去价值。

三是数据的准确性受损。由于数据的不完整，使数据所描述的特征不能与客观对象的真实特征相符。比如统计一个项目的焊接施工质量问题，由于数据的缺失，使焊接的统计合格率不能真实反映焊接的质量状况。

四是数据的一致性受损。由于存在数据缺失的情况，使前后数据进行比较的时候，缺少一致性，反映在时间上、属性上、数量上、方法上等，造成信息不一致。

5.1　不能完整记录数据

———

不能完整记录数据，是指在观察、测量、试验等产生数据的过程中，存在部分数据没有被记录的情形。数据不能被完整记录是由多种原因造成，有些是限于各种条件而无法记录完整，有些是由于人为的因素而无法记录完整等。

数据收集方案存在缺陷，直接导致数据收集的不完整现象发生。发生这样的情形原因主要有以下两种。

1. 没有明确数据收集的时间和时间段

数据收集应当明确具体的收集时间和时间段，否则会发生两种情况，一是没有明确具体的收集时间，则收集数据的随意性太强。二是没有明确时间段，则可能收集数据的时间太短，使数据量明显缺失；也可能收集数据的时间太长，不仅增加大量成本，而且存在数据的时效性、可比性等方面产生困难。

举例：

某质量管理小组开展活动，小组的现状调查过程如下：

为进一步提高预制型橡胶跑道铺装质量，小组对于对周边的体育场馆（包含自己所在公司）中橡胶跑道铺装进行检查，通过调查共计 600 个点，现场一次验收平均合格率仅为 81%，分类统计存在的质量缺陷见表 5-1。

表 5-1　现状调查表

序号	检查项目	检查数量/个	合格点数/个	合格率/%
1	面层开裂		53	53
2	面层不平整		57	57
3	面层鼓包起泡	100	92	92
4	标志线偏差		94	94
5	面层污染		96	96
6	其他		94	94
	总计	600	486	81

初看这张现状调查表，似乎现状调查的过程是完全正确的。但仔细分析，小组在进行现状调查的时候没有明确具体的调查时间段，因此存在着这样的问题：

（1）通常来说，一个城市的体育场馆以及周边的体育场馆不会在同一时间进行施工，况且具有橡胶跑道的体育场馆在同一个城市中的数量是不多的，那就说明有的体育场馆施工的时间已经在很久以前了。这样就造成了现状调查的结果不是当前的，而是很多年以前的跑道情况了。

（2）橡胶跑道本身的质量提升很快，其施工技术发展也很快。数年前的橡胶跑道与当前的橡胶跑道在工艺上、质量上有差异；数年前的施工技术与当前的施工技术也不能在同一个平台上进行对比。

（3）随着时间的推移，以前施工的橡胶跑道的面层本来就会发生开裂、污染等情况，因此，小组收集到的"面层开裂""面层不平整""面层鼓包起泡""面层污染"等质量问题，可能是本身老化、缺少保养的结果。

因此，本例是典型的数据记录不完整情况。

2. 采集的数据指标有遗漏

数据指标是指在数据方案中所规定的、用来描述调研对象的量化特征。不同的调研目的、不同的调研对象，有着不同的数据指标。就具体的一次调研过程来说，数据指标应当能够全面反映调研对象的特征。在设计数据采集方案时，如果数据指标有遗漏，则会造成某一类别的数据整体缺失，调研工作将无法完成。

举例：

某实体公司要求营销部门进行调研，期望通过最近一年来的产品销售情况分析，为下一步的产品功能改进、销售网络布局等提供决策依据。为此，营销部门制定了调研所用的数据指标，见表 5-2。

表 5-2　营销部门制定的数据指标

用户数据	用户的存量
	用户的增量
	用户的联系方法

	购买频次
行为数据	满意度
	销售量
业务数据	销售单价
	销售高峰时间段

针对营销部门制定的数据指标，实体公司的决策层表示不满意，认为数据指标存在着缺陷，有些关键指标没有列入，比如：用户的流动率、用户的主要来源、人均消费情况等。于是，营销部门邀请了外部第三方，重新制定了数据指标，见表 5-3。

表 5-3　第三方制定的数据指标

	用户的来源地
	用户的联系方法
	用户的流失率
	用户的新增量
用户数据	用户的存量
	用户的爱好
	用户的平均年龄
	用户的使用爱好
	用户的学历
	购买频次
	购买前的主要关注点
行为数据	逗留实体商铺的时间
	用户关注的产品摆放位置
	用户的满意度
	各产品的销售单价
	不同产品的销售总量
业务数据	各产品的销售总价
	销售高峰时间段
	广告产品的销售单价和总量
	人均消费情况

从第三方指定的数据指标与营销部门制定的数据指标对比来看，第三方的数据指标显然更完整，能够为决策者提供更多的决策信息。

数据指标的设定没有统一的模式，关键是能够通过指标的收集，来完整反映出客观对象的所有特征。在上述的例子中，仅针对用户的满意度还可以开发新的数据指标，比如用 HEART 模型来表示，见表 5-4。

表 5-4　衡量用户满意度的 HEART 模型

用户的愉悦感	用于衡量用户的满意度,在功能层面可表现为评分、好评、一般、分享、推荐、持续推荐等
用户的参与度	用于衡量用户的参与程度,在功能层面可以表现为评论、好评、一般、分享、推荐等
用户的接受程度	用于衡量用户的接受程度,在功能层面可以表现为观摩、欣赏、采购、迟疑、退货等
用户的留存率	用于衡量用户的接受程度,在功能层面表现为新、老用户的留下来数量等
任务完成率	用于衡量用户的不同路径完成程度,在功能层面表现为产品功能的使用情况、完成时间等

关于数据指标的设定,还有其他建模方式,比如 GASM 模型,它可以衡量产品、功能的数据表现,见表 5-5。

表 5-5　GASM 模型

目标	用户目标、产品目标、业务目标
行动	针对目标的具体优化策略、路径、方法
表现	显示优化策略等有用与否、有用程度的信号
指标	与显示信号相对应的关键指标,可设定若干个

针对网络页面开发的数据模型有 PULSE 模型,见表 5-6。

表 5-6　PULSE 模型

网络页面浏览量	反映开发页面的受欢迎程度
正常运行时间	反映页面的业务稳定性指标
响应时间	反映页面对于用户的工作响应效率
七天内活跃的用户	反映页面受到关注和应用的持续性
收益	反映页面的开发使用价值

针对手机 App 开发的数据模型有 AARRR 模型,它关注的是业务的增长,见表 5-7。

表 5-7　AARRR 模型

拉新	通过了解并定位目标用户人群,明确传达产品的核心价值,尽可能吸引用户
激活	在吸收新的用户后,在一定的时间段内引导用户再次使用产品,通常可以用用户的平均使用时长、平均每天启动次数等
留存	通过产品的价值来真实留住用户,反映产品的吸引力、用户对于产品的忠实度
变现	通过适当的方法从用户身上获得营业收入
传播	通过提升用户的体验,采取各种奖励的方法,激励用户把产品分享给新的潜在用户

Data missing mainly refers to the loss of some or all data in all data sets that should be collected and recorded completely due to intentional or unintentional circumstances. The consequence of missing data is relatively serious. It makes the data set unable to correctly and completely reflect the characteristics of the objective object, causing a series of problems in the data：

First, the integrity of data is damaged. The situation of missing data may be that all data records are missing, or some data records are missing, such as incomplete data attributes, incomplete data entries, incomplete data processes, etc. The use value of incomplete data will be greatly reduced.

Second, the validity of data is damaged. The complete data can effectively describe the objective object, but the missing data is likely to lead to the failure of the data set, such as the lack of time-marked data, which will lose value in the analysis.

Third, the accuracy of data is damaged. Due to the incompleteness of the data, the characteristics described by the data cannot be consistent with the real characteristics of the objective object. For example, when calculating the welding quality of a construction project, due to the lack of data, the statistical qualification rate of welding cannot truly reflect the welding quality status.

Fourth, data consistency is damaged. Due to the lack of data, there is a lack of consistency in time, attribute, quantity and method when comparing the data before and after, resulting in inconsistent information.

5.1   Unable to Record Data Completely

Incomplete data recording means that some data are not recorded in the process of observation, measurement and test. The data cannot be recorded completely due to various reasons, some are limited to various conditions and some are not recorded completely due to human factors.

The data collection scheme has defects, which directly leads to incomplete data collection. Two main reasons for this situation are as follows：

1. There is no clear time and time period for data collection

The specific collection time and time period should be specified for data collection, otherwise two situations will occur. First, there is no specific collection time, the data collection is too arbitrary. Second, if there is no clear time period, the time for collecting data may be too short, resulting in a significant lack of data; It may also take too long to collect data, which not only increases a lot of costs, but also has difficulties in timeliness and comparability of data.

Case:

A quality management team carried out activities, and the status survey process of the team is as follows:

In order to further improve the pavement quality of prefabricated rubber runways, the team inspected the pavement of rubber runways in the surrounding stadiums and gymnasiums (including its own company). Through the investigation of a total of 600 points, the average passing rate of on-site acceptance was only 81%. The quality defects in the classification statistics are as shown in Table 5-1:

Table 5-1　Status survey results

No.	Inspection items	Inspection numbers	Qualified points	Qualified rate/%
1	Surface cracking	100	53	53
2	Surface uneven		57	57
3	Surface blistering		92	92
4	Mark line deviation		94	94
5	Surface pollution		96	96
6	The others		94	94
Total		600	486	81

At first glance, it seems that the process of current situation investigation is completely correct. However, after careful analysis, the team did not specify the investigation period when conducting the status survey, so there are such problems:

(1) Generally speaking, the construction of stadiums and gymnasiums in a city and its surrounding areas will not be carried out at the same time. Moreover, the number of stadiums and gymnasiums with rubber runways in the same city is very small, which means that the construction time of some stadiums and gymnasiums has been a long time ago. Thus, the result of the status survey is not the current situation, but the runway situation many years ago.

(2) The quality of rubber runway itself has been improved rapidly, and its construction technology has also developed rapidly. There are differences in technology and quality between the rubber runway a few years ago and the current rubber runway; The construction technology several years ago cannot be compared with the current construction technology on the same platform.

(3) As time goes by, the surface of the previously constructed rubber runway would have cracked and polluted. Therefore, the quality problems such as "surface cracking" "surface uneven" "surface blistering" and "surface pollution" collected by the team may be the result of aging and lack of maintenance.

Therefore, this example is a typical case of incomplete data recording.

2. The collected data indicators are missing

Data indicators refer to the quantitative characteristics specified in the data plan to describe the research object. Different research purposes and objects have different data indicators. For a specific survey process, data indicators should fully reflect the characteristics of the survey object. When designing the data collection scheme, if the data indicators are omitted, the overall data of a certain category will be missing, and the research work will not be completed.

Case：

An entity company requires the marketing department to conduct research and expects to provide decision-making basis for the next step of product function improvement and sales network layout through the analysis of product sales in the last year. For this reason, the marketing department has developed the data indicators for the research, as shown in Table 5-2.

Table 5-2   Data indicators made by the marketing department

User data	User stock
	User increment
	User's contact method
Behavioral data	Purchase frequency
	Satisfaction degree
Business data	Sales volume
	Sales unit price
	Sales peak period

For the data indicators developed by the marketing department, the decision-making level of the entity company expressed dissatisfaction and believed that the data indicators had defects, and some key indicators were not included, such as user turnover rate, main sources of users, and per capita consumption. Therefore, the marketing department invited an external third party to reformulate the data indicators, as shown in Table 5-3.

Table 5-3   Data indicators developed by the third party

User data	User's origin
	User's contact method
	User churn rate
	User increment
	User stock
	User's hobbies
	Average age of users
	User's usage hobbies
	User's educational background

Behavioral data	Purchase frequency
	Main concerns before purchase
	Time spent in physical stores
	Location of products that users pay attention to
	Satisfaction degree
Business data	Sales unit price of each product
	Total sales of different products
	Total sales price of each product
	Sales peak period
	Sales unit price and total amount of advertising products
	Per capita consumption

From the comparison between the data indicators specified by the third party and the data indicators formulated by the marketing department, the data indicators of the third party are obviously more complete and can provide more decision-making information for decision makers.

There is no unified model for the setting of data indicators. The key is to fully reflect all characteristics of objective objects through the collection of indicators. In the above example, new data indicators can also be developed only for user satisfaction, such as the HEART model, as shown in Table 5-4.

Table 5-4　HEART model for measuring user satisfaction

Happiness	It is used to measure user satisfaction, which can be expressed as scoring, praise, average, sharing, recommendation, continuous recommendation, etc. at the functional level
Engagement	It is used to measure the degree of user participation, which can be expressed as comments, favorable comments, general comments, sharing, recommendation, etc. at the functional level
Adoption	It is used to measure the acceptance of users, which can be expressed as observation, appreciation, purchase, hesitation, return, etc. at the functional level
Retention	It is used to measure the acceptance of users, which is represented by the number of new and old users at the functional level
Task success	It is used to measure the degree of completion of different paths of users, which is reflected in the use of product functions, completion time, etc. at the functional level

As for the setting of data indicators, there are other modeling methods, such as GASM model, which can measure the data performance of products and functions. See Table 5-5.

Table 5-5　GASM model

| Goal | User objectives, product objectives and business objectives |
| Action | Specific optimization strategies, paths and methods for objectives |

Signal	Display the signal of whether the optimization strategy is useful or not and how useful it is
Metric	Several key indicators corresponding to the display signal can be set

The data model developed for web pages includes PULSE model, as shown in Table 5-6.

Table 5-6   PULSE model

Page view	Reflect the popularity of the development page
Uptime	Business stability indicators reflecting the page
Latency	Reflect the efficiency of page response to users
Seven days active	Reflect the page's attention and application continuity
Earning	Reflect the development and use value of the page

The data model developed for mobile App includes AARRR model, which focus on business growth, as shown in Table 5-7.

Table 5-7   AARRR model

Acquisition	By understanding and positioning the target user, clearly convey the core value of the product and try to attract users
Activation	After absorbing new users, guide users to use the product again within a certain period of time, usually using the average use time of users, the average number of starts per day, etc.
Retention	To truly retain users through the value of the product, reflecting the attractiveness of the product and users' loyalty to the product
Revenue	Obtain business income from users through appropriate methods
Referral	By improving user experience and adopting various incentive methods, users are encouraged to share products with new potential users

5.2　数据采集的过程存在缺陷

———

在第 2 章中介绍了数据的来源，如观察法、调查法、试验法等，这些数据的来源也是数据采集的方法。当方法确定以后，在具体的执行过程中，由于各种缺陷的存在会导致数据缺失的发生。

5.2.1 没有区分定性数据和定量数据

定性数据是表示客观对象性质、规定客观对象类别的文字表述型数据。定量数据是以量化属性，即通过数值并结合恰当的单位来对客观对象进行描述的数据。在进行数据采集的过程中，如果没有对定性数据和定量数据进行区分，那么即使数据采集到位，但由于数据不能完整反映客观对象的状态、特征，相当于缺失了相应的数据。

举例：

某养殖基地着手开发一种保鲜剂，用于鸡肉在运输、存储过程中的保鲜。本次研发主要是以茶多酚配方为主要成分，另外有某种添加剂、被膜剂、浸泡时间，共四种因子。每种因子各有三个水平。

为此，试验人员进行了正交试验。第一次试验由实习生进行，试验结果见表 5-8。

表 5-8　保鲜剂正交试验结果

试验序号	因素 A 茶多酚浓度/%	因素 B 添加剂/g	因素 C 被膜剂类别	因素 D 浸泡时间/min	实验结果
1	1（10）	1（30）	3（3）	2（30）	保鲜效果好
2	2（20）	1（30）	1（1）	1（20）	效果一般
3	3（30）	1（30）	2（2）	3（10）	效果一般
4	1（10）	2（40）	2（2）	1（20）	效果尚可
5	2（20）	2（40）	3（3）	3（10）	保鲜效果好
6	3（30）	2（40）	1（1）	2（30）	效果一般
7	1（10）	3（60）	1（1）	3（10）	效果尚可
8	2（20）	3（60）	2（2）	2（30）	保鲜效果好
9	3（30）	3（60）	3（3）	1（20）	效果尚可

从表 5-8 可以看出，由于试验人员缺乏经验，在记录试验结果时，用定性数据"效果一般、效果尚可、保鲜效果好"来描述。由于这些用语不能精确反映试验结果，等于失去了试验数据，试验人员无法进行下一步的分析。

于是，该基地负责人安排富有经验的试验人员进行第二次正交试验，整个试验过程与第一次完全相同，只是试验的结果改为"保鲜时间 h"。试验结果见表 5-9。

表 5-9　保鲜剂正交试验结果

试验序号	因素 A 茶多酚浓度/%	因素 B 添加剂/g	因素 C 被膜剂类别	因素 D 浸泡时间/min	实验结果 保鲜时间/h
1	1（10）	1（30）	3（3）	2（30）	35.1
2	2（20）	1（30）	1（1）	1（20）	30.2
3	3（30）	1（30）	2（2）	3（10）	29.6
4	1（10）	2（40）	2（2）	1（20）	32.7
5	2（20）	2（40）	3（3）	3（10）	36.3
6	3（30）	2（40）	1（1）	2（30）	30.0
7	1（10）	3（60）	1（1）	3（10）	34.7
8	2（20）	3（60）	2（2）	2（30）	35.9
9	3（30）	3（60）	3（3）	1（20）	33.8
位级 1 之和	35.1+32.7+ 34.7＝102.5	35.1+30.2+ 29.6＝94.9	30.2+30.0+ 34.7＝94.9	30.2+32.7+ 33.8＝96.7	
位级 2 之和	30.2+36.3+ 35.9＝102.4	32.7+36.3+ 30.0＝99.0	29.6+32.7+ 35.9＝98.2	35.1+30.0+ 35.9＝101.0	
位级 3 之和	29.6+30.0+ 33.8＝93.4	34.7+35.9+ 33.8＝104.4	35.1+36.3+ 33.8＝105.2	29.6+36.3+ 34.7＝100.6	
极差 R	9.1	9.5	10.3	4.3	
主次顺序	C ＞ B ＞ A ＞ D				
优水平	A_1	B_3	C_3	D_2	
优组合	$A_1B_3C_3D_2$				

从表 5-9 可以看出，当试验人员用定量数据记录试验结果后，可以顺利进行后续的分析工作，并最终得出结论：本次保鲜剂的优组合为 $A_1B_3C_3D_2$。

5.2.2 采集数据的时间和频数不符合要求

数据具有时效性，如果不能及时采集数据，会错失时机，造成数据缺失。造成数据缺失主要有以下五个原因。

1. 采集数据提前

有些数据的生成是需要有足够时间的，如果时间未到，那么结果是不准确的。

举例：

某实验室进行菌落的培养，试验人员在 37 ℃的恒温条件下进行试验。试验人员在 24 h、48 h、72 h 分别进行观察，记录的结果见表 5-10。

表 5-10 菌落培养结果

培养时间/h	观察到的菌落
24	大肠杆菌
48	大肠杆菌、真菌
72	大肠杆菌、真菌、霉菌

从表 5-10 可以看到，试验人员在不同的时间观察到的结果是有区别的。假如试验人员把 24h 观察的结果作为全部实验的结论，那就造成数据缺失。

2. 采集数据滞后

有些数据在过了一定的时间后，就不再出现。

举例：

每年 3—4 月份，在长江口滩涂上的候鸟，如白头鹤、小天鹅等开始向北方迁徙。如果到了夏天去观察这些侯鸟的有关情况，就错失了良机。

3. 采集数据的周期过长或过短

有些数据的采集，必须在适当的周期内才会出现。

举例：

表 5-11 是某疾控中心提供的相关细菌、病毒在干燥无生命物体表面的存活时间表。

表 5-11 相关细菌、病毒在干燥无生命物体表面的存活时间表

细菌名称	存活时间	病毒名称	存活时间
不动杆菌属	3 天~5 个月	腺病毒	7 天~3 个月
百日咳杆菌	3 天~5 天	SARS 病毒	72 小时~96 小时
白喉杆菌	7 天~6 个月	HAV 病毒	2 小时~60 天
大肠杆菌	1.5 小时~16 个月	HSV，1 型和 2 型	4.5 小时~8 周
结核分枝杆菌	1 天~4 个月	流感病毒	1 天~2 天
淋球菌	1 天~3 天	脊髓灰质炎病毒	1 天~8 周

由表 5-11 可知，实验人员要观察到相关的细菌、病毒，应当在恰当的周期内进行。如果过了相关的时间段，那么有关数据就不存在。

4. 时点值与时段值混淆

时点值是在某一个具体时点上的数据，而时段值是经过一个时间跨度的数据，或者是在一个时段上所累积的数据。如果两者混淆，则数据采集失真。

举例

某企业人力资源部门统计的员工数据，见表 5-12。

表 5-12　员工数据	单位：人
××年 12 月 31 日员工数量	1086
××年 12 月员工离职数量	25

显然，1086 人是员工在 12 月 31 日的时点数据；25 人是员工在整个 12 月份的时段数据。

5. 采集数据的频数不够

有些数据的采集，需要有足够的频数才能保证数据的完整性。在实践中，出于各种原因造成有的数据频数不足而影响后续的分析。

举例：

某质量管理小组开展活动，课题是提高智能型诱导风机的一次调试成功率。小组在现状调查过程中，通过某一天的观察，采集到的数据见表 5-13。

表 5-13　智能型诱导风机的一次调试缺陷统计表　　单位：次

序号	缺陷类型	出现频率	累计频数
1	风量达不到设计要求	2	2
2	噪声达不到设计要求	7	9
3	设备运转不正常	5	14
4	设备控制不正常	6	20

从表 5-13 发现，出现问题最多的是"噪声达不到设计要求"，发生频次为 7 次；其次是"设备控制不正常"，发生频数为 6 次。

但质量人员要求增加观察天数，现状调查由一天改为七天，观察结果见表 5-14。

表 5-14　智能型诱导风机的一次调试缺陷统计表　　单位：次

序号	缺陷类型	出现频率	累计频数
1	风量达不到设计要求	5	5
2	噪声达不到设计要求	12	17
3	设备运转不正常	15	32
4	设备控制不正常	38	70

由表 5-14 可知，"设备控制不正常"的发生频数最高，达到 38 次，显然高于其他的问题项，因此，小组选择"设备控制不正常"为本次活动的症结。

从本例中可以看出，由于数据采集的频数不够多所造成的数据缺失，会导致管理决策的失误。实质上，大数定理本身就需要数据的频数足够多才能准确判断。

5.2.3 采集数据所使用的仪器设备有问题

现在大量的数据采集，仅仅依靠人工已经远远不能满足要求，因此，数据采集工作需要仪器设备等辅助完成。数据采集的过程，是把客观对象的各种特征数据（包括几何量、物理量、化学量等各方面）通过各种传感元器件作适当转换后，再经过信号调整、量化、编码、传输等步骤，传送到终端进行数据收集、归类、整理、存储记录的过程。

1. 使用仪器设备录入数据的方式

（1）数据分析工具录入，比如 Smartbi，在数据分析、数据可视化等方面有较好应用。

（2）普通以太网模式，通过以太网开发包的设备，对接计算机的以太网接口，利用相关的软件，对设备进行的网络数据传输。

（3）数据采集卡，通过与生产设备的相关 I/O 点与对应的传感器进行连接，采集相应的过程信息，包括生产运行、故障等参数。

（4）TCP/ IP 协议的以太网模式。采集模式内容丰富，可实现远程控制。对于生产加工、质量控制有较好作用。

（5）组态软件采集，通过 PLC 控制类的设备，对非数控类组态软件进行相关信息的读取，将读取的 I/O 点信息存入数据库中。它是工业自动化领域的应用工具，为工厂实时数据采集、传输等提供可能，实现智能生产管控。

2. 因仪器设备的原因使数据缺失的情形

（1）接触式采集与非接触式采集的仪器设备故障

仪器设备采集数据分接触式采集和非接触式采集两种。接触式采集主要指通过各种传感器来采集数据，包括光敏、气敏、力敏、磁敏、声敏等不同类别的传感器。非接触式采集是利用各种信号来采集数据，比如 RFID（Radio Frequency Identification，射频识别，适用于精确度较高的场所），通过射频信号自动识别目标对象，获取相关的数据信息。一旦这些仪器设备发生故障，则数据收集缺失。

数据采集设备故障的情形有很多，比如 ISA 总线、PCI 总线等接口形式的 A/D 采集卡，安装麻烦，受到计算机限制多，可扩展性查，容易受到电磁干扰，导致采集的数据失真。一些高速及超高速数据采集系统的应用，比如 SOC（system on chip，系统级芯片），它采用嵌入式微处理器的方案，由早期的 A/D 器件、标准单片机组成应用系统，发展到在单芯片上实现完整的数据采集与分析，但由于漏电、电参数漂移、静电击穿、强电流击穿、材料本身等各种原因而失效，使数据采集缺失。

（2）模拟信号数据转化为数字信号数据失败

模拟数据是由传感器采集得到的连续变化的值，数字数据是模拟数据量化后得到的离散值。大量的数据采集系统是模拟量的测量设备，测得的信号数据是模拟数据。模拟信号与数字信号可以相互转换，根据不同的需要，把模拟量数据转换为数字量后，再经过计算机处理得到数字数据。在模拟数据转化为数字数据的过程中，由于各种原因会导致转换失败，比如数据文本格式问题、错误值出现、匹配数据区域缺少需要查询的值、模拟数字转换器的模块损坏等，由此导致数字数据缺失。

另外，一些数字化处理过程中本身有数据缺失问题，比如在 CAD 制图、测图、设计等工作中，对于图形、图像的数字化处理过程中之直接发生数据缺失。

（3）数据存储失败

数据存储失败的情形主要有：①存储器损坏等，导致某个是时间段的数据收集或保存失败；②数据保存过程中，相关程序发生错误；③操作过程中出现失误，使保存数据失败；④中电脑病毒；⑤数据备份失效等。

5.2.4 客观条件限制

数据收集是需要有相应的条件，比如有的需要有适当的温度、湿度等环境条件，有的需要有相应的技术手段支持等。由于客观条件受到限制而造成数据缺失的情形非常多，简单说明如下。

1. 环境温度不符合要求

在化学反应中，温度等环境条件是很有必要的。比如氯气，如果在常温下，它与氢氧化钠反应生成氯化钠和次氯酸钠，如果试验人员想要获得氯酸钠的数据，没有适当的加热温度就不能得到相应的结果。

2. 环境湿度不符合要求

在化学反应中，湿度等条件有时会直接影响试验结果。比如，化学方程式：

$$H_2S + O_2 = S + H_2O$$

硫化氢在燃点以下是可逆反应，而且与水汽的湿度有关。当减小湿度的时候，化学平衡向正方向移动，此时要收集硫化氢的数据就少了。而当湿度增大的时候，化学平衡向逆方向移动，此时要收集硫的数据就少了。

3. 器材的条件受限

器材对于收集数据是极其重要的，不同的器材能够达到不同的数据精度。比如采集图像时，对拍摄器材有分辨率的要求，对于感光器有像素的要求，对于镜头有极限分辨率的要求。如果相关要求达不到，则达标数据无法收集。

4. 限于条件只记录部分数据

限于各种条件而无法收集全部数据的情形非常多。比如：不使用卫星遥感技术，则测绘的范围受到限制；不花大的成本进行人口普查，则只能收集到片面的人口数据；在信息行业，如果不能利用各种平台广泛收集数据，则无法形成大数据的数据库等。

5. 无法拿到专业机构信息数据

专业机构的信息数据往往是某个组织对于数据收集的极好补充。如果无法拿到专业机构的信息数据，比如各类统计部门、信息中心、专业期刊等所提供的数据，那么相关的数据缺失。

6. 因数据的隐匿性导致无法收集

有些数据具有隐匿性，需要借助一定的设备才能收集。比如红外线、紫外线等，不借助于专业的仪器，是不能把隐匿的数据显现出来。

5.2.5 人为缺失数据

1. 主观上的失误导致数据缺失

主观失误的行为多发生新手身上，常见的有：紧张、情绪不稳定、不熟悉数据采集规范和流程、事前准备不充分、遇到紧急情况或疑难情形不知所措、环境变化而缺少应对办法、时间安排太紧凑而没能全部完成采集数据等。

2. 认知上的局限导致数据缺失

做好数据采集工作，应当对于收集对象、方法、过程等有完整清晰的认识。有时候，数据采集人员对于自己的角色定位不准、缺少相关专业知识、智力差异、经验欠缺等而形成认知上的局限性；不同的数据采集人员对于同一个客观对象在认知上有不同的理解；思维谨慎的人员与思维不谨慎人员之间形成认知的局限；对于语言文字信息的分析判断能力导致认知局限等，从而导致数据缺失。另外，如果客观对象是新生事物，对其定义本就处于一个模糊、探索的阶段，也会形成认知局限，使数据采集缺失。

3. 刻意隐瞒数据

刻意隐瞒数据导致数据缺失的情形也时有发生，比如：市场调查时，被调查人员不配合；调查时涉及商业机密、隐私等，不便透露相关信息；出于各种不同的目的，对于数据予以隐藏等。

4. 人工录入数据时失误导致数据缺失

人工录入数据的方式有：①手工直接录入方式，即利用人工用纸张进行记录数据；②用人工的方式按键盘来录入数据；③通过手工条码扫描的方式进行数据录入等。

人工录入数据时，会由于疲劳、情绪波动影响、多个事务同时开展等因素，从而导致数据缺失。在条码录入时，如果条码标签被损坏、遗失，导致扫描仪器无法识别数据，使数据无法录入。在条码扫描时，由于扫描仪有一定的扫描范围，不在扫描范围内的数据无法识别。数据录入人员遗漏扫描相关信息使数据缺失等。

5.2　Defects in the Process of Data Collection

In Chapter 2, the sources of data are introduced, such as observation method, investigation method, test method. These quantitative sources are also data collection methods. When the method is determined, data loss will occur due to various defects in the specific implementation process.

5.2.1 There is no distinction between qualitative data and quantitative data

Qualitative data is wordenunciated data that represents the nature of objective objects and specifies the category of objective objects. Quantitative data is data that describes objective objects with quantitative attributes, that is, by combining numerical values with appropriate units. In the process of data collection, if there is no distinction between qualitative data and quantitative data, then even if the data is collected in place, because the data cannot fully reflect the status and characteristics of the objective object, it is equivalent to missing the corresponding data.

Case：

A breeding base started to develop a preservative for chicken during transportation and storage. This research and development is mainly based on the tea polyphenol formula as the main ingredient, in addition, there are some additives, film coating agent, soaking time, a total of four factors. Each factor has three levels.

For this reason, the test personnel carried out orthogonal test. The first test was conducted by interns. See Table 5-8 for the test results.

Table 5-8　Orthogonal test result of preservative

Test number	Factor A Tea polyphenol concentration/%	Factor B Additive/g	Factor C Type of film coating	Factor D Soaking time/min	Test result
1	1（10）	1（30）	3（3）	2（30）	Good preservation effect
2	2（20）	1（30）	1（1）	1（20）	General effect
3	3（30）	1（30）	2（2）	3（10）	General effect
4	1（10）	2（40）	2（2）	1（20）	Effect is acceptable
5	2（20）	2（40）	3（3）	3（10）	Good preservation effect
6	3（30）	2（40）	1（1）	2（30）	General effect
7	1（10）	3（60）	1（1）	3（10）	Effect is acceptable
8	2（20）	3（60）	2（2）	2（30）	Good preservation effect

Test number	Factor A Tea polyphenol concentration/%	Factor B Additive/g	Factor C Type of film coating	Factor D Soaking time/min	Test result
9	3（30）	3（60）	3（3）	1（20）	Effect is acceptable

It can be seen from Table 5-8 that due to the lack of experience of the test personnel, the qualitative data "general effect, effect is acceptable and good preservation effect" are used to describe the test results. Because these terms can not accurately reflect the test results, it means that the test data is lost, and the tester cannot conduct the next analysis.

Therefore, the person in charge of the base arranged experienced test personnel to carry out the second orthogonal test. The whole test process was the same as the first one, but the test result was changed to "preservation time h". See Table 5-9 for test results.

Table 5-9   Orthogonal test result of preservative

Test number	Factor A Tea polyphenol concentration/%	Factor B Additive/g	Factor C Type of film coating	Factor D Soaking time/min	Test result Preservation time/h
1	1（10）	1（30）	3（3）	2（30）	35.1
2	2（20）	1（30）	1（1）	1（20）	30.2
3	3（30）	1（30）	2（2）	3（10）	29.6
4	1（10）	2（40）	2（2）	1（20）	32.7
5	2（20）	2（40）	3（3）	3（10）	36.3
6	3（30）	2（40）	1（1）	2（30）	30.0
7	1（10）	3（60）	1（1）	3（10）	34.7
8	2（20）	3（60）	2（2）	2（30）	35.9
9	3（30）	3（60）	3（3）	1（20）	33.8
Sum of level 1	35.1+32.7+ 34.7＝102.5	35.1+30.2+ 29.6＝94.9	30.2+30.0+ 34.7＝94.9	30.2+32.7+ 33.8＝96.7	
Sum of level 2	30.2+36.3+ 35.9＝102.4	32.7+36.3+ 30.0＝99.0	29.6+32.7+ 35.9＝98.2	35.1+30.0+ 35.9＝101.0	
Sum of level 3	29.6+30.0+ 33.8＝93.4	34.7+35.9+ 33.8＝104.4	35.1+36.3+ 33.8＝105.2	29.6+36.3+ 34.7＝100.6	
Range R	9.1	9.5	10.3	4.3	
Primary and secondary order	C > B > A > D				
Optimal level	A_1	B_3	C_3	D_2	
Optimal combination	$A_1B_3C_3D_2$				

From Table 5-9, it can be seen that after the test results are recorded with quantitative data, the subsequent analysis can be carried out smoothly, and the final conclusion is that the optimal combination of this preservative is $A_1B_3C_3D_2$.

5.2.2 The time and frequency of collecting data do not meet the requirements

Data is time-sensitive. If you cannot collect data in time, you will miss the opportunity and cause data loss. Five main causes lead to data loss.

1. Collect data in advance

The generation of some data needs enough time. If the time is not up, the result is inaccurate.

Case:

A laboratory carries out colony culture, and the test personnel conduct the test under the constant temperature condition of 37℃. The test personnel observed at 24 hours, 48 hours and 72 hours respectively, and the recorded results are shown in Table 5-10.

Table 5-10　Bacterial culture results

Bacterial culture time	Observed bacterial colonies
24 hour	Escherichia coli
48 hour	Escherichia coli, fungus
72 hour	Escherichia coli, fungus, mold

It can be seen from Table 5-10 that the results observed by the test personnel at different times are different. If the test personnel take the results of 24-hour observation as the conclusion of all the experiments, it will cause data loss.

2. The collected data lags behind

Some data will not appear after a certain period of time.

Case:

From March to April every year, migratory birds, such as white-headed cranes and small swans, on the tidal flats of the Yangtze River estuary begin to migrate to the north. If we go to observe the situation of these migratory birds in summer, you will miss a good opportunity.

3. The period of collecting data is too long or too short

Some data must be collected in an appropriate period.

Case:

Table 5-11 shows the survival time of related bacteria and viruses on the surface of dry inanimate objects provided by a CDC.

Table 5-11　Survival time of related bacteria and viruses on the surface of dry inanimate objects

Bacterial name	Survival time	Virus name	Survival time
Acinetobacter	3 days ~5 months	Adenovirus	7 days ~3 months
Pertussis bacilli	3 days ~5 months	SARS virus	72 hours ~96 hours
Diphtheria	7 days ~6 months	HAV virus	2 hours ~60 days

Bacterial name	Survival time	Virus name	Survival time
Escherichia coli	1.5hours ~16 months	HSV, type 1 and 2	4.5hours ~8 weeks
Mycobacterium tuberculosis	1day ~4 months	influenza virus	1day ~2days
Gonococcus	1day ~ 3days	Poliovirus	1day ~8 weeks

It can be seen from Table 5-11 that the laboratory personnel should observe the relevant bacteria and viruses within an appropriate period. If the relevant time period has passed, the relevant data will not exist.

4. Confusion between the time point value and the time period value

The time point value is the data at a specific time point, while the time period value is the data over a time span, or the data accumulated in a time period. If the two are confused, the data acquisition is distorted.

Case:

See Table 5-12 for the employee data collected by the human resources department of an enterprise.

Table 5-12　Employee data

Unit: people

Number of employees at December 31st. ××year	1086
Number of employees leaving in December, ××year	25

Obviously, 1086 people are the data of employees on December 31; 25 people are the data of employees in the whole period of December.

5. The frequency of data collection is not enough

Some data collection requires sufficient frequency to ensure data integrity. In practice, due to various reasons, the frequency of some data is insufficient, which affects the subsequent analysis.

Case:

A quality management team carried out activities, and their activity topic is to improve the success rate of the first commissioning of intelligent induction fans. The data collected by the team during the current situation investigation is shown in Table 5-13.

Table 5-13　Statistics of defects in the first commissioning of intelligent induction fans

No.	Defect type	Occurrence frequency / times	Cumulative frequency / times
1	The air volume cannot meet the design requirements	2	2
2	The noise cannot meet the design requirements	7	9
3	Abnormal operation of equipments	5	14
4	Abnormal control of equipments	6	20

From Table 5-13, it is found that the most common problem is "the noise can not meet the design requirements", and the frequency is 7 times; The second is "abnormal control of equipments", which occurs 6 times.

However, the quality personnel required to increase the number of observation days, and the status survey was changed from one day to seven days. See Table 5-14 for the observation results.

Table 5-14 Statistics of defects in the first commissioning of intelligent induction fans

No.	Defect type	Occurrence frequency / times	Cumulative frequency / times
1	The air volume cannot meet the design requirements	5	5
2	The noise cannot meet the design requirements	12	17
3	Abnormal operation of equipment	15	32
4	Abnormal control of equipments	38	70

It can be seen from Table 5-14 that the frequency of "abnormal controle of equipments" is the highest, reaching 38 times, which is obviously higher than other problem items. Therefore, the team chose "abnormal control of equipments" as the crux of this activity.

It can be seen from this example that the lack of data caused by the insufficient frequency of data collection will lead to mistakes in management decisions. In essence, the theorem of large numbers itself requires sufficient frequency of data to accurately judge.

5.2.3 There are problems with the instruments and equipment used to collect data

At present, a large number of data acquisition can not meet the requirements only by manual work. Therefore, the data acquisition work needs the assistance of instruments and equipment to complete. The process of data acquisition is the process of transferring various characteristic data of the objective object (including geometric quantity, physical quantity, chemical quantity and other aspects) to the terminal for data collection, classification, sorting, storage and recording after proper conversion through various sensing elements, and then through signal adjustment, quantification, coding, transmission and other steps.

1. The way of using instruments and equipments to logging data

(1) Input with data analysis tools, such as Smartbi, has good applications in data analysis, data visualization, etc.

(2) In the normal Ethernet mode, connect the equipment of the Ethernet development package to the Ethernet interface of the computer, and use the relevant software to carry out network data transmission for the equipment.

(3) The data acquisition card is connected with the relevant I/O points of the production equipment and the corresponding sensors to collect the corresponding process information, including production operation, fault and other parameters.

(4) Ethernet mode of TCP/IP protocol. The collection mode is rich in content and can realize remote control. It plays a good role in production, processing and quality control.

（5）The configuration software collects the relevant information of the non-NC configuration software through the PLC control equipment, and stores the read I/O point information into the database. It is an application tool in the field of industrial automation, providing the possibility for real-time data collection and transmission in factories, and realizing intelligent production control.

2. Data missing due to instrument and equipment

（1）Instrument and equipment failure of contact acquisition and non-contact acquisition

The data collected by instruments and equipment can be divided into two types: contact acquisition and non-contact acquisition. Contact acquisition mainly refers to collecting data through various sensors, including light sensitive, gas sensitive, force sensitive, magnetic sensitive, acoustic sensitive and other different types of sensors. Non-contact acquisition uses various signals to collect data, such as RFID（Radio Frequency Identification, applicable to places with high accuracy）, and automatically identifies target objects through radio frequency signals to obtain relevant data information. Once these instruments and equipment fail, data collection is missing.

There are many cases of data acquisition equipment failure, such as the A/D acquisition card in the form of interface such as ISA bus and PCI bus, which is difficult to install, is subject to many computer restrictions, scalability check, and is prone to electromagnetic interference, resulting in the distortion of the collected data. Some applications of high-speed and ultra-high-speed data acquisition systems, such as SOC（system on chip）, adopt the scheme of embedded microprocessor. The application system is composed of early A/D devices and standard single chip computers. It has developed to achieve complete data acquisition and analysis on a single chip. However, it fails due to various reasons such as leakage, electrical parameter drift, electrostatic breakdown, strong current breakdown, and material itself. This leads to the loss of collected data.

（2）Conversion of analog signal data to digital signal data failure

The analog data is the continuously changing value acquired by the sensor, and the digital data is the discrete value obtained after the analog data is quantized. A large number of data acquisition systems are analog measurement equipment, and the measured signal data is analog data. Analog signal and digital signal can be converted to each other. According to different needs, analog data can be converted to digital data, and then digital data can be obtained through computer processing. In the process of converting analog data to digital data, the conversion will fail due to various reasons, such as data text format problems, error values, lack of values to be queried in the matching data area, and damage to the module of the analog-to-digital converter, resulting in the loss of digital data.

In addition, some digital processing processes have their own data loss problems, such as the direct data loss in the digital processing of graphics and images in CAD drawing, mapping, design and other work.

（3）Data storage failure

Data storage failures mainly include: ①The data collection or storage of a certain period of time fails due to memory corruption; ②In the process of data saving, errors occurred in relevant programs; ③Failure to save data due to errors during operation; ④Attacked by computer virus; ⑤Data backup failure, etc.

5.2.4 Limitations of objective conditions

Data collection requires corresponding conditions, such as appropriate temperature, humidity and other environmental conditions, and corresponding technical support. There are many cases of data missing due to the restriction of objective conditions. The brief description is as follows.

1. The ambient temperature does not meet the requirements

In chemical reactions, environmental conditions such as temperature are necessary. For example, if chlorine reacts with sodium hydroxide to generate sodium chloride and sodium hypochlorite at normal temperature, if the tester wants to obtain the data of sodium chlorate, the corresponding results cannot be obtained without proper heating temperature.

2. The ambient humidity does not meet the requirements

In chemical reactions, humidity and other conditions sometimes directly affect the test results. For example, chemical equation:

$$H_2S + O_2 = S + H_2O$$

Hydrogen sulfide is a reversible reaction below the ignition point and is related to the humidity of water vapor. When the humidity is reduced, the chemical balance moves to the positive direction, and there is less data to collect hydrogen sulfide. When the humidity increases, the chemical balance moves in the opposite direction, and there is less data to collect.

3. Equipment conditions are limited

Equipment is very important for data collection. Different equipment can achieve different data accuracy. For example, when collecting images, there are requirements for the resolution of the shooting equipment, the pixel of the photosensitive device, and the limit resolution of the lens. If the relevant requirements are not met, the compliance data cannot be collected.

4. Only part of the data is recorded because of limited conditions

There are many cases where all data cannot be collected due to various conditions. For example, if satellite remote sensing technology is not used, the scope of surveying and mapping will be limited; If the census is not carried out at a large cost, only one-sided population data can be collected; In the information industry, if the data cannot be collected widely by various platforms, the database of big data cannot be formed.

5. Unable to obtain the information and data of professional institutions

The information and data of professional institutions are often an excellent supplement for data collection by an organization. If the information data of professional institutions, such as the data provided by various statistical departments, information centers, professional journals, etc., cannot be obtained, the relevant data is missing.

6. Data cannot be collected due to its concealment

Some data have concelment and can only be collected with the help of certain equipment, such as infrared ray, ultraviolet ray, etc. Hidden data cannot be displayed without the help of professional instruments.

5.2.5 Human caused missing data

1. Subjective errors lead to data missing

Subjective errors often occur in novices, including nervousness, emotional instability, lack of familiarity with data collection specifications and procedures, inadequate preparation in advance, confusion in emergency or difficult situations, lack of response measures due to environmental changes, and too tight schedule to complete data collection.

2. Cognitive limitations lead to data missing

In order to do a good job of data collection, collectors should have a complete and clear understanding of the collection object, method, process, etc. Sometimes, data collectors have cognitive limitations due to their inaccurate role positioning, lack of relevant professional knowledge, intellectual differences, and lack of experience; Different data collectors have different cognitive understanding of the same objective object; There are cognitive limitations between people with cautious thinking and people with careless thinking; The ability to analyze and judge language and text information leads to cognitive limitations, which leads to data loss. In addition, if the objective object is a new thing, its definition is in a vague and exploratory stage, which will also form cognitive limitations and make data collection missing.

3. Deliberately conceal data

Deliberate concealment of data leads to data loss. For example, when conducting market research, the respondents do not cooperate; The investigation involves trade secrets, privacy, etc., and it is inconvenient to disclose relevant information; Data is hidden for various purposes.

4. Errors in manual data entry lead to data loss

Manual data entry methods include: ①manual direct entry. Manual paper is used to record data; ②Enter data manually by pressing the button; ③Data entry is performed through manual barcode scanning.

When entering data manually, data will be missing due to fatigue, the impact of mood fluctuations, and the simultaneous development of multiple transactions. If the barcode label is damaged or lost, the scanning instrument cannot recognize the data and the data cannot be entered. During barcode scanning, data outside the scanning range cannot be recognized because the scanner has a certain scanning range. Data entry personnel missed scanning relevant information, resulting in data missing, etc.

5.3　删失数据

———

在数据缺失的各种情形中，有一类数据会由于各种原因被停止观察或不能被观察到，这种状况下的数据就属于删失数据。

5.3.1 删失数据的含义

删失数据（censored data）是指在可靠性分析和生存分析中，由于某种原因被截断了的数据，它也是某一个时刻的数据观察值（等同于截尾试验的数据）。删失数据按时间延伸方向分，可以分为右删失、左删失、区间删失和任意删失。

1. 右删失

对于某个产品来说，如果它的可靠性或寿命的精确值未知，而只知道它的可靠性或寿命大于等于某一个观测值，则称这个观测值在这一点右删失（right censored）。比如：在对某种型号的灯泡进行寿命检验中，发现在 500 小时检验时灯泡仍然在工作，则可以断定灯泡的寿命大于 500 小时，可以记为"500+"，这种数据称为"右删失数据"。

2. 左删失

如果某个产品的可靠性或寿命的精确值未知，而只知道它小于等于某一个观测值，则称这个观测值在这一点左删失（left censored）。比如：在对某种型号的灯泡进行寿命检验中，发现在 600 小时检验时灯泡已经失效，但具体何时失效却不知道，则可以肯定它的寿命小于 600 小时，可以记为"600-"，这种数据称为"左删失数据"。

3. 区间删失

如果某个产品的可靠性或寿命的精确值未知，而只知道它在某个区间内，则称这个观测值在这一区间删失（interval censored）。比如：在对某种型号的灯泡进行寿命检验中，可以肯定它的寿命超过 500 小时，但小于 550 小时，则称这种数据为区间删失数据。

4. 任意删失

任意删失（arbitrary censored）是各种删失都适用的数据记录存放、表示方法，即对于右删失、左删失、区间删失全部适用。

任意删失数据的表示方法，见表 5-15。

表 5-15　任意删失数据的表示方法

观测值类型	开始栏输入	结束栏输入
精确时间	失效时间	失效时间
右删失	时间（在此之后失效）	*
左删失	*	时间（在此之前失效）
区间删失	失效发生区间的开始时间	失效发生区间的结束时间

举例：

有 100 个零件，第 8 天有 15 个零件正被使用，对剩下的 85 个零件进行观测。第 10 天发现有 4 个已经失效，2 个恰好在第 10 天失效。则工作表结构见表 5-16。

表 5-16 零件删失数据工作表

序	C1	C2	C3
	开始	结束	频率
1	8	*	15
2	*	10	4
3	10	10	2
4	10	*	79

5.3.2 右删失的类型

右删失在寿命数据中比较常见。根据试验安排的研究周期结束的判断准则不同，右删失可以分为时间删失（Ⅰ型删失）、失效删失（Ⅱ型删失）、随机删失（Ⅲ型删失）三种基本类型。

1. 时间删失（Ⅰ型删失）

在进行产品的寿命检验中，由于观察的时间、费用等会受到一定的限制，由此规定：试验在一定的时间范围内进行，到最终时刻就停止观测。这个最终时刻通常在试验开始前就确定。这样，样品的寿命只有在小于或等于事先给定的时间范围内才能被观测到。此时获得的数据称为时间删失（time censoring），也称为Ⅰ型删失或定时删失。比如：某型号机器的寿命时间删失在1000 小时，说明该型机器到了 1000 小时仍在工作，其寿命超过 1000 小时，但机器的准确失效时间不知道。

2. 失效删失（Ⅱ型删失）

事先规定个体失效的固定数目，在进行产品的寿命检验时，观察持续到固定数目的个体失效为止。此时获得的数据称为失效删失（failure censoring），也称为Ⅱ型删失或定数删失。失效删失试验常常用于设备寿命水平的测试研究，测试时所有的试验设备同时开始运行，直到有事先给定数目的设备失效时终止测试。如果要坚持观察到所有的设备都失效，可能需要相当漫长的时间来等待，而通过失效删失检验，可节省大量时间和费用。比如：某种型号机器的寿命试验中，其寿命失效删失在第 10 件机器，则意味着第 10 件以后失效的机器寿命数据都是删失的。

3. 随机删失（Ⅲ型删失）

在进行产品的寿命检验时，不是根据事先规定的时间或失效个数来停止试验，而是根据另一个随机变量的取值来决定试验的每个样品停止与否，这种情况下获得的观察数据称为随机删失（random censoring），也称为Ⅲ型删失。随机删失是试验处于比较复杂的情况，造成失效的干扰因素很多，事先也没有计划好失效时的情形。往往是在试验的过程中，受到了某个干扰因素的影响，从而提前离开了试验或终止试验观察。比如：在某次寿命检验时，试验地点发生了变化而终止了试验；两栋商务楼的新建 VAV 空调系统，分别在正常运行后的第 60 个月和第 75 个月更新了其他的空调系统，则这两栋商务楼的 VAV 新建空调系统的寿命分别为 60+月和 75+月。

Ⅰ型删失和Ⅱ型删失统称为单一删失（singly censored），Ⅲ型删失也称为多删失（multiply censored）。

5.3.3 截尾试验

截尾试验是一种寿命试验，是指在产品试验的过程中，一旦有部分产品失效就停止试验的一种试验方法。在截尾试验中获得的失效数据称为截尾样本。

1. 定时截尾试验

定时截尾试验，又称 I 型截尾试验，是指试验在事先规定的时间停止。试验停止时所获得的失效数据个数是随机变量。比如对于 100 个产品样本进行寿命试验，事先规定到 50 小时就停止试验，假设失效数为 5 个，则该失效数是随机变量。恰当地规定试验停止时间是定时截尾试验的关键。

2. 定数截尾试验

定数截尾试验，又称 II 型截尾试验，是指试验在事先规定地失效个数达到时就停止试验。试验停止时的时间是随机变量。比如对于 100 个产品样本进行寿命试验，事先规定到第 30 个产品（30% 的产品）发生失效时就停止试验，则该停止试验的时间是随机变量。恰当地规定失效比例是定数截尾试验的关键。

5.3.4 删失数据的应用

删失数据在产品可靠型分析、生存分析中有着广泛的应用。利用不同的删失数据，可以进行数据的分布确认、分布的参数估计、不同寿命的分布分析、拟合、分析置信区间等。

举例 1：参数估计

某质量小组研究某种型号的温度传感器的可靠性，小组随机抽取了 150 个温度传感器进行寿命加速试验，事先规定失效数目为 15，测得的 15 个产品的寿命数据为（单位：日）：90，170，280，360，550，730，920，1100，1300，1600，2000，2400，2900，3400，4000。小组已经找到最适合这组数据的参数分布模型是威布尔分布。现在小组需要估计该型温度传感器寿命为 3600 日的概率，估计 90% 的温度传感器寿命为多少，这批抽样的温度传感器寿命服从威布尔分布的形状参数 β 是否等于 0.5。

具体分析：

利用 Minitab 软件构建工作表，输入 15 个寿命数据，第 16 个为右删失且频数为 135，见图 5-1。

	C1	C2	C3	C
	寿命	删失	频数	
1	90	1	1	
2	170	1	1	
3	280	1	1	
4	360	1	1	
5	550	1	1	
6	730	1	1	
7	920	1	1	
8	1100	1	1	
9	1300	1	1	
10	1600	1	1	
11	2000	1	1	
12	2400	1	1	
13	2900	1	1	
14	3400	1	1	
15	4000	1	1	
16	4000	0	135	

图 5-1 工作表

选择"统计"中的"可靠性/生存"，选择"分布分析（右删失）"中的"参数分布分析"。在

"删失"中选择"失效删失在"，框中输入"16"。在"估计"中，选择"极大似然"，在"估计下列百分比的百分位数"中，输入"10"，在"估计这些寿命的概率"中输入"3600"。在"检验"中，"检验形状（Weibull斜率）或尺度等于"中输入"0.5"。

经过软件分析后，得到温度传感器的寿命数据参数分布分析的结果输出，见图5-2。

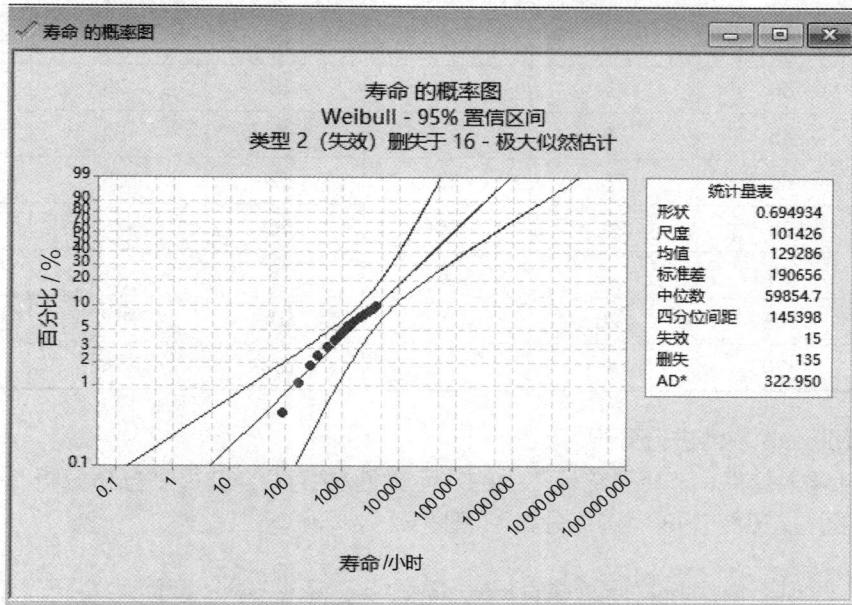

图5-2　寿命数据参数分布分析的结果输出

结果分析：

威布尔分布的形状参数 β 的点估计值为 0.694 934，它的 95% 置信区间为（0.422 158，1.143 96），尺度参数 α 的点估计值为 101 426，它的 95% 置信区间为（16 988，605 557）。

从百分位数表格看，在百分比 10 处，百分位数为 3979.19，说明 90% 的温度传感器寿命为 3979.19 小时，它的 95% 的置信区间为（1920.63，8244.17）。

从生存概率表看，温度传感器寿命为 3600 日的概率为 0.906 397，它的 95% 置信区间为（0.849 508，0.942 501）。

在对形状参数 β 等于 0.5 的检验时，P 值等于 0.195，显然 P 值小于 0.5，因此拒绝原假设，即这批抽样的温度传感器寿命服从威布尔分布的形状参数 β 不等于 0.5。

对于威布尔分布形状参数 β 的检验在实践中经常用到，比如为了区分数据是来自何种的失效状态，若 $\beta < 1$ 则属于早期失效，若 $\beta = 1$ 则属于随机失效，若 $\beta > 1$ 则属于耗损失效。

举例2：求最佳拟合分布

有某型加工机械需要安装加工零件以进行加工产品，这种零件有新、旧两种方法生产。为了用数据来说明哪种零件寿命更长，试验人员各取 100 个进行寿命试验，相关试验的数据见表5-17。显然，新、旧两种方法试验的时间间隔是不相同的。在表中列出了新、旧方法的各个时间区间的开始值、结束值，用观测到的数据记录两种方法对应的加工后成品的个数。现在求最佳拟合分布。

表 5-17　新旧两种方法加工的零件寿命试验数据　　　　单位：天

序号	C1	C2	C3	C4	C5	C6
	开始 1	结束 1	频率 1	开始 2	结束 2	频率 2
1	0	15	2	0	13	1
2	15	30	4	13	26	1
3	30	45	8	26	39	1
4	45	60	12	39	52	8
5	60	75	23	52	65	20
6	75	90	16	65	78	14
7	90	105	7	78	91	9
8	105	120	2	91	104	2
9	120	*	1	104	*	1

现在使用 Minitab 软件进行分析。

把数据全部输入软件，选择"统计"中的"可靠性/生存"，选择"分布分析（任意删失）"中的"分布 ID 图"，见图 5-3 "选择分布 ID 图"。

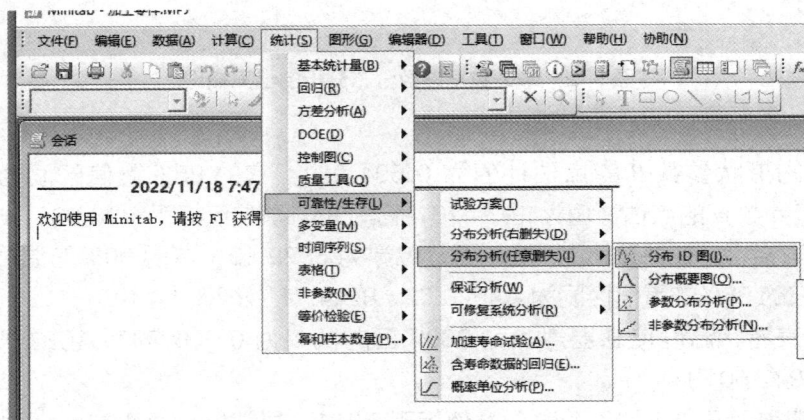

图 5-3　选择分布 ID 图

在"分布 ID 图－任意删失"中，"初始变量"选择"开始 1 开始 2"，"结尾变量"选择"结束 1 结束 2"，"频率列"选择"频率 1 频率 2"，见图 5-4 "分布 ID 图－任意删失"。

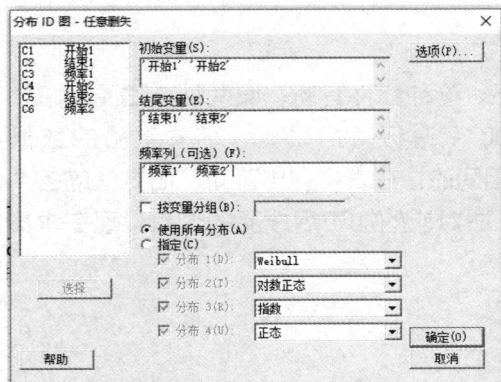

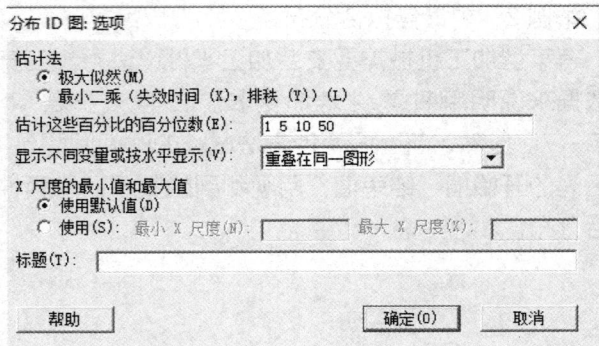

图 5-4　分布 ID 图—任意删失　　　　　图 5-5　分布 ID 图：选项

点击"选项"，在"估计法"中选择"极大似然"，见图5-5"分布 ID 图：选项"。

在全部对话框"确定"后，在会话框有如下输出内容，见图5-6"会话框输出"部分内容。

拟合优度

分布	Anderson-Darling（调整）
Weibull	0.670
对数正态	0.890
指数	1.995
对数 Logistic	0.634
3 参数 Weibull	0.602
3 参数对数正态	0.642
2 参数指数	2.012
3 参数对数 Logistic	0.616
最小极值	0.515
正态	0.637
Logistic	0.622

拟合优度

分布	Anderson-Darling（调整）
Weibull	0.851
对数正态	1.070
指数	2.401
对数 Logistic	0.845
3 参数 Weibull	0.797
3 参数对数正态	0.841
2 参数指数	2.418
3 参数对数 Logistic	0.865
最小极值	0.730
正态	0.844
Logistic	0.878

图 5-6 "会话框输出"部分内容

在图形框，出现了三张图形，见图5-7"分布 ID 图"。

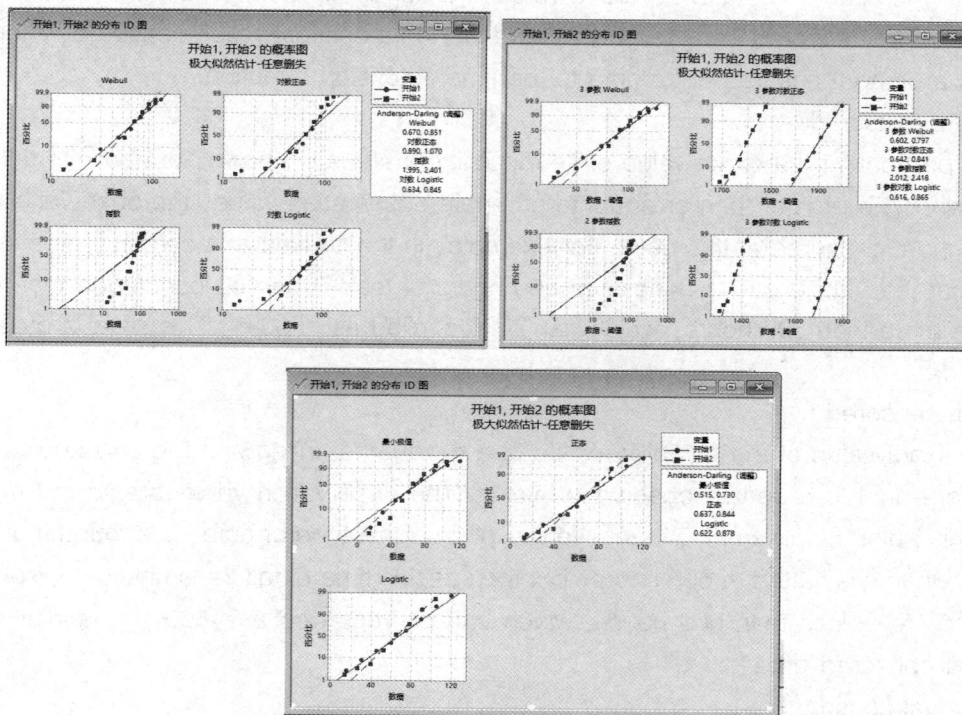

图 5-7 分布 ID 图

结果分析：

（1）新、旧两组数据的变化程度都不算剧烈，所以考虑在正态分布、logistic 分布、极值分布三个分布中选择一个。

（2）从会话框 Anderson-Darling 拟合优度值可以看出，两组数据中最小极值分布的 Anderson-Darling 拟合优度值为：0.515、0.730，而且都是各分布中最小的值。

（3）从图5-7 最后一张图中可以看出，两组数据的最小极值分布都是拟合最好的分布。所以，最小极值分布是最适合本例的分布。

5.3　Censored Data

In various cases of missing data, there is a kind of data that will be stopped or cannot be observed for various reasons, and the data in this case is deleted data.

5.3.1 Meaning of censored data

Censored data refers to the data truncated for some reason in reliability analysis and survival analysis. It is also the data observation value at a certain time (equivalent to the data of censoring test). The censored data can be divided into right censored, left censored, interval censored and arbitrary censored according to the time extension direction.

1. Right censored

For a product, if the exact value of its reliability or life is unknown, but only its reliability or life is known to be greater than or equal to a certain observation value, the observation value is said to be right censored at this point. For example, in the life test of a certain type of bulb, if it is found that the bulb is still working after 500 hours of test, it can be concluded that the life of the bulb is more than 500 hours, which can be recorded as "500+". This data is called "right censored data".

2. Left censored

If the exact value of the reliability or life of a product is unknown, it is only known that it is less than or equal to a certain observation value. The observation value is said to be left censored at this point. For example, in the life test of a certain type of bulb, it is found that the bulb has failed when it is tested at 600 hours, but the specific time of failure is unknown, so it can be sure that its life is less than 600 hours, which can be recorded as "600 −", and this data is called "left censored data".

3. Interval censored

If the precise value of the reliability or life of a product is unknown, it is only known that it is within a certain range. The observed value is called interval censored at this point. For example, in the life test of a certain type of bulb, it can be sure that its life is more than 500 hours, but less than 550 hours, then this kind of data is called interval censored data.

4. Arbitrary censored

Arbitrary censored is a data record storage and representation method applicable to all kinds of censored data, that is, right censored data, left censored data and interval censored data are all applicable.

See Table 5−15 for the representation method of arbitrarily censored data.

Table 5-15　Representation method of arbitrarily censored data

Types of observed value	Input of start column	Input of end column
Exact time	Failure time	Failure time
Right censored	Time (expires after)	*
Left censored	*	Time (expires before)
Interval censored	Start time of failure occurrence interval	End time of failure occurrence interval

Case：

There are 100 parts, 15 parts are being used on the 8th day, and the remaining 85 parts are observed. On the 10th day, four of them were found to have failed, and two just failed on the 10th day. See Table 5-16 for the structure of the worksheet.

Table 5-16　Part censored data worksheet

No.	C1	C2	C3
	Start	End	Frequency
1	8	*	15
2	*	10	4
3	10	10	2
4	10	*	79

5.3.2 Types of right censored data

The right censored data is more common in the life data. According to the different criteria for judging the end of the research cycle of the experimental arrangement, the right censored data can be divided into three basic types: Time censoring (Type I censoring), Failure censoring (Type II censoring) and Random censoring (Type III censoring).

1. Time censoring (Type I censoring)

In the life test of products, due to the time and cost of observation will be limited, it is stipulated that the test should be carried out within a certain time range and the observation shall be stopped at the final time. The final time is usually determined before the test. In this way, the life of the sample can be observed only in a time range less than or equal to a given time range. The data obtained at this time is called time censoring, also known as Type I censoring or Timing censoring. For example, the life time censoring of a type of machine is 1000 hours, which means that the machine is still working after 1000 hours. Its life is greater than 1000 hours, but the accurate failure time of the machine is unknown.

2. Failure censoring (Type II censoring)

A fixed number of individual failures shall be specified in advance. During the life test of the product, it shall be observed until a fixed number of individual failures. The data obtained at this time is called failure censoring, also known as type II censoring or fixed number censoring. Failure censoring test is often used to test and study the service life level of equipment. During the test, all test equipment begin to operate at the same time until a given number of equipment fail. It may take a long time to observe that all equipment fails, and a lot of time and cost can be saved by failure censoring inspection. For example, in the life test of a type of machine, the life failure is censored in the 10th machine, which means that the life data of machines that fail after the 10th machine are censored.

3. Random censoring (Type Ⅲ censoring)

During the life test of products, the test is not stopped according to the predetermined time or the number of failures, but whether each sample of the test is stopped or not is determined according to the value of another random variable. The observed data obtained in this situation is called random censoring, also known as Type Ⅲ censoring. Random censoring means that the test is in a complex situation. There are many interference factors causing failure, and the failure is not planned in advance. Usually, in the process of the test, it is affected by some interference factor, so it leaves the test or terminates the test observation in advance. For example, in a life test, the test was terminated due to the variation of the test site; the newly constructed VAV air conditioning system in two commercial buildings were updated by other air conditioning system after the 60th and 75th month normal operation, then the lives of the newly constructed air conditioning system in the two commercial buildings are 60^+ month and 75^+ month respectively.

Type Ⅰ censoring and Type Ⅱ censoring are collectively referred to as single censored, and Type Ⅲ deletion is also referred to as multiply censored.

5.3.3 Censored test

Censored test, also known as truncated test, is a kind of life test, which refers to a test method that stops the test once some products fail during the product test. The failure data obtained in the censoring test is called the censored sample.

1. Fix time censored test

Fix time censored test, also known as type Ⅰ truncation test, means that the test is stopped at a predetermined time. The number of failure data obtained when the test is stopped is a random variable. For example, for life test of 100 product samples, it is required to stop the test after 50 hours in advance. Assuming that the number of failures is 5, the number of failures is a random variable. Properly specifying the test stop time is the key to the fix time censored test.

2. Fix number censored test

Fix number censored test, also known as type Ⅱ truncation test, refers to stopping the test when the number of failures specified in advance reaches. The time when the test stopped is a random variable. For example, for the life test of 100 product samples, it is specified in advance that the test will be stopped when the 30th product (30% of the product) fails, and the time to stop the test is a random variable. Properly specifying the failure ratio is the key to the fix number censored test.

5.3.4 Application of censored data

Censored data is widely used in product reliability analysis and survival analysis. Using different censored data, we can confirm the distribution of data, estimate the parameters of distribution, analyze the distribution of different lives, and fit and analyze the confidence interval.

Case1: Paramenter estimation
A quality team studied the reliability of a certain type of temperature sensor. The team ran-

domly selected 150 temperature sensors for life acceleration test. The number of failures was 15, and the measured life data of 15 products was (unit: day):90, 170, 280, 360, 550, 730, 920, 1100, 1300, 1600, 2000, 2400, 2900, 3400, 4000. The team has found that the most suitable parameter distribution model for this group of data is Weibull distribution. Now the team needs to estimate the probability that the life of this type of temperature sensor is 3600 days, and estimate the life of 90% of temperature sensors. The team needed to estimate whether the shape parameter β of the sampled temperature sensors that obey Weibull distribution is equal to 0.5.

Analyze as follows:

Use Minitab software to build a worksheet, input 15 lifedata, and the 16th one is right censored and the frequency is 135, as shown in Figure 5-1.

↓	C1	C2	C3
	Life data	Censored	Frequency
1	90	1	1
2	170	1	1
3	280	1	1
4	360	1	1
5	550	1	1
6	730	1	1
7	920	1	1
8	1100	1	1
9	1300	1	1
10	1600	1	1
11	2000	1	1
12	2400	1	1
13	2900	1	1
14	3400	1	1
15	4000	1	1
16	4000	0	135

Figure 5-1　Worksheet

Select "Reliability/Survival" in the "Statistics". Choose "Parametric Distribution Analysis" in the" Distribution Analysis (Right Censoring)". Input "10" into the "Estimate percentiles for these additional percents", and input "3600" into the "Estimate probabilities for these times (values)". Select "Censored" into "Use censoring columns", or select "Failure censor at" and input "16" into the frame. Input "0.5" into the "Test scale (Weibull or expo) or location (other dists) equal to".

After software analysis, the result output of life data parameter distribution analysis of temperature sensor is obtained, as shown in Figure 5-2.

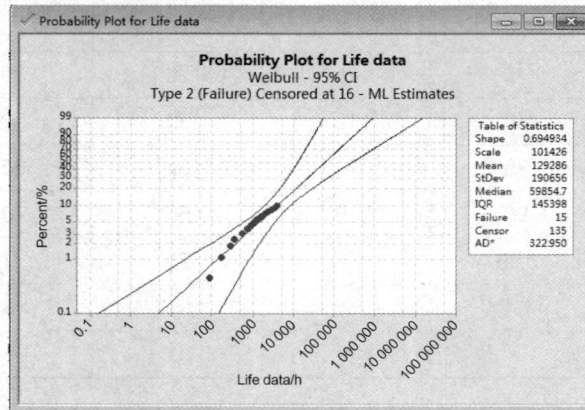

Figure　5-2　Result output of life data parameter distribution analysis

Result analysis:

The point estimate value of the shape parameter β of Weibull distribution is 0.694 934. Its 95% confidence interval is(0.422 158,1.143 96); The point estimate value of scale parameter α is 101 426, and its 95% confidence interval is (16 988,605 557).

From the percentile table, at percentage 10, the percentile is 3979.19, indicating that 90% of the temperature sensor life is 3979.19 hours, and its 95% confidence interval is(1920.63, 8244.17).

From the survival probability table, the probability of the temperature sensor life of 3600 days is 0.906 397, and its 95% confidence interval is(0.849 508,0.942 501).

When test "the shape parameter β is equal to 0.5", the P value is equal to 0.195, obviously the P value is less than 0.5, so the null assumption is rejected, that is, the shape parameter β of Weibull distribution for the life of the temperature sensors sampled in this batch is not equal to 0.5.

The test for the shape parameter β of Weibull distribution is often used in practice. For example, we can distinguish which failure state the data comes from. If $\beta < 1$, it is early failure; if $\beta = 1$, it is random failure; if $\beta > 1$, it is wearout failure.

Case 2:Find the best fit distribution

A certain type of processing machine needs to install processing parts to process products. This kind of parts can be produced by new and old methods. In order to use data to explain which parts have a longer service life, the test personnel take 100 each for life test. See Table 5-17 for relevant test data. Obviously, the time interval between the new and old methods is different. The start and end values of each time interval of the new and old methods are listed in the table, and the number of finished products corresponding to the two methods is recorded with the observed data. Now we need to find the best fit distribution.

Table 5-17 Life test data of parts processed by new and old methods   Unit: days

No.	C1	C2	C3	C4	C5	C6
	Start1	End1	Frequency1	Start2	End2	Frequency2
1	0	15	2	0	13	1
2	15	30	4	13	26	1
3	30	45	8	26	39	1
4	45	60	12	39	52	8
5	60	75	23	52	65	20
6	75	90	16	65	78	14
7	90	105	7	78	91	9
8	105	120	2	91	104	2
9	120	*	1	104	*	1

Now use Minitab software for analysis.

Input all the data into the software. Select "Reliability/Survival" in "Statistics". Choose

"Distribution ID Plot" in "Distribution Analysis (Arbitrary Censoring)". See Figure 5-3 "Select Distribution ID Plot".

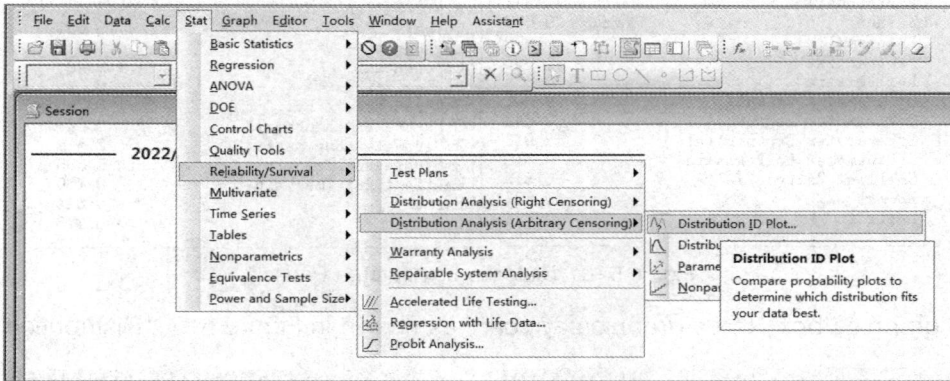

Figure 5-3   Distribution ID Plot

In "Distribution ID Plot—Arbitrary Censoring", Select "Start1, Start 2" into "Start variables", select "End1, End2" into "End variables", and select "Frequency1, Frequency 2" into "Frequency columns (optional)". See Figure 5-4 "Distribution ID Plot—Arbitrary Censoring".

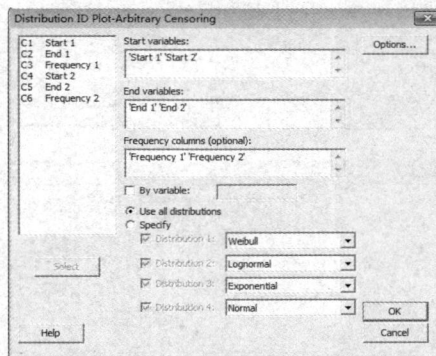

Figure 5-4   Distribution ID Plot—Arbitrary Censoring

Click "Options". Select "Maximum Likelihood" in the "Estimation Method", as shown in Figure 5-5 "Distribution ID Plot：Options".

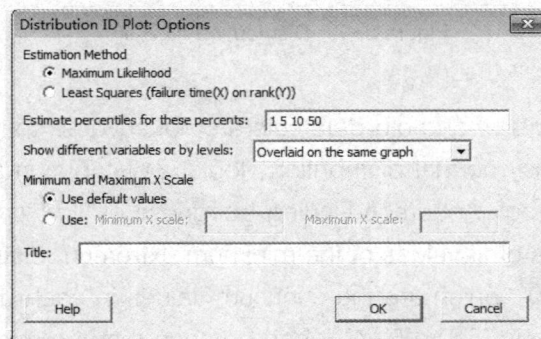

Figure 5-5   Distribution ID Plot：Options

After all dialog boxes are "OK", the following output contents are displayed in the dialog box, as shown in Figure 5-6 "Dialogue box output".

```
Goodness-of-Fit                          Goodness-of-Fit

                    Anderson-Darling                        Anderson-Darling
Distribution              (adj)          Distribution              (adj)
Weibull                   0.670          Weibull                   0.851
Lognormal                 0.890          Lognormal                 1.070
Exponential               1.995          Exponential               2.401
Loglogistic               0.634          Loglogistic               0.845
3-Parameter Weibull       0.602          3-Parameter Weibull       0.797
3-Parameter Lognormal     0.642          3-Parameter Lognormal     0.841
2-Parameter Exponential   2.012          2-Parameter Exponential   2.418
3-Parameter Loglogistic   0.616          3-Parameter Loglogistic   0.865
Smallest Extreme Value    0.515          Smallest Extreme Value    0.730
Normal                    0.637          Normal                    0.844
Logistic                  0.622          Logistic                  0.878
```

Figure 5-6　Dialogue box output（Part）

In the graphics box, three graphics appear, as shown in Figure 5-7 "Distribution ID Plot".

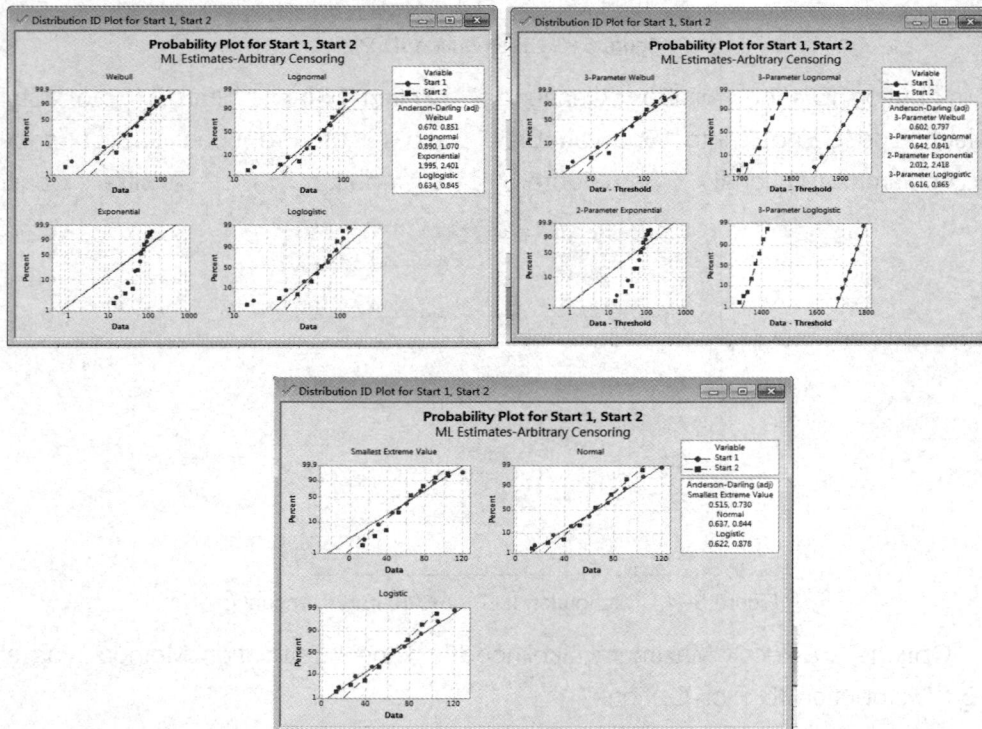

Figure 5-7　Distribution ID Plot

Result analysis：

（1）The changes of the new and old data sets are not severe, so it is considered to choose one of the three distributions：normal distribution, logistic distribution and extreme distribution.

（2）From the dialogue box Anderson-Darling goodness of fit value, it can be seen that the Anderson-Darling goodness of fit values of the minimum extremum distribution in the two groups of data are 0.515 and 0.730, which are the minimum values in each distribution.

（3）From the last figure in Figure 5-7, it can be seen that the minimum extremum distribution of the two groups of data is the best fitting distribution. Therefore, the minimum extremum distribution is the most suitable distribution for this example.

5.4　数据资料遗失

数据资料遗失是由各种原因导致的，比如数据储存设备损坏、数据管理人员工作责任心不强、误操作、环境条件干扰等引起。数据资料遗失会对于数据分析产生较大的影响。

举例：

某小组开展活动。小组在现状调查中找到的症结为冷剪机的"曲轴轴瓦使用寿命短"，小组在原因分析中找到的一调末端原因为"端盖与曲轴之间间隙大"。

小组分析后认为，端盖与曲轴之间有间隙给轴向串动留有空间。小组通过受力分析得知，冷剪机运行一段时间后由于剪口端受力较大，曲轴在一个动作后会使轴的轴向有一个推力从而造成轴瓦磨损。接着小组进行现场试验，通过对压盖增加垫片组的方法，以减少冷剪机端盖与曲轴之间的间隙，并进行了相关的数据统计。

小组在采集完数据以后，由于保管的原因，有部分数据缺失，因此只有 16 组数据，见表 5-18。

表 5-18　冷剪机端盖与曲轴之间的间隙测试 16 组数据

序号	垫片数量 /（1mm·个$^{-1}$）	端盖与轴间隙 /mm	曲轴振动值 /（mm·s^{-1}）	序号	垫片数量 /（1mm·个$^{-1}$）	端盖与轴间隙 /mm	曲轴振动值 /（mm·s^{-1}）
1	5	95	7.2	9	45	55	7.8
2	10	90	7.3	10	50	50	7.6
3	15	85	7.4	11	55	45	7.3
4	20	80	7.3	12	60	40	7.2
5	25	75	7.3	13	65	35	7.4
6	30	70	7.6	14	70	30	7.5
7	35	65	7.3	15	75	25	7.1
8	40	60	7.2	16	80	20	7.4

小组根据表 5-18 的数据，绘制散点图，见图 5-8。

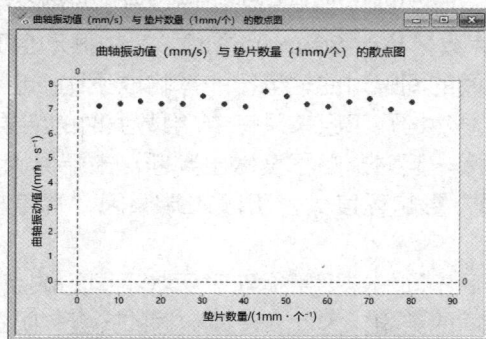

图 5-8　冷剪机端盖与曲轴之间的间隙对曲轴振动影响散点图

从图 5-7 观察，曲轴振动值与垫片数量之间几乎没有相关关系。因此，小组认为末端原因"端盖与曲轴之间间隙大"对于症结"曲轴轴瓦使用寿命短"没有大的影响，由此判断该末端原因为非主要原因。

由于存在数据缺失的情形，因此，有关负责人要求小组重新做试验，而且数据总量要达到30 组。于是，小组进行了第二次试验，得到了相关数据统计表，见表 5-19。

表 5-19　冷剪机端盖与曲轴之间的间隙测试 30 组数据

序号	垫片数量 /（1mm·个⁻¹）	端盖与轴间隙 /mm	曲轴振动值 /（mm·s⁻¹）	序号	垫片数量 /（1mm·个⁻¹）	端盖与轴间隙 /mm	曲轴振动值 /（mm·s⁻¹）
1	5	95	7.2	16	80	20	7.4
2	10	90	7.3	17	82	18	6.8
3	15	85	7.4	18	84	16	6.6
4	20	80	7.3	19	86	14	6.6
5	25	75	7.3	20	88	12	6.3
6	30	70	7.6	21	89	11	6.3
7	35	65	7.3	22	91	9	6.3
8	40	60	7.2	23	92	8	5.7
9	45	55	7.8	24	93	7	5.4
10	50	50	7.6	25	94	6	5.1
11	55	45	7.3	26	95	5	4.8
12	60	40	7.2	27	96	4	4.5
13	65	35	7.4	28	97	3	3.8
14	70	30	7.5	29	98	2	3.4
15	75	25	7.1	30	99	1	2.1

根据表 5-19 的数据，小组制作了对应的散点图，见图 5-9。

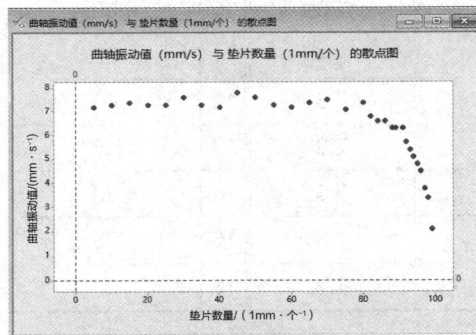

图 5-9　冷剪机端盖与曲轴之间的间隙对曲轴振动影响散点图

由散点图可以看出，当垫片数量超过 80 mm/个时，曲轴振动值开始发生明显的向下弯曲趋势，表明冷剪机端盖与曲轴之间的间隙对曲轴运行的影响成正比，也说明端盖间隙对曲轴的使用寿命有较大影响。这一现象，显然推翻了原来只有 16 组数据时的结果。

最终，小组会同相关负责人一起对影响程度做出判断：末端原因"端盖与曲轴之间间隙大"对症结"曲轴轴瓦使用寿命短"影响程度大。所以末端原因"端盖与曲轴之间间隙大"为主要原因。

从本例中可以看出，由前后不同的数据量得到了两张不同的散点图，而且得出的结论正好相反，给最终的决策带来截然不同的结果。因此，注重数据的齐全，防止数据遗失是极其重要的。

5.4　Data Missing

Data missing is caused by various reasons, such as damage to data storage equipment, weak sense of responsibility of data management personnel, misoperation, interference of environmental conditions, etc. Data loss will have a great impact on data analysis.

Case：

A quality management team carries out activities. The crux found by the team in the status survey is "short service life of crankshaft bearing bush" of the cold shear, and the cause found by the team in the cause analysis is "large gap between end cover and crankshaft".

After analysis, the team believed that there was a gap between the end cap and the crankshaft to leave space for axial string movement. The team learned from the stress analysis that after the cold shear operated for a period of time, due to the large stress on the shear end, the crankshaft will have a thrust on the axial direction of the shaft after an action, which will cause bearing bush wear. Then the team carried out field tests to reduce the gap between the end cover of the cold shear and the crankshaft by adding shims to the gland, and carried out relevant data statistics.

After the team collected the data, due to storage reasons, some data were missing, so there were only 16 groups of data. See Table 5−18 "Clearance test table between the end cover of cold shear and crankshaft".

Table 5−18　Clearance test table between the end cover of cold shear and crankshaft

No.	Number of shims /（1mm · piece^{-1}）	Clearance between end cover and shaft/mm	Crankshaft vibration value /（mm · s^{-1}）	No.	Number of shims /（1mm · piece^{-1}）	Clearance between end cover and shaft/mm	Crankshaft vibration value /（mm · s^{-1}）
1	5	95	7.2	9	45	55	7.8
2	10	90	7.3	10	50	50	7.6
3	15	85	7.4	11	55	45	7.3
4	20	80	7.3	12	60	40	7.2
5	25	75	7.3	13	65	35	7.4
6	30	70	7.6	14	70	30	7.5
7	35	65	7.3	15	75	25	7.1
8	40	60	7.2	16	80	20	7.4

According to the data in Table 5−18, the team drew a scatter diagram, as shown in

Figure 5-8.

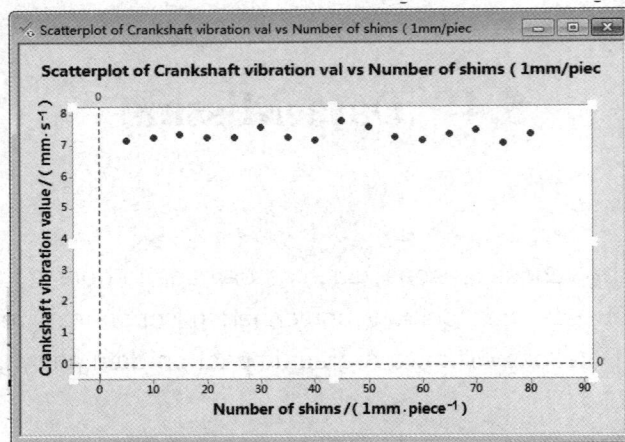

Figure 5-8 Scatter diagram of influence of clearance between end cover
and crankshaft of cold shear on crankshaft vibration

From Figure 5-8, there is almost no correlation between the crankshaft vibration value and the number of shims. Therefore, the team believes that the end cause "large gap between end cap and crankshaft" has no significant impact on the crux "short service life of crankshaft bearing bush", so it is judged that the end cause is not the main cause.

Due to the lack of data, the relevant person in charge asked the group to conduct the test again, and the total number of data should reach 30 groups. Therefore, the team conducted the second test and obtained the relevant statistical data table, as shown in Table 5-19.

Table 5-19 30 sets of data for clearance test between end cover and crankshaft of cold shear

No.	Number of shims / (1mm · piece^{-1})	Clearance between end cover and shaft/mm	Crankshaft vibration value / (mm · s^{-1})	No.	Number of shims / (1mm · piece^{-1})	Clearance between end cover and shaft/mm	Crankshaft vibration value / (mm · s^{-1})
1	5	95	7.2	16	80	20	7.4
2	10	90	7.3	17	82	18	6.8
3	15	85	7.4	18	84	16	6.6
4	20	80	7.3	19	86	14	6.6
5	25	75	7.3	20	88	12	6.3
6	30	70	7.6	21	89	11	6.3
7	35	65	7.3	22	91	9	6.3
8	40	60	7.2	23	92	8	5.7
9	45	55	7.8	24	93	7	5.4
10	50	50	7.6	25	94	6	5.1
11	55	45	7.3	26	95	5	4.8
12	60	40	7.2	27	96	4	4.5

No.	Number of shims / (1mm · piece⁻¹)	Clearance between end cover and shaft/mm	Crankshaft vibration value / (mm · s⁻¹)	No.	Number of shims / (1mm · piece⁻¹)	Clearance between end cover and shaft/mm	Crankshaft vibration value / (mm · s⁻¹)
13	65	35	7.4	28	97	3	3.8
14	70	30	7.5	29	98	2	3.4
15	75	25	7.1	30	99	1	2.1

According to the data in Table 5-19, the team made the corresponding scatter chart, as shown in Figure 5-9.

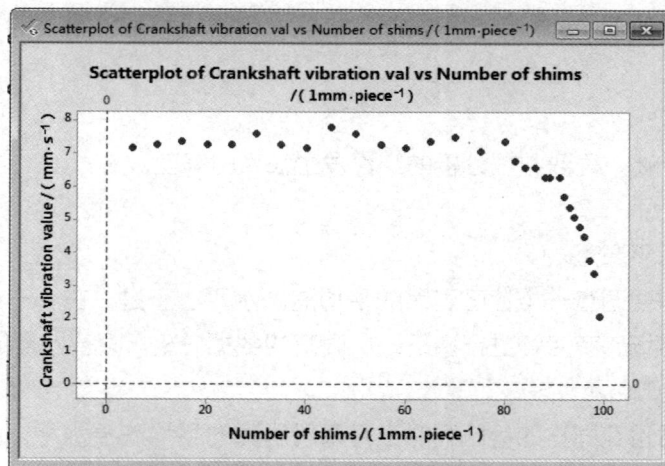

Figure 5-9   Scatter diagram of influence of clearance between end cover and crankshaft of cold shear on crankshaft vibration

It can be seen from the scatter diagram that when the number of shims exceeds 80mm/piece, the vibration value of the crankshaft starts to bend downward obviously, indicating that the gap between the end cap of the cold shear and the crankshaft has a direct impact on the operation of the crankshaft, which also shows that the end cap gap has a great impact on the service life of the crankshaft. This phenomenon clearly overturned the original results when there were only 16 sets of data.

Finally, the team together with relevant responsible persons made a judgment on the degree of impact, that is, the end cause "large gap between the end cover and the crankshaft" has a great impact on the crux "short service life of the crankshaft bearing bush". So the end cause "large gap between end cover and crankshaft" is the main reason.

It can be seen from this example that two different scatter diagrams are obtained by drawing different data volumes before and after, and the conclusions are exactly the opposite, bringing different results to the final decision. Therefore, it is extremely important to pay attention to the completeness of data and prevent data loss.

5.5　数据错用

数据错用是数据缺失的一个特殊情形。当数据错用发生的时候，相当于间接发生了数据缺失。数据错用可以分几种类型，比如数值错误、论证错误、感知错误等引起。

5.5.1 数值错用的情形

数值错用是指对于数据中的数字使用发生了错误，有的是不理解数字的具体含义，有的是样本数值理解错误等。

1. 样本的内涵错用

样本的内涵发生变化，导致数据分析的结论发生很大的偏差。

举例：人均消费水平调查

有一个调查小组需要调查一个社区居民的平均消费水平。于是，在周一至周五的白天，调查小组到该社区附近的菜场、商店等进行调研。从调研的数据分析，调查小组认为该社区的消费水平折合为人均月消费 1000 元。

另有一个调查小组也在同样的社区同时进行调研居民的平均消费水平。由于很多上班族都是下班后开始消费，因此该调查小组不仅调查了白天的消费情况，还调查了晚上的消费情况。同时，调查小组还针对上班族进行了网上购物的消费调查。从调研的数据分析，调查小组认为该社区的消费水平折合为人均月消费 1200 元。具体见表 5-20。

表 5-20　消费水平调查表

组名	调查时间	调查内容	人均消费结果/（元·月$^{-1}$）
第一组	周一至周五白天	菜场、商店	1000
第二组	周一至周五白天、傍晚	菜场、商店、网购	1200

在这个例子中，第一个调查小组显然缺失了上班族的晚上消费数据，缺失了网上消费数据。

2. 统计口径的变化

统计口径是统计数据时所采用的相关标准，可以理解为开展数据统计工作所遵循的指标体系，它包括统计的方式、统计的范围等指标。如果统计口径发生变化，那么数据分析也发生偏差。

举例：统计公司产值

有一家大型集团公司发展势头良好，产值逐年增加。现要求相关人员统计其各子公司的产值，以作为制定新一轮发展计划和相关管理举措的依据。统计的范围为全部子公司。经过对全部

12 家子公司的产值数据分析，发现该集团的人均年产值为 400 万元，而上一年度人均年产值为 500 万元，同比下降了 20%。这样的结论，似乎与企业的发展势头相悖，那么问题出在哪儿呢？

经过深度分析，调查人员发现，该集团今年从市场上新控股了 2 家亏损企业，而这 2 家企业由于前期的管理不善，发展严重滞后。今年刚并入集团公司，业务稍有起色，但与其他成熟的子公司相比，差距甚远。何况，新并入的 2 家公司中，员工数量比较多。这样，虽然集团公司原有的 10 家子公司发展良好，但由于新入企业的状况，就把人均产值降了下来。具体数据见表 5-21。

表 5-21　产值调查表

年份	调查子公司	子公司情况	总人数/人	人均产值/万元
今年	12 家子公司	10 家在增长，新入 2 家发展滞后	5800	400
去年	10 家子公司	10 家在增长	5000	500

在本例中，去年的产值统计中 2 家亏损企业的数据是缺失的。如果在分析的时候，把去年 2 家公司的产值状况也纳入到去年的统计数据中，这样得出的结论就正确。

3. 差异不显著的排序结果

排序是一种分类的过程，它是把各种无序的数据按照一定的要求、规律进行排列，把无序的数据变为有序的序列。排序现在在很多行业都有应用。

举例：高校排名

世界高校的排名现在有很多种，如泰晤士世界大学排名、QS 世界大学排名等。但它们的排名结果会有一定的差异。这里摘录了 2022 年世界大学排名，见表 5-22。

表 5-22　2022 年世界大学排名

排序	泰晤士世界大学声誉排名	泰晤士世界大学排名	QS 世界大学排名
1	哈佛大学	牛津大学	麻省理工学院
2	麻省理工学院	加州理工学院	牛津大学
3	斯坦福大学	哈佛大学	斯坦福大学
4	牛津大学	斯坦福大学	剑桥大学
5	剑桥大学	剑桥大学	哈佛大学
6	加州大学伯克利分校	麻省理工学院	加州理工学院
7	普林斯顿大学	普林斯顿大学	帝国理工学院
8	耶鲁大学	加州大学伯克利分校	苏黎世联邦理工学院
9	清华大学	耶鲁大学	伦敦大学学院
10	东京大学	芝加哥大学	芝加哥大学

从三个排名来看，先后顺序各有不同，那是由学校排名的逻辑、内在关系、指标差异的显著性等方面所决定的。有些排名中多了某项指标，有的少了某项指标，因此排名有差异。

泰晤士的世界大学声誉排名更注重业内人士的主观感受，它是通过邀请制的学术问卷调查结果而得到。该声誉调查问卷仅针对有经验、发表过论文的学者，就其所属学科、所熟悉的院校的研究及教学发表意见。问卷中的主要指标是学者、研究、教学，带有一定的主观性。

泰晤士世界大学排名把实现联合国 17 项可持续发展目标作为评估的主要依据，涉及的领域包括研究、外部活动、管理、教学等方面，评估基本都是客观的数据。

QS 的世界大学排名注重的指标主要有 6 项，分别为：学术声誉占 40%，雇主声誉占 10%，单位教员论文引文数占 20%，师生比例占 20%，国际教师比例占 5%，国际学生比例占 5%。该排名把全球声誉、学术质量、国际化程度作为三个主要指标。

由此可以看出，不同的评价标准有自己的评价方法和依据，又不同程度地忽略了其他排名方法的指标，因此排名结果是有差异的。

4. 平均数不具有代表性

平均数是数据之间的均值，它代表了数据的集中趋势。用平均数来表示一组数据的情况，具有直观、简明等特点。但有时候，平均数作为指标会失去代表性。

举例：城市气温

成都位于四川盆地，年平均气温 17.9 ℃。昆明地处云贵高原，年平均气温 17.1℃。为此，有人得出结论，两个城市的气温基本一致，没有什么差异。

那么真实结果如何呢？ 把两个城市的月平均气温资料进行比较。成都的数据见表 5-23，昆明的数据见表 5-24。

表 5-23　成都市的气温　　　　　　　　　　　　　　　　　单位：℃

月份	1月	2月	3月	4月	5月	6月	7月	8月	9月	10月	11月	12月	年平均
最高	22	25	35	34	31	27	24	22	11	11	12	15	17.9
最低	14	18	25	23	21	16	13	11	4	4	4	8	

表 5-24　昆明市的气温　　　　　　　　　　　　　　　　　单位：℃

月份	1月	2月	3月	4月	5月	6月	7月	8月	9月	10月	11月	12月	年平均
最高	17	20	22	25	26	26	25	26	25	22	20	16	17.1
最低	3	5	9	12	15	18	18	18	17	13	8	4	

从上述两表可以看出，在 3 月、4 月、5 月，成都市的气温要明显高于昆明市；而在 9 月、10 月、11 月，成都市的气温要明显低于昆明市。显然"两个城市的气温基本一致"的结论，是忽视了具体月份的数据。如果分析了两个城市在春季、秋季的气温数据的结构、分布、异常值等，那么发现成都市的气温在全年的波动比较大，昆明市的气温在全年波动要小，实质是气温数据的标准差是不一样的。因此，不能省略数据的结构，轻易用平均数来代表全部数据的特征。

5. 对数据不核实

对数据不核实的现象经常发生，发生的原由是缺失了真实数据。不核实数据的情形主要有：

（1）道听途说。比如有的人在路上听说某上市公司业绩非常好，立即在二级市场购买了大量该公司的股票。但由于信息是不真实的，上市公司的股票几乎一直维持在原有的价格，略有波动。

（2）直接引用。比如有总公司的某部门负责人把下属企业相关部门递交的生产数据直接引用，没有进行核实，使企业的决策发生偏差。

（3）校对不认真。比如某技术人员手工填写数据时，填写了数字"63"，而下一道工序的技术人员把该数字误读为"68"，而且辅助技术人员也没有进行校对，直接进行相关计算而出错。

（4）判断错误。比如某个老人腹部不舒服，由于老人经常有胃酸的情形，所以家人以为是胃病，就简单服用一些胃药，但是没有效果。然后，家人把老人送到医院就诊，发现是轻度心梗，其表现的征兆的确是腹部不舒服。

5.5.2 论证逻辑错误

在各种数据引用的过程中，有些数据之间存在比较严谨的逻辑关系，如果逻辑错误，相当于中间过程的数据缺失，造成数据之间的逻辑错误。

1. 运算概念错误

运算概念错误是指人们在进行数据运算时对于数据的具体含义没有搞清楚，用自以为正确的方式进行运算，得到错误的结论。这样的一类错误相当于数据的概念错误及计算过程缺失。

举例：

某个人在经过一个十字路口时发现路边有一个显示屏，该显示屏连着分贝测量仪。每当车辆经过时，显示屏上不断变换着分贝的数值。这个人看到显示屏上的读数一会儿是 60 分贝，一会儿是 65 分贝，他告诉旁边的小孩，车辆噪声得到了很好的控制，65 分贝比 60 分贝才多了 5 个分贝。这样的结论正确吗？

看一下分贝的定义：分贝（decibel）是量度两个相同单位的数量比例的计量单位，它可以用于测量声音强度，其计算过程是以对数形式出现。声压级的分贝计算公式为

$$N_{db} = 20 \times \lg \frac{p_i}{p_0}$$

其中，p_0 = 0.000 02 帕斯卡（Pa），是参考声压；p_i 为实测声压。

显然，65 分贝与 60 分贝之间的区别，不是相差 5 分贝，而是相差了数倍。

在实际工作生活中，不能简单运算的例子还有很多。比如两杯 20 ℃的温水，加在一起不会达到 40 ℃。因此，要避免这样的错误，需要对数据的概念构成、运算过程予以正确理解，不能跳过正确运算的环节直接得出结论。

2. 归纳的逻辑错误

归纳有很多种方法，无论哪种方法，归纳必须有严谨的逻辑关系分析。有时候，由于样本的代表性、归纳的逻辑性等出了问题，那么归纳的结论也出错。

举例：

某调研公司受委托进行某市家庭使用互联网情况的调查。该公司的调研人员对 10 万户家庭通过网络调查的方式进行了调研，在收到回复后，调研人员得出结论：10 万户家庭全部用上了互联网。

在本例中，调研人员的调查方法仅限于于网络调查，导致样本数据缺失。而这种调研方法的缺陷就是逻辑错误，以网络调查的方式、且收到网络回复来证明已经使用了互联网，有点类似"没来的人请举手"。因此，把握正确的逻辑，杜绝样本缺失才是归纳的基础性工作。

3. 数据缺少代表性

在论证过程中所采用的数据必须具有代表性。有时候，由于采集数据受到各种条件的限制，使数据不具有广泛的代表性，导致论证结论错误。

举例：

某调研小组开展全球的气候变化调研，其中一项内容是全球的气候是否在变暖。调研人员共分了四组。第一组调查了 1982 到 1992 十年间的全球平均气温波动情况，发现气温大约上升了 0.2 ℃。第二组调查了 1992 到 2002 十年间的全球平均气温波动情况，发现气温变化不大。第三组调查了 2002 到 2012 十年间的全球平均气温波动情况，发现气温变化也不大，大致上升了 0.3 ℃。第四组调查了从 2012 到 2022 十年间的全球平均气温波动情况，发现气温的上升幅度在 0.3 ℃。这样，四个调查组各自得出结论，全球在十年间的平均气温上升不快。

后来，负责人把调研时间扩大到 160 年，从 1862 年到 2022 年，发现全球的平均气温波动非常明显，大致提升了 1.3 ℃，仅仅 2022 年的平均气温就比 1850—1900 年的平均温度高出 1.15±0.13 ℃。

那为什么会出现十年间的气温上升幅度不大呢？ 因为样本的调研周期太短。如果调研周期扩大到 1982 年到 2022 年的 40 年，则发现气温提升了大约在 0.8 ℃。这也说明样本的调研周期短会导致样本数据相对缺失，则样本缺少了代表性。

4. 分子或分母变化

在逻辑论证中，对于分子、分母的变化会使运算的过程发生很大的变化，导致数据运算的过程缺失。

举例：

某服装商店进行推销。第一个星期，商店推出"打五折" 策略，结果营业额得到提振。第二个星期，商店推出"买一送一"策略，结果营业额很快上升。那么，"打五折"与"买一送一"有无区别呢？ 其实是有明显区别的。

"打五折"相当于价格的分子直接缩水一半，比如单价原来为 100 元/件，那么打五折后为 50 元/件。其营业额是随着每件衣服的销售而上升。

"买一送一"是衣服的价格不变，但是其价格的分母扩大了一倍，比如单价原来为 100 元/件，那么"买一送一"后成为 100 元/2 件。其营业额是随着每 2 件衣服的销售而上升。

假定顾客购买的次数一定，那么，"买一送一"的销售速率要 2 倍于"打五折"的销售速率，要搞明白其中的逻辑关系，则要把隐藏的数据显性化。

5. 偶然性替代必然性

单次采集的数据具有偶然性，而多次的数据则会显示一定的规律，更多的数据则呈现大数定理。事实上，有些人会用偶然性来替代必然性，实质是样本数据不够。

举例:

2010 年南非世界杯的时候,章鱼保罗着实火了一把,因为它在整个足球比赛期间的预测是出奇的准。它的预测方式是:工作人员在两支队伍比赛前,把印有不同国旗的相同食物放到水族箱中,然后让章鱼保罗去选择。章鱼保罗的预测结果见表 5-25。

表 5-25　章鱼保罗的预测及真实比赛结果

比赛场次	比赛的球队	章鱼保罗的预测	真实结果
小组赛首战	德国 VS 澳大利亚	德国胜	德国以 4:0 胜
小组赛次战	德国 VS 塞尔维亚	德国负	德国以 0:1 负
小组赛末战	德国 VS 加纳	德国胜	德国以 1:0 胜
1/8 决赛	德国 VS 英格兰	德国胜	德国以 4:1 胜
1/4 决赛	德国 VS 阿根廷	德国胜	德国以 4:0 胜
半决赛	德国 VS 西班牙	德国负	德国以 0:1 负

如此高的预测准确率,是否真的能代表章鱼保罗的预测能力呢? 显然不是。其实在预测时,样本数据是存在一些奥秘的,那就是国旗的颜色。由于德国国旗是由黑、红、黄三色组成,而红色与黄色是章鱼非常喜欢的食物颜色。因此,与其说是章鱼保罗的预测,不如说是章鱼保罗选择了喜欢的食物颜色。

6. 数据间没有可比性

在数据之间进行比较是经常用到的方法。但数据之间的可比是有前提的,比如数据的单位一致、采集数据的时间长短一致、数据的内在逻辑一致等。有时候存在数据之间没有可比性的情形。

举例:

张三和李四是高中同班同学,平时的学习成绩一直是张三更好些。高考后,张三以高分进入一所高校;李四的考分稍有逊色,进入了另一所高校。两个人虽然学校不同,但所学的专业相同。在大学期间,张三读书仍旧刻苦,而李四相对轻松些。

临近大学毕业,张三与李四碰头,交流后发现,李四的学习成绩比张三好很多,两个人的部分学科成绩见表 5-26。

表 5-26　两个人的学习成绩对比(部分学科)

学科	张三的成绩	李四的成绩
高等数学	76	95
理论力学	80	92
分析化学	83	98
材料力学	74	89
结构力学	81	96

张三的功底比较扎实,学习非常认真,通常的理解,似乎张三应该比李四的学习成绩更好,

那为什么会出现不一样的结果呢？ 经过分析后发现，张三所在学校的课程内容更丰富、更难，考试的难度也更高，因此，张三的成绩显得一般。而李四所在学校的课程内容相对简单，考试难度一般，因此，成绩容易提上去。这也说明，由于样本的背景不一致，数据之间是没法对比的。产生这个问题的实质是样本的数据难度缺失了。

7. 点估计忽略了样本基数

在使用点估计时，应当对于估计的基数有准确的认识。基数不准，则估计值出现偏差。

举例：

国外某一个国家，有一年发生肺结核的病例数达到了 1000 人，发生狂犬病的病例数达到了 30 人。于是得出结论，肺结核的传染率高于狂犬病。

对于这样的点估计，是发生了样本基数的错误。因为肺结核的治愈率很高，因此防控手段相对松，样本的基数大；而狂犬病的治愈率极低，因此防控手段极紧，样本的基数少。因此，基数不准确，估计值也不准确。

8. 数据的发展趋势不可持续

惯性思维是数据处理时经常遇到的现象。其实，在时间序列预测中，也有惯性思维的影子，使用惯性思维有一个前提就是数据发展趋势的相对稳定。如果遇到了外力的作用，那么数据的发展趋势将不可持续。

举例：
有一家外贸企业成立于 2000 年，它的八年营业额发展数据见表 5-27。

表 5-27 2001 至 2008 年营业额发展数据

年份	营业额/万元	营业额增长速率/%
2001 年	10 052	—
2002 年	11 710	16.5
2003 年	13 455	14.9
2004 年	15 513	15.3
2005 年	18 585	19.8
2006 年	21 742	17.0
2007 年	25 373	16.7
2008 年	29 788	17.4

从表 5-27 可以看出，该企业自成立以来，每年营业额的增长都保持在 14% 至 29% 之间，发展势头良好。为此，企业上下管理人员深受鼓舞，认为在后面的岁月中将继续快速发展，力争成为行业龙头。于是，企业开始了加速扩张战略，提出在未来十年以后，营业额要达到 9 至 10 亿。企业在大力招兵买马的同时，在全球各地纷纷布点。一时间，各种有利因素似乎都朝着理想的状态发展。

那么，后续的发展究竟如何？ 该公司后续八年的营业额数据见表 5-28。

表 5-28 2009 至 2016 年营业额发展数据

年份	营业额/万元	营业额增长速率/%
2009 年	8847	-70.3
2010 年	7776	-12.1
2011 年	8041	3.4
2012 年	8491	5.6
2013 年	8839	4.1
2014 年	8698	-1.6
2015 年	8889	2.2
2016 年	9217	3.7

从表 5-28 可以看出，企业在后续的八年发展极其艰苦。尤其是 2009 年，企业遭遇重大冲击，原因是当年受到了全球金融风暴的重大影响，外贸企业的发展举步维艰。此后，企业进入裁员、压缩开支、压紧规模、赔偿合同订单等困难周期，到 2015 年才开始慢慢恢复，但营业额还是没有冲破亿元大关。

就一个企业发展来说，初创时期各方面的条件良好，形成了一个快速增长期。但是企业的决策层对于风险的考虑有所欠缺，所以，当全球金融风暴开始来临，给企业的发展当头一棒。另外，从一个企业的增长规律来看，大多是前期快速发展，后期就比较缓慢，因为受到了各种因素的制约，产生了规模不经济的现象。因此，不能用前期的发展趋势来代替后续的发展趋势。导致这种惯性思维的问题在于管理层做决策时对于风险数据的缺失、经济周期波动数据的缺失。

9. 数据间的反比关系

从人的主观来说，都是向往着越来越好的工作与生活，因此，在很多工作、生活的场合，人们的预期值向好是常见的事情。而在现实的工作生活中，数据之间出现反比关系的现象是很多的。其中的原因，是受到了综合因素的约束。从数据的角度去分析，是缺少了约束数据。

举例：

人们在旅游度假的时候，希望获得高品质的服务，于是服务人员延长了服务时间。与之相反的是，单个服务人员的服务效率就显得低下。

有的人在上下班高峰时，为了节省体力和时间，就以车代步。但回家以后，出于锻炼身体的需要，就挤出时间进行各种健身活动。

5.5.3 感知错误

感知错误是指人们对于客观事物产生错误的认识和判断，这里主要指由于数据的缺失、分布不同而造成不同的认识和判断。

1. 语言文字的误导

语言文字是人们得知外界信息的重要渠道。一旦语言文字中遗漏了重要数据，就会对人们的认识和判断起着误导的作用。

举例：

某公司通过向外界宣称：公司今年的营业收入同比增长 300%。如果不针对这条信息进行甄别，那么公司的生产运营等的确是取得了不俗的成绩。随后，有行业知名人士对该企业去年的生产经营情况进行调查，发现该公司在过去的三年中由于受到各种因素的影响，营业收入每况愈下，去年几乎处于停顿的状态。今年有一家上市公司对其进行了收购，开展了内部整顿，并利用上市公司的资源积极拓展业务，因此，营业收入迅速上升。

在本例中，公司对外宣称的"营业收入同比增长 300%"是真实信息，但信息中缺少了去年的营业收入情况，实质是缺少了对比的真实数据。这种情况在现实中是很多的，还比如有 A 公司，在 B 市是属于比较大的企业，感觉非常好。但到了国际上一比较，发现 A 的业绩与行业的龙头企业还相差甚远。这也是 A 公司在 B 市缺少对比的结果。

2. 坐标轴变化

人们在感知某一个客观对象时，会对其形状、大小、分布、远近等进行判断。有时候，由于参照物的变化，人们的感知也会发生变化。

举例：

图 5-10 为曲轴振动值与垫片数量之间的散点图，左图与右图的散点分布是完全相同的，区别在于左图 Y 轴是从 0 开始计，而右图 Y 轴是从 7.0 开始计。

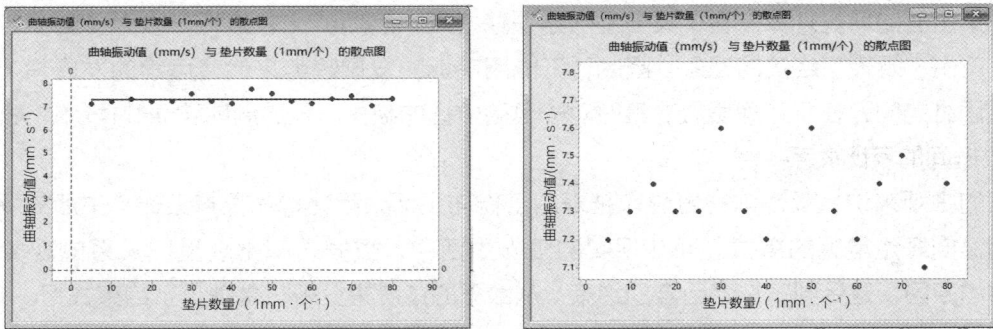

图 5-10　变换了坐标轴的散点图

从左图看，曲轴振动值与垫片数量之间的相关关系是强相关，且呈一条水平线。从右图看，曲轴振动值与垫片数量之间几乎不相关。导致这样的感知结论，是由于 Y 轴的参考线数据发生了变化。实质是相关系数数据的缺失，导致了判断不准确。

针对本例的数据，用 Minitab 计算相关系数，见图 5-11。

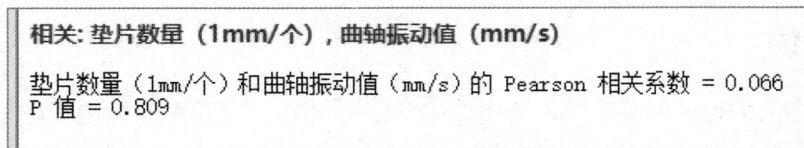

相关：垫片数量（1mm/个），曲轴振动值（mm/s）

垫片数量（1mm/个）和曲轴振动值（mm/s）的 Pearson 相关系数 = 0.066
P 值 = 0.809

图 5-11　曲轴振动值与垫片数量之间的相关系数

由图 5-11 可知，曲轴振动值与垫片数量之间的相关系数为 0.066，P 值为 0.809，结论为不相关。

5.5　Data Misuse

———

Data misuse is a special case of data loss. When data misuse occurs, it is equivalent to data loss indirectly. Data misuse can be divided into several types, such as numerical error, argument error, perception error.

5.5.1 Misuse of Numerical Values

Misuse of numerical value means that there is an error in the use of the number in the data, some of which occur because people do not understand the specific meaning of numbers, and some of which are due to wrong understanding of sample values.

1. Misuse of sample connotation

The connotation of the sample has changed, leading to great deviation in the conclusion of data analysis.

Case: Survey of per capita consumption level

A survey team needs to investigate the average consumption level of residents in a community. Therefore, during the day from Monday to Friday, the investigation team went to the vegetable market and shops near the community for investigation. From the analysis of the survey data, the survey team believes that the consumption level of the community is equivalent to 1000 *yuan* per capita per month. Another survey team also conducted a survey on the average consumption level of residents in the same community. Since many office workers start to consume after work, the survey team investigated not only the consumption during the day, but also the consumption at night. At the same time, the survey team also conducted a consumption survey of online shopping for office workers. From the analysis of the survey data, the survey team believes that the consumption level of the community is equivalent to 1200 *yuan* per capita per month. See Table 5-20 for details.

Table 5-20　Consumption level survey

Team No.	Investigation time	Investigation content	Per capita consumption results
Team 1	Daytime from Monday to Friday	Vegetable market, shops	1000 *yuan* per month
Team 2	Day and night from Monday to Friday	Vegetable market, shops, online shopping	1200 *yuan* per month

In this case, Team 1 obviously missed the evening consumption data of office workers and online consumption data.

2. Changes in statistical caliber

Statistical caliber refers to the relevant standards used in data statistics, which can be understood as the indicator system followed in data statistics. It includes indicators such as statistical method and scope. When the statistical caliber changes, the results of data analysis deviate, too.

Case: Statistics of the company's output value

There is a large group company with good development momentum, and its output value is increasing year by year. Now relevant personnel are required to count the output value of their subsidiaries as the basis for formulating a new round of development plan and relevant management measures. The scope of statistics is all subsidiaries. After analyzing the output value data of all 12 subsidiaries, it was found that the average annual output value per capita of the group was 4 million yuan, compared with 5 million yuan in the previous year, a year-on-year decrease of 20%. This conclusion seems to be contrary to the development trend of the enterprise, so what is the problem?

After in-depth analysis, the investigators found that the group had held shares in two loss-making enterprises in the market this year, and the development of these two enterprises was seriously delayed due to poor management in the early stage. This year, they have just been incorporated into the group company, and their business have improved slightly, but compared with other mature subsidiaries, it is far behind. Moreover, the number of employees in the two newly incorporated companies is relatively large. In this way, although the original 10 subsidiaries of the group company have developed well, the per capita output value has decreased due to the situation of new enterprises. See Table 5-21 for specific data.

Table 5-21 Output survey

Categroy	Subsidiaries surveyed	Situation of subsidiaries	Total number of people	Per-capita output value
This year	12 subsidiaries	10 companies are growing, and the new 2 companies are lagging behind	5800 people	4 million *yuan*
Last year	10 subsidiaries	10 companies are growing	5000 people	5 million *yuan*

In this case, the data of the two loss-making enterprises are actually missing in the output statistics of last year. If the output value of the two companies in the last year is also included in the statistical data of last year, the conclusion will be correct.

3. Ranking results with insignificant differences

Ranking is a process of classification, which arranges all kinds of unordered data according to certain requirements and rules, and turns the unordered data into an ordered sequence. Ranking is now applied in many industries.

Case: University ranking

There are many rankings of world universities, including The Times Higher Education World University Rankings and QS World University Rankings, etc. But their ranking results will be different. The ranking of world universities in 2022 is excerpted here. See Table 5-22.

Table 5-22　2022 World University Ranking

Ranking	Times Higher Education World University Reputation Rankings	Times Higher Education World University Rankings	QS World University Rankings
1	Harvard University	Oxford University	Massachusetts Institute of Technology
2	Massachusetts Institute of Technology	California Institute of Technology	Oxford University
3	Stanford University	Harvard University	Stanford University
4	Oxford University	Stanford University	Cambridge University
5	Cambridge University	Cambridge University	Harvard University
6	University of California, Berkeley	Massachusetts Institute of Technology	California Institute of Technology
7	Princeton University	Princeton University	Imperial College London
8	Yale University	University of California, Berkeley	Zurich Federal Institute of Technology
9	Tsinghua University	Yale University	University College London
10	The University of Tokyo	University of Chicago	University of Chicago

From the perspective of the three rankings, the order is different, which is determined by the logic of the school ranking, the internal relationship, and the significance of the index differences. Some rankings have more indicators, and some have less indicators, so there are differences in rankings.

The Times Higher Education World University Reputation Rankings focus more on the subjective feelings of industry insiders, which is obtained through the results of the academic questionnaire made by invitation. The reputation questionnaire is only for scholars who have experience and published papers, and give opinions on the research and teaching of their disciplines and familiar institutions. The main indicators in the questionnaire are scholars, research and teaching, with a certain degree of subjectivity.

The Times Higher Education World University Rankings take the achievement of the 17 United Nations Sustainable Development Goals as the main basis for its assessment. The areas involved include research, external activities, management, teaching and other aspects. The assessment is basically objective data.

The QS World University Rankings focus on six indicators, namely: academic reputation which accounts for 40%, employer reputation which accounts for 10%, the number of citations of faculty papers which accounts for 20%, the proportion of teachers and students which accounts for 20%, the proportion of international teachers which accounts for 5%, and the proportion of international students which accounts for 5%. The ranking takes global reputation, academic quality and degree of internationalization as three main indicators.

It can be seen from this that different evaluation criteria have their own evaluation methods and basis, and the indicators of other ranking methods are ignored to varying degrees, so the ranking results are different.

4. The average is not representative

Average is the mean value between data, which represents the concentration trend of data. It is intuitive and concise to use the average number to represent a group of data. But sometimes, the average will lose its representativeness as an indicator.

Case: Urban temperature

Chengdu is located in the Sichuan Basin, with an annual average temperature of 17.9 ℃. Kunming is located in Yunnan-Guizhou Plateau, with an annual average temperature of 17.1 ℃. For this reason, some people have concluded that the temperatures of the two cities are basically the same.

What is the real result? Compare the monthly average temperature data of the two cities. See Table 5-23 for data of Chengdu and Table 5-24 for data of Kunming.

Table 5-23　Temperature in Chengdu　　　　　Unit: ℃

Month	Jan	Feb	Mar	Apr	May	Jun	Jul	Aug	Sep	Oct	Nov	Dec	Annually average
Highest	22	25	35	34	31	27	24	22	11	11	12	15	17.9
Lowest	14	18	25	23	21	16	13	11	4	4	4	8	

Table 5-24　Temperature in Kunming　　　　　Unit: ℃

Month	Jan	Feb	Mar	Apr	May	Jun	Jul	Aug	Sep	Oct	Nov	Dec	Annually average
Highest	17	20	22	25	26	26	25	26	25	22	20	16	17.1
Lowest	3	5	9	12	15	18	18	18	17	13	8	4	

It can be seen from the above two tables that the temperature in Chengdu is significantly higher than that in Kunming in March, April and May. In September, October and November, the temperature in Chengdu is significantly lower than that in Kunming. Obviously, the conclusion that the temperatures of the two cities are basically the same ignores the data of specific months. If we analyze the structure, distribution and abnormal values of the temperature data of the two cities in spring and autumn, we can find that the temperature of Chengdu fluctuates greatly throughout the year, while the temperature of Kunming fluctuates slightly throughout the year. In fact, the standard deviation of the temperature data is different. Therefore, the structure of data cannot be omitted, and the average number can easily be used to represent the characteristics of all data.

5. No data verification

Non-verification of data often occurs because of the lack of real data. The situations of non-verifying data mainly include:

(1) Hearsay. For example, some people immediately bought a large number of shares of a listed company in the secondary market after hearing that its performance was very good on the road. However, due to the untrue information, the stock prices of listed companies have been kept at the original prices, with slight fluctuations.

(2) Direct reference. For example, the person in charge of a department of the head office directly quoted the production data submitted by the relevant departments of the subordinate

enterprises without verification, which made the enterprise's decision deviate.

(3) Proofreading is not serious. For example, when a technician filled in the data manually, he filled in the number "63", but the technician of the next process mistakenly read the number as "68", and the auxiliary technician did not proofread it, and directly made the relevant calculation and made an error.

(4) Wrong judgment. For example, an old man has abdominal discomfort. Because the old man often has stomach acid, his family thought it was stomach disease, so they simply gave him some stomach medicine, but it didn't work. Then, the family sent the old man to the hospital for treatment, and found that it was mild myocardial infarction, and its symptom is indeed abdominal discomfort.

5.5.2 The Error of Argument Logic

In the process of various data references, some data have strict logical relations. If the logic is wrong, it is equivalent to the data missing in the intermediate process, resulting in logical errors between data.

1. Wrong operation concept

Wrong operation concept refers to that people do not understand the specific meaning of data when performing data operations, and use the correct way to get the wrong conclusion. Such a kind of error is equivalent to the conceptual error of data and the lack of calculation process.

Case:

A person was passing an intersection and found a display screen on the roadside, which was connected with a decibel meter. When the vehicle passes by, the display constantly changes the decibel value. The person saw that the reading on the display screen was 60 decibels in a moment and 65 decibels awhile. He told the child next to him that the vehicle noise was well controlled. 65 decibels was only 5 decibels more than 60 decibels. Is this conclusion correct?

Take a look at the definition of decibel: decibel is a unit of measurement that measures the proportion of the quantity of two identical units. It can be used to measure the sound intensity. Its calculation process is in the form of logarithm. The decibel calculation formula of sound pressure level is

$$N_{db} = 20 \times \lg \frac{p_i}{p_0}$$

In the formula, $p_0 = 0.000\ 02$ Pascal (Pa). It is the reference sound pressure; p_i is the measured sound pressure.

Obviously, the difference between 65 dB and 60 dB is not 5 dB, but several times.

In real work and life, there are many examples that cannot be simply calculated. For example, two cups of warm water at 20 ℃ will not reach 40 ℃ when we put them together. Therefore, in order to avoid such mistakes, it is necessary to correctly understand the conceptual composition and operation process of data, and not skip the link of correct operation to directly draw conclusions.

2. Inductive logic error

There are many methods of induction, no matter which method, induction must have rigorous logical relationship analysis. Sometimes, due to problems in the representativeness of samples and the logic of induction, the conclusion of induction is also wrong.

Case:

A research company was commissioned to investigate the use of the internet by households in a city. The researchers of the company conducted a survey of 100,000 families through the internet. After receiving the reply, the researchers concluded that all 100,000 families have access to the internet.

In this case, the survey method of the investigators is limited to the network survey, resulting in the loss of sample data. The flaw of this research method is logic error. It is similar to "Please raise your hand if you haven't come yet" that the internet has been used by the way of network survey and the reply received from the network. Therefore, it is the basic work of induction to grasp the correct logic and eliminate the lack of samples.

3. Data is not representative

The data used in the demonstration process must be representative. Sometimes, the collected data is limited by various conditions, which makes the data not widely representative, leading to wrong conclusions.

Case:

A research group carried out global climate change surveys, one of which is whether the global climate is warming. The researchers were divided into four groups. The first group investigated the fluctuation of global average temperature from 1982 to 1992, and found that the temperature increased by about 0.2 ℃. The second group investigated the fluctuation of global average temperature from 1992 to 2002, and found that the temperature changed little. The third group investigated the fluctuation of global average temperature from 2002 to 2012, and found that the temperature also changed little, with an increase of approximately 0.3 ℃. The fourth group investigated the fluctuation of global average temperature from 2012 to 2022, and found that the rise of temperature was 0.3 ℃. In this way, the four survey groups concluded respectively that the average temperature of the world has not risen rapidly in the past decade.

Later, the person in charge extended the research period to 160 years, from 1862 to 2022, and found that the global average temperature fluctuated significantly, from − 0.1 ℃ to 1.3 ℃, and the average temperature in 2022 alone was 1.15 ± 0.13 ℃ higher than the average temperature in 1850−1900.

Then why did the temperature rise in the past ten years not much? Because the survey period of the sample is too short. If the research period is extended to 40 years from 1982 to 2022, it is found that the temperature has increased by about 0.8 ℃. This also shows that the short survey period of the sample will lead to the relative lack of sample data, so the sample is lack of representativeness.

4. Change of numerator or denominator

In logical argumentation, the change of numerator and denominator will cause great changes in the operation process, resulting in the loss of data operation process.

Case:

A clothing store did a sales promotion. In the first week, the store launched a "50% discount" strategy, resulting in a boost in turnover. The next week, the store launched the "buy one and get one free" strategy, resulting in a rapid increase in turnover. So, is there any difference between "50% discount" and "buy one and get one free"? There are obvious differences.

The "50% discount" is equivalent to a direct reduction of half of the numerator of the price. For example, if the unit price was originally 100 *yuan*/piece, then it would be 50 *yuan*/piece after a 50% discount. Its turnover increases with the sales of each piece of clothing.

"Buy one and get one free" means that the price of clothes remains unchanged, but the denominator of the price has doubled. For example, if the unit price was 100 *yuan* per piece, then "buy one and get one free" will become 100 *yuan*/piece. Its turnover increases with the sales of every 2 pieces of clothing.

Assuming that the number of times customers buy is constant, the sales rate of "buy one and get one free" is twice as high as the sales rate of "50% discount". To understand the logical relationship, it is necessary to make the hidden data explicit.

5. Contingency replaces necessity

The data collected in a single time is accidental, while the data collected in multiple times will show certain rules, and more data will show the theorem of large numbers. In fact, some people will use contingency to replace inevitability. In fact, the sample data is not enough.

Case：

During the 2010 World Cup in South Africa, Paul the Octopus was really popular because his prediction about matches was surprisingly accurate. Paul's prediction method was that the staff put the same food with different flags in the aquarium before the two teams competed, and then let Paul choose one flag. See Table 5-25 for the prediction results of Paul the octopus.

Table 5-25   Paul the Octopus's prediction and the real match results

Match	Team in the match	Paul's prediction	Real results
First game of group match	Germany vs Australia	Germany win	Germany won 4-0
Second game of the group match	Germany vs Serbia	Germany lost	Germany lost 0-1
Last game of the group match	Germany vs Ghana	Germany win	Germany won1-0
1/8 finals	Germany vs England	Germany win	Germany won 4-1
1/4 finals	Germany vs Argentina	Germany win	Germany won 4-0
Semi-final	Germany vs Spain	Germany lost	Germany lost 0-1

Does such high prediction accuracy really represent Paul's prediction ability? Obviously not. In fact, there are some mysteries in the sample data when forecasting, that is, the color of the national flag. Because the German flag is composed of black, red and yellow, and red and yellow are the food colors that octopus like very much. Therefore, it is not so much Paul's prediction as Paul's choice of food color.

6. No comparability between data

Comparing data is a frequently used method. However, there are preconditions for the comparability of data, such as the consistency of data units, the consistency of data acquisition time, and the consistency of data internal logic. Sometimes there is no comparability between data.

Case：

Zhang San and Li Si were high school classmates, and Zhang San's academic performance had always been better. After the college entrance examination, Zhang San entered a university with high scores; Li Si's score was inferior and he entered another university. Although the two students entered different schools, they had the same majors. During college, Zhang San still studied hard, while Li Si was relatively relaxed.

Near college graduation, Zhang San and Li Si met. After exchange, they found that Li Si's academic performance was much better than Zhang San's. See Table 5-26 for some of their academic achievements.

Table 5-26　Comparison of academic achievements between two people（some subjects）

Subject	Zhang San's achievements	Li Si's achievements
Advanced mathematics	76	95
Theoretical mechanics	80	92
Analytical chemistry	83	98
Material mechanics	74	89
Structural mechanics	81	96

Zhang San's academic background is relatively good, and he studied very hard. Generally, it seems that Zhang San should have better academic performance than Li Si, so why would there be different results? After analysis, it is found that the curriculum content of Zhang San's school is richer and more difficult, and the difficulty of the exam is also very high. Therefore, Zhang San's performance appears to be average. However, the course content of Li Si's school is relatively simple, and the exam is not so difficult, so the grades are easy to improve. This also shows that the data cannot be compared because of the inconsistent background of the samples. The essence of this problem is that the difficulty of sample data is missing.

7. Point estimation ignores sample cardinality

When using point estimation, we should have an accurate understanding of the estimated base. If the base is not correct, the estimated value will deviate.

Case：

In a certain country, the number of tuberculosis cases reached 1000 in one year, and the number of rabies cases reached 30. It was concluded that the infection rate of tuberculosis was higher than that of rabies.

For such point estimation, there is a sample cardinality error. Because the cure rate of tuberculosis is high, the prevention and control measures are relatively loose, and the sample cardinality is large; The cure rate of rabies is very low, so the means of prevention and control are extremely tight, and the cardinal number of samples is small. Therefore, the cardinal number is not accurate and the estimated value is not accurate.

8. The development trend of data is not sustainable

Inertial thinking is a common phenomenon in data processing. In fact, in time series prediction, there is also the shadow of inertia thinking. Therefore, inertia thinking cannot be said to be bad. A prerequisite for using inertia thinking is the relative stability of data development trend. If external forces are encountered, the development trend of data will not be sustainable.

Case:

A foreign trade enterprise was established in 2000. See Table 5-27 for its eight-year turn-over development data.

Table 5-27 Turnover development data from 2001 to 2008

Year	Turnover (1000 *yuan*)	Turnover growth rate/%
2001	100 520	—
2002	117 100	16.5
2003	134 550	14.9
2004	155 130	15.3
2005	185 850	19.8
2006	217 420	17.0
2007	253 730	16.7
2008	297 880	17.4

It can be seen from Table 5-27 that since its establishment, the annual turnover growth of the enterprise has maintained between 14% and 29%, with a good development momentum. To this end, the company's top and bottom managers are deeply encouraged and believe that they will continue to develop rapidly in the coming years and strive to become the industry leader. Therefore, the enterprise began to accelerate its expansion strategy and proposed that the turnover should reach 900 to 1 billion in the next ten years. While enterprises are vigorously recruiting, they are distributing their products all over the world. For a time, all kinds of favorable factors seemed to develop towards the ideal state.

Then, what is the subsequent development? See Table 5-28 for the turnover data of the company in the following eight years.

Table 5-28 Turnover development data from 2009 to 2016

Year	Turnover (1000 *yuan*)	Turnover growth rate/%
2009	88 470	−70.3
2010	77 760	−12.1
2011	80 410	3.4
2012	84 910	5.6
2013	88 390	4.1
2014	86 980	−1.6
2015	88 890	2.2
2016	92 170	3.7

It can be seen from Table 5-28 that the development of enterprises in the following eight years is extremely difficult. In particular, in 2009, enterprises suffered a major impact because of the major impact of the global financial crisis that year, and the development of foreign trade enterprises was difficult. Since then, enterprises have entered a difficult cycle of layoffs, spending reduction, scale reduction, compensation for contracts and orders, and only began to recover slowly in 2015, but the turnover has not broken the 100-million-yuan mark.

As far as the development of an enterprise is concerned, the conditions in all aspects of the start-up period are good, forming a period of rapid growth. However, the decision-making level of the enterprise lacks consideration of risk. Therefore, when the global financial crisis begins to come, it will give a blow to the development of the enterprise. In addition, from the perspective of the growth law of an enterprise, most of them developed rapidly in the early stage, but were relatively slow in the later stage. Because of the constraints of various factors, the phenomenon of diseconomies of scale occurred. Therefore, the development trend in the early stage cannot replace the subsequent development trend. The problems that lead to this kind of inertia thinking are the lack of risk data and the lack of economic cycle fluctuation data when the management makes decisions.

9. Inverse relationship between data

From the subjective point of view, people are yearning for better and better work and life. Therefore, in many work and life occasions, it is common for people's expectations to improve. In real work and life, there are many phenomena of inverse relationship between data. The reason is restricted by comprehensive factors. From the perspective of data, it is lack of constraint data.

Case:

When people are on holiday, they want to get high-quality service, so the service staff extended the service time. On the contrary, the service efficiency of individual service personnel is low.

In order to save energy and time, some people use cars instead of walking during rush hours. But when some people go home, they squeeze out time for various fitness activities for the purpose of exercising.

5.5.3 Perception Error

Perception error refers to people's wrong understanding and judgment of objective things. Here, it mainly refers to different understanding and judgment caused by the lack and distribution of data.

1. Misleading language

Language is an important channel for people to learn about the outside world. Once important data is omitted from the language, it will mislead people's understanding and judgment.

Case:

A company announced to the outside world that its operating revenue this year increased by 300% year on year. If this information is not screened, the company's production and operation have indeed achieved good results. Subsequently, a well-known person in the industry investigated the production and operation of the company last year, and found that the company's operating income has been deteriorating in the past three years due to various factors, and was almost at a standstill last year. This year, a listed company acquired it, carried out internal rectification, and actively expanded its business using the resources of the listed company. Therefore, the operating income rose rapidly.

In this case, the company's "300% year-on-year increase in operating revenue" is true information, but the information lacks the operating revenue of last year, which is essentially the lack of real data for comparison. There are many such situations in reality. For example, company A is a relatively large enterprise in city B, which feels very good. However, in the international comparison, it is found that A's performance is far from that of the leading enterprises in the industry. This is also the result of company A's lack of comparison in city B.

2. Coordinate axis change

When people perceive an objective object, they will judge its shape, size, distribution, distance, etc. Sometimes, people's perception will also change due to the change of reference objects.

Case：

Figure 5－10 shows the scatter diagram between the crankshaft vibration value and the number of shims. The scatter distribution in the left and right figures is the same. The difference is that the Y-axis in the left figure is calculated from 0, while the Y-axis in the right figure is calculated from 7.0.

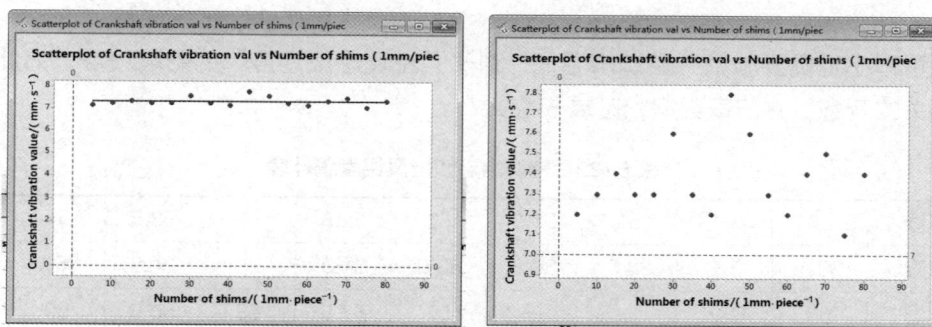

Figure 5－10　Scatter diagram with transformed coordinate axis

The left figure shows that the correlation between the crankshaft vibration value and the number of shims is strong, illustrated by a horizontal line. From the right figure, there is almost no correlation between the crankshaft vibration value and the number of shims. The reason for this perception conclusion is that the reference line data of the Y-axis has changed. The essence is the lack of correlation coefficient data, resulting in inaccurate judgment.

For the data of this case, use Minitab to calculate the correlation coefficient, as shown in Figure 5－11.

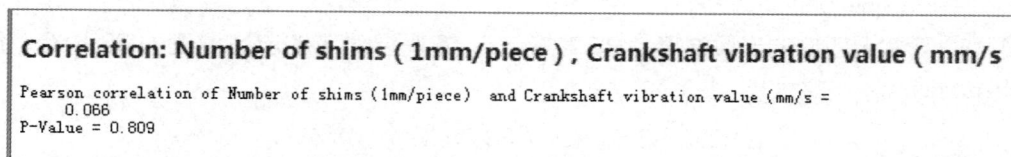

Correlation: Number of shims (1mm/piece) , Crankshaft vibration value (mm/s

Pearson correlation of Number of shims (1mm/piece) and Crankshaft vibration value (mm/s = 0.066
P-Value = 0.809

Figure 5－11　Correlation coefficient between crankshaft vibration value and number of shims

It can be seen from Figure 5－11 that the correlation coefficient between the crankshaft vibration value and the number of shims is 0.066, and the P value is 0.809. The conclusion is irrelevant.

5.6 缺失数据的弥补

――

数据集合中有数据缺失的情况发生，对于该数据集的分析肯定会带来影响，有的影响甚至是颠覆性的。这里所指的"缺失数据的弥补"有一个大前提，那就是数据的产生处于稳定的状况中，人们能够根据数据产生的过程，预测其接下来的趋势。在这样的前提下，如果发生有中间少量数据的缺失，就不至于产生特别大的影响。

5.6.1 均值数据的插补

均值插补是根据缺失值所在位置，取相邻两个值的均值来进行插补的方法。在插补法中，利用中位数、众数等进行插补是相同的原理。

举例：

某质量管理小组需要验证"空调水温低"对于"空调出风温度"的影响程度。为此，小组到现场进行相关的测量，并记录数据，见表 5-29。

表 5-29　空调水温和出风温度统计表 单位：℃

空调水温	50	—	51	51.5	52	52.5	—	53.5	54	54.5
出风温度	33.1	—	33.6	33.8	33.9	34.3	—	34.9	35.1	35.4
空调水温	55	55.5	—	—	57	57.5	58	58.5	59	59.5
出风温度	35.6	35.8	—	—	36.4	36.6	36.9	37.3	37.7	37.8
空调水温	60	60.5	61	61.5	—	—		63.5	64	64.5
出风温度	37.9	38.0	38.4	38.8				40	40.1	40.3

从表 5-29 可以看出，小组在收集数据的过程中，有多处数据缺失。为此，小组决定采用均值插补的方式，对数据进行估计。

（1）小组发现，对于水温的插补比较简单，相邻数据都相差 0.5 ℃，则 50 ℃与 51 ℃之间插入 50.5 ℃，其余水温数据以此类推。

已知：50 ℃水温对应的出风温度为 33.1 ℃，51 ℃水温对应的出风温度为 33.6 ℃，那么，50.5 ℃水温对应的出风温度为

$$（33.1+33.6）/2 = 33.35 ℃ ≈ 33.4 ℃$$

同理，53 ℃水温对应的出风温度，经过计算为 34.6 ℃。

（2）在水温 55.5 ℃与 57 ℃之间，插入水温 56 ℃、56.5 ℃。

已知：55.5 ℃水温对应的出风温度为 35.8 ℃，57 ℃水温对应的出风温度为 36.4 ℃，那么，对应的出风温度需要插入 2 个值。

第一个值是 56 ℃水温对应的出风温度，为

$$35.8 + （36.4 - 35.8）/3 = 36 ℃$$

第二个值是 56.5 ℃水温对应的出风温度，为

$$36.4 - （36.4 - 35.8）/3 = 36.2 ℃$$

（3）在水温 61.5 ℃与 63.5 ℃之间，插入水温 62 ℃、62.5 ℃、63 ℃。

已知：61.5 ℃水温对应的出风温度为 38.8 ℃，63.5 ℃水温对应的出风温度为 40 ℃，那么，对应的出风温度需要插入 2 个值。

第一个值是 62 ℃水温对应的出风温度，为

$$38.8 + （40 - 38.8）/4 = 39.1 ℃$$

第二个值是 62.5 ℃水温对应的出风温度，为

$$39.1 + （40 - 38.8）/4 = 39.4 ℃$$

第三个值是 63 ℃水温对应的出风温度，为

$$39.4 + （40 - 38.8）/4 = 39.7 ℃$$

这样，经过整理，小组得到了均值插补后的统计数据，见表 5-30。

表 5-30　均值插补后的统计数据　　　　　　　　　　　　单位：℃

空调水温	50	50.5	51	51.5	52	52.5	53	53.5	54	54.5
出风温度	33.1	33.4	33.6	33.8	33.9	34.3	34.6	34.9	35.1	35.4
空调水温	55	55.5	56	56.5	57	57.5	58	58.5	59	59.5
出风温度	35.6	35.8	36	36.2	36.4	36.6	36.9	37.3	37.7	37.8
空调水温	60	60.5	61	61.5	62	62.5	63	63.5	64	64.5
出风温度	37.9	38.0	38.4	38.8	39.1	39.4	39.7	40	40.1	40.3

5.6.2 直接删除缺失数据

直接删除法就是把缺失的数据予以忽略、删除，然后直接引用其他数据进行分析。

举例：

继续上一个例子，根据表 5-29 中的数据，把其中的缺失值直接删去，得到表 5-31。也就是说，把表 5-31 的数据视为连续测量得到的数据，不再做任何的插补。

表 5-31　删除空缺值后的统计数据　　　　　　　　　　　单位：℃

空调水温	50	51	51.5	52	52.5	53.5	54	54.5
出风温度	33.1	33.6	33.8	33.9	34.3	34.9	35.1	35.4
空调水温	55	55.5	57	57.5	58	58.5	59	59.5
出风温度	35.6	35.8	36.4	36.6	36.9	37.3	37.7	37.8
空调水温	60	60.5	61	61.5	63.5	64	64.5	
出风温度	37.9	38.0	38.4	38.8	40	40.1	40.3	

5.6.3 建立回归方程弥补

建立回归方程法，是先根据已有数据建立回归方程，再根据回归方程来计算缺失之处的数据以作弥补。

继续上一个例子。先根据表 5–29 的数据，建立回归方程。

把表 5–29 的数据输入 Minitab，运用回归分析，会话框出现相关数据，见图 5–12。

系数

项	系数	系数标准误	T 值	P 值	方差膨胀因子
常量	8.182	0.351	23.29	0.000	
空调水温（摄氏度）	0.49736	0.00613	81.11	0.000	1.00

回归方程

出风温度（摄氏度） = 8.182 + 0.49736 空调水温（摄氏度）

异常观测值的拟合和诊断

观测值	出风温度（摄氏度）	拟合值	残差	标准残差	
18	38.0000	38.2715	-0.2715	-2.26	R
21	40.0000	39.7636	0.2364	2.05	R

R 残差大

图 5–12 回归分析输出数据

根据图 5–12 可知，出风温度（ y ）与空调水温（ x ）之间可以建立的线性回归方程为：

$$y = 8.182 + 0.497\ 36\ x$$

下一步，根据该回归方程可以进行相应数据的测算。

当空调水温为 50.5 ℃时，出风口温度为 8.182+0.497 36 × 50.5 = 33.3 ℃；

当空调水温为 53 ℃时，出风口温度为 8.182+0.497 36 × 53 = 34.5 ℃；

当空调水温为 56 ℃时，出风口温度为 8.182+0.497 36 × 56 = 36.0 ℃；

当空调水温为 56.5 ℃时，出风口温度为 8.182+0.497 36 × 56.5 = 36.3 ℃；

当空调水温为 62 ℃时，出风口温度为 8.182+0.497 36 × 62 = 39.0 ℃；

当空调水温为 62.5 ℃时，出风口温度为 8.182+0.497 36 × 62.5 = 39.3 ℃；

当空调水温为 63 ℃时，出风口温度为 8.182+0.497 36 × 63 = 39.5 ℃。

5.6.4 数据替换

1. 用固定值替换

用固定值替换是指当数据发生缺失以后，缺失之处的数据用其他固定值来替换。这种方法通常适用于数据变化幅度不大的情形。用固定值替换会导致数据间的方差放大。

举例：

某调研小组对某建筑公司工人的上年度工资进行调研，部分调研数据见表 5–32。

表 5–32 部分工人的上年度工资

序号	工人姓名	年度工资/元
1	王 ××	87 600
2	施 ××	92 300

序号	工人姓名	年度工资/元
3	马××	—
4	张××	79 500
5	秦××	97 600
6	陆××	69 700
7	焦××	—
8	干××	—
9	沈××	75 300
10	全××	93 400

从表 5-32 得知，工人马××、焦××、干××的上年度工资未能调查到。为了使后续分析工作顺利开展，调研人员决定用固定值来替换。他们查阅了上年度建筑行业工人的年度平均工资情况，得到数据为 85 200 元。于是，工人马××、焦××、干××的上年度工资全部用 85 200 元替换。

2. 临近值替换

用临近值替换，就是用缺失值最近的数值来替换，并根据需要，取最小的临近值或最大的临近值。

举例：

继续上一个例子，由表 5-32 得知，工人马××、焦××、干××的上年度工资未能调查到。于是，调研人员决定用临近值来替换。出于谨慎的考虑，调研人员决定采用较小的临近值来替换缺失值。马××的上年度工资取 min(92 300，79 500)，取值 79 500 元。焦××的上年度工资取 min(69 700，75 300)，取值 69 700 元。这样，就得到了工人的上年度工资表，见表 5-33。

表 5-33　工人的上年度工资

序号	工人姓名	年度工资/元
1	王××	87 600
2	施××	92 300
3	马××	79 500
4	张××	79 500
5	秦××	97 600
6	陆××	69 700
7	焦××	69 700
8	干××	69 700
9	沈××	75 300
10	全××	93 400

5.6.5 极大似然估计法

极大似然估计法是指在缺失类型为随机缺失的条件下，假设模型对于完整的样本是正确、适用的，则通过观测数据的边际分布对未知参数进行极大似然估计，这种方法可以称为忽略缺失值的极大似然估计。

对于极大似然的参数估计实际中常采用的计算方法是期望值最大化（Expectation Maximization，简称 EM）。EM 算法最早由邓普斯特（Dempster）于 1977 年提出，它是通过迭代计算来进行极大似然估计或计算后验概率分布。它的使用前提是样本容量大，并且有效样本的数量能够保证估计值是渐近无偏并服从正态分布。

EM 的思路是：先用估计值来替换缺失值，接着对新构成的完整数据进行参数估计，然后根据参数估计值再反过来估计缺失值。

EM 算法有两个步骤，步骤一是求期望值，即在给定观测数据的条件下，求出缺失值的条件期望，并用计算出的条件期望对缺失数据进行插补。步骤二是进行极大似然估计，对插补后的完整数据集进行参数的极大似然估计。

由于 EM 人工计算工作量相当大，现在通常使用计算机软件来进行。比如 SPSS 软件中，通过选择"缺失值分析"的相关内容，其中就有 EM 的计算。

5.6.6 不做处理

针对缺失数据进行插补等处理，是把主观估计值予以弥补的权宜之计，不一定完全符合客观事实。因为当主观插补数据进入到数据集后，实际上是改变了原有的数据集内涵，有时候分析结果也未必令人满意。

在这样的情况下，对于缺失数据不做任何处理也是一个选项，那么分析人员可以在掌握真实数据的情况下，对于数据进行相应的分析。

5.6.7 电脑缺失数据

1. 常见的电脑数据缺失

在电脑使用过程中，经常会遇到电脑逻辑问题导致数据缺失。这样的情形有以下七种：

（1）在电脑上误删除数据；

（2）格式化，即对 U 盘、内存卡等进行了格式化，导致数据缺失；

（3）复制粘贴时出现问题，没有把数据真正复制粘贴好；

（4）U 盘、硬盘的分区丢失；

（5）重装系统时，C 盘数据丢失；

（6）突然断电、强行拔出电源等行为，且没有 UPS 电源，导致数据缺失；

（7）遭遇计算机病毒，使数据缺失。

针对这些问题，可以使用数据恢复工具来帮助数据恢复，包括：删除恢复、格式化恢复、硬盘恢复、移动存储设备恢复、深度恢复等。

2. 电脑数据的不可逆

在电脑操作中，有时候不慎使数据不可逆。这样的情形有以下四种：

（1）电脑硬件损坏。电脑硬件是重要的数据存储设备，一旦损坏严重，通常很难恢复。

（2）遭遇病毒。有些电脑病毒专门破坏文件或截取信息，平时必须做好病毒防范工作。

（3）物理删除。在电脑操作中，如果是误删除、格式化等是属于逻辑删除，数据有恢复的可能。而物理删除时把数据从存储介质上抹去，则肯定不可逆。

（4）数据覆盖。在电脑操作中，数据一旦被覆盖，则原有数据无法找回。比如数据在分区中丢失，还有可能找回。如果丢失数据的分区继续使用，则写入新数据后，原来丢失的数据再也无法找回。

5.6   Remedy for Missing Data

_____

If there is missing data in the data set, the analysis of the data set will certainly have an impact, some of which may even be subversive. There is a major premise for the "compensation of missing data" mentioned here, that is, the data generation is in a stable state, and people can predict the next trend of the data according to the process of data generation. Under this premise, if there is a small amount of data missing in the middle, it will not have a particularly big impact.

5.6.1 Interpolation of Mean Data

Mean value interpolation is a method of taking the mean value of two adjacent values according to the location of the missing value to carry out difference interpolation. In the interpolation method, it is the same principle to use the median, mode, etc. for interpolation.

Case:

A quality management team needs to verify the impact of "low air conditioning water temperature" on "air conditioning outlet temperature". To this end, the team went to the site to carry out relevant measurements and record the data, as shown in Table 5-29.

Table 5-29   Statistics of air conditioning water temperature and air outlet temperature   Unit: ℃

Air conditioning water temperature	50	—	51	51.5	52	52.5	—	53.5	54	54.5
Air outlet temperature	33.1	—	33.6	33.8	33.9	34.3	—	34.9	35.1	35.4
Air conditioning water temperature	55	55.5	—	—	57	57.5	58	58.5	59	59.5
Air outlet temperature	35.6	35.8	—	—	36.4	36.6	36.9	37.3	37.7	37.8
Air conditioning water temperature	60	60.5	61	61.5	—	—	—	63.5	64	64.5
Air outlet temperature	37.9	38.0	38.4	38.8	—	—	—	40	40.1	40.3

It can be seen from Table 5-29 that there are many missing data in the process of data collection. To this end, the team decided to use the mean interpolation method to estimate the data.

(1) The team found that the interpolation of water temperature is relatively simple, and the difference between adjacent data is 0.5 ℃, so 50.5 ℃ is inserted between 50 ℃ and 51 ℃, and the rest of the water temperature data is the same.

It is known that the outlet temperature corresponding to 50 ℃ water temperature is 33.1 ℃, and the outlet temperature corresponding to 51 ℃ water temperature is 33.6 ℃, so the outlet temperature corresponding to 50.5 ℃ water temperature is：

$$(33.1+33.6)/2 = 33.35 ℃ ≈ 33.4 ℃$$

Similarly, the outlet temperature corresponding to 53 ℃ water temperature is calculated as 34.6 ℃.

（2）Insert the water temperature of 56 ℃ and 56.5 ℃ between 55.5 ℃ and 57 ℃. It is known that the outlet temperature corresponding to 55.5 ℃ water temperature is 35.8℃, and the outlet temperature corresponding to 57 ℃ water temperature is 36.4 ℃. Therefore, two values need to be inserted into the corresponding outlet temperature.

The first value is the outlet temperature corresponding to 56 ℃ water temperature, which is：

$$35.8 + (36.4 - 35.8)/3 = 36 ℃$$

The second value is the outlet temperature corresponding to 56.5 ℃ water temperature, which is：

$$36.4 - (36.4 - 35.8)/3 = 36.2 ℃$$

（3）Insert water temperature of 62 ℃, 62.5 ℃ and 63 ℃ between 61.5 ℃ and 63.5 ℃.It is known that the outlet temperature corresponding to 61.5 ℃ water temperature is 38.8 ℃, and the outlet temperature corresponding to 63.5 ℃ water temperature is 40 ℃, so two values need to be inserted into the corresponding outlet temperature.

The first value is the outlet temperature corresponding to the water temperature of 62 ℃, which is：

$$38.8 + (40 - 38.8)/4 = 39.1 ℃$$

The second value is the outlet temperature corresponding to the water temperature of 62.5 ℃, which is：

$$39.1 + (40 - 38.8)/4 = 39.4 ℃$$

The third value is the outlet temperature corresponding to the water temperature of 63 ℃, which is：

$$39.4 + (40 - 38.8)/4 = 39.7 ℃$$

In this way, after sorting, the team obtained the statistical data after mean interpolation, as shown in Table 5-30.

Table 5-30　Statistical data after mean interpolation　　　　　　　Unit：℃

Air conditioning water temperature	50	50.5	51	51.5	52	52.5	53	53.5	54	54.5
Air outlet temperature	33.1	33.4	33.6	33.8	33.9	34.3	34.6	34.9	35.1	35.4
Air conditioning water temperature	55	55.5	56	56.5	57	57.5	58	58.5	59	59.5
Air outlet temperature	35.6	35.8	36	36.2	36.4	36.6	36.9	37.3	37.7	37.8
Air conditioning water temperature	60	60.5	61	61.5	62	62.5	63	63.5	64	64.5

Air conditioning water temperature	50	50.5	51	51.5	52	52.5	53	53.5	54	54.5
Air outlet temperature	37.9	38.0	38.4	38.8	39.1	39.4	39.7	40	40.1	40.3

5.6.2 Delete missing data directly

The direct deletion method is to ignore and delete the missing data, and then directly reference other data for analysis.

Case:

Continue with the case in 5.6.1. According to the data in Table 5-29, delete the missing values directly to get Table 5-31. In other words, the data in Table 5-31 is regarded as the data obtained from continuous measurement, and no interpolation is required.

Table 5-31　Statistics after deleting vacancy values　　　　　Unit: ℃

Air conditioning water temperature	50	51	51.5	52	52.5	53.5	54	54.5
Air outlet temperature	33.1	33.6	33.8	33.9	34.3	34.9	35.1	35.4
Air conditioning water temperature	55	55.5	57	57.5	58	58.5	59	59.5
Air outlet temperature	35.6	35.8	36.4	36.6	36.9	37.3	37.7	37.8
Air conditioning water temperature	60	60.5	61	61.5	63.5	64	64.5	
Air outlet temperature	37.9	38.0	38.4	38.8	40	40.1	40.3	

5.6.3 Establish regression equation to remedy

The establishment of regression equation method is to establish the regression equation based on the existing data, and then calculate the missing data based on the regression equation to make up for it.

Continue with the example in 5.5.6. First, the regression equation is established according to the data in Table 5-29.

Input the data in Table 5-29 into Minitab, and use regression analysis to show the relevant data in the dialog box, as shown in Figure 5-12.

```
Coefficients

Term                        Coef  SE Coef  T-Value  P-Value   VIF
Constant                   8.182    0.351    23.29    0.000
Water temperature (℃)   0.49736  0.00613    81.11    0.000  1.00

Regression Equation

Air outlet temperature (℃) = 8.182 + 0.49736 Water  temperature  (℃)

Fits and Diagnostics for Unusual Observations

         Air outlet
Obs     temperature (℃)     Fit     Resid  Std Resid
 18         38.0000      38.2715  -0.2715    -2.26   R
 21         40.0000      39.7636   0.2364     2.05   R

R  Large residual
```

Figure 5-12　Regression analysis output data

According to Figure 5-12, the linear regression equation that can be established between the air outlet temperature (y) and the air conditioning water temperature (x) is:

$$y = 8.182 + 0.497\,36x$$

Next, the corresponding data can be calculated according to the regression equation.

When the air conditioning water temperature is 50.5 ℃, the air outlet temperature is 8.182 + 0.497 36× 50.5 = 33.3 ℃;

When the air conditioning water temperature is 53 ℃, the air outlet temperature is 8.182 + 0.497 36× 53 = 34.5 ℃;

When the air conditioning water temperature is 56, the air outlet temperature is 8.182 + 0.497 36× 56 = 36.0 ℃;

When the air conditioning water temperature is 56.5, the air outlet temperature is 8.182 + 0.497 36× 56.5 = 36.3 ℃;

When the air conditioning water temperature is 62, the air outlet temperature is 8.182 + 0.497 36× 62 = 39.0 ℃;

When the air conditioning water temperature is 62.5, the air outlet temperature is 8.182 + 0.49736× 62.5 = 39.3 ℃;

When the air conditioning water temperature is 63, the air outlet temperature is 8.182 + 0.497 36× 63 = 39.5 ℃.

5.6.4 Data replacement

1. Replace with fixed value

Replace with fixed values means that when data is missing, the missing data is replaced with other fixed values. This method is usually applicable to the situation where the data change is not large. Replacing with a fixed value will cause the variance between the data to be enlarged.

Case：

A survey team investigated the wages of workers in a construction company in the previous year. Some survey data are shown in Table 5-32.

Table 5-32　Annual wages of some workers in the prerious year

No.	Name of worker	Annual salary/*yuan*
1	Wang ××	87 600
2	Shi ××	92 300
3	Ma ××	—
4	Zhang ××	79 500
5	Qin ××	97 600
6	Lu ××	69 700
7	Jiao ××	—
8	Gan ××	—
9	Shen ××	75 300
10	Quan ××	93 400

According to Table 5-32, the annual wages of workers Ma ××, Jiao ×× and Gan ×× have not been investigated. In order to make the follow-up analysis work smoothly, the researchers decided to replace it with fixed value. They looked up the average annual wage of workers in the construction industry in the previous year and got the data of 85 200 *yuan*. Therefore, the annual wages of workers Ma ××, Jiao ×× and Gan ×× were all replaced by 85 200 *yuan*.

2. Replace with adjacent values

Replace with adjacent values is to replace the missing value with the nearest value, and take the minimum or maximum neighboring value as required.

Case：

Continuing with the case in 5.6.1, we can see from Table 5-32 that the annual wages of workers Ma ××, Jiao ×× and Gan ×× have not been investigated. Therefore, the researchers decided to replace it with the neighboring value. Due to careful consideration, the researchers decided to replace the missing value with a smaller neighboring value. Ma ××'s annual salary is min (92 300, 79 500), which is 79 500 *yuan*. Jiao ××'s annual salary is min (69 700, 75 300), which is 69 700 yuan. In this way, the annual wage table of workers is obtained, as shown in Table 5-33.

Table 5-33　Annual wages of some workers in the previous year

No.	Name of worker	Annual salary/*yuan*
1	Wang ××	87 600
2	Shi ××	92 300
3	Ma ××	79 500
4	Zhang ××	79 500
5	Qin ××	97 600
6	Lu ××	69 700
7	Jiao ××	69 700
8	Gan ××	69 700
9	Shen ××	75 300
10	Quan ××	93 400

5.6.5 Maximum likelihood estimation method

Maximum likelihood estimation method refers to the estimation of unknown parameters through the marginal distribution of observation data, assuming that the model is correct and applicable for the complete sample under the condition that the missing type is random missing. This method can be called the maximum likelihood estimation of ignoring missing values.

For the parameter estimation of maximum likelihood, the commonly used calculation method is expectation maximization (EM). The EM algorithm was first proposed by Dempster in 1977. It uses iterative calculation to perform maximum likelihood estimation or calculate a posterior probability distribution. The premise of its use is that the sample size is large, and the number of effective samples can ensure that the estimated value is asymptotically unbiased and subject to normal distribution.

The idea of EM is to replace the missing value with the estimated value, then estimate the parameters of the newly formed complete data, and then estimate the missing value according to the parameter estimation value.

EM algorithm has two steps. The first step is to find the expected value, that is, under the condition of given observation data, find the conditional expectation of the missing value, and use the calculated conditional expectation to interpolate the missing data. The second step is to perform maximum likelihood estimation for the parameters of the interpolated complete data set.

Due to the considerable workload of EM manual calculation, computer software is usually used now. For example, in SPSS software, by selecting the relevant content of "missing value analysis", the calculation of EM is included.

5.6.6 No treatment

Interpolation and other processing for missing data is an expedient way to make up for the subjective estimate, which does not necessarily conform to the objective facts. Because when subjective imputation data enters the dataset, it actually changes the original dataset connotation, and sometimes the analysis results are not satisfactory.

In this case, it is also an option not to deal with the missing data, so the analyst can analyze the data based on the real data.

5.6.7 Computer data missing

1. Common computer data missing

In the process of computer use, we often encounter computer logic problems that lead to data loss. Such situations include the following seven types:

(1) Delete data on the computer by mistake;

(2) Format, that is, format the USB flash disk and memory card, resulting in data loss;

(3) There was a problem when copying and pasting. The data was not copied and pasted properly;

(4) The partition of U disk and hard disk is missing;

(5) When reinstalling the system, C disk data is lost;

(6) Sudden power failure, forced unplugging and other behaviors, and no UPS power supply, resulting in data loss;

(7) Encounter computer virus, resulting in data loss.

To solve these problems, data recovery tools can be used to help data recovery, including: delete recovery, format recovery, hard disk recovery, removable storage device recovery, deep recovery, etc.

2. Irreversibility of computer data

In computer operation, sometimes the data is not reversible due to carelessness. Such situations include the following four types:

(1) The computer hardware is damaged. Computer hardware is an important data storage device. Once it is seriously damaged, it is usually difficult to recover.

(2) Encounter virus. Some computer viruses specially destroy files or intercept information, so we must do a good job of virus prevention at ordinary times.

(3) Physical deletion. In computer operation, if it is deleted or formatted by mistake, it is logical deletion, and the data may be recovered. The physical deletion of data from the storage media is certainly irreversible.

(4) Data coverage. In computer operation, once the data is overwritten, the original data cannot be retrieved. For example, if data is lost in the partition, it may be retrieved. If the partition with lost data continues to be used, the original lost data cannot be retrieved after the new data is written.

Chapter 6
Data Risk

第 6 章
数据的风险

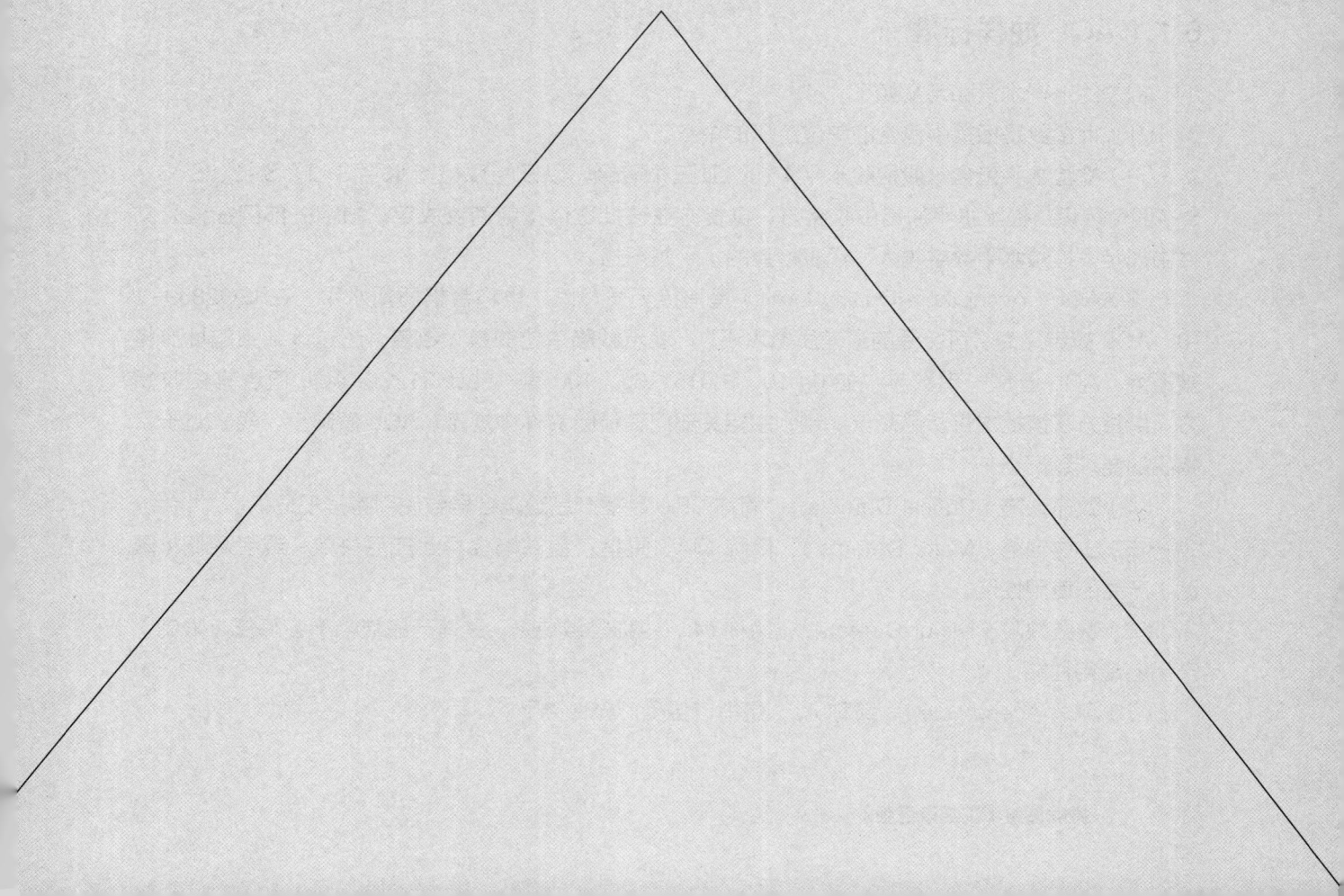

数据在采集、应用的过程中都会存在风险，因此，清楚数据风险之所在，可以更好地为使用数据服务。数据采集的风险是指在获取数据的过程中所面对的风险，比如抽样时采集的不是随机数据；采集到假性数据；采集过程中受到干扰（政策、环境、人为）；限于条件而无法获取人、机、料、法、环、测等方面的数据；测量、试验不准确；可靠性中主动缺失数据，否则成本太大；当样本大于5%时数据为非独立样本。数据应用的风险是指在数据应用的过程中所面对的风险，比如由于过程能力不足而产生风险、假设检验的风险等。

6.1 抽样的风险

——

抽样风险是指由抽样方案的合理性、抽样过程的可控制程度、样本容量及其代表性等所构成的风险。抽样风险的核心是样本不能代表总体的风险。通常来说，抽样风险与样本容量的大小成反比，即样本容量越小，抽样风险越大；样本容量越大，抽样风险越小。

6.1.1 AQL 抽样标准

相关的一些术语和定义如下：

（1）批量数是指批中包含的单位产品的个数。

（2）检验水平包含一般检验水平Ⅰ、Ⅱ、Ⅲ三个等级，以及特殊检验水平 S-1、S-2、S-3、S-4四个等级。检验水平通常由收货方、供货方在该批次供货前商议决定，由相关部门在检验文件中规定。检验水平数字越大，代表检查的水平越严格。

（3）AQL（Acceptable Quality Limit）是指接受质量限，也称合格质量水平，在 ISO2859-1中，AQL 被定义为"可接受的最差质量水平"，表示缺陷单位的最大数量，超过该数量的批次将被拒绝。AQL 值是一串数字，如 0.010、0.015、65、100 等，共计 21 个。AQL 值通常由收货方、供货方在该批次供货前商议决定，由相关部门在检验文件中规定。AQL 值越小，代表对于合格品的要求越严格。

（4）致命缺陷（Critical Defects），简称 CR，是指产品缺陷会导致使用者呈现危机。

（5）主要缺陷（Major Defects），简称 MA，是指产品缺陷会导致产品失效，或者降低（限制）产品的使用性能。

（6）次要缺陷（Minor Defects），简称 MI，也称轻微缺陷，是指产品缺陷不会降低（限制）产品的使用性能。

（7）AC（Aceeptable），又称 Ac，是指可接受、允许接受。

（8）RE（Reject），又称 Re，是指拒收。

（9）抽样检验，是指从总体中随机抽取一定数量的样本，并对随机抽取的样本进行质量检验，然后把检验结果与接受准则进行比较，从而确定总体是否合格或需要进行抽检后进行判定的一种质量检验方法。

抽样检验的优点是费用低、有利于对产品进行仔细全面检查等。抽样检验的缺点是有风险，因为产品品质是制造出来的，而不是检验出来的，有可能导致误判。

（10）Lot，又称批次，是指用同一批号的原料，在同一班次、同一生产线上生产出来的同一品种和同一规格的产品。

（11）一次抽样，又称单次抽样，是指根据抽样检验方案，每次从检验批中随机抽取样本，经检验后可以判定是否合格。

（12）二次抽样，是指第一次抽样结果不能判定，必须进行再次抽样检验。

（13）多次抽样，是指二次抽样的延续。

一个完整的检验项目包含三个要素：批量数、检验水平、AQL 值。由批量数和检验水平确定了检验的两个关键：一是在这一批次产品中，需要抽检多少个产品；二是缺陷产品达到多少个，该批次产品会被拒绝。由批量数和检验水平，可以确定样本量字码，样本量字码表见表 6-1。使用样本量字码与 AQL 值进行检索，就得到相应的抽样方案。

表 6-1　样本量字码表

批量	特殊检验水平				一般检验水平		
	S-1	S-2	S-3	S-4	I	II	III
2~8	A	A	A	A	A	A	B
9~15	A	A	A	A	A	B	C
16~25	A	A	B	B	B	C	D
26~50	A	B	B	C	C	D	E
51~90	B	B	C	C	C	E	F
91~150	B	B	C	D	D	F	G
151~280	B	C	D	E	E	G	H
281~500	B	C	D	E	F	H	J
501~1200	C	C	E	F	G	J	K
1201~3200	C	D	E	G	H	K	L
3201~10 000	C	D	F	G	J	L	M
10 001~35 000	C	D	F	H	K	M	N
35 001~150 000	D	E	G	J	L	N	P
150 001~500 000	D	E	G	J	M	P	Q
500 000 及其以上	D	E	H	K	N	Q	R

举例：

批量数在 1201 至 3200 个之间，检验水平确定为 II（可以称为"抽样标准为 AQL Level II"），那么查样本量字码表，得到样本量字码为 K。

6.1.2 检验水平的选择原则

（1）在没有特殊规定的情况下，一般选择检验水平Ⅱ。

（2）对检查费用进行比较，如果每个样品的检查费用为 a，处理一个判批次为不合格样品的费用为 b，那么，检查水平应当遵循：如果 $a > b$，选择检验水平Ⅰ；如果 $a = b$，选择检验水平Ⅱ；如果 $a < b$，选择检验水平Ⅲ。

（3）四个特殊检验水平 S–1、S–2、S–3、S–4，可以用于样本量相对较小，而且能容许较大抽样风险的情形，比如检验费用极高或者破坏性检验的情形。

（4）辨别能力：Ⅰ < Ⅱ < Ⅲ；S–1 < S–2 < S–3 < S–4

6.1.3 检验一次抽样方案

检验一次抽样方案是指从批量为 N 的一个产品批中，随机抽取容量为 n 的样本，对样本中的产品逐一检验。如果其中的不合格品数 d 不超过合格判定数 c，则接受该产品批；如果不合格品数 d 超过合格判定数 c，则拒收该产品批。

正常检验一次抽样方案见表 6–2。

表 6–2　正常检验一次抽样方案

样本量字码	样本量	0.010	0.015	0.025	0.040	0.065	0.10	0.15	0.25	0.40	0.65	1.0	1.5	2.5	4.0	6.5	10	15	25	40	65	100	150	250	400	650	1000
		Ac Re	Ac Re	Ac Re	Ac Re	Ac Re	Ac Re	Ac Re	Ac Re	Ac Re	Ac Re	Ac Re	Ac Re	Ac Re	Ac Re	Ac Re	Ac Re	Ac Re	Ac Re	Ac Re	Ac Re	Ac Re	Ac Re	Ac Re	Ac Re	Ac Re	Ac Re
A	2	↓	↓	↓	↓	↓	↓	↓	↓	↓	↓	↓	↓	↓	↓	↓	↓	0 1	1 2	2 3	3 4	5 6	7 8	10 11	14 15	21 22	30 31
B	3	↓	↓	↓	↓	↓	↓	↓	↓	↓	↓	↓	↓	↓	↓	↓	0 1	1 2	2 3	3 4	5 6	7 8	10 11	14 15	21 22	30 31	44 45
C	5	↓	↓	↓	↓	↓	↓	↓	↓	↓	↓	↓	↓	↓	↓	0 1	1 2	2 3	3 4	5 6	7 8	10 11	14 15	21 22	30 31	44 45	↑
D	8	↓	↓	↓	↓	↓	↓	↓	↓	↓	↓	↓	↓	↓	0 1	1 2	2 3	3 4	5 6	7 8	10 11	14 15	21 22	30 31	44 45	↑	↑
E	13	↓	↓	↓	↓	↓	↓	↓	↓	↓	↓	↓	↓	0 1	1 2	2 3	3 4	5 6	7 8	10 11	14 15	21 22	30 31	44 45	↑	↑	↑
F	20	↓	↓	↓	↓	↓	↓	↓	↓	↓	↓	↓	0 1	1 2	2 3	3 4	5 6	7 8	10 11	14 15	21 22	30 31	44 45	↑	↑	↑	↑
G	32	↓	↓	↓	↓	↓	↓	↓	↓	↓	↓	0 1	1 2	2 3	3 4	5 6	7 8	10 11	14 15	21 22	30 31	44 45	↑	↑	↑	↑	↑
H	50	↓	↓	↓	↓	↓	↓	↓	↓	↓	0 1	1 2	2 3	3 4	5 6	7 8	10 11	14 15	21 22	30 31	44 45	↑	↑	↑	↑	↑	↑
J	80	↓	↓	↓	↓	↓	↓	↓	↓	0 1	1 2	2 3	3 4	5 6	7 8	10 11	14 15	21 22	30 31	44 45	↑	↑	↑	↑	↑	↑	↑
K	125	↓	↓	↓	↓	↓	↓	↓	0 1	1 2	2 3	3 4	5 6	7 8	10 11	14 15	21 22	30 31	44 45	↑	↑	↑	↑	↑	↑	↑	↑
L	200	↓	↓	↓	↓	↓	↓	0 1	1 2	2 3	3 4	5 6	7 8	10 11	14 15	21 22	30 31	44 45	↑	↑	↑	↑	↑	↑	↑	↑	↑
M	315	↓	↓	↓	↓	↓	0 1	1 2	2 3	3 4	5 6	7 8	10 11	14 15	21 22	30 31	44 45	↑	↑	↑	↑	↑	↑	↑	↑	↑	↑
N	500	↓	↓	↓	↓	0 1	1 2	2 3	3 4	5 6	7 8	10 11	14 15	21 22	30 31	44 45	↑	↑	↑	↑	↑	↑	↑	↑	↑	↑	↑
P	800	↓	↓	↓	0 1	1 2	2 3	3 4	5 6	7 8	10 11	14 15	21 22	30 31	44 45	↑	↑	↑	↑	↑	↑	↑	↑	↑	↑	↑	↑
P	1250	↓	↓	0 1	1 2	2 3	3 4	5 6	7 8	10 11	14 15	21 22	30 31	44 45	↑	↑	↑	↑	↑	↑	↑	↑	↑	↑	↑	↑	↑
R	2000	↓	0 1	1 2	2 3	3 4	5 6	7 8	10 11	14 15	21 22	30 31	44 45	↑	↑	↑	↑	↑	↑	↑	↑	↑	↑	↑	↑	↑	↑

⇩ —— 使用箭头下面的第一个抽样方案，如果样本等于或超过批量，则执行100%验收。

⇧ —— 使用箭头上面的第一个抽样方案。

Ac —— 接受数

Re —— 拒收数

根据已经确定的检验水平、AQL 水准，制定抽样方案。

举例：

批量数在 1201 至 3200 个之间，检验水平确定为Ⅱ，得到样本量字码为 K，样本量为 125。

现在把 AQL 设定为：主要缺陷 2.5（%），次要缺陷 4.0（%），即检验标准按 AQL2.5，4.0 抽样。

查表 6-2，得到：如果主要缺陷的产品不超过 7 个，同时次要缺陷的产品不超过 10 个，则该批次产品被接受。如果主要缺陷的产品达到 8 个，或者次要缺陷的产品达到 11 个，则拒收该批次产品。

举例：

批量数在 1201 至 3200 个之间，检验水平确定为 Ⅱ，得到样本量字码为 K。

现在 AQL 设定为 4.0。

查表 6-2，从"K 行"到"4.0 列"交叉处得到：Ac = 10，Re = 11。但是，根据注释，当表中箭头向上时，使用箭头上面的第一个抽样方案。因此，向上箭头对应的第一个格子为"Ac21，Re22"。由此格子向左行看去，得到样本量为 315。

举例：

批量数为 40，检验水平为 Ⅱ，得到样本量字码为 D。

现在 AQL 设定为 0.15.

查表 6-2，由于 D 行与"0.15 列"交叉没有数字，则由"0.15 列"箭头继续向下，查到第一个格子为"Ac0，Re1"。由该格子向左行看，得到样本量为 80。

现在的问题是：批量数为 40，显然小于 80。根据注释：如果样本等于或者超过批量，则执行 100% 检验。因此，该方案为 100% 全部检验。当检验结果为 0 缺陷时，接受该批次产品；当检验结果有 1 个缺陷时，拒收该批次产品。

在应用时，如果 AQL≤10，则可以用不合格品百分数表示质量水平，也可以用每百单位产品不合格数表示质量水平。如果 AQL>10，就不能用不合格品百分数表示质量水平，而是用每百单位产品不合格数表示。

6.1.4 质量管理中的抽样风险

1. 生产方风险与使用方风险

在质量管理中，抽样风险可以反映为根据样本错误地拒绝或接受特定批次的概率。由于抽样检验中样本的随机性，且样本只是批的一小部分，所以存在错判的可能。

当批质量符合要求却不被接受时，由生产方承担的风险称为生产方风险（错判概率 α）。

当批质量不符合要求却被接受时，由使用方承担的风险称为使用方风险（漏判概率 β）。

使用方根据其对于产品的质量水平，结合能接受的风险程度来确定 AQL 水准。表 6-3 为不同水准对应的使用方风险概率。

表6-3 不同水准对应的使用方风险概率
（对一次抽样方案以不合格品百分数表示，适合于不合格品百分数检验）

样本量字码	样本量	接受质量限（AQL）														
		0.010	0.015	0.025	0.040	0.065	0.10	0.15	0.25	0.40	0.65	1.0	1.5	2.5	4.0	6.5	10
A	2															68.4	69.0 *
B	3														51.6	54.1 *	57.6 *
C	5													36.9	37.3 *	39.8 *	58.4 *
D	8											25.0	25.2 *	27.0 *	40.6	53.8
E	13										16.2	16.4 *	17.5 *	26.8	36.0	44.4
F	20										10.9	11.0 *	11.8 *	18.1	24.5	30.4	41.5
G	32									6.94	7.01 *	7.50 *	11.6	15.8	19.7	27.1	34.0

样本量字码	样本量	接受质量限（AQL）														
		0.010	0.015	0.025	0.040	0.065	0.10	0.15	0.25	0.40	0.65	1.0	1.5	2.5	4.0	6.5	10
H	50								4.50	4.54 *	4.87 *	7.56	10.3	12.9	17.8	22.4	29.1
J	80						2.84	2.86 *	3.07 *	4.78	6.52	8.16	11.3	14.3	18.6	24.2
K	125					1.83	1.84 *	1.97 *	3.08	4.20	5.27	7.29	9.24	12.1	15.7	21.9
L	200					1.14	1.16 *	1.24 *	1.93	2.64	3.31	4.59	5.82	7.60	9.91	13.8
M	315				0.728	0.735 *	0.788 *	1.23	1.68	2.11	2.92	3.71	4.85	6.33	8.84
N	500			0.459	0.464 *	0.497 *	0.776	1.06	1.33	1.85	2.34	3.06	4.00	5.60
P	800		0.287	0.290 *	0.311 *	0.485	0.664	0.833	1.16	1.47	1.92	2.51	3.51
Q	1250	0.184	0.186 *	0.199 *	0.311	0.425	0.534	0.741	0.940	1.23	1.61	2.25
R	2000	0.116 *	0.124 *	0.194	0.266	0.334	0.463	0.588	0.769	1.00	1.41

注1：在使用方风险质量处，预期10%的批会被接受。

注2：所有表值均基于二项分布。

注3：标注 * 表示该值适合于供选择的分数接受数一次抽样方案。

2. 构成风险的原因

（1）样本容量不足，影响了抽样误差大小。

（2）仪器设备等没有校准，仪器设备有偏差；对仪器设备的误读误判；准确度不够，检测能力差。

（3）抽样方案中，样本不能代表质量状况。比如重复抽样的误差比不重复抽样的误差大。

（4）总体中各单位标志值的差异程度大，制程能力差，CPK 达不到要求；标准偏差大，制程波动大。

举例：

批量3000件，不良品率 p =5%，按照 AQL=0.65，那么抽样 n = 125 件，AC = 2 的时候，抽到3件或3件以上不良品的概率是多少？

具体分析：

$$P = \frac{(np)^d}{d!} e^{-np}, N \geqslant 10\,n, \text{且 } p < 0.1$$

经过计算，

抽到0件不良品的概率 P（0）= 0.19%；

抽到1件不良品的概率 P（1）= 1.2%；

抽到2件不良品的概率 P（2）= 3.76%；

抽到3件或3件以上不良品的概率 P（3）= 1 - P（0）- P（1）- P（2）= 94.85%。

说明，采用 AQL = 0.65 进行抽样，产品被拒收的概率很高。

6.1.5 正常、加严、加宽检验

1. 正常、加严、加宽检验的转移规则

基本的转移规则和程序见图6-1。

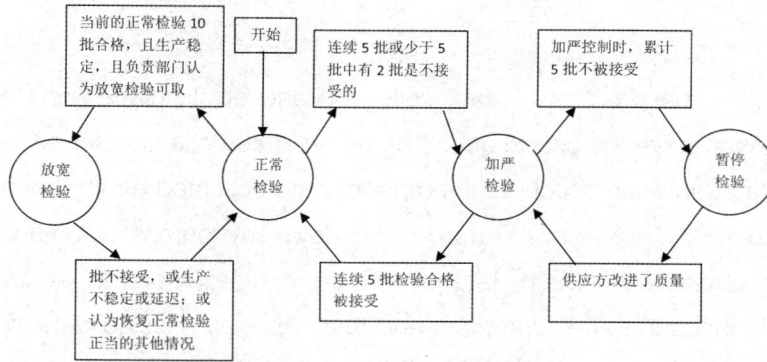

图 6-1　转移规则和程序图

2. 正常到加严

当正在采用正常检验时，只要初次检验中连续 5 批或少于 5 批中，有 2 批是不可接受的，则转移到加严检验。

3. 加严到正常

当正在采用加严检验时，如果初次检验的接连 5 批已被认为是可接受的，应回复正常检验。

4. 正常到放宽

当正在采用正常检验时，如果下列各条件均满足，应转移到放宽检验：

一是当前的转移得分至少是 30 分；

二是生产稳定；

三是相关负责部门认为放宽检验可取。

5. 放宽到正常

当正在执行放宽检验时，如果初次检验出现下列任一情况，应恢复正常检验。

一是一个批未被接受；

二是生产不稳定或延迟；

三是认为恢复正常检验是正当的其他情况。

6. 暂停检验

如果在初次加严检验的一系列连续批中，未接受批的累计数达到 5 批，应暂时停止检验，直到供方为改进所提供产品或服务的质量已采取行动，而且负责部门承认此行动可能有效时，才能恢复检验程序。恢复检验应从使用加严检验开始。

6.1.6 OC 曲线

OC（Operating Characteristics）曲线是进行风险比较的有效方式。OC 曲线是批接受概率 $P_a(p)$ 与批质量水平（批不合格品率）P 的关系曲线。每一个抽样方案都有一条对应的 OC 曲线。OC 曲线表示了一个抽样方案对一个批质量的辨识能力。正常的 OC 曲线见图 6-2。

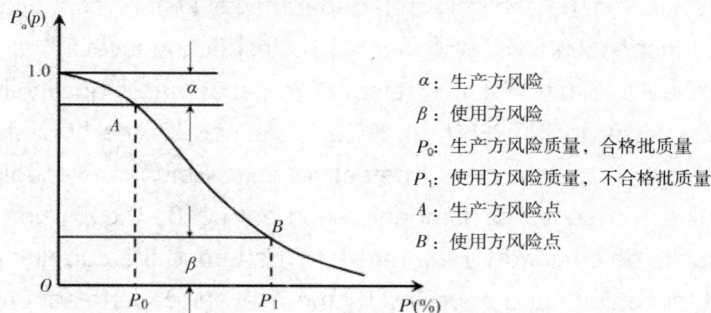

α：生产方风险
β：使用方风险
P_0：生产方风险质量，合格批质量
P_1：使用方风险质量，不合格批质量
A：生产方风险点
B：使用方风险点

图 6-2　正常的 OC 曲线

There are risks in the process of data collection and application. Therefore, knowing the risk of data can better serve the use of data. The risk of data collection refers to the risk faced in the process of data acquisition, such as the random data collected during sampling; False data collected; The acquisition process is disturbed (policy, environment, human); Due to limited conditions, it is impossible to obtain data on man, machine, material, method, environment and measurement; Inaccurate measurement and test; Actively missing data in reliability, otherwise the cost is too high; When the sample is greater than 5%, the data is non-independent. The risk of data application refers to the risk faced in the process of data application, such as the risk caused by insufficient process capability, the risk of hypothesis testing, etc.

6.1　Sampling Risk

———

Sampling risk refers to the risk composed of the rationality of the sampling plan, the controllability of the sampling process, the sample size and its representativeness. The core of sampling risk is that the sample cannot represent the overall risk. Generally speaking, the sampling risk is inversely proportional to the size of the sample size, that is, the smaller the sample size, the greater the sampling risk; The larger the sample size, the smaller the sampling risk.

6.1.1 AQL Sampling Standard

Some relevant terms and definitions are as follows.

(1) Batch number. It refers to the number of unit products included in the batch.

(2) The inspection level. It includes three levels of general inspection level I, II and III, and four levels of special inspection level S-1, S-2, S-3 and S-4. The inspection level is usually determined by the receiving party and the supplier through negotiation before the delivery of this batch, and is specified by the relevant departments in the inspection documents. The higher the number of inspection level, the stricter the inspection level.

(3) Acceptable Quality Limit(AQL). It refers to the acceptable quality limit, also known as the acceptable quality level. In ISO2859-1, AQL is defined as the "worst acceptable quality level", indicating the maximum number of defective units. Batches exceeding this number will be rejected. AQL value is a series of numbers, such as 0.010, 0.015, 65, 100, etc., totaling 21. AQL value is usually determined by the receiving party and the supplier through negotiation before the supply of this batch, and specified by the relevant department in the inspection document. The smaller the AQL value is, the stricter the requirements for qualified products.

(4)Critical Defects (CR). It means that product defects will cause users to present a crisis.

(5)Major Defects (MA). It means that product defects will lead to product failure or reduce (limit) the use performance of the product.

(6)Minor Defects (MI), also known as minor defect, means that the product defect will not reduce (limit) the service performance of the product.

(7)AC(Acceptable), also known as Ac, means acceptable and allow acceptance.

(8)RE(Reject), also known as Re, refers to rejection.

(9)Sampling inspection, refers to a quality inspection method that randomly selects a certain number of samples from the population, carries out quality inspection on the randomly selected samples, and then compares the inspection results with the acceptance criteria to determine whether the population is qualified or needs to be determined after random inspection.

The advantages of sampling inspection are low cost, convenient for careful and comprehensive inspection of products, etc. The disadvantage of sampling inspection is that it is risky, because the product quality is produced rather than inspected, which may lead to misjudgment.

(10)Lot, also known as batch, refers to the products of the same variety and specification produced from the same batch of raw materials in the same shift and on the same production line.

(11)One-time sampling, also known as single sampling, refers to the random sampling from the inspection lot each time according to the sampling inspection scheme, and whether it is qualified can be determined after inspection.

(12)Secondary sampling means that the results of the first sampling cannot be determined and must be re-sampled for inspection.

(13)Multiple sampling refers to the continuation of secondary sampling.

A complete inspection item contains three elements: batch number, inspection level, and AQL value. Two key points of inspection are determined by batch number and inspection level: first, how many products need to be sampled in this batch of products; second, the number of defective products will be rejected. The sample size code can be determined from the batch number and inspection level. See Table 6-1 for the sample code table. Use the sample size word code and AQL value to retrieve the corresponding sampling plan.

Table 6-1　Sample size word code table

Batch	Special inspection level				General inspection level		
	S-1	S-2	S-3	S-4	I	II	III
2~8	A	A	A	A	A	A	B
9~15	A	A	A	A	A	B	C
16~25	A	A	B	B	B	C	D
26~50	A	B	B	C	C	D	E
51~90	B	B	C	C	C	E	F
91~150	B	B	C	D	D	F	G

Batch	Special inspection level				General inspection level		
	S-1	S-2	S-3	S-4	I	II	III
151~280	B	C	D	E	E	G	H
281~500	B	C	D	E	F	H	J
501~1200	C	C	E	F	G	J	K
1201~3200	C	D	E	G	H	K	L
3201~10 000	C	D	F	G	J	L	M
10 001~35 000	C	D	F	H	K	M	N
35 001~150 000	D	E	G	J	L	N	P
150 001~500 000	D	E	G	J	M	P	Q
500 000 or more	D	E	H	K	N	Q	R

Case:

If the number of batches is between 1201 and 3200, and the inspection level is determined to be II (which can be called "AQL Level II"), then check the sample size word code table, and the sample size word code is K.

6.1.2 Selection principle of inspection level

1. In the absence of special provisions, inspection level II is generally selected.

2. Compare the inspection cost. If the inspection cost of each sample is a and the cost of handling a batch of unqualified samples is b, then the inspection level should follow the rules as bellow: if $a > b$, select inspection level I; if $a = b$, select inspection level II; if $a < b$, select inspection level III.

3. The four special inspection levels S-1, S-2, S-3 and S-4 can be used in cases where the sample size is relatively small and the sampling risk is relatively large, such as the case of extremely high inspection cost or destructive inspection.

4. Discrimination: I < II < III; S-1 < S-2 < S-3 < S-4.

6.1.3 One-time sampling plan for inspection

The one-time inspection sampling scheme refers to randomly sampling samples with a capacity of n from a product batch with a batch size of N, and inspecting the products in the sample one by one. If the number of nonconforming products d does not exceed the number of qualified products c, the product batch shall be accepted; If the number of nonconforming products d exceeds the number of qualified products c, the product batch will be rejected.

See Table 6-2 for a sampling plan for normal inspection.

Table 6-2　One-time sampling plan for normal inspection

code	Sample size	0.010	0.015	0.025	0.040	0.065	0.10	0.15	0.25	0.40	0.65	1.0	1.5	2.5	4.0	6.5	10	15	25	40	65	100	150	250	400	650	1000
		AcRe	AcRe	AcRe	AcRe	AcRe	AcRe	AcRe	AcRe	AcRe	AcRe	AcRe	AcRe	AcRe	AcRe	AcRe	AcRe	AcRe	AcRe	AcRe	AcRe	AcRe	AcRe	AcRe	AcRe	AcRe	AcRe
A	2															0 1			1 2	2 3	3 4	5 6	7 8	10 11	14 15	21 22	30 31
B	3														0 1			1 2	2 3	3 4	5 6	7 8	10 11	14 15	21 22	30 31	44 45
C	5													0 1			1 2	2 3	3 4	5 6	7 8	10 11	14 15	21 22	30 31	44 45	
D	8											0 1			1 2	2 3	3 4	5 6	7 8	10 11	14 15	21 22	30 31	44 45			
E	13										0 1			1 2	2 3	3 4	5 6	7 8	10 11	14 15	21 22	30 31	44 45				
F	20									0 1			1 2	2 3	3 4	5 6	7 8	10 11	14 15	21 22							
G	32								0 1			1 2	2 3	3 4	5 6	7 8	10 11	14 15	21 22								
H	50							0 1			1 2	2 3	3 4	5 6	7 8	10 11	14 15	21 22									
J	80						0 1			1 2	2 3	3 4	5 6	7 8	10 11	14 15	21 22										
K	125					0 1			1 2	2 3	3 4	5 6	7 8	10 11	14 15	21 22											
L	200				0 1			1 2	2 3	3 4	5 6	7 8	10 11	14 15	21 22												
M	315			0 1			1 2	2 3	3 4	5 6	7 8	10 11	14 15	21 22													
N	500		0 1			1 2	2 3	3 4	5 6	7 8	10 11	14 15	21 22														
P	800	0 1			1 2	2 3	3 4	5 6	7 8	10 11	14 15	21 22															
P	1250	0 1		1 2	2 3	3 4	5 6	7 8	10 11	14 15	21 22																
R	2000		1 2	2 3	3 4	5 6	7 8	10 11	14 15	21 22																	

⇓ ———— Use the first sampling plan below the arrow.
　　　　 If the sample is equal to or more than the batch size, 100% acceptance is required.

⇑ ———— Use the first sampling plan above the arrow.

Ac ----- Accepted number

Re ----- Rejection number

Develop sampling plan according to the determined inspection level and AQL level.

Case：

The number of batches is between 1201 and 3200, the inspection level is determined as Ⅱ, and the sample size code is K, and the sample size is 125.

Now the AQL is set as：2.5 (%) for major defects and 4.0 (%) for minor defects, that is, the inspection standard is based on AQL 2.5 and 4.0.

According to Table 6-2, if there are no more than 7 products with major defects and no more than 10 products with minor defects, this batch of products will be accepted. If there are 8 products with major defects or 11 products with minor defects, the batch of products will be rejected.

Case：

The number of batches is between 1201 and 3200, the inspection level is determined as Ⅱ, and the sample size is K.

Now AQL is set to 4.0.

According to Table 6-2, from the intersection of "K row" to "4.0 column", Ac = 10, Re = 11. However, according to the note, when the arrow in the figure is upward, the first sampling scheme above the arrow is used. Therefore, the first cell corresponding to the up arrow is "Ac21, Re22". Looking to the left of the grid, the sample size is 315.

Case：

The batch number is 40, the inspection level is Ⅱ, and the sample size code is D.

Now AQL is set to 0.15.

According to Table 6-2, since there is no number at the intersection of row D and "0.15 column", the arrow "0.15 column" continues downward, and the first grid is "Ac0, Re1". From this grid to the left, the sample size is 80.

The problem is that the number of batches is 40, which is obviously less than 80. According to the note: if the sample is equal to or more than the batch size, perform 100% inspection. Therefore, the scheme is 100% full inspection. When the inspection result is 0 defect, accept the batch of products; If there is one defect in the inspection result, the batch of products will be rejected.

In application, if AQL ≤ 10, the quality level can be expressed by the percentage of nonconforming products, or by the number of nonconforming products per 100 units. If AQL>10, the quality level cannot be expressed by the percentage of nonconforming products, but by the number of nonconforming products per 100 units.

6.1.4 Sampling risk in quality management

1. Producer's risk and user's risk

In quality management, sampling risk can be reflected as the probability of wrongly rejecting or accepting a specific batch according to the sample. Because of the randomness of the samples in the sampling inspection and the samples are only a small part of the batch, there is the possibility of misjudgment.

When the batch quality meets the requirements but is not accepted, the risk borne by the manufacturer is called the producer's risk (misjudgment probability α).

When the batch quality does not meet the requirements but is accepted, the risk borne by the user is called the user risk (misjudgment probability β).

The user determines the AQL level according to the quality level of the product and the acceptable risk level. Table 6-3 shows the user risk probability corresponding to different levels.

Table 6-3　User risk probability corresponding to different levels

(A sampling plan is expressed as the percentage of nonconforming products, which is suitable for the percentage inspection of nonconforming products)

Character code	sample size	Acceptance quality limit (AQL)															
		0.010	0.015	0.025	0.040	0.065	0.10	0.15	0.25	0.40	0.65	1.0	1.5	2.5	4.0	6.5	10
A	2															68.4	69.0 *
B	3														51.6	54.1 *	57.6 *
C	5													36.9	37.3 *	39.8 *	58.4 *
D	8												25.0	25.2 *	27.0 *	40.6	53.8
E	13											16.2	16.4 *	17.5 *	26.8	36.0	44.4
F	20										10.9	11.0 *	11.8 *	18.1	24.5	30.4	41.5
G	32									6.94	7.01 *	7.50 *	11.6	15.8	19.7	27.1	34.0
H	50								4.50	4.54 *	4.87 *	7.56	10.3	12.9	17.8	22.4	29.1
J	80							2.84	2.86 *	3.07 *	4.78	6.52	8.16	11.3	14.3	18.6	24.2
K	125						1.83	1.84 *	1.97 *	3.08	4.20	5.27	7.29	9.24	12.1	15.7	21.9
L	200					1.14	1.16 *	1.24 *	1.93	2.64	3.31	4.59	5.82	7.60	9.91	13.8	
M	315				0.728	0.735 *	0.788 *	1.23	1.68	2.11	2.92	3.71	4.85	6.33	8.84		

Character code	sample size	Acceptance quality limit (AQL)															
		0.010	0.015	0.025	0.040	0.065	0.10	0.15	0.25	0.40	0.65	1.0	1.5	2.5	4.0	6.5	10
N	500			0.459	0.464 *	0.497 *	0.776	1.06	1.33	1.85	2.34	3.06	4.00	5.60			
P	800		0.287	0.290 *	0.311 *	0.485	0.664	0.833	1.16	1.47	1.92	2.51	3.51				
Q	1250	0.184	0.186 *	0.199 *	0.311	0.425	0.534	0.741	0.940	1.23	1.61	2.25					
R	2000	0.116 *	0.124 *	0.194	0.266	0.334	0.463	0.588	0.769	1.00	1.41						

Note 1: At the risk quality department of the user, it is expected that 10% of the batches will be accepted.

Note 2: All table values are based on binomial distribution.

Note 3: The mark * indicates that the value is suitable for the sampling scheme of the acceptable number of points for selection.

2. Causes of risk

(1) Insufficient sample size affects the sampling error.

(2) The instruments and equipment are not calibrated, and the instruments and equipment have deviations; misreading and misjudgment of instruments and equipment; insufficient accuracy, poor detection ability.

(3) In the sampling plan, the sample cannot represent the quality status. For example, the error of repeated sampling is larger than that of non-repeated sampling.

(4) In general, the difference of the mark value of each unit is large, the process capacity is poor, and the CPK cannot meet the requirements; large standard deviation and large process fluctuation.

Case:

The batch is 3000 pieces, the defective rate $p = 5\%$, AQL $=0.65$. What is the probability of sampling 3 or more defective products when $n = 125$ and AC $=2$?

Analyze as follows:

$$P = \frac{(np)^d}{d!} \mathrm{e}^{-np}, \ N \geqslant 10n, \text{and } p < 0.1$$

After calculation,

the probability of picking 0 defective products is $P(0) = 0.19\%$;

the probability of picking 1 defective products is $P(1) = 1.2\%$;

the probability of picking 2 defective products is $P(2) = 3.76\%$;

the probability of picking 3 or more defective products $P(3) = 1 - P(0) - P(1) - P(2) = 94.85\%$.

Therefore, if AQL $=0.65$ is used for sampling, the probability of product rejection is high.

6.1.5 Normal, Tightening and Widening Inspection

1. Transfer rules for normal, tightened and widened inspection
See Figure 6-1 for basic transfer rules and procedures.

2. Normal to tightened
When normal inspection is being used, as long as two of the five consecutive batches or less in the initial inspection are unacceptable, it will be transferred to tightened inspection.

3. Tightened to normal
When the tightened inspection is being used, if the five consecutive batches of the initial in-

spection have been considered acceptable, the normal inspection shall be resumed.

4. Normal to relaxed

When normal inspection is being used, if the following conditions are met, it shall be transferred to relaxed inspection:

First, the current transfer score is at least 30 points;

Second, stable production;

Third, the relevant responsible departments think it is advisable to relax the inspection.

5. Relax to normal

When the relaxation inspection is being carried out, if any of the following conditions occur in the initial inspection, the normal inspection shall be resumed.

First, a batch was not accepted;

Second, production is unstable or delayed;

Third, other circumstances where the resumption of normal testing is justified.

6. Suspension of inspection

If the cumulative number of unaccepted batches reaches 5 in a series of consecutive batches of the initial strict inspection, the inspection shall be suspended temporarily until the supplier has taken actions to improve the quality of the products or services provided, and the responsible department acknowledges that this action may be effective, then the inspection procedure can be resumed. The recovery inspection shall start from the use of the stricter inspection.

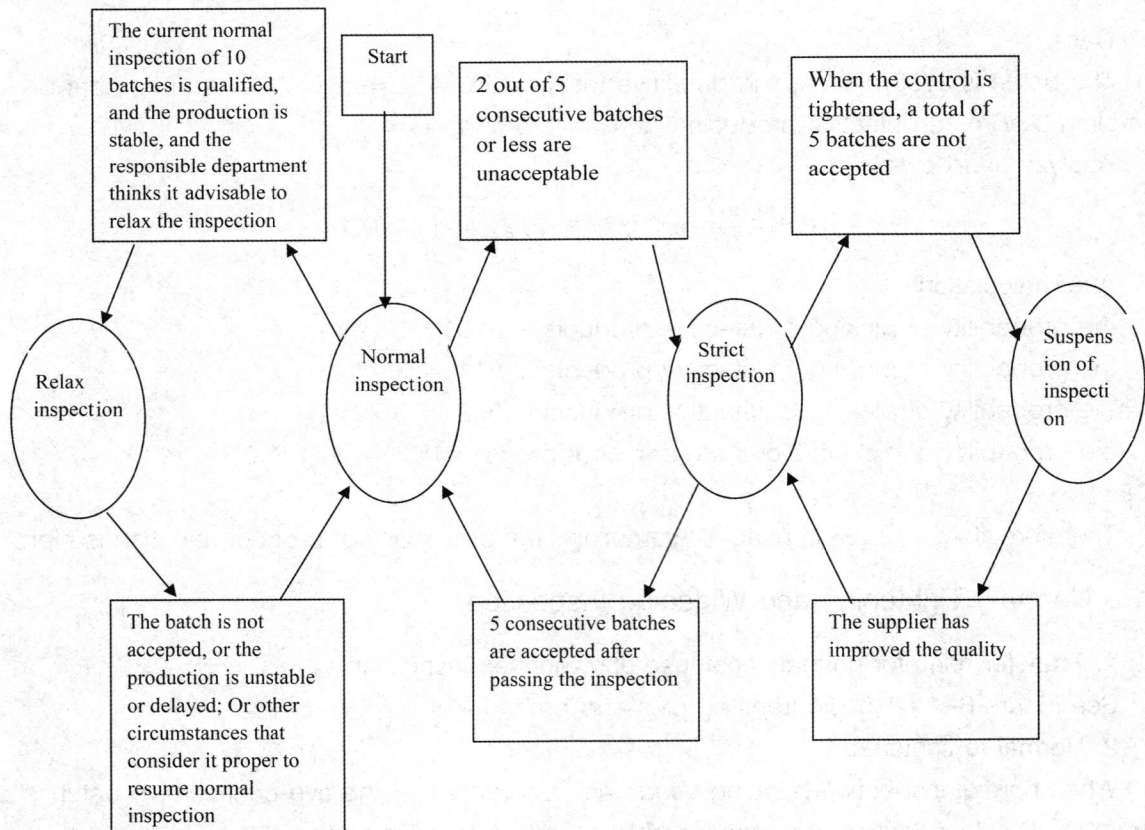

Figure 6-1　Transfer rules and procedures diagram

6.1.6 OC curve

Operating Characteristics (OC) curves is an effective way to compare risks. OC curve is the relation curve between the batch acceptance probability $P_a(p)$ and the batch quality level (batch nonconforming product rate) P. Each sampling plan has a corresponding OC curve. OC curve shows the ability of a sampling scheme to identify the quality of a batch. See Figure 6-2 for normal OC curve.

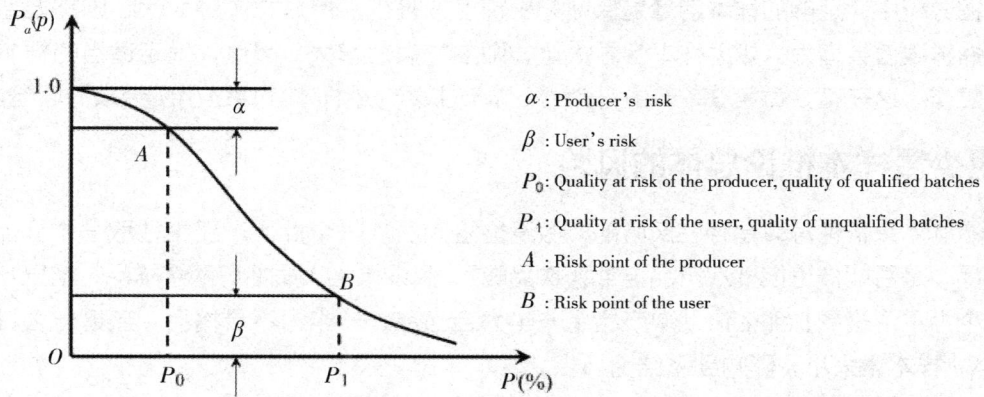

α : Producer's risk

β : User's risk

P_0: Quality at risk of the producer, quality of qualified batches

P_1: Quality at risk of the user, quality of unqualified batches

A : Risk point of the producer

B : Risk point of the user

Figure 6-2   Normal OC curve

6.2　样本容量过小的风险

在数据分析中，足够的样本容量是保证分析精度的基础。但是样本量太大，则需要增加很多成本，非抽样误差会增大。如果样本容量太小，则抽样误差增大。因此，确定适当的样本容量是很重要的工作，这在第3章第3.2节中有介绍。本节主要介绍样本容量过小所带来的风险。

6.2.1 用少量样本推论总体的风险

用少量样本来推论总体是存在风险的。假设检验人员从一个批次产品中抽取一个样品，结果为不合格品，是否能推论该批次产品全部为次品呢？ 假设检验人员抽取两个样品，结果一个为合格品，一个为不合格品，那么该批次产品是一半为合格品、一半为不合格品，还是什么其他情形呢？ 显然，样本量太小，影响到了结论。

事实上，随着样本量的增加，对总体参数的估计精度会不断提高。不同的样本容量有不同的样本平均值分布，示意图见图6-3。

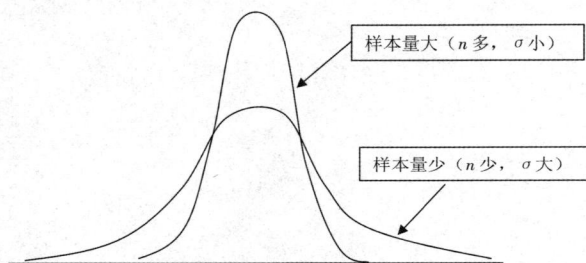

图6-3　不同样本量的样本平均值分布示意图

图6-3中，σ代表（总体）固有的标准差，n是样本量。可以看出，样本量大，则误差小。

现在以δ代表两个总体平均值的实际差值。δ/σ，称为检验灵敏度或灵敏系数，是用样本平均值的差值除以标准差。下面看样本容量（n）对检验概率的影响，见图6-4。

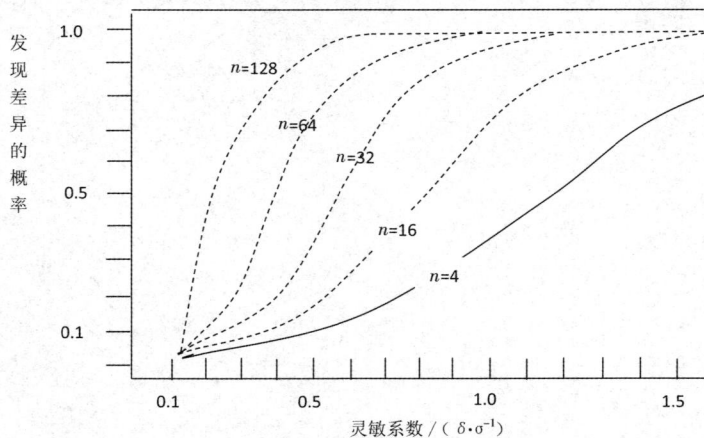

图6-4　样本容量（n）对检验概率的影响

从图 6-4 可以看出，在相同的灵敏系数下，随着样本容量的增大，发现差异的概率增大。

6.2.2 样本量少的相关分析

在样本量少的情况下进行相关分析是不正确的。如果成对的数据太少，比如少于 30 组，可以根据数据绘制成散点图，也可以计算出相关系数，但真正的相关性是难以判断的。

举例：

某质量管理小组进行主要原因的确认。末端原因为"进水温度高"，症结为"加工材料表面的气泡数多"。为此，小组进行了试验，并在不同的时间、不同的进水温度下抽取了 15 组数据进行分析，数据见表 6-4。

表 6-4   温度与材料气泡的数据

序号	温度/℃	气泡数/（个·平方米$^{-1}$）
1	84.5	81
2	85.3	82
3	85.5	91
4	90.0	80
5	90.2	85
6	90.3	87
7	91.1	84
8	91.2	92
9	91.3	86
10	92.0	90
11	93.3	91
12	94.4	92
13	94.9	93
14	96.2	95
15	98.3	96

根据表 6-4，可以制作散点图，见图 6-5。

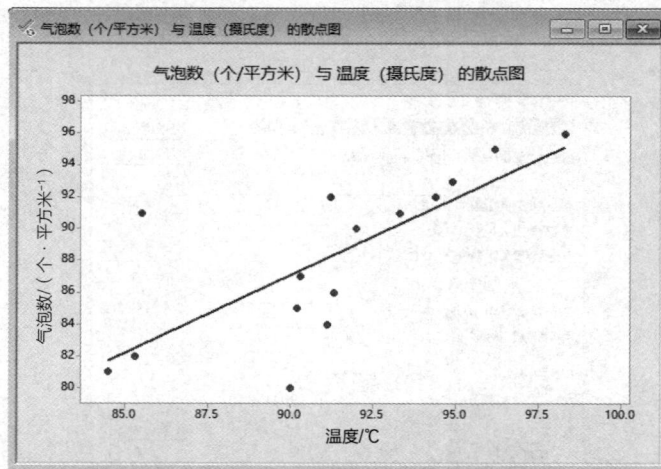

图 6-5   温度与材料气泡的散点图

从图6-5看，温度与材料气泡之间似乎有一条明显的回归线，根据计算，两者的相关系数达到0.772，那么应该为强相关。但是，如果使用Minitab中的"相关系数"进行检验，发现这个相关系数的 P 值为0.000，显然没有任何相关。

6.2.3 按属性抽样验收

对于采购方来说，当采购大量零件等材料时，需要评估其过程能力（ C_{pk} ），用控制图检查过程参数的受控情况等，但往往会受制于条件。因此，抽样验收是更有效的一种方法。

按照属性抽样，需要制定AQL和RQL，还需要制定批次大小、生产者风险 α 和使用者风险 β 。

举例：

现在采购方采购5000个零件，希望通过属性抽样检验来确定需要检验零件的数量，以及确定样本中允许有多少个不合格品。现设定AQL=1.5%，RQL=5.0%， α =0.05， β =0.1。

使用Minitab软件。

选择软件中"统计"中的"质量工具"，选择其中的"按属性抽样验收"，见图6-6。

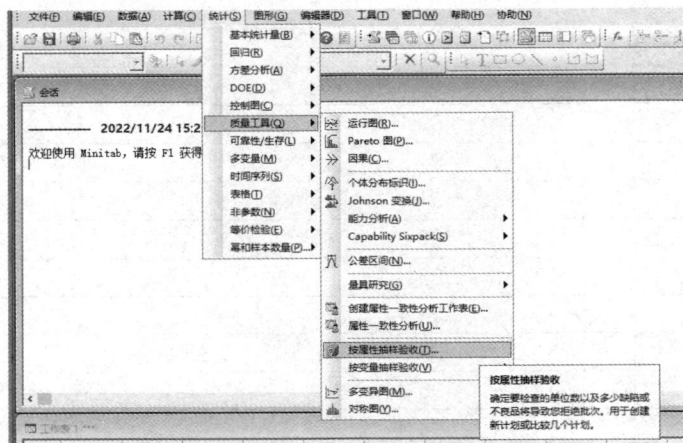

图6-6 选择"按属性抽样验收"

点击后，在会话框中输入AQL=1.5%，RQL=5.0%， α =0.05， β =0.1；"批次大小"中输入5000，出现对话框见图6-7。

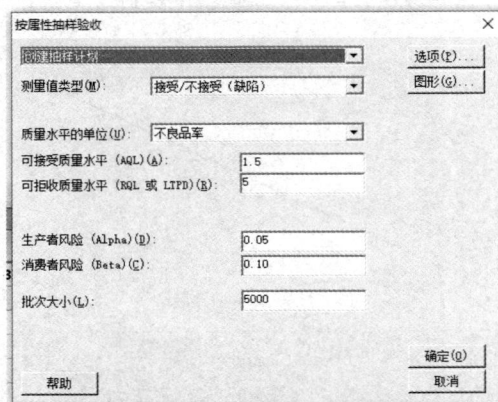

图6-7 按属性抽样会话框

点击确定后，会话框输出内容见图6-8。

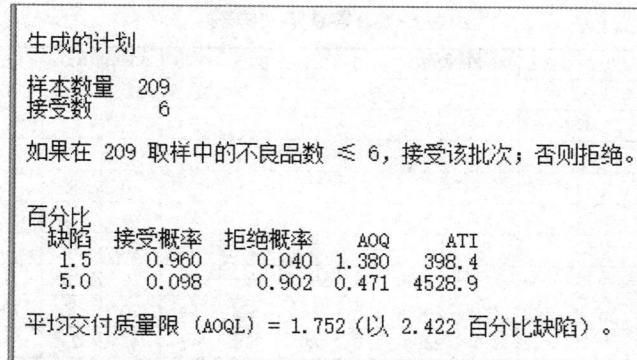

```
生成的计划

样本数量   209
接受数      6

如果在 209 取样中的不良品数 ≤ 6，接受该批次；否则拒绝。

百分比
  缺陷   接受概率   拒绝概率    AOQ    ATI
  1.5    0.960     0.040     1.380   398.4
  5.0    0.098     0.902     0.471  4528.9

平均交付质量限（AOQL）= 1.752（以 2.422 百分比缺陷）。
```

图6-8　会话框输出

图形框输出内容见图6-9。

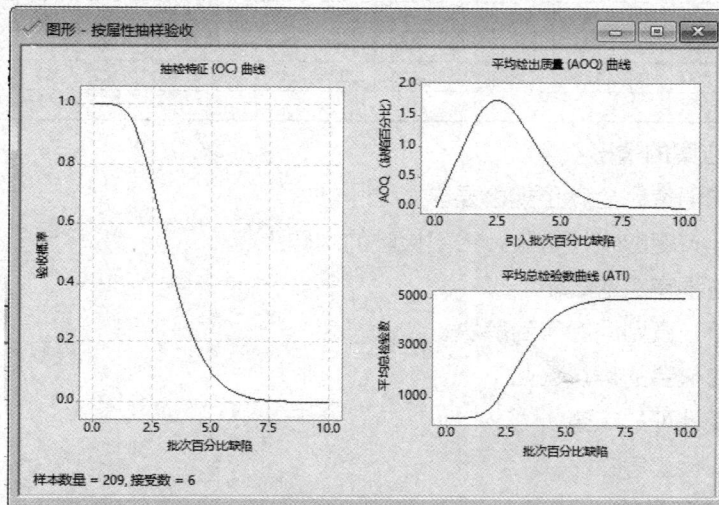

图6-9　图形框输出内容

由此可以得到：需要检验的零件为 209 个。如果所检验的 209 个零件中的不合格品不超过 6 个，则可以接受整批零件。如果有 7 个及以上的不合格品，则拒绝接受整批次零件。

6.2.4 零缺陷抽样方案

1. 零缺陷抽样概念

执行零缺陷抽样方案是一种经济的方法，在 AQL 值相同的情况下，零缺陷抽样方案的风险概率与执行 GB/T 2828.1—2012 相类似，但抽样量大大减少。

零缺陷抽样方案是指无论批量和样本大小，其抽样检验的接受数 Ac＝0。与 GB/T 2828.1—2012 相比，零缺陷抽样方案不存在加严、放宽、正常检验之间的转移规定，也没有二次抽样、多次抽样计划。实施零缺陷抽样检验方案时，其抽样方案的样本量可以采用执行正常检验一次抽样方案中的样本量。

2. 零缺陷抽样表

实际上，为了操作简便，需要对零缺陷抽样方案中的样本量与正常抽样方案中的样本量之间进行转化。其转化是基于二项分布的公式。按照 GB 2828 抽样表转化的零缺陷抽样表见

表 6-5。

表 6-5　零缺陷抽样表

接受质量限（AQL）	0.010	0.015	0.025	0.040	0.065	0.10	0.15	0.25	0.40	0.65	1.0	1.5	2.5	4.0	6.5	10.0
批量数	样本数															
2 to 8	*	*	*	*	*	*	*	*	*	*	*	*	5	3	2	2
9 to 15	*	*	*	*	*	*	*	*	*	*	13	8	5	3	2	2
16 to 25	*	*	*	*	*	*	*	*	*	20	13	8	5	3	3	2
26 to 50	*	*	*	*	*	*	*	*	32	20	13	8	5	5	5	3
51 to 90	*	*	*	*	*	*	80	50	32	20	13	8	7	6	5	4
91 to 150	*	*	*	*	*	125	80	50	32	20	13	12	11	7	6	5
151 to 280	*	*	*	*	200	125	80	50	32	20	20	19	13	10	7	6
281 to 500	*	*	*	315	200	125	80	50	48	47	29	21	16	11	9	7
501 to 1200	*	800	500	315	200	125	80	75	73	47	34	27	19	15	11	8
1201 to 3200	1250	800	500	315	200	125	120	116	73	53	42	35	23	18	13	9
3201 to 10 000	1250	800	500	315	200	192	189	116	86	68	50	38	29	22	15	9
10 001 to 35 000	1250	800	500	315	300	294	189	135	108	77	60	46	38	29	15	9
35 001 to 150 000	1250	800	500	490	476	294	218	170	123	96	74	56	40	29	15	9
150 001 to 500 000	1250	800	750	715	476	345	270	200	156	119	90	64	40	29	15	9
500 001 以上	1250	1200	1200	715	556	435	303	244	189	143	102	64	40	29	15	9

注："＊"代表需要百分百全检

3. 适用零缺陷抽样方案的情形

（1）对于产品的期望完全满足规格要求。

（2）对于一些非关键的特性检验，希望检验的次数少。

（3）需要满足各种审查的要求。

（4）临时的检验，直到问题得到解决。

（5）对于仓库内材料的确认检验。

（6）供应商接受认证的审查。

（7）一般的外观检查等。

4. 零缺陷抽样方案举例

继续上一个例子，创建 C=0 的计划。由于 Minitab 软件中 AQL 始终大于 0，因此，设定 AQL = 0.1，RQL=5.0%。另外，$\alpha=0.5$，$\beta=0.05$（α 的值设得高，且可以是任意值；另，$\alpha<1-\beta$）；则出现会话框见图 6-10。

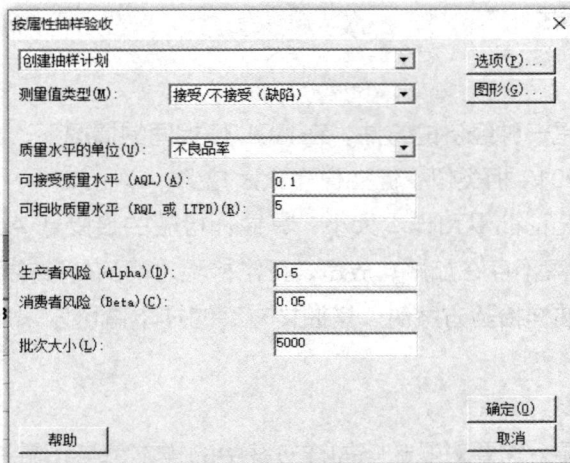

图 6-10　按属性抽样 C=0 会话框

点击确定后，出现会话框内容，见图 6-11。

```
生成的计划

样本数量    59
接受数       0

如果在 59 取样中的不良品数 ≤ 0，接受该批次；否则拒绝。

百分比
缺陷    接受概率    拒绝概率    AOQ      ATI
 0.1     0.943       0.057      0.093    342.2
 5.0     0.048       0.952      0.240   4760.4

平均交付质量限（AOQL）= 0.611（以 1.667 百分比缺陷）。
```

图 6-11　会话框输出

另有图形框输出，见图 6-12。

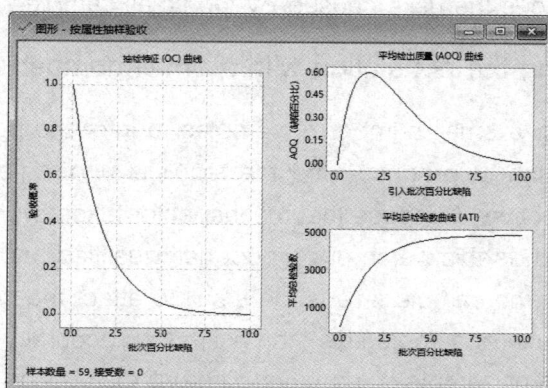

图 6-12　图形框输出

从输出内容可以看出，样本数量为 59，比按属性抽样大为减少。接受 0.1% 不合格率批次的概率为 94.3%；接受 5% 不合格率批次的概率为 4.8%，也意味着 95.2% 的批次被拒收。

根据零缺陷计划，如果所检验的 59 个零件中不合格品不超过 0 个，可以接受整批次；如果有 1 个或更多不合格品，则拒收整批次。

6.2　Risk of Too Small Sample Size

In data analysis, sufficient sample size is the basis to ensure the analysis accuracy. However, if the sample size is too large, it will increase a lot of costs, and the non-sampling error will increase. If the sample size is too small, the sampling error will increase. Therefore, it is very important to determine the appropriate sample size, which is introduced in Section 2 of Chapter 3. This section mainly introduces the risks caused by too small sample size.

6.2.1 The risk of inferring the population with a small number of samples

There is a risk of using a small number of samples to infer the population. If the inspector takes a sample from a batch of products and the result is unqualified, can we infer that the batch of product is all defective? Suppose that the inspector takes two samples, one is qualified and the other is unqualified. Is this batch of product half qualified and half unqualified, or what other situation? Obviously, the sample size is too small to affect the conclusion.

In fact, with the increase of sample size, the estimation accuracy of overall parameters will continue to improve. Different sample sizes have different distribution of sample mean values, as shown in Figure 6-3.

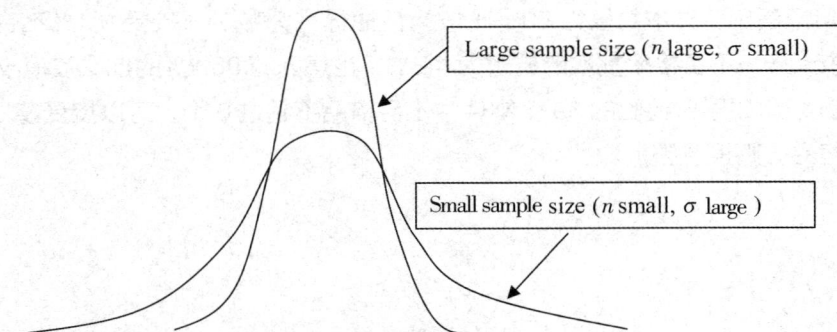

Figure 6-3　Distribution diagram of sample mean value of different sample sizes

In Figure 6-3, σ represents the inherent standard deviation (of population), and n is the sample size. It can be seen that if the sample size is large, the error is small.

Now with δ represents the actual difference between the two population averages. δ / σ is called test sensitivity or sensitivity coefficient, which is the difference of the average value of the sample divided by the standard deviation. Next, observe the effect of sample size (n) on the test probability, as shown in Figure 6-4.

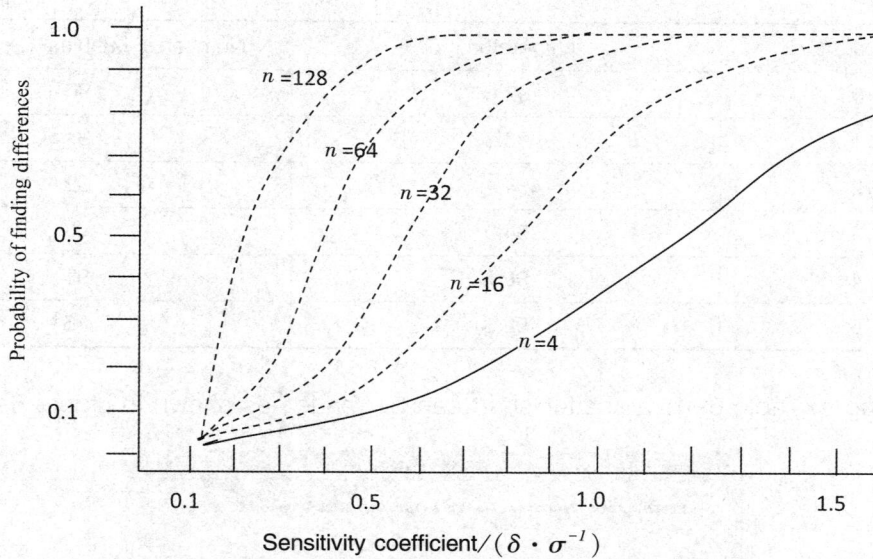

Figure 6-4 Influence of sample size (*n*) on the test probability

It can be seen from Figure 6-4 that under the same sensitivity coefficient, the probability of finding differences increases with the increase of sample size.

6.2.2 Correlation analysis with small sample size

It is incorrect to carry out correlation analysis with a small sample size. If there are too few pairs of data, such as less than 30pairs, we can draw a scatter chart based on the data and calculate the correlation coefficient, but the true correlation is difficult to judge.

Case:

A quality management team confirms the main causes. The end cause is "high water inlet temperature", and the crux is "too many bubbles on the surface of processed materials". For this reason, the team conducted a test and extracted 15 groups of data for analysis at different time and different inlet water temperatures. See Table 6-4 for the data.

Table 6-4 Data of temperature and material bubbles

No.	Temperature/℃	Number of bubbles/(pcs · m^{-2})
1	84.5	81
2	85.3	82
3	85.5	91
4	90.0	80
5	90.2	85
6	90.3	87
7	91.1	84
8	91.2	92
9	91.3	86

No.	Temperature/℃	Number of bubbles/(pcs · m^{-2})
10	92.0	90
11	93.3	91
12	94.4	92
13	94.9	93
14	96.2	95
15	98.3	96

According to Table 6-4, a scatter chart can be made, as shown in Figure 6-5.

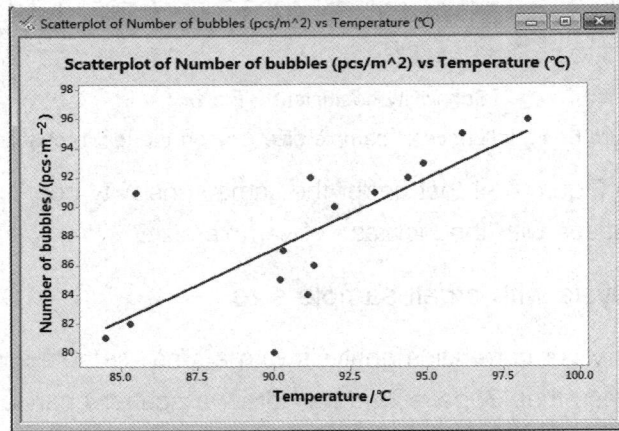

Figure 6-5   Scatter plot of temperature and material bubbles

From Figure 6-5, there seems to be an obvious regression line between temperature and material bubbles. According to calculation, the correlation coefficient of the two is 0.772, so it should be strong correlation. However, if the "correlation coefficient" in Minitab is used for inspection, it is found that the P value of this correlation coefficient is 0.000.

6.2.3 Sampling acceptance by attributes

For the purchaser, when purchasing a large number of parts and other materials, it is necessary to evaluate its process capability (C_{pk}) and check the control of process parameters with control chart, etc. but it is often subject to conditions. Therefore, sampling acceptance is a more effective method.

According to attribute sampling, AQL and RQL need to be drawn up, as well as the batch size, producer risk α and user risk β.

Case:

Now the purchaser purchases 5000 parts, and hopes to determine the quantity of parts to be inspected through attribute sampling inspection, and determine how many unqualified products are allowed in the sample. Now set AQL=1.5%, RQL=5.0%, α =0.05, β =0.1.

Use Minitab to analyze.

Choose "Quality tools" in the "Statistics", then select "Acceptance Sampling by Attributes", as shown in Figure 6-6.

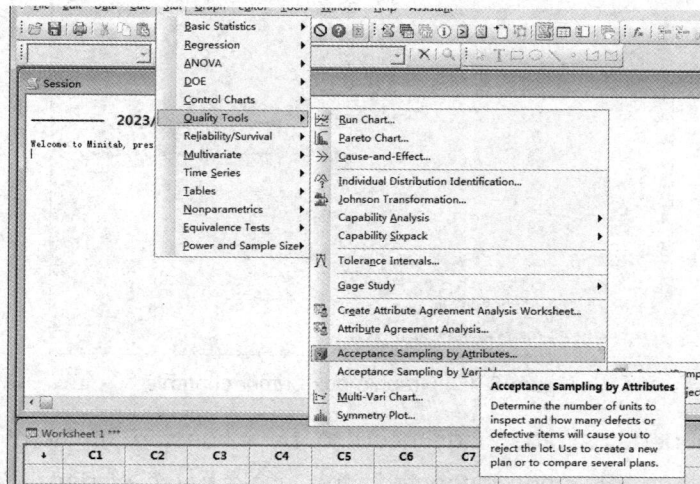

Figure 6-6   Select "Acceptance Sampling by Attributes"

After clicking, input "AQL =1.5%, RQL =5.0%, α =0.05, β =0.1" into the dialogue box; and input "5000" into the "Lot size". The dialogue box appears as shown in Figure 6-7.

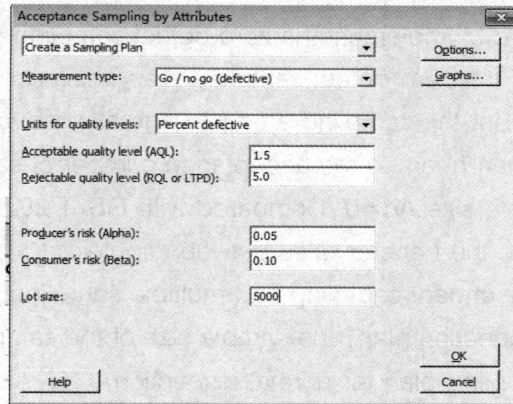

Figure 6-7   "Acceptance Sampling by Attributes" dialogue box

After clicking OK, the output content in the dialog box is shown in Figure 6-8.

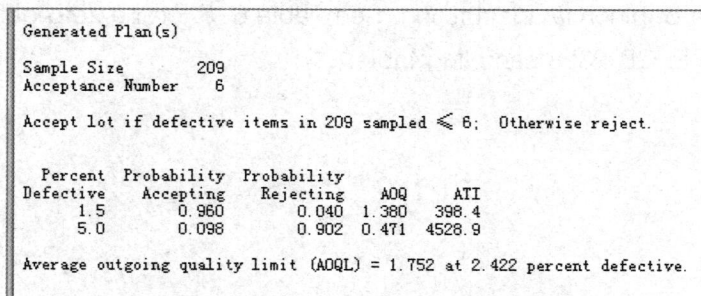

Figure 6-8   Dialogue box output

See Figure 6-9 for graphic frame output.

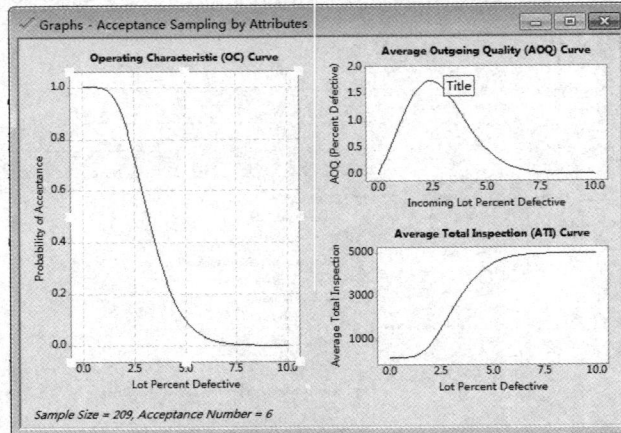

Figure 6-9　Graphic box output contents

It can be concluded that 209 parts need to be inspected. If there are no more than 6 unqualified products in 209 parts inspected, the whole batch of parts can be accepted. If there are 7 or more unqualified products, the whole batch of products will be rejected.

6.2.4 Zero defect sampling plan

1. Concept of zero defect sampling

It is an economic method to implement the zero defect sampling plan. Under the same AQL value, the risk probability of the zero defect sampling plan is similar to that of the implementation of GB/T 2828.1-2012, but the sampling amount is greatly reduced.

Zero defect sampling plan refers to the acceptance number of sampling inspection regardless of batch size and sample size Ac=0. Compared with GB/T 2828.1-2012, the zero-defect sampling plan does not have the transfer rules between tightened, relaxed and normal inspection, nor does it have the secondary sampling and multiple sampling plans. When implementing the zero-defect sampling inspection plan, the sample size of the sampling plan can be the sample size in the one-time sampling plan for normal inspection.

2. Zero defect sampling table

In fact, in order to simplify the operation, it is necessary to convert the sample size in the zero defect sampling plan to the sample size in the normal sampling plan. Its transformation is based on the formula of binomial distribution. See Table 6-5 for the zero-defect sampling table converted according to GB 2828 sampling table.

Table 6-5　Zero defect sampling table

Acceptance qualitylimit(AQL)	0.010	0.015	0.025	0.040	0.065	0.10	0.15	0.25	0.40	0.65	1.0	1.5	2.5	4.0	6.5	10.0
Batch number	Sample number															
2 to 8	*	*	*	*	*	*	*	*	*	*	*	*	5	3	2	2
9 to 15	*	*	*	*	*	*	*	*	*	*	13	8	5	3	2	2
16 to 25	*	*	*	*	*	*	*	*	*	20	13	8	5	3	3	2
26 to 50	*	*	*	*	*	*	*	*	32	20	13	8	5	5	5	3
51 to 90	*	*	*	*	*	*	80	50	32	20	13	8	7	6	5	4
91 to 150	*	*	*	*	*	125	80	50	32	20	13	12	11	7	6	5
151 to 280	*	*	*	*	200	125	80	50	32	20	20	19	13	10	7	6
281 to 500	*	*	*	315	200	125	80	50	48	47	29	21	16	11	9	7
501 to 1200	*	800	500	315	200	125	80	75	73	47	34	27	19	15	11	8
1201 to 3200	1250	800	500	315	200	125	120	116	73	53	42	35	23	18	13	9
3201 to 10 000	1250	800	500	315	200	192	189	116	86	68	50	38	29	22	15	9
10 001 to 35 000	1250	800	500	315	300	294	189	135	108	77	60	46	38	29	15	9
35 001 to 150 000	1250	800	500	490	476	294	218	170	123	96	74	56	40	29	15	9
150 001 to 500 000	1250	800	750	715	476	345	270	200	156	119	90	64	40	29	15	9
500 001 以上	1250	1200	1200	715	556	435	303	244	189	143	102	64	40	29	15	9

Note：" * " means 100% inspection is required

3. The case of applying the zero defect sampling plan

（1）The expectation of the product fully meets the specification requirements.

（2）For some non-critical characteristics, it is hoped that the number of inspections will be less.

（3）The requirements of various reviews need to be met.

（4）Temporary inspection until the problem is solved.

（5）Confirm and inspect the materials in the warehouse.

（6）Suppliers are subject to certification review.

（7）General appearance inspection, etc.

4. Example of zero defect sampling plan

Continue the case in 6.9 and create a plan with C=0. Since AQL in Minitab software is always greater than 0, set AQL=0.1, RQL=5.0%. In addition, $\alpha = 0.5$, $\beta = 0.05$（ The value of α is set high and can be any value; in addition, $\alpha < 1-\beta$). The dialog box appears as shown in Figure 6-10.

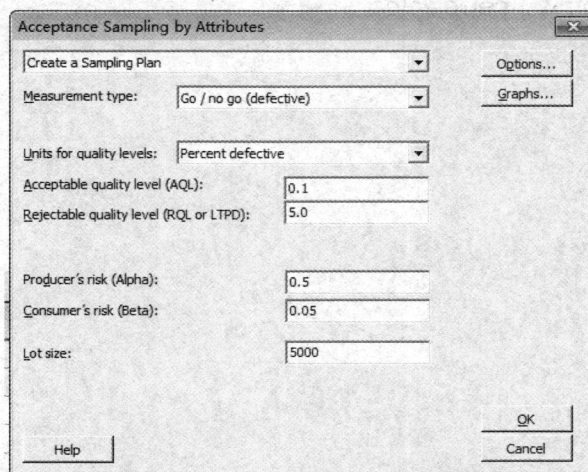

Figure 6-10　Dialogue box of "Acceptance Sampling by Attributes C=0"

After clicking OK, the dialog box will appear, as shown in Figure 6-11.

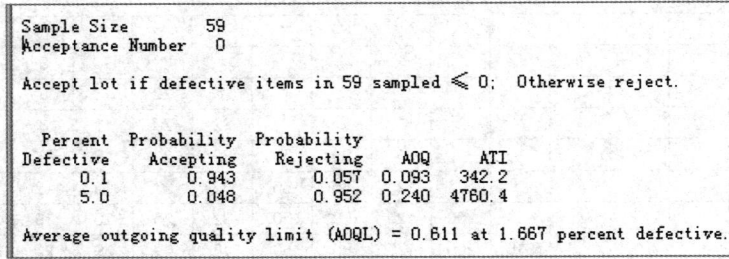

```
Sample Size        59
Acceptance Number   0

Accept lot if defective items in 59 sampled ≤ 0;  Otherwise reject.

 Percent  Probability  Probability
Defective  Accepting   Rejecting    AOQ    ATI
   0.1      0.943        0.057      0.093  342.2
   5.0      0.048        0.952      0.240 4760.4

Average outgoing quality limit (AOQL) = 0.611 at 1.667 percent defective.
```

Figure 6-11   Dialogue box output

There is also graphic box output, as shown in Figure 6-12.

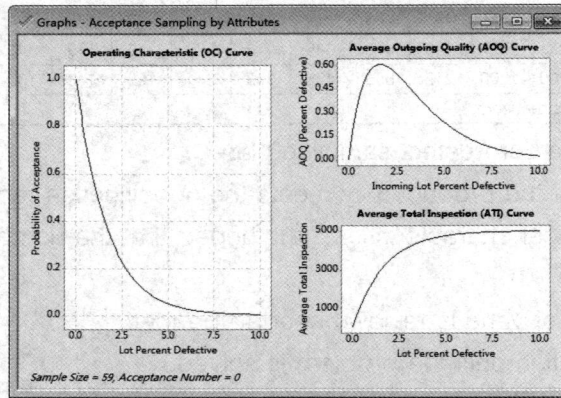

Figure 6-12   Graphics box output

From the output, we can see that the number of samples is 59, which is much less than sampling by attributes. The probability of accepting batches with 0.1% unqualified rate is 94.3%. The probability of accepting a batch with a 5% rejection rate is 4.8%, which also means that 95.2% of the batch will be rejected.

According to the zero defect plan, if there are no more than 0 unqualified products in the 59 parts inspected, the whole batch can be accepted. If there is one or more nonconforming products, the whole batch will be rejected.

6.3 应用统计方法不适宜

在数据分析过程中，应用统计方法强调适宜性。如果应用不适宜，就有可能造成数据分析不正确等。应用统计方法不适宜的情形非常多，本节只能作部分介绍。

6.3.1 抽样后进行分析的结论是否正确

通过抽样进行数据分析是常有的事情。在实践中会遇到这样的情形：抽样方法是恰当的，但对于抽样的数据如何进行分析却存在着问题。

举例：

在质量管理小组活动中，经常有小组从作业工人中随机选代表进行现场操作的情形。现在有一个班组共 50 个工人，其中技能好的有 40 人，技能差的有 10 人。在抽样之前，质量管理小组只是从介绍中知晓该班组的整体操作水平是较好的，并不清楚有 10 人操作技能差的事情。现在小组从中抽取 5 人组成一个小组取参加测试，那么，这个测试结果是否能够准确地反映该班组操作水平属于较好的情况？

该小组认为，从 50 个工人中随机抽取 5 人进行操作，能够正确反映班组的作业水平的，而且能够验证班组成员的操作水平都是较好的。

正确的分析过程如下：

本例中抽取样本的方法属于超几何分布，即从一个总体 N 中，不放回地随机抽 n 个产品，这 n 个产品中，技能差的个数 X 服从超几何分布。

如果随机变量 X 的概率密度函数为

$$P(x = i) = \frac{\binom{c}{i} \cdot \binom{N-C}{n-i}}{\binom{N}{n}}$$

其中，$i = 0, 1, \cdots, n$，N 为总体数，C 为总体中具有某个特征的单位数，在本例中为技能差的工人。n 是从 N 中抽取的样本数，i 为样本中具有某特征的单位数。则称 X 服从超几何分布。

因此

$$P(x = 0) = \frac{\binom{10}{0} \cdot \binom{50-10}{5-0}}{\binom{50}{5}}$$

$$= \frac{\dfrac{10!}{0! \ 10!} \cdot \dfrac{40!}{5! \ 35!}}{\dfrac{50!}{5! \ 45!}}$$

$$= 0.311$$

从分析的数据看，样本中没有技能差的工人的概率为 0.311，意味着在随机抽取的 5 名工人中，存在技能差的工人。因此，不能认为在抽取的 5 个工人中，都具有较好的操作水平。

6.3.2 由频率占比判断而导致分析结果失误

在质量管理实践中，分析哪个质量问题是主要问题，经常采取的方法是通过不同质量问题的占比大小来进行判断。但有时候这样的方法会有失偏颇。

举例：

某公司在质量检查中发现，所生产的某型机器总是发生噪声问题。经过长期的现场观察，发现在噪声发生的时候，总是有导流管堵塞、螺栓松动、金属盖板裂纹等问题出现。为了调查清楚究竟是哪个问题主要导致了噪声的产生，检查人员经过调查获得了相关数据，整理成表 6-6。

表 6-6　导致机器噪声的问题　　　　　　　　　单位：次

导致机器噪声的问题	发生次数	出现噪声的次数
导流管堵塞	480	440
螺栓松动	310	220
金属盖板裂纹	760	540

根据表 6-6 的数据，检查人员进行了三个问题出现频率的相应占比计算，分别为：

"导流管堵塞"占"噪声"的比例为 440/480 = 0.917；

"螺栓松动"占"噪声"的比例为 220/310 = 0.710；

"金属盖板裂纹"占"噪声"的比例为 540/760 = 0.711。

于是检查人员通过三个质量问题的频率比较后得出结论："导流管堵塞"是导致机器噪声的最主要问题。

这样的结论似乎无懈可击。但是复检人员在复查过程中，发现检查人员的结论不正确。理由如下：

整个机器产生噪声及相应问题的分析，应当用贝叶斯公式来进行计算。

贝叶斯公式的表述为：如果时间 A1，A2，…，An，两两互不相容，

$$P(Ai) > 0,$$

$$\sum_{i=1}^{\infty} Ai = \Omega$$

则对任意满足 $P(B) > 0$ 的事件 B，有

$$P(Ai/B) = \frac{P(Ai)P(B/Ai)}{\sum_{i=1}^{\infty} P(Ai)P(B/Ai)}$$

其中，P（Ai）为先验概率，P（Ai/B）为后验概率。

复检人员提出，贝叶斯公式可用于机器的故障分析。现在用 A 表示噪声，Di 表示三个质量问题。那么，问题的总数为 T = 480+310+760 = 1550。

P（D1）= 480/1550 = 0.31；

P（D2）= 310/1550 = 0.20；

P（D3）= 760/1550 = 0.49；

P（A／D1）= 440/480 = 0.917；

P（A／D2）= 220/310 = 0.710；

P（A／D3）= 540/760 = 0.711。

$$\sum_{i=1}^{3} P\,(\text{Ai})\,P\,(\text{B／Ai}) = 0.31 \cdot 0.917 + 0.20 \cdot 0.710 + 0.49 \cdot 0.711$$

$$= 0.2843+0.1420+0.3484 = 0.7747$$

所以，P（D1/A）= 0.2843/0.7747 = 0.367；

P（D2/A）= 0.1420/0.7747 = 0.183；

P（D3/A）= 0.3484/0.7747 = 0.450。

由此可以看出，"金属盖板裂纹"出现的概率是 0.450，故复检人员判断，"金属盖板裂纹"是导致噪声的最大问题。

6.3.3 隐藏着的相关性

举例：

A 公司是一家著名的饮料生产企业。月初，A 公司在各大媒体上发布广告，定于月底在一个广场上举行最佳饮料评选活动，欢迎全体市民参与。到了月底，活动如期进行。在广场中央的很多桌子上，摆放着 8 种不同企业生产的饮料，为避嫌，所有的饮料都倒入式样一致的瓶子中，瓶身上除了各自的编号 1 至 8，没有其他任何标记。参与活动的市民在一一品尝 8 种饮料后，把自己认为最佳的饮料编号投入投票箱。投票环节结束后，工作人员清点票箱，结果显示：A 公司生产的饮料得票最高，获得了最佳饮料称号。

这个活动的过程，看似参与品尝活动的市民是随机来到现场的人员，与 A 公司没有丝毫关联，但实质上该活动隐藏着一个"相关性"的秘密，那就是当天出席活动的大多数人员，本身是A 公司饮料的爱好者，而且熟悉 A 公司饮料的口味。而不喜欢 A 公司饮料的人员则对于 A 公司举办的活动缺少兴趣，这部分人员大多不会参加。因此，这个相关性就是"喜欢 A 公司饮料的市民"的投票与"最佳饮料"的结果是紧密相关的。

6.3.4 应用不同分布曲线的情形

在统计分析过程中，引用恰当的分布曲线可以很好地与数据进行拟合等工作，为分析结论的正确性提供依据。引用分布曲线的不同，会有着不同的结论。

举例：

在一个机械加工企业中，有多台某型加工机械。在最近的一个月内共发生了 25 次机械故障，维修人员对于这些故障机械的维修时间（单位：分钟）为：48，20，111，38，50，47，28，79，45，98，22，52，67，51，96，35，43，40，53，64，76，87，48，54，53。

现在需要求出这个机械故障模型的参数估计，然后用该模型求出故障平均维修时间的估计值和置信区间。

分析 1：

为了便于分析，把机械的维修时间看成是"寿命的生存时间"。现应用 Minitab 软件，先建立工作表，部分工作表见图 6-13。

图 6-13　工作表（部分）

在 Minitab 软件"统计"中选择"可靠性/生存"，选择"分布分析（右删失）"中的"分布 ID 图"，见图 6-14。

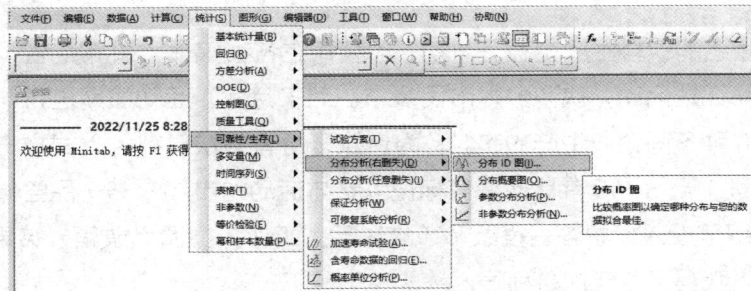

图 6-14　分布 ID 图

在会话框"分布 ID 图 – 右删失"中，"变量"选择"维修时间"，点击"指定"，见图 6-15。

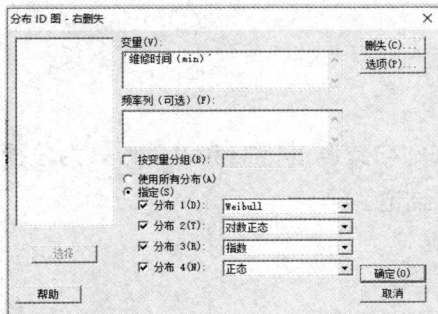

图 6-15　分布 ID 图 – 右删失会话框

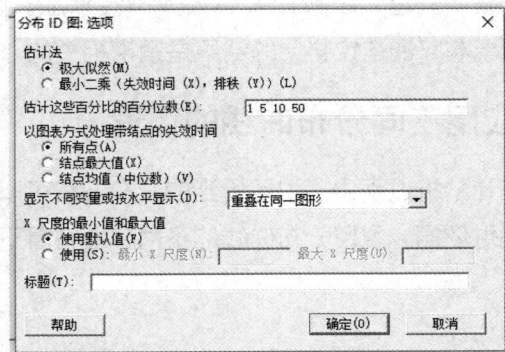

图 6-16　分布 ID 图:选项会话框

点击"选项",在"估计法"中选择"极大似然",见图6-16。

全部确定后,会话框输出内容见图6-17。

拟合优度

分布	Anderson-Darling(调整)
Weibull	1.062
对数正态	0.872
指数	4.717
正态	1.245

百分位数表格

分布	百分比	百分位数	标准误	95% 正态置信区间 下限	上限
Weibull	1	10.8299	3.28230	5.97925	19.6157
对数正态	1	19.2500	3.13962	13.9830	26.5008
指数	1	0.564829	0.112966	0.381660	0.835906
正态	1	2.50620	8.88647	-14.9110	19.9234
Weibull	5	20.2574	4.31709	13.3408	30.7599
对数正态	5	25.6924	3.33882	19.9154	33.1453
指数	5	2.88268	0.576537	1.94785	4.26616
正态	5	18.2356	7.08059	4.35788	32.1133
Weibull	10	26.7108	4.67104	18.9598	37.6306
对数正态	10	29.9670	3.42624	23.9509	37.4942
指数	10	5.92126	1.18425	4.00105	8.76303
正态	10	26.6209	6.22955	14.4112	38.8306
Weibull	50	55.0819	4.98996	46.1209	65.7841
对数正态	50	51.5718	4.36928	43.6814	60.8875
指数	50	38.9549	7.79097	26.3222	57.6504
正态	50	56.2	4.61615	47.1525	65.2475

平均故障时间间隔表格

分布	均值	标准误	95% 正态置信区间 下限	上限
Weibull	56.3239	4.6723	47.8720	66.2679
对数正态	56.4129	4.9892	47.4348	67.0903
指数	56.2000	11.2400	37.9748	83.1719
正态	56.2000	4.6161	47.1525	65.2475

图6-17 会话框输出内容

图形框输出内容见图6-18。

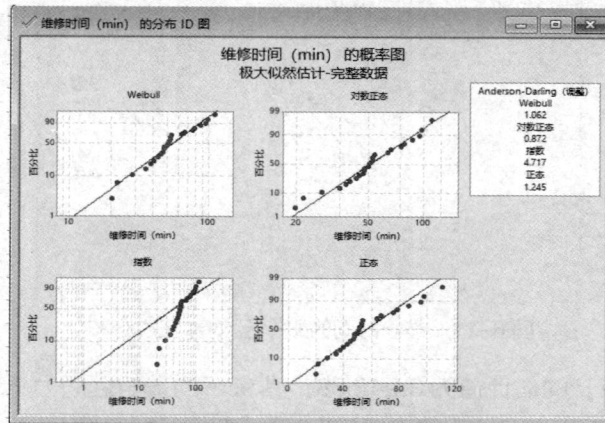

图6-18 图形框输出内容

从图6-17拟合优度Anderson-Darling值中可以看到,对数正态的值0.872最小,结合图6-18中的图形,可以认为"对数正态分布"最拟合这组数据。

下面进行参数估计。

在"统计"中选择"可靠性/生存",选择"分布分析(右删失)"中的"参数分布分析"。

在"参数分布分析 - 右删失"对话框中,"变量"中选择"维修时间",在"假定分布"中选择"对数正态",见图6-19。

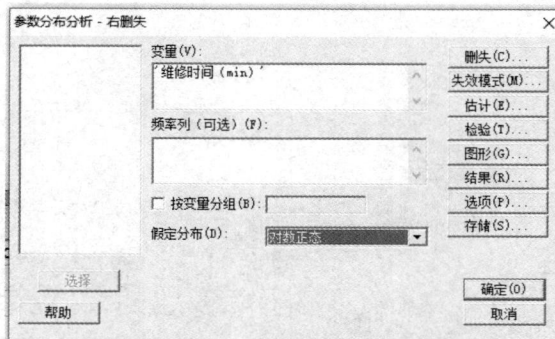

图6-19 参数分布分析 - 右删失会话框

点击"估计","估计法"中选择"极大似然"。全部确定后，会话框输出内容（部分）见图6-20。

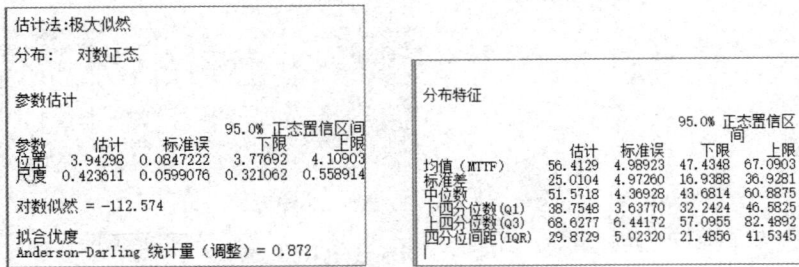

图6-20 会话框输出内容（部分）

图形框输出为"维修时间的概率图 对数正态分布"，见图6-21。

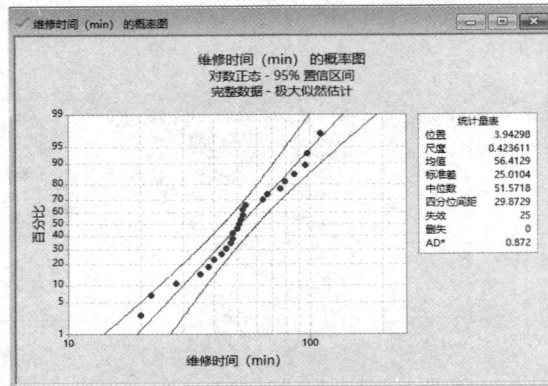

图6-21 维修时间的概率图 对数正态分布

可以看出，位置参数 μ 的估计值为3.942 98，其95%的置信区间为（3.776 92，4.109 03）；尺度参数 σ 的估计值为0.423 611，其95%的置信区间为（0.321 062，0.558 914）；平均维修时间 MTTF 估计值为56.4129，其95%的置信区间为（47.4348，67.0903）。

分析2：

现在参数估计选择"指数分布"。

在"统计"中选择"可靠性/生存"，选择"分布分析（右删失）"中的"参数分布分析"。

在"参数分布分析 – 右删失"会话框中，"变量"中选择"维修时间"，在"假定分布"中选择"指数"。"估计"中的"估计法"选择"极大似然"。全部确定后，会话框输出内容（部分）见图6-22。

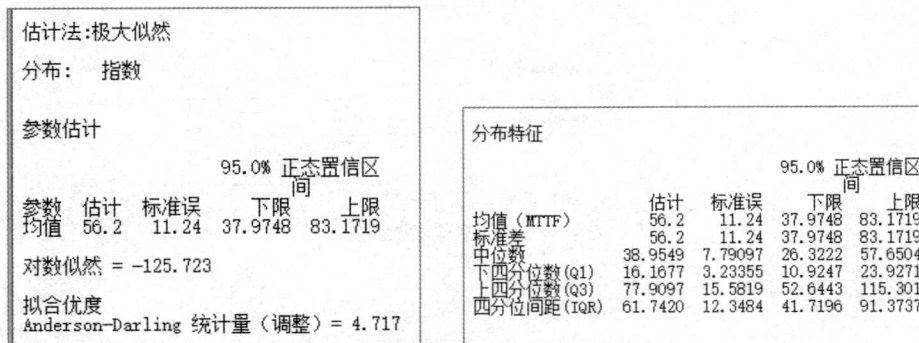

图6-22 会话框输出内容（部分）

图形框输出为"维修时间的概率图 指数分布",见图6-23。

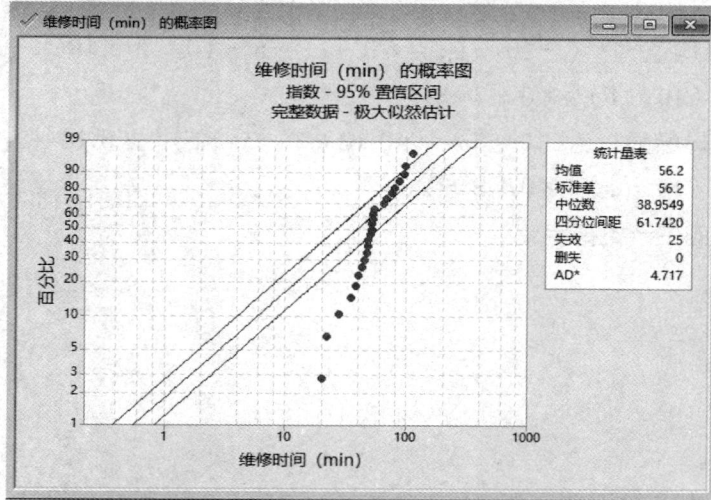

图6-23 维修时间的概率图 指数分布

平均维修时间为56.2，其95%的置信区间为（37.9748，83.1719）。

通过"对数正态"拟合及分析的结果与"指数"拟合及分析的结果的比较，发现两者的平均维修时间虽然接近，但95%的置信区间相差大。可见，应用适宜的分布曲线很重要。

6.3.5 应用期望值不适宜

平均值是很多管理活动中常用的方法，但是也存在应用平均值不适宜的情形。

举例：

有A、B两条生产线，各自生产产品的误差数据见表6-7。现在需要判断哪条生产线的生产更稳定。

表6-7 产品误差数据

A 产品误差/‰	−10	−5	0	5	10
A 出现误差的概率	0	0.18	0.54	0.18	0.10
B 产品误差/‰	−10	−5	0	5	10
B 出现误差的概率	0.18	0.10	0.24	0.30	0.18

分析1：用平均值来分析

A产品的误差平均值为0，B产品的误差平均值为0，表明A与B没有区别。

分析2：用期望值来分析

$E（A）=（−10）×0+（−5）×0.18+0×0.54+5×0.18+10×0.10 = 1$；

$E（B）=（−10）×0.18+（−5）×0.10+0×0.24+5×0.30+10×0.18 = 1$。

表明A与B没有区别。

分析3：用离差的平方平均值来分析

计算离差的平方平均值公式为

$$\sum_{i=1}^{n} (X_i - EX)^2 p(X_i)$$

A 的离差平方平均值为：$(-10-1)^2 \times 0 + (-5-1)^2 \times 0.18 + (0-1)^2 \times 0.54 + (5-1)^2 \times 0.18 + (10-1)^2 \times 0.1 = 18$；

B 的离差平方平均值为：$(-10-1)^2 \times 0.18 + (-5-1)^2 \times 0.10 + (0-1)^2 \times 0.24 + (5-1)^2 \times 0.30 + (10-1)^2 \times 0.18 = 45$。

因此，A 生产线的生产更稳定。

6.3   Inappropriate Application of Statistical Method

———

In the process of data analysis, statistical methods are used to emphasize suitability. If the application is not appropriate, it may cause incorrect data analysis, etc. There are many cases where the application of statistical methods is not suitable, and this section can only give a partial introduction.

6.3.1 Whether the conclusion of analysis after sampling is sorrect

It is common to analyze data through sampling. In practice, we will encounter such situations: the sampling method is appropriate, but there are problems in how to analyze the sampled data.

Case:

In the activities of the quality management team, the team often selects representatives randomly from the workers to carry out on-site operations. There are now 50 workers in a team, including 40 with good skills and 10 with poor skills. Before sampling, the quality management team only knew from the introduction that the overall operation level of the team was good, and did not know that there were 10 people with poor operation skills. Now the team selects five people from the team to form a team to take part in the test. Then, whether the test result accurately reflect the operation level of the team is good?

The team believes that 5 workers randomly selected from 50 workers can correctly reflect the operation level of the team and verify that the operation level of the team members is good.

The correct analysis process is as follows.

The method of sampling in this example belongs to hypergeometric distribution, that is, from a total N, randomly select n products without putting them back. Among these n products, the number X of poor skills follows hypergeometric distribution.

If the probability density function of random variable X is

$$P(x = i) = \frac{\binom{c}{i} \cdot \binom{N - C}{n - i}}{\binom{N}{n}},$$

in which, $i = 0, 1, \ldots, n$; N is the total number, C is the number of units with a certain characteristic in the population, and in this case, the workers with poor skills. N is the number of samples extracted from N, and i is the number of units with certain characteristics in the sample. Then X is said to obey hypergeometric distribution. Therefore:

$$P(x=0) = \frac{\binom{10}{0} \cdot \binom{50-10}{5-0}}{\binom{50}{5}}$$

$$= \frac{\dfrac{10!}{0! \ 10!} \cdot \dfrac{40!}{5! \ 35!}}{\dfrac{50!}{5! \ 45!}}$$

$$= 0.311$$

From the analysis data, the probability of no workers with poor skills in the sample is 0.311, which means that there are workers with poor skills among the five randomly selected workers. Therefore, it cannot be considered that all the five workers selected have good operation level.

6.3.2 Analysis result error caused by frequency ratio judgment

In the practice of quality management, to analyze which quality problem is the main problem, the often adopted method is to judge by the proportion of different quality problems. But sometimes this method will be biased.

Case:

During the quality inspection of a company, it was found that a certain type of machine produced by the company always has noise problems. Through long-term on-site observation, it is found that when the noise occurs, there are always problems such as blockage of the guide pipe, loose bolts and cracks in the metal cover plate, etc. In order to find out which problem mainly caused the noise, the inspectors obtained relevant data through investigation and compiled them into Table 6-6.

Table 6-6　Problems that cause machine noise

Problems that cause machine noise	Number of occurrences	Number of noises
Blocked diversion pipe	480	440
Loose bolts	310	220
Cracks in metal cover plate	760	540

According to the data in Table 6-6, the inspectors calculated the corresponding proportion of the frequency of three problems, which are:

The ratio of "Blocked diversion pipe" to "noise" is 440/480 = 0.917;

The ratio of "Loose bolts" to "noise" is 220/310 = 0.710;

The ratio of "Cracks in metal plate" to "noise" is 540/760 = 0.711.

Then the inspector concluded through the frequency comparison of the three quality problems that the "Blocked diversion pipe" is the main problem causing the machine noise.

Such a conclusion seems to be impeccable. However, the reinspection personnel found that the conclusions of the inspectors were incorrect during the review. The reasons are as follows:

Bayesian formula should be used to calculate the noise generated by the whole machine and the corresponding problems.

The expression of Bayesian formula is: if the timeA1, A2, $\cdots$, An are incompatible with each other,

P (Ai) > 0,

$$\sum_{i=1}^{\infty} Ai = \Omega$$

Then, for any event B that satisfies P(B) > 0, there is:

$$P \text{ (Ai/B)} = \frac{P(\text{Ai})P(\text{B/Ai})}{\sum_{i=1}^{\infty} P(\text{Ai})P(\text{B/Ai})}$$

In the formula, P (Ai) is a prior probability, P (Ai/B) is a posterior probability.

The reinspection personnel proposed that Bayesian formula can be used for machine fault analysis. Now, A represents noise and Di represents three quality problems. Then, the total number of questions is T = 480+310+760 = 1550.

P (D1) = 480/1550 = 0.31;

P (D2) = 310/1550 = 0.20;

P (D3) = 760/1550 = 0.49;

P (A/D1) = 440/480 = 0.917;

P (A/D2) = 220/310 = 0.710;

P (A/D3) = 540/760 = 0.711.

$$\sum_{i=1}^{3} P(\text{Ai})P(\text{B/Ai}) = 0.31 \cdot 0.917 + 0.20 \cdot 0.710 + 0.49 \cdot 0.711$$

$$= 0.2843 + 0.1420 + 0.3484 = 0.7747$$

Therefore, P (D1/A) = 0.2843/0.7747 = 0.367;

P (D2/A) = 0.1420/0.7747 = 0.183;

P (D3/A) = 0.3484/0.7747 = 0.450.

It can be seen from this that the probability of "crack on metal cover plate" is 0.450, so the reinspection personnel judged that "crack on metal cover plate" is the biggest problem causing noise.

6.3.3 Hidden correlation

Case:

Company A is a famous beverage manufacturer. At the beginning of the month, Company A published advertisements in various major media and was scheduled to hold the best beverage selection activity in a square at the end of the month. All citizens are welcome to participate. By the end of the month, the activity was on schedule. On many tables in the center of the square, there are eight kinds of drinks produced by different enterprises. In order to avoid suspicion, all drinks are poured into bottles with the same style. There are no other marks on the bottles except their respective numbers 1 to 8. After tasting eight kinds of drinks one by one, the citizens who participated in the activity put their best drink number into the voting box. After the voting session, the staff counted the ticket boxes, and the results showed that the beverage produced by Company A received the highest number of votes and won the title of the best beverage.

In the process of the event, it seems that the citizens participating in the tasting activity are random people who come to the scene, and have nothing to do with company A, but in fact, the event hides a secret of "relevance", that is, most of the people attending the event that day are themselves fans of company A's drinks, and are familiar with the taste of company A's drinks. The people who don't like the drinks of Company A are not interested in the activities held by Company A, and most of them will not participate. Therefore, the correlation is that the vote of "citizens who like company A's drinks" is closely related to the result of "the best drink".

6.3.4 Application of different distribution curves

In the process of statistical analysis, the proper distribution curve can be used to fit the data well and provide a basis for the correctness of the analysis conclusion. Different conclusions will be drawn according to different distribution curves.

Case:

In a machining enterprise, there are multiple processing machines of a certain type. A total of 25 mechanical failures have occurred in the last month, and the maintenance time (unit: minutes) for these mechanical failures is: 48, 20, 111, 38, 50, 47, 28, 79, 45, 98, 22, 52, 67, 51, 96, 35, 43, 40, 53, 64, 76, 87, 48, 54, 53.

Now we need to find out the parameter estimation of the mechanical failure model, and then use the model to find out the estimated value and confidence interval of the mean maintenance time.

Analysis 1:

In order to facilitate analysis, the maintenance time of machinery is regarded as "the survival time of life". Now set up the worksheet in the Minitab, part of the worksheet is shown in Figure 6-13.

↓	C1	C2
	Repair time (min)	
1	48	
2	20	
3	111	
4	38	
5	50	
6	47	
7	28	
8	79	
9	45	
10	98	
11	22	
12	52	
13	67	
14	51	
15	96	
16	35	

Figure 6-13 Part of the worksheet

In Minitab software, choose "Reliability/Survival" in "Statistics", select "Distribution ID plot" in the "Distribution Analysis(Right Censoring)", as shown in Figure 6-14.

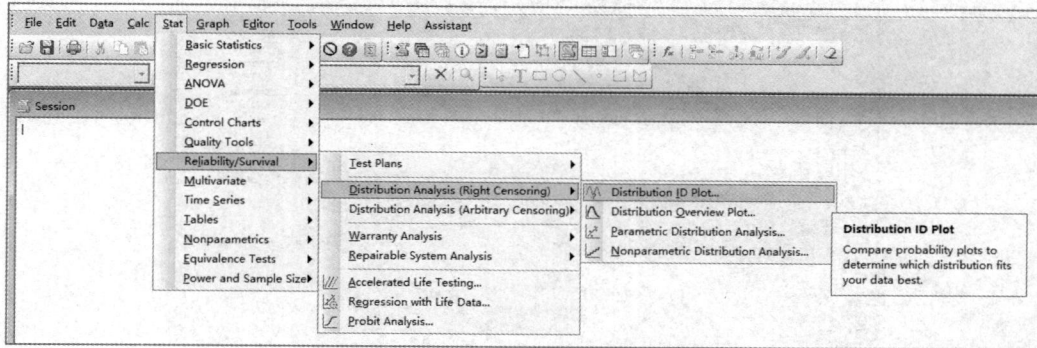

Figure 6-14　Distribution ID Plot

In the dialogue box of "Distribution ID Plot-Right Censoring", choose "Repair time" in the "Variables"; then Click "Specify", as shown in Figure 6-15.

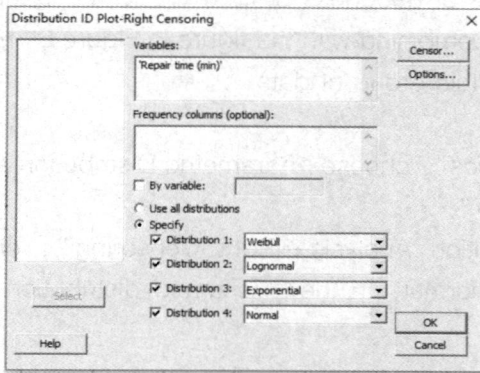

Figure 6-15　Distribution ID Plot-Right Censoring

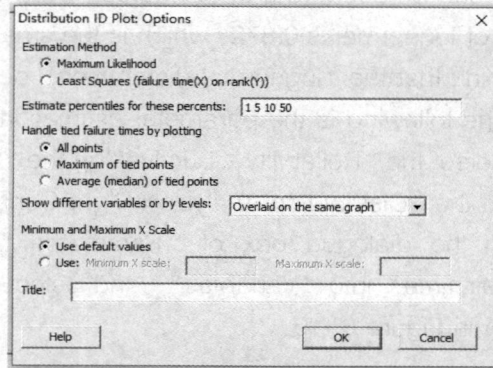

Figure 6-16　Distribution ID Plot: Options

Click "Options". Choose "Maximum Likelihood" in "Estimation Method", as shown in Figure 6-16.

After clicking all "OK", the output contents will appear in the dialogue box as shown in Figure 6-17.

Figure 6-17　Output contents of dialogue box

The output content of the graphic box is shown in the Figure 6-18.

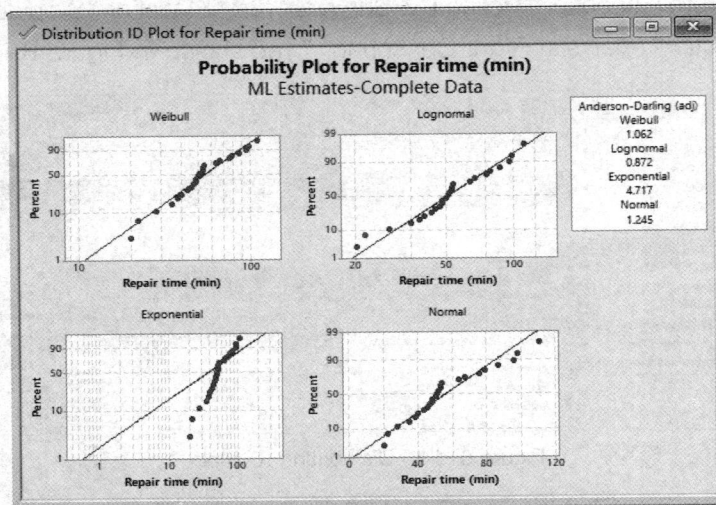

Figure 6-18   Output contents of the graphic box

It can be seen from the Anderson-Darling value of goodness of fit in Figure 6-17 that the value of lognormal is 0.872, which is the smallest. Combining with the figure in Figure 6-18, we can think that the "lognormal distribution" best fits this group of data.

The following is the parameter estimation.

Select the "Reliability / Survival" in the "Statistics"; choose "Parametric Distribution Analysis" in the "Distribution Analysis (Right Censoring)".

In the dialogue box of "Parametric Distribution Analysis—Right Censoring", choose "Repair time" into "Variables", and select "Lognormal" in the "Assumed distribution", as shown in Figure 6-19.

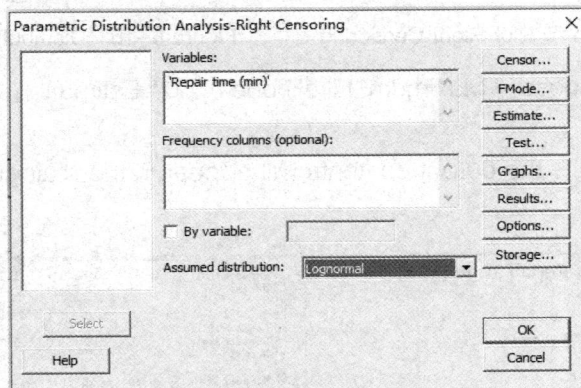

Figure 6-19   Parametric Distribution Analysis—Right Censoring

Click "Estimate". Choose "Maximum likelihood" in the "Estimation Method". After all are confirmed, the output content (part) of the session box is shown in Figure 6-20.

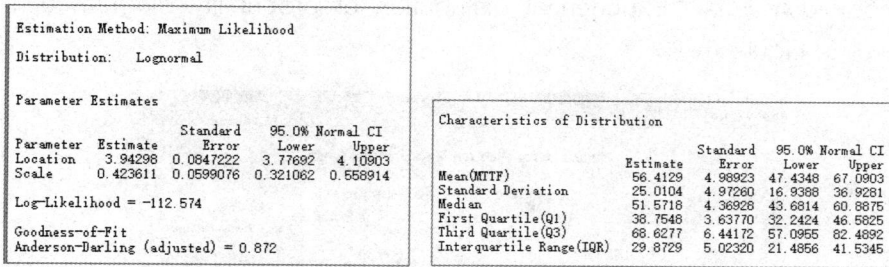

```
Estimation Method: Maximum Likelihood

Distribution:  Lognormal

Parameter Estimates

                    Standard    95.0% Normal CI
Parameter  Estimate   Error    Lower    Upper
Location   3.94298  0.0847222  3.77692  4.10903
Scale      0.423611 0.0599076  0.321062 0.558914

Log-Likelihood = -112.574

Goodness-of-Fit
Anderson-Darling (adjusted) = 0.872
```

```
Characteristics of Distribution

                              Standard   95.0% Normal CI
                     Estimate   Error    Lower    Upper
Mean(MTTF)            56.4129   4.98923  47.4348  67.0903
Standard Deviation   25.0104   4.97260  16.9388  36.9281
Median               51.5718   4.36928  43.6814  60.8875
First Quartile(Q1)   38.7548   3.63770  32.2424  46.5825
Third Quartile(Q3)   68.6277   6.44172  57.0955  82.4892
Interquartile Range(IQR) 29.8729 5.02320 21.4856 41.5345
```

Figure 6-20　Dialogue box output content（part）

The graphic box output is "Probability Plot for Repair Time Lognormal Distribution", as shown in Figure 6-21.

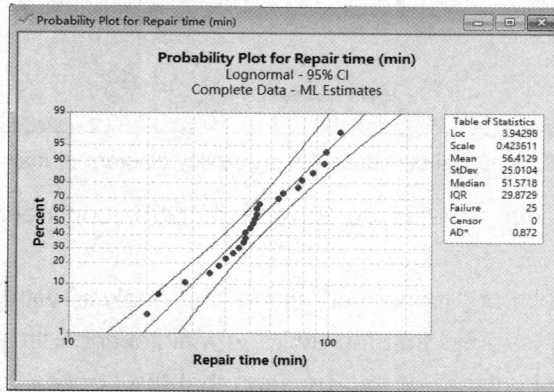

Figure 6-21　Probability Plot for Repair Time Lognormal Distribution

It can be seen that the estimated value of the position parameter μ is 3.942 98, and its 95% confidence interval is（3.776 92, 4.109 03）. The estimated value of scale parameter σ is 0.423 611, and its 95% confidence interval is（0.321 062, 0.558 914）; The estimated value of MTTF is 56.4129, with a 95% confidence interval of（47.4348, 67.0903）.

Analysis 2：

Now select "exponential distribution" for parameter estimation.

Select the "Reliability / Survival" in the "Statistics"; choose "Parametric Distribution Analysis" in the "Distribution Analysis（Right Censoring）".

In the dialogue box of "Parametric Distribution Analysis—Right Censoring", choose "Repair time" into "Variables", and select "Exponential" in the "Assumed distribution". Choose "Maximum likelihood" in the "Estimation Method". After all are confirmed, the output content（part）of the dialogue box is shown in Figure 6-22.

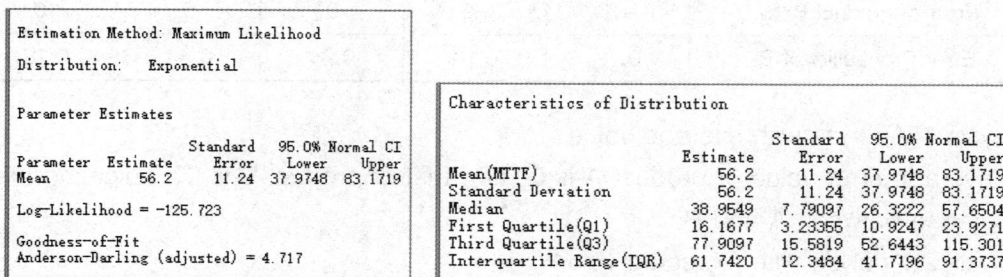

```
Estimation Method: Maximum Likelihood

Distribution:  Exponential

Parameter Estimates

                    Standard   95.0% Normal CI
Parameter  Estimate  Error    Lower    Upper
Mean         56.2    11.24   37.9748  83.1719

Log-Likelihood = -125.723

Goodness-of-Fit
Anderson-Darling (adjusted) = 4.717
```

```
Characteristics of Distribution

                            Standard   95.0% Normal CI
                   Estimate   Error    Lower    Upper
Mean(MTTF)           56.2    11.24   37.9748  83.1719
Standard Deviation   56.2    11.24   37.9748  83.1719
Median              38.9549  7.79097 26.3222  57.6504
First Quartile(Q1)  16.1677  3.23355 10.9247  23.9271
Third Quartile(Q3)  77.9097 15.5819  52.6443 115.301
Interquartile Range(IQR) 61.7420 12.3484 41.7196 91.3737
```

Figure 6-22　Dialogue box output content（part）

Graphic box output is "Exponential distribution of probability diagram of maintenance time", as shown in Figure 6-23.

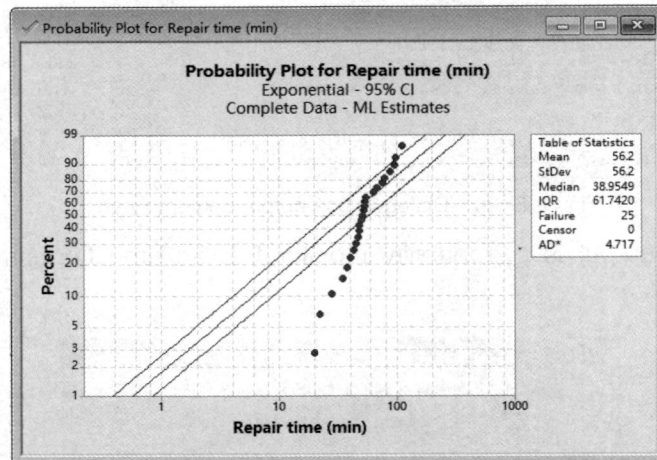

Figure 6-23 Exponential distribution of probability diagram of maintenance time

The average maintenance time is 56.2, and its 95% confidence interval is (37.9748, 83.1719).

By comparing the results of "lognormal" fitting and analysis with the results of "exponential" fitting and analysis, it is found that the average maintenance time of the two is close, but the 95% confidence interval is large. It can be seen that the application of appropriate distribution curve is very important.

6.3.5 Unsuitable expected value of application

Average value is a common method in many management activities, but there are also cases where the application of average value is not appropriate.

Case:

There are two production lines A and B, and the error data of their products are shown in Table 6-7. Now we need to determine which production line is more stable.

Table 6-7 Product error data

Error of product A/‰	−10	−5	0	5	10
Error Probability of A	0	0.18	0.54	0.18	0.10
Error of product B/‰	−10	−5	0	5	10
Error Probability of B	0.18	0.10	0.24	0.30	0.18

Analysis 1: Analyze with average value

The average error value of product A is 0, and that of product B is 0, indicating that there is no difference between A and B.

Analysis 2: Analyze with expected value

$E(A) = (-10) \times 0 + (-5) \times 0.18 + 0 \times 0.54 + 5 \times 0.18 + 10 \times 0.10 = 1;$

$E(B) = (-10) \times 0.18 + (-5) \times 0.10 + 0 \times 0.24 + 5 \times 0.30 + 10 \times 0.18 = 1.$

It indicates that there is no difference between A and B.

Analysis 3:Use the square mean of the deviation to analyze

The formula for calculating the square mean of the deviation is

$$\sum_{i=1}^{n} (X_i - EX)^2 p(X_i)$$

The mean square of the deviation of A is: $(-10-1)^2 \times 0 + (-5-1)^2 \times 0.18 + (0-1)^2 \times 0.54 + (5-1)^2 \times 0.18 + (10-1)^2 \times 0.1 = 18$;

The mean square of the deviation of B is: $(-10-1)^2 \times 0.18 + (-5-1)^2 \times 0.10 + (0-1)^2 \times 0.24 + (5-1)^2 \times 0.30 + (10-1)^2 \times 0.18 = 45$。

Therefore, the operation of production line A is more stable.

6.4 过程能力不足

在管理实践中，过程能力不足（$C_p < 1$）的时候，说明产品的质量水平低，就产生了相应的产品质量数据风险。

6.4.1 过程能力

1. 过程能力的含义

过程能力（Process Capability，简称 PC）是指过程在加工产品质量方面的能力，表现在过程稳定（受控）的程度上。过程能力通常用质量特性值分布的 6 倍标准差来表示，记为 6σ。有关过程能力有以下特点：

（1）过程稳定时，质量特性 X 通常服从正态分布 $N(\mu, \sigma^2)$，其中标准差 σ 的大小表示过程稳定的程度。σ 越小，过程越稳定；

（2）稳定过程的 99.73% 的产品质量特性值散布在区间 $[\mu - 3\sigma, \mu + 3\sigma]$ 内；

（3）过程能力 PC $= 6\sigma$。

2. 过程能力指数

过程能力指数 C_P（Process Capability Index，简称 PCI）是用来度量一个过程满足标准要求的程度。它的指标值可以视为过程的潜在能力，即当过程的平均值可以调至规格中心或目标值时，过程符合规格的能力。有关过程能力指数有以下特点：

（1）规范限 T =（LSL，USL）（规范限是标准要求，通常是顾客的要求）。LSL 是公差下限（下偏差）；USL 是公差上限（上偏差）；

（2）规范中心：M =（LSL+USL）/2；

（3）公差（规范限的宽度）T = USL−LSL；

（4）规范中心 M 与受控过程中心（即正态均值）μ 重合时，过程能力指数定义为：

$C_P =$ 标准要求/过程能力 = USL−LSL/ 6σ = T/6σ。

过程能力的等级与状态见表 6−8。

表 6−8　过程能力的等级与状态

等级	$C_P > 1.67$	$1.33 < C_P \leq 1.67$	$1.00 < C_P \leq 1.33$	$0.67 < C_P \leq 1.00$	$C_P \leq 0.67$
状态	过程能力非常充足	过程能力充足	过程能力尚可	过程能力不足	过程能力严重不足

过程能力等级对应的合格品率及不合格品率见表 6−9。

表 6-9　过程能力等级对应的合格品率及不合格品率

C_P	公差范围	合格品率/%	不合格品率 / 10^{-6}
0.67	±2σ	95.56	444 432
1.00	±3σ	99.73	2700
1.33	±4σ	99.9934	66
1.67	±5σ	99.999 936	0.64
2.00	±6σ	99.999 999 8	0.002

3. 实际过程能力指数C_{pk}

由于实际数据的分布中心与公差中心会发生偏离，即使过程能力充足（$C_P = 1.33$），操作者也严格按照操作规定进行作业，但也会出现不合格品大大超过 $C_P = 1.33$ 时的预测数，因此提出了实际过程能力指数 C_{pk}。$\hat{\sigma}$ 为实际的标准偏差。

（1）$C_{PL} = (\mu - \text{LSL})/3\hat{\sigma}$，为单侧下限过程能力指数；

（2）$C_{PU} = (\text{USL} - \mu)/3\hat{\sigma}$，为单侧上限过程能力指数；

（3）$C_{pk} = \min\{C_{PL}, C_{PU}\}$，为实际过程能力指数；

（4）$C_P = \text{USL} - \text{LSL}/6\hat{\sigma} = T/6\hat{\sigma}$，为潜在过程能力指数。

4. 过程性能指数P_{pk}

过程性能，也称为长期过程能力（Process Performance），记为 P_p。其定义也是过程标准差的 6 倍。过程性能实际上是对未被证明处于统计受控状态的过程特性输出的统计度量。过程性能同时考虑了偶然因素、异常因素所引起的变异，所以实际反映了长期的变异。使用过程性能时应谨慎。

（1）$P_{PL} = (\bar{x} - \text{LSL})/3\hat{\sigma}$，为单侧下限过程性能指数；

（2）$P_{PU} = (\text{USL} - \bar{x})/3\hat{\sigma}$，为单侧上限过程性能指数；

（3）$P_{pk} = \min\{P_{PL}, P_{PU}\}$，为实际过程性能指数；

（4）$P_p = \text{USL} - \text{LSL}/6\hat{\sigma} = T/6\hat{\sigma}$，为潜在过程性能指数。

6.4.2 过程能力不足的影响

1. 影响过程能力的因素

影响过程能力主要有人、机、料、法、环、测等方面的因素。现在也有提出影响过程能力的 9M 因素（市场 Market，资金 Money，管理 Management，人 Men，激励 Motivation，材料 Material，机器和机械化 Machine and Mechanization，现代信息方式 Modern Information Methods，产品规格要求 Mounting Product Requirement）。当过程能力不足的时候，往往可以从以上因素中寻找问题，比如：操作者情绪波动、人员更换、机器移动、机械老化、修理不当、测量设备或量具没有校准、原材料变化、供应商变动、温度湿度变化等。

2. 过程能力不足的后果

过程能力不足的影响是较大的，最直接的后果就是不合格产品的产生，此时需要调整过程加工的分布中心以减少偏移量，或者提高过程能力以减少分散程度。

采取的措施包括：

（1）在不影响产品质量的情况下，适当放宽产品的公差范围；

（2）分析过程质量低的原因，改进过程，如采用控制图进行过程控制，改进工艺方法，修订操作规程，优化工艺参数；

（3）检修、改造设备，采用高精度设备；

（4）加强过程监控，进行全数检验或者加严检验；

（5）当过程能力非常不足时，要采取紧急措施停止生产，检查原因，对过程进行更改，对产品全数检验或加严检验等。

举例：

加工产品的重量，共分 10 组，每组 10 个数据，汇总后数据见表 6-10。

表 6-10　加工产品的重量

第一组	第二组	第三组	第四组	第五组	第六组	第七组	第八组	第九组	第十组
3909	2543	2455	2364	3178	2636	1632	2181	2910	1282
3089	2006	2728	2635	1999	2180	2003	2543	4362	366
2182	2634	3180	3277	2725	3088	1274	3820	3457	563
2540	2910	2001	2270	3271	3547	2180	1635	2541	1458
3458	3275	2004	1819	2363	1817	1637	734	1090	3364
3635	2541	2547	1092	2726	2814	2729	2367	2543	4279
3814	2907	3096	1817	2543	3092	1816	2189	2449	2185
2637	1632	1906	4183	1270	913	1911	1999	3094	2002
2542	2543	1814	3456	1095	2909	1726	2728	2546	1726
2728	1819	2183	3184	1819	2545	2182	2185	2907	3633

先检验这组数据是否符合正态分布。

用 Minitab 软件进行分析，进行正态性检验。见图 6-24。

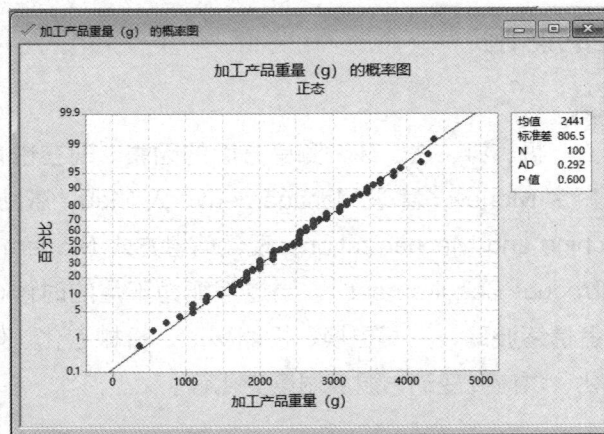

图 6-24　数据的正态性检验

由图 6-24 可知，该组数据平均值为 2441，标准差为 806.5，最大值为 4279，最小值为 366，P 值为 0.600，符合正态分布。

下面进行过程能力分析。

选择 Minitab 软件，"统计"中的"质量工具"，选择"能力分析"中的"正态"，见图6-25。

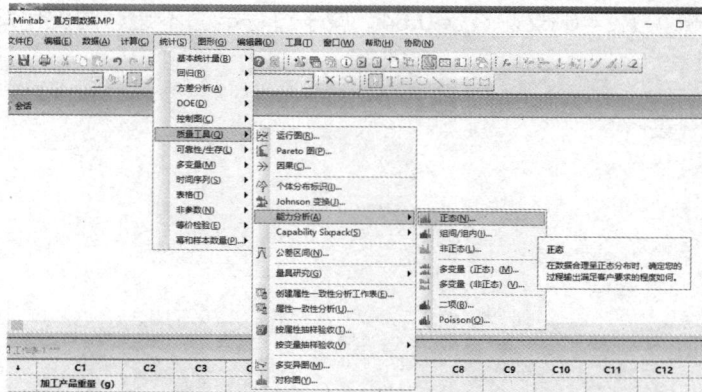

图 6-25　选择"正态"

在会话框中，"单列"选择"产品重量"，子组大小选择 10，"规格下限"设为"750"（假定），"规格上限"设定为"3750"（假定），见图 6-26。

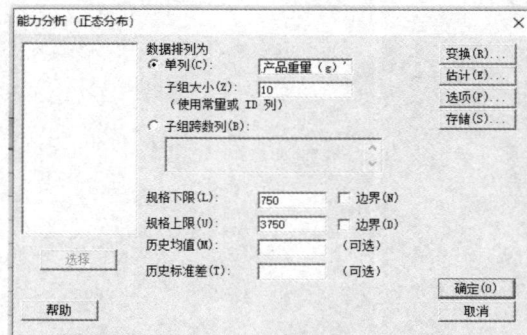

图 6-26　能力分析（正态分布）

点击"确定"后，有"过程能力报告"输出，见图 6-27。

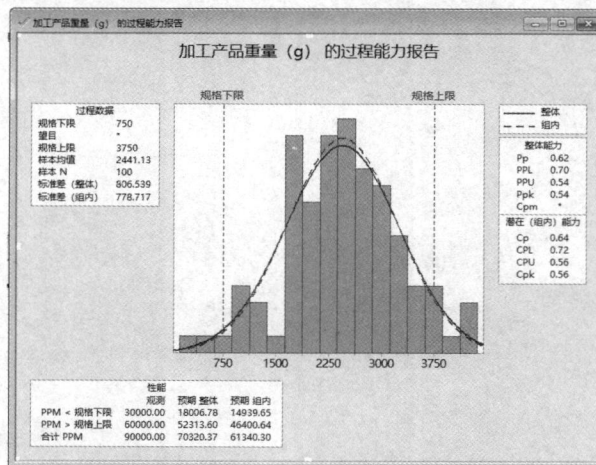

图 6-27　过程能力报告

从图 6-27 看到，整体的不合格率为 70 320.37PPM，超下限的不合格率为 18 006.78PPM，超上限的不合格率为 52 313.60PPM。说明这批产品中，有 7.03％的不合格品。在不合格品中，74.39％（52 313/70 320 ＝ 0.7439）为没有达到上限而导致不合格。这就要求产品管理者把生产过程进行调整，以满足产品上限的要求。同时看到，组内的不合格率为 61 340.30PPM，好于整体的不合格率，说明在子组内的生产相对稳定。

从过程能力来分析，P_P =0.62，C_p =0.64，属于过程能力不足，由此也解释了产品数据不理想的原因。

6.4 Insufficient Process Capability

In the management practice, when the process capability is insufficient($C_p < 1$), it indicates that the product quality level is low, and the corresponding product quality data risk is generated.

6.4.1 Process Capability

1. Meaning of process capability

Process capability (PC) refers to the capability of the process in the quality of processed products, which is shown in the degree of process stability (control). The process capability is usually expressed by 6 times the standard deviation of the distribution of quality characteristic value, which is recorded as 6 σ . Relevant process capabilities have the following characteristics:

(1)When the process is stable, the quality characteristic X usually follows the normal distribution N(μ , σ^2). The size of the standard deviation σ indicates the stability of the process. The smaller the σ, the more stable the process.

(2)99.73% of the product quality characteristic values of the stable process are scattered in the range of [$\mu -3\sigma$, $\mu +3\sigma$].

(3)Process capability PC $=6\sigma$.

2. Process capability index

Process capability index(PCI) C_P is used to measure the degree to which a process meets the standard requirements. Its index value can be regarded as the potential capability of the process, that is, the ability of the process to meet the specification when the average value of the process can be adjusted to the specification center or target value. Relevant process capability indexes have the following characteristics:

(1)Specification tolerance T = (LSL, USL) (The specification tolerance is a standard requirement, usually a customer requirement). LSL is the lower tolerance limit (lower deviation); USL is the upper tolerance limit(upper deviation).

(2)Specification Center: M =(LSL+USL)/2.

(3)Tolerance (width of specification limit) T = USL–LSL.

(4)When the specification center M coincides with the controlled process center (i.e. normal mean value) μ , the process capability index is defined as:

C_P = standard requirements/process capability = USL–LSL/ 6σ = T/ 6σ

See Table 6–8 for the level and status of process capability.

Table 6-8 Level and status of process capability

Level	$C_P > 1.67$	$1.33 < C_P \leqslant 1.67$	$1.00 < C_P \leqslant 1.33$	$0.67 < C_P \leqslant 1.00$	$C_P \leqslant 0.67$
Status	Process capability is very sufficient	Sufficient process capability	Process capability is acceptable	Insufficient process capability	Process capability is seriously insufficient

See Table 6-9 for qualified product rate and unqualified product rate corresponding to process capability level.

Table 6-9 Qualified product rate and unqualified product rate
corresponding to process capability level

C_P	Tolerance range	Qualified product rate/%	Unqualified product rate /10^{-6}
0.67	$\pm 2\sigma$	95.56	444 432
1.00	$\pm 3\sigma$	99.73	2700
1.33	$\pm 4\sigma$	99.9934	66
1.67	$\pm 5\sigma$	99.999 936	0.64
2.00	$\pm 6\sigma$	99.999 999 8	0.002

3. Actual process capability index C_{pk}

Because the distribution center of the actual data will deviate from the tolerance center, even if the process capacity is sufficient ($C_P = 1.33$), and the operator operates in strict accordance with the operation regulations, but the number of nonconforming products will greatly exceed the predicted number at $C_P = 1.33$. Therefore, the actual process capability index C_{pk} is proposed. $\hat{\sigma}$ is the actual standard deviation.

(1) $C_{PL} = (\mu - LSL)/3\hat{\sigma}$ is the unilateral lower limit process capability index;

(2) $C_{PU} = (USL - \mu)/3\hat{\sigma}$ is the unilateral upper limit process capability index;

(3) $C_{pk} = \min\{C_{PL}, C_{PU}\}$ is the actual process capability index;

(4) $C_P = USL - LSL/6\hat{\sigma} = T/6\hat{\sigma}$ is the potential process capability index.

4. Process performance index P_{pK}

Process performance, also known as long-term process performance, is recorded as P_p. Its definition is also 6 times of the process standard deviation. Process performance is actually a statistical measurement of the output of process characteristics that have not been proved to be in a statistically controlled state. The process performance can consider the variation caused by accidental factors and abnormal factors at the same time, so it actually reflects the long-term variation. We should use process performance carefully.

(1) $P_{PL} = (\bar{x} - LSL)/3\hat{\sigma}$ is the unilateral lower limit process performance index;

(2) $P_{PU} = (USL - \bar{x})/3\hat{\sigma}$ is the unilateral upper limit process performance index;

(3) $P_{pk} = \min\{P_{PL}, P_{PU}\}$ is the actual process performance index;

(4) $P_P = USL - LSL/6\hat{\sigma} = T/6\hat{\sigma}$ is the potential process performance index.

6.4.2 Impact of Insufficient Process Capability

1. Factors affecting process capability

The main factors affecting the process capability are men, machine, material, method, environment and measurement. Now there are also 9M (market, money, management, men, motivation, material, machine and mechanization, modern information methods, and mounting product requirements) factors that affect the process capability. When the process capability is insufficient, problems can often be found from the above factors, such as operator mood fluctuation, personnel replacement, machine movement, mechanical aging, improper repair, calibration of measuring equipment or measuring tools, raw material change, supplier change, temperature and humidity change, etc.

2. Consequences of insufficient process capability

The impact of insufficient process capability is large, and the most direct consequence is the generation of nonconforming products. At this time, it is necessary to adjust the distribution center of process processing to reduce the offset, or improve the process capability to reduce the dispersion.

Measures taken include the following items.

(1) Relax the tolerance range of the product properly without affecting the product quality.

(2) Analyze the causes of low process quality and improve the process, such as using control charts for process control, improving process methods, revising operating procedures, and optimizing process parameters。

(3) Overhaul and reform equipment, and use high-precision equipment.

(4) Strengthen process monitoring and conduct full inspection or strict inspection.

(5) When the process capacity is very insufficient, emergency measures should be taken to stop production, check the reasons, change the process, and conduct full or strict inspection of products.

Case：

The weight of processed products is divided into 10 groups with 10 data for each group. The summarized data are shown in Table 6-10 "Weight of processed products".

Table 6-10 Weight of processed products

Graup 1	Graup 2	Graup 3	Graup 4	Graup 5	Graup 6	Graup 7	Graup 8	Graup 9	Graup 10
3909	2543	2455	2364	3178	2636	1632	2181	2910	1282
3089	2006	2728	2635	1999	2180	2003	2543	4362	366
2182	2634	3180	3277	2725	3088	1274	3820	3457	563
2540	2910	2001	2270	3271	3547	2180	1635	2541	1458
3458	3275	2004	1819	2363	1817	1637	734	1090	3364
3635	2541	2547	1092	2726	2814	2729	2367	2543	4279
3814	2907	3096	1817	2543	3092	1816	2189	2449	2185
2637	1632	1906	4183	1270	913	1911	1999	3094	2002

Graup 1	Graup 2	Graup 3	Graup 4	Graup 5	Graup 6	Graup 7	Graup 8	Graup 9	Graup 10
2542	2543	1814	3456	1095	2909	1726	2728	2546	1726
2728	1819	2183	3184	1819	2545	2182	2185	2907	3633

First, check whether this group of data conforms to the normal distribution.

Minitab software is used for analysis of normality test. Select "Normality Test" in "Basic Statistics". The result is seen in Figure 6-24.

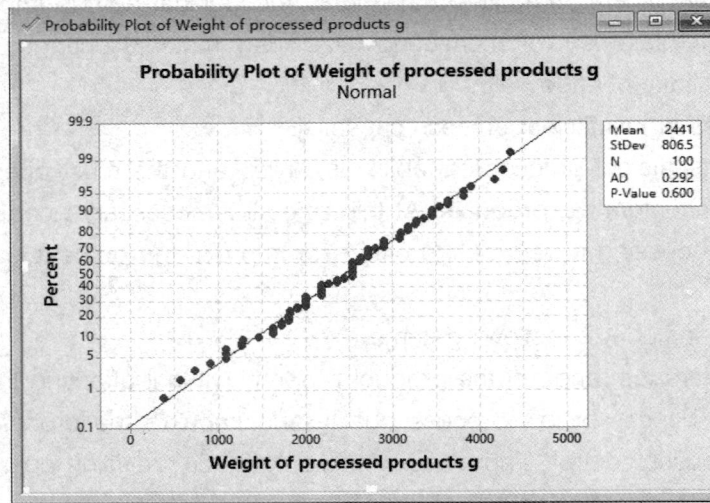

Figure 6-24　Normality test of data

It can be seen from Figure 6-24 that the average value of this group of data is 2441, the standard deviation is 806.5, the maximum value is 4279, the minimum value is 366, and the P value is 0.600, which conforms to the normal distribution.

The following is the process capability analysis.

Select Minitab software. Select "Quality Tools" in "Statistics", and choose "Normal" in "Capability Analysis", as shown in Figure 6-25.

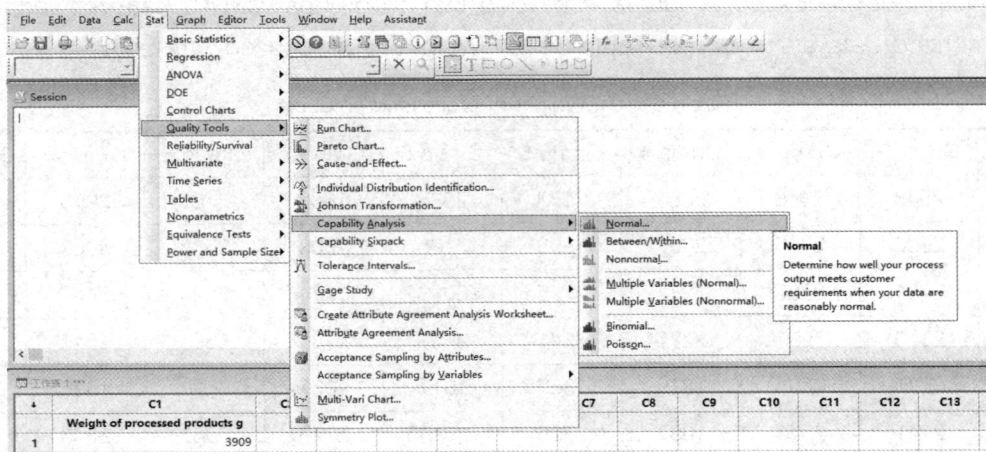

Figure 6-25　Select "Normal"

In the dialogue box, select "Weight of processed products" into "Single column". Select

"10" in the "Subgroup size". Set the "Lower specification limit" as "750" (assume) and the "Upper specification limit" as "3750" (assume), as shown in Figure 6-26.

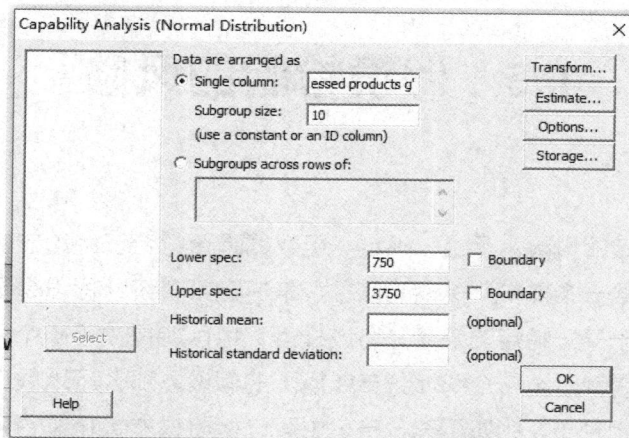

Figure 6-26　Capability analysis (normal distribution)

After clicking "OK", the "Process Capability Report" will be output, as shown in Figure 6-27.

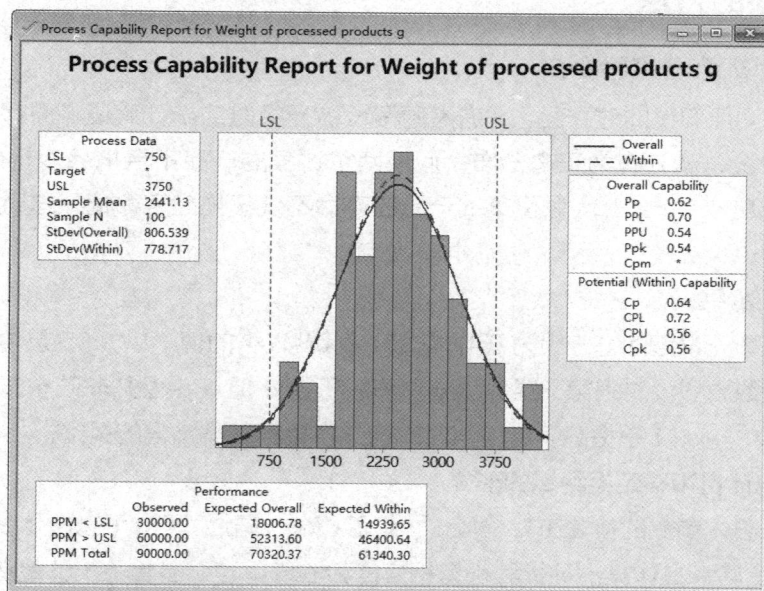

Figure 6-27　Process capability report

It can be seen from Figure 6-27 that the overall unqualified rate is 70 320.37PPM, the unqualified rate exceeding the lower limit is 18 006.78PPM, and the unqualified rate exceeding the upper limit is 52 313.60PPM. It indicates that 7.03% of the products are unqualified. Among the nonconforming products, 74.39% (52 313/70 320=0.7439) failed to reach the upper limit, resulting in nonconformity. This requires the product manager to adjust the production process to meet the requirements of the product ceiling. At the same time, we can see that the unqualified rate in the group is 61 340.30PPM, which is better than the overall unqualified rate, indicating that the production in the sub group is relatively stable.

Analyze from process capability, P_P =0.62, C_p =0.64, it is due to insufficient process capacity, which also explains the reason why the product data is not ideal.

6.5 假设检验的风险

假设检验是指在规定的风险水平上，确定一组数据是否符合已经给定假设的一种统计方法。或者说，假设检验是根据一个样本所提供的信息，来决定某个统计假设是否应该被拒绝的方法，这个统计假设可能是关于某一特定的统计分布或模型，也可能是关于一个分布的某些参数值，比如平均值或方差。如果是针对某一个统计分布或模型的假设，那么运用假设检验方法可以检验总体数据是否符合某种特定分布，比如正态分布；也可以检验样本数据是否随机等。如果是针对总体参数的假设，那么运用假设检验方法可以检验某一个总体的参数，比如平均值或方差是否符合规定的要求；也可以检验两个或多个总体之间的差异。

6.5.1 假设检验的方法

1. 提出原假设 H_0 和备择假设 H_1

备择假设也称为原假设的对立面，或者是原假设被拒绝时的统计假设。原假设是根据实际问题提出的，如：假设总体分布为正态分布；假设两个过程的产品不合格品率相同，原假设 H_0：$P_1 = P_2$、备择假设 H_1：$P_1 \neq P_2$；假设某批产品的寿命不低于规定寿命值，这时原假设 H_0：$\mu \geq \mu_0$，备择假设为 H_1：$\mu < \mu_0$。

2. 确定显著性水平 α

显著性水平是事先设定的。一般为 0.1、0.05、0.01 较小的值，以 0.05 最为常见。显著性水平的用意是：在拒绝原假设时犯第一类错误的概率小于该 α 值。α 值并非越小越好，因为当 α 值越小时，置信概率（$1-\alpha$）是提高了，但同时就增加了犯第二类错误的风险。

3. 选择检验统计量和确定拒绝域的形式

对于单个正态总体参数的检验有 μ 检验、t 检验、x^2 检验；两个正态总体参数的检验有：μ 检验、t 检验、F 检验。具体某一种检验都有相应的检验统计量。当检验统计量选择后，可以根据该检验统计量的分布以及选定的显著性水平，确定拒绝域的形式，拒绝域是指检验统计量取值的一个区域，如果样本的检验统计量值处于这个区域，则应当拒绝原假设。根据原假设的不同，假设检验分为单侧检验和双侧检验，相应的拒绝域形式也不同。拒绝域的边界值为临界值，拒绝域以外的区域为接受区域。

当原假设 H_0：$\mu \leq (\geq) \mu_0$，或 $\sigma^2 \leq (\geq) \sigma_0^2$ 时，一般为单侧检验；当原假设 H_0：$\mu = \mu_0$，或 $\sigma^2 = \sigma_0^2$ 时，一般为双侧检验。然后根据给定的检验统计量的分布，查表确定临界值，从而确定具体的拒绝域。对于 μ 检验，根据确定的显著性水平 a，查标准正态分布的分位表来确定临界值。单侧检验直接按 a 查出 μ_{1-a}；双侧检验要按 $a/2$ 查出 $\mu_{1-a/2}$。对于 t 检验、x^2 检验、F 检验，确定临界值时应分别查 t 分布的分位数表、x^2 分布的分位数表、F 分布的分位数表；查表时，应考虑是单侧检验还是双侧检验、给定的显著性水平 α、依据样本统计量的自由度 f。单个总体假设检验的自由度为 $n-1$；两个总体假设检验的自由度为 n_1+n_2-2；F 检验中 n_1 为分子自由度，n_2 为分母自由度。

4. 收集样本数据

计算所确定的检验统计量值。将计算出的检验统计量值与在相应表中查出的拒绝域临界值进行比较：如果其落入拒绝域，则拒绝原假设；否则做出保留原假设的结论。

举例：P 值检验

某公司有一批物资长期存放在仓库内，由于该仓库的存储条件不理想，仓库管理员认为这批物资中会有一部分不能正常使用。根据企业的管理要求，如果有 60% 及以上的物资完好无损，那么这批物资经过清理后还可以继续使用。反之，如果物资的完好率小于 60%，那么综合考虑成本后，这批物资只能进入处理程序。于是仓库管理员请公司的质检人员进行检查。质检人员随机抽查了 200 件物资进行检查，发现有 115 件物资完好无损。

仓库管理人员得知检查数据后，认为物资的完好率 = 115/200 = 57.5%，显然小于 60% 的管理要求，因此，请示上级把该仓库物资予以处理。这样的处理方式是否妥当呢？

下面进行重新计算，取 $\alpha = 0.05$，

由于抽样的样本为 200，因此，可以认为仓库物资的总量是大样本，那么把 $n=200$ 视为大样本，计算 $\widehat{P}$ 的近似正态分布来进行检验。

原假设为 H_0：物资予以清理。

$$\widehat{P} = 115/200 = 0.575$$

$$\frac{\widehat{P}-P_0}{\sqrt{P_0(1-P_0)/n}} = \frac{0.575-0.60}{\sqrt{0.60(1-0.60)/200}} = -0.72$$

查正态分布表，得到临界值 $Z(0.05) = 1.65$。

显然，$-0.72 > -1.65$。

因此，拒绝原假设。那么，这批物资应当值得清理后继续使用。

6.5.2 第一类错误和第二类错误

由于假设检验是利用有限的样本数据来对总体假设作出判断，因此不可避免地会存在判断错误。当原假设为真而被拒绝，即"弃真"时称为第一类错误。犯第一类错误的概率称为第一类风险，在检验中犯第一类错误概率不能超过的界限称为显著性水平 α。当原假设不真而予以接受，即"存伪"时，称为第二类错误。犯第二类错误的概率称第二类风险，记为 β。

1. 第一类错误

产生第一类错误的原因有：样本中存在极端的数值，使原假设看上去不成立；采用的标准比较宽松，使原假设为真但被拒绝。

举例：

假设有一个家具生产企业，每加工生产的 100 件家具里有 5 件是不合格品，那么不合格率是5%。现在有质检部门需要检查家具生产质量，质检人员事先不知道该企业有 5% 的不合格率。质

检人员采取的方法是用假设检验来验证。

第一步，根据统计学关于小概率事件的定义，即概率小于 5% 的事件在一次试验中不会发生，质检人员假设该企业家具的不合格率不超过 5%，那么在这个假设下，一次抽样抽不到不合格品的家具。

第二步，因为 5% 是统计学上的小概率事件，但毕竟在 100 件家具里面的确有 5 件不合格品，因此，一次抽样抽到次品的概率为 5%，还是存在可能的。于是质检人员开始抽样，他们随机抽取了一件家具，结果正好抽中了一件不合格的家具。

第三步，根据抽到一件不合格品家具的情况，质检人员认为，原假设是 H_0：不合格率 <5%，也就是说，在只有 5 个不合格品家具的情况下，一次抽样是抽不到不合格品的。但现在却真实地抽到了一件不合格品，因此，拒绝原假设。质检人员认为在 100 件家具里的不合格品超过 5 件。

分析：本例中由于抽中不合格品，质检按人员就拒绝原假设，也就是原假设为真而拒绝原假设，质检人员犯了第一类错误。犯第一类错误的概率就是 5 件不合格品家具所占的比例。因为"低于 5% 的事件是小概率事件"是人为规定，那就存在 5% 的可能性犯错。在原假设为真的情况下，质检人员抽中不合格品的概率就是 5%，也说明犯错的概率等于小概率事件发生的概率。

2. 第二类错误

产生第二类错误的原因：检验设计不灵敏，使原假设看上去成立；样本数据变异过大，使原假设成立，实质原假设为假；处理效应本身比较小，使原假设为假但被接受。

举例：

与上一个例子相仿。假设有一个家具生产企业，每加工生产的 100 件家具里有 10 件是不合格品，那么不合格率是 10%。现在质检部门需要检查家具生产质量，质检人员事先不知道该企业有 10% 的不合格率。质检人员采取的方法是用假设检验来验证。

第一步，质检人员提出了原假设是 H_0：不合格率 < 5%，即 100 件家具里面的不合格品小于 5 件。

第二步，质检人员开始抽样检查，他们随机抽取了一件家具，结果正好抽中了一件合格的家具。因为 100 件家具中，有 90 件是事实上的合格品，因此，质检人员抽中合格家居的概率是 90%。

第三步，根据抽到一件合格品家具的情况，质检人员认为，原假设是 H_0：不合格率<5%，也就是说，在只有 5 个不合格品家具的情况下，一次抽样是抽不到不合格品的。现在一次抽样的结果是抽到了一件合格品，因此，接受原假设。质检人员认为在 100 件家具里的不合格品小于 5 件。

分析：本例中，由于一次抽样抽到了合格品，质检人员就不能拒绝原假设，也就是原假设为"假"但接受了原假设，质检人员犯了第二类错误。犯第二类错误的概率就是 90 件合格品家具所占的比例，即达到了 90%。

在本次假设检验中，事实上的不合格率是大于 10%，如果质检人员要拒绝原假设 H_0：不合格率<5%，那么需要在一次抽样中抽到不合格品家具。只要质检人员在一次抽样中抽到合格品家具，那么就犯错，犯错的概率就是抽中正品的概率。

3. 第一类错误和第二类错误的关系

（1）α 可以称为检验水准，$1-\beta$ 称为检验效能（或称为把握度 power of test）。在实践中，要权衡第一类错误与第二类错误哪个更重要，以此来决定检验水准 α 的大小。

（2）当样本的数量一定时，α 越小，则 β 越大；反之，α 越大，则 β 越小。这是因为犯第一类错误的概率是根据检验水准人为设定的。在实践中，可以通过选定 α 的值来控制 β 的大小。

（3）由于 α 和 β 分别代表了不同风险下的显著性水平，因此，$\alpha + \beta$ 不一定等于 1。

（4）当其他条件不变时，α 与 β 不会同时增加或减少。只有在增加样本数量的情况下，α 与 β 才同时减少，这是因为样本数量的增加会减小标准误差，那么总体的图形显得细瘦，犯错的概率就小。反之，在减少样本数量的情况下，α 与 β 才同时增加。

（5）当其他条件不变时，如果要求犯第一类错误的概率小，那么犯第二类错误的概率就大；同样，当其他条件不变时，如果要求犯第二类错误的概率小，那么犯第一类错误的概率就大。

6.5　Risk of Hypothesis Test

———

Hypothesis test refers to a statistical method to determine whether a group of data conforms to the given assumptions at the specified risk level. In other words, hypothesis test is a method to determine whether a statistical hypothesis should be rejected based on the information provided by a sample. This statistical hypothesis may be about a specific statistical distribution or model, or it may be about some parameter values of a distribution, such as the average value or variance. If it is a hypothesis for a certain statistical distribution or model, the hypothesis test method can be used to test whether the population data conforms to a specific distribution, such as normal distribution; It can also test whether the sample data is random. If the hypothesis is based on the population parameters, the hypothesis test method can be used to test whether the parameters of a certain population, such as the average value or variance, meet the specified requirements; Differences between two or more populations can also be examined.

6.5.1 Methods of hypothesis test

1. Put forward the null hypothesis H_0 and the alternative hypothesis H_1

The Alternative hypothesis is also called the opposite of the null hypothesis, or the statistical hypothesis when the null hypothesis is rejected. The null hypothesis is put forward according to the actual problems, such as the assumption that the population distribution is normal. Assume that the nonconforming product rate of the two processes is the same, and the null hypothesis is $H_0: p_1 = p_2$, the alternative hypothesis $H_1: p_1 \neq p_2$. Assuming that the life of a batch of products is not less than the specified life value, at this time, the null hypothesis $H_0: \mu \geqslant \mu_0$, the alternative hypothesis $H_1: \mu < \mu_0$, etc.

2. Determine significance level α

The significance level is preset. Generally, it is the smaller value of 0.1, 0.05 and 0.01, and 0.05 is the most common value. The significance level means that the probability of making the first type of error when rejecting the original hypothesis is less than the α Value. It is not the smaller the α value, the better. Because when the α value becomes smaller, the confidence probability $(1-\alpha)$ has increased, at the same time, it increases the risk of making the type II error.

3. Select the test statistics and determine the forms of rejection regions

The tests for single normal population parameters are μ test, t test and x^2 test. The test for two normal population parameters are μ-test, t-test, and F-test. A specific test has corresponding test statistics. When the test statistic is selected, the form of rejection field can be determined according to the distribution of the test statistic and the selected significance level. The rejection region refers to a region where the test statistic values are taken. If the sample's test statistic values are in this region, the null hypothesis should be rejected. According to the different null hypotheses, hypothesis test can be divided into unilateral test and bilateral test, and the corresponding forms of rejection regions are also different. The boundary value of the rejection region is the critical value, and the region outside the rejection domain is the acceptance region.

When the null hypothesis is $H_0: \mu \leqslant (\geqslant) \mu_0$, or $\sigma^2 \leqslant (\geqslant) \sigma_0^2$, it is generally the single side test. When the null hypothesis $H_0: \mu = \mu_0$, or $\sigma^2 = \sigma_0^2$, it is generally the two sided test. Then, according to the distribution of the given test statistics, the critical value is determined by looking up the table, and the specific rejection region is determined. For μ test, according to the determined significance level a, check the quantile table of standard normal distribution to determine the critical value. In the one side test, we find out μ_{1-a} based on a directly. In the two sided test, we find out $\mu_{1-a/2}$ based on $a/2$. For t test, x^2 test, F test, when determining the critical value, the quantile tables of t distribution, x^2 distribution and F distribution shall be checked respectively. When looking up the table, we should consider whether it is one side test or two sided test, the given significance level a and the degree of freedom f according to the sample statistics. The degree of freedom for a single population hypothesis test is $n-1$; The degree of freedom for two populations hypothesis test is $n_1 + n_2 - 2$. In F test, n_1 is the molecular degree of freedom, and n_2 is the denominator degree of freedom.

4. Collect sample data

Calculate the determined test statistic value. Compare the calculated test statistic value with the critical value of the rejection region found in the corresponding table: if it falls into the rejection region, reject the null hypothesis; Otherwise, make a conclusion that the null hypothesis is retained.

Case: P value test

A company has a batch of materials stored in the warehouse for a long time. Because the storage conditions in the warehouse are not ideal, the warehouse keeper thinks that some of these materials cannot be used normally. According to the management requirements of the enterprise, if 60% or more of the materials are intact, these materials can still be used after cleaning. On the contrary, if the availability rate of materials is less than 60%, the materials can only enter the processing procedure after considering the cost comprehensively. So the warehouse keeper asked the company's quality inspectors to check. The quality inspector randomly inspected 200 pieces of materials and found 115 pieces of materials in good condition.

After knowing the inspection data, the warehousekeeper think that the availability rate of materials = 115/200 = 57.5%, it is obviously less than 60% of management requirements. Therefore, they ask the superior to dispose the materials in the warehouse. Whether such a treatment is appropriate?

Then recalculate, take $\alpha = 0.05$.

Since the sample size is 200, it can be considered that the total quantity of warehouse materials is a large sample, so $n = 200$ is considered as a large sample. We calculate the approximate normal distribution of $\hat{P}$ for testing.

The null hypothesis is H_0: Materials to be cleaning up.

$$\widehat{P} = 115/200 = 0.575$$

$$\frac{\widehat{P} - P_0}{\sqrt{P_0(1 - P_0)/n}} = \frac{0.575 - 0.60}{\sqrt{0.60(1 - 0.60)/200}} = -0.72$$

Check the normal distribution table to get the critical value $Z(0.05) = 1.65$,

Obviously, $-0.72 > -1.65$,

So, the null hypothesis is rejected. Then, this batch of materials should be worth cleaning up and continuing to be used.

6.5.2 Type I error and type II error

Because hypothesis testing uses limited sample data to judge the overall hypothesis, there will inevitably be judgment errors. When the null hypothesis is true and rejected, that is, "abandon the truth", it is called the type I error. The probability of making the type I error is called the first type of risk, and the limit that the probability of making the type I error cannot exceed in the test is called the significance level α. When the null hypothesis is not true and accepted, that is, "false", it is called the type I error. The probability of making type II error is called the second type of risk, which is recorded as β.

1. Type I error

The reasons for the type I error are as follows: there are extreme values in the sample, which makes the original hypothesis seem invalid; The standard adopted is relatively loose, making the null hypothesis true but rejected.

Case:

Suppose there is a furniture manufacturer, and 5 of every 100 pieces of furniture processed and produced are unqualified products, then the unqualified rate is 5%. Now there are quality inspection departments that need to check the production quality of furniture. The quality inspection personnel do not know that the enterprise has a 5% disqualification rate in advance. The method adopted by the quality inspector is to verify with hypothesis test.

Step 1: According to the definition of small probability events in statistics, that is, events with a probability of less than 5% will not occur in a single test. The quality inspector assumes that the unqualified rate of the furniture of the enterprise does not exceed 5%. Under this assumption, a single sampling is a piece of furniture with no unqualified products.

Step 2: because 5% is a small probability event in statistics, but after all, there are 5 unqualified products in 100 pieces of furniture, so the probability of sampling defective products at one time is 5%, which is still possible. So the quality inspectors began to sample, they randomly selected a piece of furniture, and the result was just a piece of unqualified furniture.

Step 3: According to the situation of a piece of furniture with unqualified products, the quality inspector thinks that the null hypothesis is H_0: the unqualified rate is less than 5%, that is to say, when there are only 5 unqualified products, the unqualified products cannot be sampled at one time. But now we have actually picked up a nonconforming product, so we reject the null hypothesis. The quality inspector thinks that there are more than 5 unqualified pieces in 100 pieces of furniture.

Analysis: In this example, because the unqualified products are selected, the quality inspection personnel reject the null hypothesis, that is, the null hypothesis is true and it is rejected. The quality inspection personnel make the type I error. The probability of making the type I error is the proportion of five pieces of unqualified furniture. Because "less than 5% of events are low probability events" is artificially stipulated, there is a 5% probability of making mistakes. If the null hypothesis is true, the probability that the quality inspector will select the unqualified products is 5%, which means that the probability of making mistakes is equal to the probability of small probability events.

2. Type II error

The reasons for the type II error: the test design is not sensitive, which makes the null hypothesis seem to be valid; The sample data has too much variation, which makes the null hypothesis tenable and the null hypothesis in essence false; The treatment effect itself is relatively small, making the null hypothesis false but accepted.

Case:

It is similar to the above example. Suppose there is a furniture manufacturer, and 10 pieces of each 100 pieces of furniture processed and produced are unqualified, then the unqualified rate is 10%. Now the quality inspection department needs to check the production quality of furniture. The quality inspection personnel do not know that the enterprise has a 10% failure rate in advance. The method adopted by the quality inspector is to verify with hypothesis test.

Step 1: step, the quality inspector put forward the null hypothesis that H_0: the unqualified rate is less than 5%, that is, the unqualified rate of 100 pieces of furniture is less than 5 pieces.

Step 2: the quality inspectors start sampling inspection. They randomly select a piece of furniture, and the result is just a qualified piece of furniture. Because 90 of the 100 pieces of furniture are actually qualified products, the probability that quality inspectors will select qualified furniture is 90%.

Step 3: according to the condition of a piece of qualified furniture, the quality inspector thinks that the null hypothesis is H_0: the unqualified rate is less than 5%, that is to say, when there are only 5 unqualified products, no unqualified products can be sampled at one time. Now, the result of a sampling is that a qualified product has been selected. Therefore, the null hypothesis is accepted. The quality inspector thinks that there are less than 5 unqualified pieces in 100 pieces of furniture.

Analysis: In this example, because qualified products are sampled at one time, the quality inspector cannot reject the null hypothesis, that is, the null hypothesis is "false", but it is accepted. The quality inspector makes the type II error. The probability of making the type II error is the proportion of 90 pieces of qualified furniture, that is, 90%.

In this hypothesis test, the actual unqualified rate is more than 10%. If the quality inspector wants to reject the null hypothesis H_0: unqualified rate<5%, then it is necessary to sample unqualified furniture in one sampling. As long as the quality inspector picks up the qualified furniture in a sampling, he or she will make mistakes. The probability of making mistakes is the probability of picking up the authentic furniture.

3. The relationship between the first type of error and the second type of error

（1）α can be called inspection level. $1-\beta$ is called test efficiency（or power of test）. In practice, we should weigh the first type of error against the second type of error to determine the size of the inspection level α.

（2）When the number of samples is fixed, the smaller the α, the larger the β; conversely, the larger the α, the smaller the β. This is because the probability of making the first type of error is artificially set according to the inspection level. In practice, we can select the value of α to control the size of β.

（3）because α and β represent the significance level under different risks, so, $\alpha+\beta$ is not necessarily equal to 1.

（4）When other conditions remain unchanged, α and β will not increase or decrease at the same time. Only when the number of samples is increased, α and β will decrease at the same time.This is because the increase in the number of samples will reduce the standard error, so the overall figure looks thin and the probability of making mistakes is small. On the contrary, if the number of samples is reduced, α And β Only increase at the same time.

（5）When other conditions remain unchanged, if the probability of making the first type of error is required to be small, the probability of making the second type of error is large; Similarly, when other conditions remain unchanged, if the probability of making the second type of error is required to be small, then the probability of making the first type of error is large.

6.6　数据人员职业技能的影响

——

数据人员是指从事与数据有关的人员，包括数据采集的策划、数据采集的执行、数据的处理分析、数据的应用等所有环节上的相关人员。显然，数据人员的职业技能对于数据分析的最终结果会产生影响。

6.6.1 数据人员的职业技能

几乎各领域各行业都有数据人员，因此，数据人员应当具备必要的职业技能，以提高数据工作质量。

1. 基本的数学和逻辑能力

在大多数的情况下，数据人员所用到的数学技能不是很复杂，主要有：

（1）基本的数学运算、各种数学概念的理解；

（2）基础的逻辑知识；

（3）对于各类信息的敏感度和管理信息的能力；

（4）做好数据相关活动的计划、预算编制能力等。

2. 统计分析能力

统计分析能力是数据人员的核心能力，主要有：

（1）必要的统计学基础知识，特别是抽样、描述性统计、统计推论、统计预测、统计决策等；

（2）概率论专业知识；

（3）清楚各类数据分布的情形和相应分析；

（4）掌握必要的统计软件使用。

3. 测量和试验等专业能力

（1）掌握基础的测量知识；

（2）熟悉测量工作流程和工作内容；

（3）掌握基本的测量仪器设备的使用、保养等知识；

（4）掌握基础的试验知识；

（5）了解各种试验设计方法，尤其是单因子、双因子试验设计方法。

4. 计算机应用能力

（1）掌握基本的计算机知识和计算机应用能力；

（2）掌握建模的相关知识；

（3）熟悉模拟的相关知识；

（4）对于数据分析师要掌握必要的计算机编程语言；

（5）专业人士应掌握数据的可视化能力；

（6）了解大数据分析的相关知识。

5. 必要的软实力

（1）沟通与倾听能力；

（2）关注细节，做好数据的记录与保存；

（3）独立分析判断能力，不受外界的干扰；

（4）解决问题的能力，能够达到数据分析的目标；

（5）不断学习的能力，掌握跨界知识；

（6）高素养的培养，做到职业、敬业、专业。

6.6.2 关注数据的潜在风险

数据废气（Data Exhaust）是指被废弃的数据。一个人在网站上浏览了哪些页面，在某个页面上停留的时间长度、在具体文本上输入了什么信息、鼠标光标停留的具体位置，这些都构成了一个人的网上数字轨迹，这些都是数字废气。

这些看似无用的数字废气，具有很大的利用价值。如果数字废气得到开发利用和循环利用，则企业可以形成新的竞争优势。比如谷歌公司通过长期的观察研究发现，人们在搜索某个关键词的时候，如果打开了某个页面却找不到有效信息，于是使用者就不得不回到搜索页面重新开始搜索，直至在某一个页面的某一个具体链接上找到所需要的资料。为此，谷歌公司敏锐地捕捉到了这种"数据废气"的潜在价值，直接把使用者最终找到资料的具体链接给予前置，这样就大大提高了用户的搜索效率，也进一步提高了谷歌的搜索价值。

但是，反过来思考，数据废气也无意中泄露了使用者的大量数据，比如消费习惯、个人爱好、工作性质、急于解决问题的内容、希望实施的计划、就医记录、教育过程等。尤其是物联网时代的到来，数字废气作为一种新型的数据，应当得到有效保护和利用，预防相关风险的产生。

6.6.3 预测的风险

利用数据进行预测、决策是管理活动的重要内容，但预测的风险也紧紧相随。黑天鹅事件是预测风险的代表。缘由是这样的：在 17 世纪之前，澳大利亚的黑天鹅一直未被发现。当时的欧洲人普遍认为天鹅一向都是白色的。但随着澳大利亚第一只黑天鹅的出现，这个不可动摇的信念被否决了。黑天鹅的存在意味着不可预测的重大、稀有事件的发生。纽约大学特聘教授纳西姆·尼古拉斯·塔勒布（Nassim Nicholas Taleb）在他的代表作《黑天鹅》中提到，未来发生的事情是纯随机的，人类要是根据过去的经验对未来的事情进行预测，那是不靠谱的。塔勒布还举了一个简单的例子：如果一只火鸡，每天都去收集主人喂它的饲料多少的数据，那么一定时间之后，这只火鸡认为能够预测明天主人喂多少吃的给它。直到感恩节的前一天晚上，这只火鸡还做出了第二天的预测。结果是第二天该火鸡的命运就突然发生了重大的转折……。这个转折，就是典型的黑天鹅事件。

从中可以看到，对于数据的分析利用，必须考虑各种风险的存在。固有的思维、惯性思维有时候会束缚住正确的结论。

6.6 Influence of Professional skills of data personnel

——

Data personnel refer to those engaged in data related activities, including data acquisition planning, data acquisition implementation, data processing and analysis, and data application. Obviously, the professional skills of data personnel will have an impact on the final results of data analysis.

6.6.1 Professional skills of data personnel

There are data personnel in almost all fields and industries. Therefore, data personnel should have the necessary professional skills to improve the quality of data work.

1. Basic mathematical and logical abilities

In most cases, the mathematical skills used by data personnel are not very complex, mainly including：

(1)Basic mathematical operations and understanding of various mathematical concepts；

(2)Basic logical knowledge；

(3)Sensitivity to all kinds of information and ability to manage information；

(4)Ability to plan and budget data related activities.

2. Statistical analysis ability

Statistical analysis ability is the core ability of data personnel, mainly including：

(1)Necessary basic statistical knowledge, especially sampling, descriptive statistics, statistical inference, statistical prediction, statistical decision-making；

(2)Probability theory expertise；

(3)Understand the distribution of various data and the corresponding analysis；

(4)Master the use of necessary statistical software.

3. Professional abilities such as measurement and test

(1)Master basic measurement knowledge；

(2)Be familiar with the measurement workflow and work content；

(3)Master the use and maintenance of basic measuring instruments and equipment；

(4)Master basic test knowledge；

(5)Understand various test design methods, especially single factor and double factor test design methods.

4. Computer application ability

(1)Master basic computer knowledge and computer application ability；

(2) Master relevant knowledge of modeling;

(3) Be familiar with simulation related knowledge;

(4) The data analyst should master the necessary computer programming language;

(5) Professionals should master the visualization ability of data;

(6) Understand big data analysis.

5. Necessary soft power

(1) Communication and listening ability;

(2) Pay attention to details and keep records of data;

(3) Independent analysis and judgment ability, free from external interference.

(4) Ability to solve problems and achieve the goal of data analysis.

(5) The ability to continuously learn and master cross-border knowledge.

(6) Cultivate high attainment and be professional, dedicated and specialized.

6.6.2 Focus on the potential risks of data

Data exhaust refers to the discarded data. What pages a person browses on a website, the length of time he stays on a page, what information he enters in specific text, and the specific location where the mouse cursor stays constitute a person's online digital trajectory, which are all digital exhaust.

These seemingly useless digital exhausts are of great use value. If digital exhaust is developed and recycled, enterprises can form new competitive advantages. For example, Google has found through long-term observation and research that when people search for a keyword, if they open a page but cannot find effective information, users will have to go back to the search page and start searching again until they find the required information on a specific link of a page. For this reason, Google has keenly captured the potential value of this "data exhaust", and directly put the specific link of the information that the user finally finds in the front, which greatly improves the user's search efficiency and further improves Google's search value.

However, on the other hand, the data exhaust also unintentionally discloses a large amount of user data, such as consumption habits, personal hobbies, types of work, content eager to solve problems, plans to implement, medical records, education process, etc. Especially with the advent of the internet of things era, the data exhaust, as a new type of data, should be effectively protected and utilized to prevent related risks.

6.6.3 The risk of prediction

The use of data for forecasting and decision-making is an important part of management activities, but the risks of forecasting are closely related. A black swan event is a representative of risk prediction. The reason is this: Before the 17th century, the black swan in Australia had not been found. At that time, Europeans generally believed that swans were always white. But with the appearance of Australia's first black swan, this unshakable belief was rejected. The existence of black swans means the occurrence of unpredictable major and rare events. In his representative work Black Swan, Nassim Nicholas Taleb, a distinguished professor of New York University, mentioned that future events are purely random, and it is not reliable for humans to predict future events based on past experience. Taleb also gave a simple example: if a turkey collects data on how much the master feeds it every day, then after a certain period of time, the turkey thinks it can predict how much the master will feed it tomorrow. Until the night before Thanksgiving, the turkey made a prediction for the next day. As a result, the fate of the turkey suddenly took a significant turn the next day. This turning point is a typical black swan event.

It can be seen that for the analysis and utilization of data, various risks must be considered. Inherent thinking and inertial thinking sometimes fetters the correct conclusion.

6.7 非线性数据和多元数据的存在

——

人们为了分析数据方便，把非线性数据简化为线性数据；把数据简化为一元数据、二元数据。而真实的客观世界中，非线性数据和多元数据无处不在。比如，经济的增长不可能呈线性上升趋势；加工厂生产一个零件，就会碰到长度、宽度、高度、弧度、重量等多元指标。因此，对于非线性数据和多元数据的存在要有客观认识。

6.7.1 非线性数据

在数据分析中，如果忽视了非线性数据的存在，那么，数据的风险就高，分析结论就不准确。这里介绍几个主要的非线性模型。

1. 半对数模型

半对数模型是指在因变量和解释变量中，其中的一个变量为对数形式，另一个变量为线性模型。如果因变量为对数形式，称为对数–线性模型（log-line model）；如果解释变量为对数形式的，则称为线性–对数模型（line-log model）。

对数–线性模型的表达式为

$$\ln Y_t = \beta_0 + \beta_1 x_t + u_t$$

在这个模型中，斜率的含义是 Y 的百分比变动，即斜率度量的是解释变量 X 变动一个单位所引起的因变量 Y 的百分比变动（也可以理解为 Y 的增长率）。因此，这个模型通常称为增长模型，通常可以用于所关心的经济变量的增长率。

线性–对数模型的表达式为

$$Y_t = \beta_0 + \beta_1 \ln x_t + u_t$$

在这个模型中，Y 的绝对变动量等于 β_1 乘以 X 的相对变动量。该模型通常用于研究解释变量每变动 1% 引起的因变量的绝对变动量的多少问题。

2. 双曲线函数模型

双曲线函数模型的表达式为

$$Y_t = \beta_0 + \beta_1 \left(\frac{1}{X_t} \right) + u_t$$

这个模型中，只有自变量 X 为非线性。当 X 趋向无穷大时，Y 趋向 β_0，意味着 Y 将无限接近其渐进线（$Y = \beta_0$）。该模型通常可用于描述恩格尔曲线、菲利普斯曲线。

3. 多项式回归模型

对于一个多元数据的模型：

$$y_i = \beta_0 + \beta_1 x_{i1} + \cdots + \beta_m x_i + \varepsilon_{im}, \ i = 1, 2, \cdots, n$$

多项式回归模型常用于描述生产成本。

在进行自变量的选择时，可以从以下几个方面来考虑：

一是从理论上判断，该自变量是否应该作为因变量的解释变量，如果是，则考虑放入模型。

二是采用 t 检验，如果 t 检验表示该自变量的系数估计值为显著，则考虑放入模型。

三是计算偏倚，如果自变量放进模型后，其他变量的系数估计值发生显著变化，则考虑放入模型。

四是计算标准误差，如果把自变量放入模型后，回归方程的标准误差最小，则考虑放入模型。

6.7.2 多元数据

从数学的角度，如果个体的观测数据能表述为 P 维欧几里得空间的点，这样的数据就是多源数据。多元数据的客观存在，需要用多元统计的方法来进行相关分析，以解决多个对象、多个指标情况下的数据处理，找到多个变量之间的统计规律性。无论在什么领域，如果忽视了多元数据的存在，那么数据的风险是巨大的。以下简要介绍多元正态分布的相关内容。

1. 多元正态分布的含义

设 μ 是 p 维常数向量，p 阶方阵 $\Sigma > 0$。若 p 维随机向量 X 的密度函数表达为

$$f(x) = (2\pi)^{-p/2} \mid \Sigma \mid^{-1/2} \exp\{-1/2(x-\mu)'\Sigma^{-1}(x-\mu)\}$$

则称 X 服从 p 维正态分布，记为 $X \sim N_p(\mu, \Sigma)$。

多元正态分布具有如下性质：

多元正态分布的均值向量为 $E(X) = \mu$；

多元正态分布的协方差阵为 $V(X) = \Sigma$；

多元正态随机向量的任何线性变换为正态随机向量。

2. 多元正态分布的参数估计

（1）样本均值向量 $\bar{X}$

令 $\bar{x}_j = \dfrac{1}{n}\sum\limits_{i=1}^{n} X_{ij}$，$j = 1, 2, \cdots, p$，

则

$$\bar{X} = \frac{1}{n}\sum_{i=1}^{n} X_{(i)} = (\bar{x}_1, \bar{x}_2, \cdots, \bar{x}_P)'$$

称为样本均值向量。

其中的 $(\bar{x}_1, \bar{x}_2, \cdots, \bar{x}_P)'$ 是行向量转置的形式。

举例：

某公司有 10 名管理人员接受管理相关知识的培训，培训完成后进行考试，成绩见表 6-11。

表6-11　管理人员考试成绩

学员	管理学 x_1	组织行为学 x_2	市场营销 x_3
学员1	85	93	87

学员	管理学 x_1	组织行为学 x_2	市场营销 x_3
学员2	89	87	85
学员3	92	95	76
学员4	78	91	94
学员5	91	84	83
学员6	87	90	75
学员7	85	88	93
学员8	96	79	89
学员9	77	89	95
学员10	84	92	81

使用 Minitab 软件进行分析。在建立工作表后，在"统计"中选择"基本统计量"，再选择"显示描述性统计量"。在会话框"变量"中选择输入"x_1，x_2，x_3"，见图6-28。

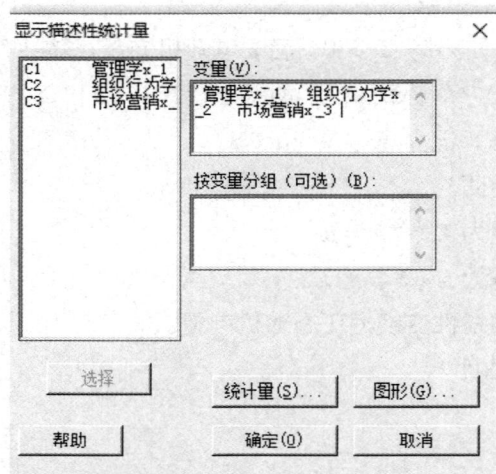

图6-28　显示描述性统计量

点击"统计量"，在"显示描述性统计量：统计量"中选择"均值"，见图6-29。

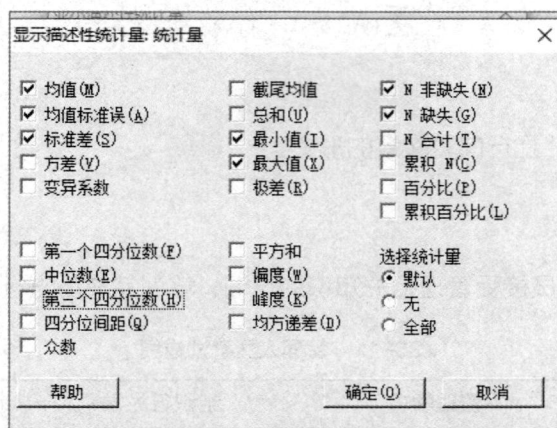

图6-29　显示描述性统计量：统计量

在全部"确定"后，会话框出现相关内容，见图6-30。

```
欢迎使用 Minitab，请按 F1 获得有关帮助。|

描述性统计量：管理学x¯_1，组织行为学x¯_2，市场营销x¯_3

                              均值标
变量              N   N*   均值   准误   标准差   最小值   最大值
管理学x¯_1        11   0   86.27  1.71   5.68    77.00   96.00
组织行为学x¯_2    11   0   89.18  1.39   4.60    79.00   95.00
市场营销x¯_3      11   0   85.91  2.05   6.79    75.00   95.00
```

图6-30 会话框输出内容

由图6-30可知，样本均值向量为

$$\bar{X} = \begin{bmatrix} 86.27 \\ 89.18 \\ 85.91 \end{bmatrix}$$

（2）样本协方差阵 S

样本协方差阵：

$$S = (S_{ij}) = \frac{1}{n-1} \sum_{i=1}^{n} (X_{(i)} - \bar{X})(X_{(i)} - \bar{X})'$$

其中，S 的 i 行 j 列元素为

$$S_{ij} = \frac{1}{n-1} \sum_{k=1}^{n} (X_{ki} - \bar{x}_i)(X_{kj} - \bar{x}_j)$$

S_{ij} 是随机变量 X_I 与 X_J 的样本协方差。

S_{ii} 是对角线上的元素

$$S_{ii} = \sum_{a=1}^{n} (x_{ai} - \bar{x}_i)^2 / (n-1), \quad i = 1, 2, \cdots, p$$

S_{ii} 是 X_i 的样本方差。

举例：

继续上面的例子。在 Minitab 软件建立工作表后，在"统计"中选择"基本统计量"，再选择"协方差"。在会话框"变量"中选择输入" x_1, x_2, x_3 "，见图6-31。

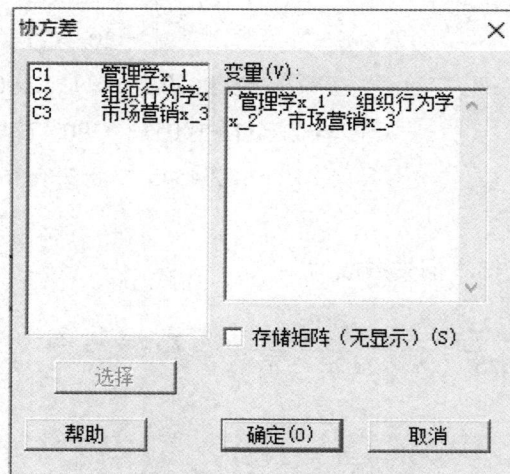

图6-31 协方差

在全部确定后，会话框出现相关内容，见图6-32。

协方差: 管理学x_1, 组织行为学x_2, 市场营销x_3

	管理学x_1	组织行为学x_2	市场营销x_3
管理学x_1	32.2182		
组织行为学x_2	-12.3545	21.1636	
市场营销x_3	-20.2727	-8.1818	46.0909

图6-32　会话框输出内容

由会话框可以看出，输出数据为对角线左下半的内容。由于样本协方差阵 S 呈对称形状，故

$$S = \begin{bmatrix} 32.2182 & -12.3545 & -20.2727 \\ -12.3545 & 21.1636 & -8.1818 \\ -20.2727 & -8.1818 & 46.0909 \end{bmatrix}$$

在该矩阵的对角线上三个数据即为样本方差 S_{ii}，即"32.2182""21.1636""46.0909"分别是管理学、组织行为学、市场营销三门课程的样本方差。

（3）样本离差阵 A

样本离差阵：

$$A = (a_{ij}) = (n-1)S$$
$$= \sum_{i=1}^{n}(X_{(i)} - \bar{X})(X_{(i)} - \bar{X})$$

把 A 记为 $(a_{ij})_{p \times p}$，那么，
A 的元素为

$$a_{ij} = \sum_{k=1}^{n}(X_{ki} - \bar{x}_i)(X_{kj} - \bar{x}_j)，\quad i, j = 1, 2, \cdots, p$$

举例：

继续上面的例子。

$$A = (n-1)S$$
$$= (10-1) \times \begin{bmatrix} 32.2182 & -12.3545 & -20.2727 \\ -12.3545 & 21.1636 & -8.1818 \\ -20.2727 & -8.1818 & 46.0909 \end{bmatrix}$$
$$= \begin{bmatrix} 289.9638 & -111.1905 & -182.4543 \\ -111.19051 & 90.4724 & -73.6362 \\ -182.4543 & -73.6362 & 414.8181 \end{bmatrix}$$

（4）样本相关阵 R

随机变量 X_i 与 X_j 之间的样本相关系数为

$$r_{ij} = \frac{S_{ij}}{\sqrt{S_{ii}}\sqrt{S_{jj}}} = \frac{a_{ij}}{\sqrt{a_{ii}}\sqrt{a_{jj}}}，\quad i, j, = 1, 2, \cdots, p$$

则矩阵

$$R = (r_{ij})_{p \times p}$$

为随机向量 X 的样本相关阵。

举例:

继续上面的例子。在 Minitab 软件建立工作表后,在"统计"中选择"基本统计量",再选择"相关"。在会话框"变量"中选择输入"x_1,x_2,x_3"。选择"显示 P 值",见图 6-33。

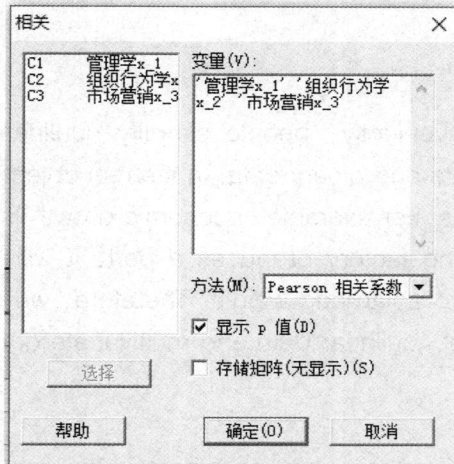

图 6-33 "相关"会话框

在点击"确定"后,会话框出现相关内容,见图 6-34。

图 6-34 会话框输出内容

在图 6-34 中,得到了 x_1,x_2,x_3 中两两变量之间的相关系数。由于样本相关阵对角线的值都是"1",因此,本例的样本相关阵为

$$R = \begin{bmatrix} 1.000 & -0.473 & -0.526 \\ -0.473 & 1.000 & -0.262 \\ -0.526 & -0.262 & 1.000 \end{bmatrix}$$

在图 6-34 中,还得到了相关系数是否为 0 的 P 值。

P 值检验的原假设为: $H_0 : \rho_{ij} = 0, H_1 : \rho_{ij} \neq 0$。

当 $P < 0.05$ 时,拒绝原假设,即 X_i 与 X_j 一定相关;当 $P \geq 0.05$ 时,不能拒绝原假设,即不能确定 X_i 与 X_j 之间相关,也可以说 X_i 与 X_j 不相关。

本例中,x_1,x_2 之间的相关系数为 -0.473,为弱的负相关,检验的 P 值为 0.142,大于 0.05,不能确定 x_1,x_2 之间的相关。x_1,x_3 之间的相关系数为 -0.526,为负相关,检验的 P 值为 0.096,稍大于 0.05,说明 x_1,x_3 之间的相关不紧密。x_2,x_3 之间的相关系数为 -0.262,为较弱的负相关,检验的 P 值为 0.436,明显大于 0.05,说明 x_2,x_3 之间的不相关。

6.7 Multivariate Data Exist

In order to analyze data conveniently, people simplify nonlinear data into linear data and simplify data into univariate data and binary data. In the real objective world, nonlinear data and multivariate data are ubiquitous. For example, economic growth is unlikely to show a linear upward trend; When a processing factory produces a part, it will encounter multiple indicators such as length, width, height, radian and weight. Therefore, we should have an objective understanding of the existence of nonlinear data and multivariate data.

6.7.1 Nonlinear data

In data analysis, if the existence of nonlinear data is ignored, the risk of data will be high and the analysis conclusion will be inaccurate. Here are some main nonlinear models.

1. Semi logarithmic model

Semi logarithmic model refers to the dependent variable and explanatory variable, one of which is in logarithmic form, and the other is a linear model. If the dependent variable is in logarithmic form, it is called log-line model; If the interpretation variable is in logarithmic form, it is called line-log model.

The expression of the log-line model is

$$\ln Y_t = \beta_0 + \beta_1 x_t + u_t$$

In this model, the meaning of slope is the percentage change of Y, that is, the slope measures the percentage change of dependent variable Y caused by a unit change of explanatory variable X (also known as the growth rate of Y). Therefore, this model is usually called the growth model and can be used for the growth rate of the economic variables concerned.

The expression of the line-log model is

$$Y_t = \beta_0 + \beta_1 \ln x_t + u_t$$

In this model, the absolute change of Y is equal to β_1 times the relative change of X. This model is usually used to study the absolute change of dependent variable caused by every 1% change of explanatory variable.

2. Hyperbolic function model

The expression of hyperbolic function model is

$$Y_t = \beta_0 + \beta_1 \left(\frac{1}{X_t} \right) + u_t$$

In this model, only the independent variable X is nonlinear. When X tends to infinity, Y tends to β_0, which means that Y will be infinitely close to its asymptotic line ($Y = \beta_0$). This model can usually be used to describe Engel curve and Phillips curve.

3. Polynomial regression model

For a multivariate data model,

$$y_i = \beta_0 + \beta_1 x_{i1} + \cdots + \beta_m x_i + \varepsilon_{im} , \ i = 1, 2, \cdots, n$$

Polynomial regression models are often used to describe production costs.

When selecting the independent variable, the following aspects can be considered:

First, theoretically judge whether the independent variable should be used as the explanatory variable of the dependent variable, and if so, consider putting it into the model.

The second is to use t test. If the t-test indicates that the estimated value of the coefficient of the independent variable is significant, it is considered to be put into the model.

The third is calculation bias. If the estimated values of coefficients of other variables change significantly after the independent variables are put into the model, they should be put into the model.

The fourth is to calculate the standard error. If the standard error of the regression equation is the smallest after the independent variable is put into the model, it should be considered to put into the model

6.7.2 Multivariate data

From a mathematical point of view, if the individual observation data can be expressed as points in P-dimensional Euclidean space, such data is multi-source data. The objective existence of multivariate data requires the use of multivariate statistical methods for correlation analysis to solve the data processing in the case of multiple objects and indicators and find the statistical regularity among multiple variables. In any field, if the existence of multivariate data is ignored, the risk of data is huge. The following is a brief introduction to the content of multivariate normal distribution.

1. Meaning of multivariate normal distribution

Set up μ is a p-dimensional constant vector, square matrix Σ of order $p > 0$. If the density function of the p-dimensional random vector X is expressed as:

$$f(x) = (2\pi)^{-p/2} \ | \ \Sigma \ | \ ^{-1/2} \exp \{ -1/2(x - \mu)' \Sigma^{-1}(x - \mu) \}$$

X is said to obey the p-dimensional normal distribution, which is recorded as $X \sim N_p(\mu, \Sigma)$.

The multivariate normal distribution has the following properties:

The mean vector of multivariate normal distribution is $E(X) = \mu$;

The covariance matrix of multivariate normal distribution is $V(X) = \Sigma$;

Any linear transformation of multivariate normal random vector is normal random vector.

2. Parameter estimation of multivariate normal distribution

(1) Sample mean vector $\bar{X}$

Let $\bar{x}_j = \dfrac{1}{n} \displaystyle\sum_{i=1}^{n} X_{ij}$, $j = 1, 2, \cdots, p$,

then

$$\bar{X} = \frac{1}{n} \sum_{i=1}^{n} X_{(i)} = (\bar{x}_1, \bar{x}_2, \cdots, \bar{x}_P)'$$

It is called sample mean vector.

Where $(\bar{x}_1, \bar{x}_2, \cdots, \bar{x}_p)'$ is the transposed form of the row vector.

Case:

A company has 10 management personnel who have received training on management related knowledge. After the training, they will take an examination. See Table 6-11 for the results.

Table 6-11 Examination result of management personnel

Student	Management x_1	Organizational Behavior x_2	Marketing Management x_3
student 1	85	93	87
student 2	89	87	85
student 3	92	95	76
student 4	78	91	94
student 5	91	84	83
student 6	87	90	75
student 7	85	88	93
student 8	96	79	89
student 9	77	89	95
student 10	84	92	81

Minitab software was used for analysis. After creating a worksheet, select "Basic Statistics" in "Statistics", and then select "Display Descriptive Statistics". Select input "x_1, x_2, x_3" in the session box "Variable". See Figure 6-28.

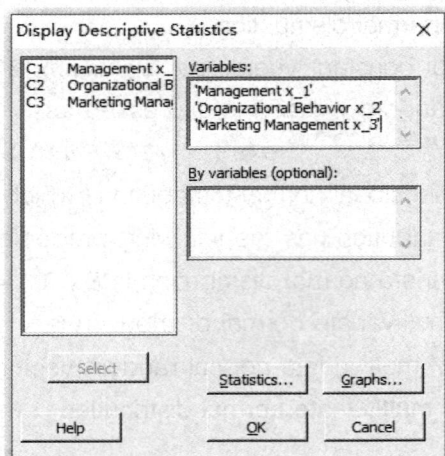

Figure 6-28 Display description statistics

Click "Statistics". Select "Mean" in "Display Descriptive Statistics: Statistics", as shown in Figure 6-29.

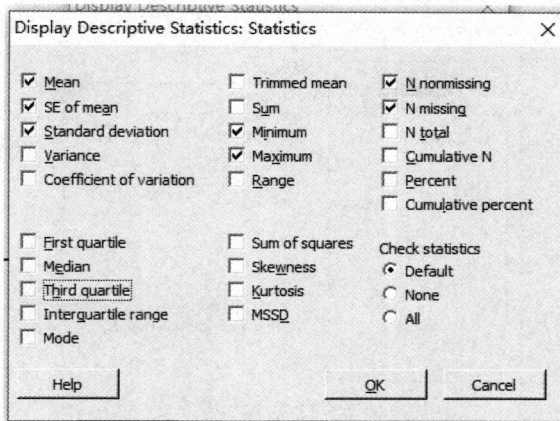

Figure 6-29　Display descriptive statistics：statistics

After clicking all "OK", the relevant content appears in the dialog box, as shown in Figure 6-30.

Descriptive Statistics: Management x_1, Organizational Behavior x_2, Marketing Management x_3

Variable	N	N*	Mean	SE Mean	StDev	Minimum	Maximum
Management x_1	11	0	86.27	1.71	5.68	77.00	96.00
Organizational Behavior	11	0	89.18	1.39	4.60	79.00	95.00
Marketing Management x_3	11	0	85.91	2.05	6.79	75.00	95.00

Figure 6-30　Content in the dialog box

As shown in Figure 6-30, the sample mean vector is

$$\bar{X} = \begin{bmatrix} 86.27 \\ 89.18 \\ 85.91 \end{bmatrix}$$

（2）Sample covariance matrix S.

The sample covariance matrix：

$$S = (S_{ij}) = \frac{1}{n-1} \sum_{i=1}^{n} (X_{(i)} - \bar{X})(X_{(i)} - \bar{X})'$$

Where, the elements in row i and column j of S are

$$S_{ij} = \frac{1}{n-1} \sum_{k=1}^{n} (X_{ki} - \bar{x_i})(X_{kj} - \bar{x_j})$$

S_{ij} is the sample covariance of the random variables X_I and X_J.

S_{ii} is an element on the diagonal

$$S_{ii} = \sum_{a=1}^{n} (x_{ai} - \bar{x_i})^2 / (n-1), \ i = 1, 2, \cdots, p$$

S_{ii} is the sample variance of X_i.

Case：

Continue with the above example. After creating the worksheet in Minitab software, select "Basic Statistics" in "Statistics", and then select "Covariance". Input " x_1, x_2, x_3 " in the "Variables" of the dialogue box, as shown in Figure 6-31.

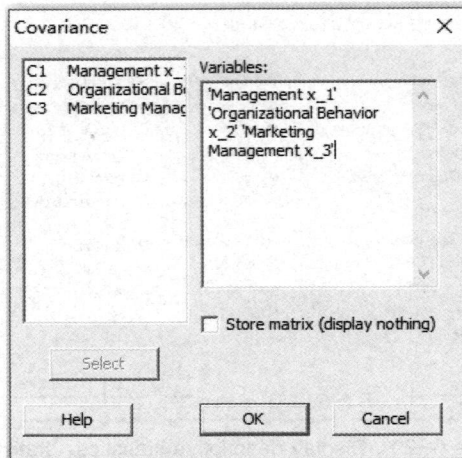

Figure 6-31　Covariance

After all are confirmed, the dialog box will display relevant contents, as shown in Figure 6-32.

Covariances: Management x_1, Organizational Behavior x_2, Marketing Management x_3

	Management x_1	Organizational B	Marketing Manage
Management x_1	32.2182		
Organizational B	-12.3545	21.1636	
Marketing Manage	-20.2727	-8.1818	46.0909

Figure 6-32　Output contents of the dialogue box

It can be seen from the dialog box that the output data is the lower left half of the diagonal. Since the sample covariance matrix S is symmetric, so

$$S = \begin{bmatrix} 32.2182 & -12.3545 & -20.2727 \\ -12.3545 & 21.1636 & -8.1818 \\ -20.2727 & -8.1818 & 46.0909 \end{bmatrix}$$

Three data on the diagonal of the matrix is the sample variance S_{ii}. It means that "32.2182" "21.1636" "46.0909" is the sample variance of management, organizational behavior and marketing respectively.

(3) Sample deviation matrix A

Sample deviation matrix:

$$A = (a_{ij}) = (n-1)S$$
$$= \sum_{i=1}^{n} (X_{(i)} - \bar{X})(X_{(i)} - \bar{X})$$

Mark A as $(a_{ij})_{p \times p}$, then,

the elements of A are

$$a_{ij} = \sum_{k=1}^{n} (X_{ki} - \bar{x}_i)(X_{kj} - \bar{x}_j) \ i, j = 1, 2, \cdots, p$$

Case:

Continue with the above example.

$$A = (n-1)S$$

$$= (10-1) \times \begin{bmatrix} 32.2182 & -12.3545 & -20.2727 \\ -12.3545 & 21.1636 & -8.1818 \\ -20.2727 & -8.1818 & 46.0909 \end{bmatrix}$$

$$= \begin{bmatrix} 289.9638 & -111.1905 & -182.4543 \\ -111.19051 & 90.4724 & -73.6362 \\ -182.4543 & -73.6362 & 414.8181 \end{bmatrix}$$

（4）Sample correlation matrix R

The sample correlation coefficient between random variables X_i and X_j is

$$r_{ij} = \frac{S_{ij}}{\sqrt{S_{ii}} \sqrt{S_{jj}}} = \frac{a_{ij}}{\sqrt{a_{ii}} \sqrt{a_{jj}}} , \quad i, j, = 1, 2, \cdots, p$$

Then matrix

$$R = (r_{ij})_{p \times p}$$

is the sample correlation matrix of random vector X.

Case：

Continue with the above example. After creating the worksheet in Minitab software, select "Basic Statistics" in "Statistics", and then select "Correlation". Input " x_1, x_2, x_3 " in the session box "Variable". Select "Display p-values", as shown in Figure 6-33.

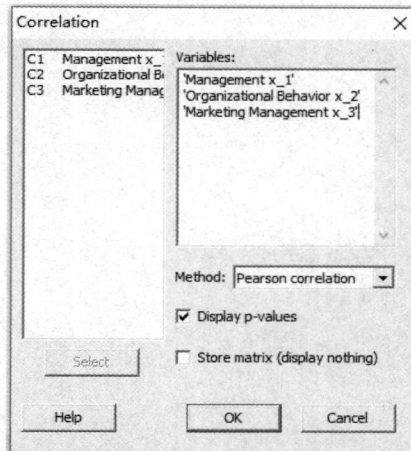

Figure 6-33　"Correlation" dialog box

After clicking "OK", relevant contents will appear in the dialog box, as shown in Figure 6-34.

Figure 6-34　Output contents of the dialog box

In Figure 6-34, the correlation coefficients between two variables in x_1, x_2, x_3 are obtained.

Since the diagonal value of the sample correlation matrix is "1", the sample correlation matrix in this example is

$$R = \begin{bmatrix} 1.000 & -0.473 & -0.526 \\ -0.473 & 1.000 & -0.262 \\ -0.526 & -0.262 & 1.000 \end{bmatrix}$$

In Figure 6-34, the P value of whether the correlation coefficient is 0 is also obtained.

The null hypothesis of the P value test is: $H_0 : \rho_{ij} = 0$, $H_1 : \rho_{ij} \neq 0$.

When $P < 0.05$, reject the null hypothesis, i.e. X_i and X_j must be relevant. When $P \geqslant 0.05$, then null hypothesis cannot be rejected, that is, the correlation between X_i and X_j cannot be determined, it can also be said that X_i and X_j are not related.

In this example, the correlation coefficient between x_1 and x_2 is −0.473, which is a weak negative correlation. The p value tested is 0.142, greater than 0.05, then we cannot determine the correlation between x_1 and x_2. The correlation coefficient between x_1 and x_3 is −0.526. It is negative correlation. The p value tested is 0.096. It is slightly greater than 0.05, which indicates that the correlation between x_1 and x_3 is not tight. The correlation coefficient between x_2 and x_3 is −0.262. It is weak negative correlation. The p value tested is 0.436. It is significantly greater than 0.05, which indicates the irrelevance between x_2 and x_3.

参考文献 ／Bibliography

[1]鲍劲松，刘天元，郑宇.工业智能：方法与应用[M]. 北京：电子工业出版社，2022.

[2]车品觉. 数据的本质[M]. 北京：联合出版公司，2017.

[3]陈根. 数字孪生[M]. 北京：电子工业出版社，2020.

[4]程正兴. 数据拟合[M]. 西安：西安交通大学出版社，1986.

[5]杜子芳. 多元统计分析[M]. 北京：清华大学出版社，2016.

[6]方匡南. 数据科学[M]. 北京：电子工业出版社，2018.

[7]黄丽. 算子代数上的完全保持映射[M]. 西安：西安交通大学出版社，2022.

[8]霍雨佳，周若平，钱晖中. 大数据科学[M]. 成都：电子科技大学出版社，2017.

[9]季生平. 管理心得[M]. 成都：四川大学出版社，2018.

[10]季生平. 建筑企业管理分析[M]. 北京：中国原子能出版社，2021.

[11]季生平. 简明建筑工程质量管理小组实务问答[M]. 北京：中国市场出版社，2022.

[12]季生平. 质量管理统计应用[M]. 西安：世界图书出版社，2023.

[13]季生平. Statistical application in quality control[M]. 西安：世界图书出版社，2024.

[14]金勇进，杜子芳，蒋妍. 抽样技术[M]. 北京：中国人民大学出版社，2012.

[15]李建华. 测量学[M]. 上海：上海交通大学出版社，2008.

[16]李文林. 数学史概论[M]. 北京：高等教育出版社，2021.

[17]罗泰晔. 数据赋能[M]. 北京：机械工业出版社，2022.

[18]马逢时，吴诚鸥，蔡霞. 基于 Minitab 的现代实用统计[M]. 北京：中国人民大学出版社，2013.

[19]茆诗松，周纪芗，陈颖. 试验设计[M]. 北京：中国统计出版社，2012.

[20]孙希文. 数理逻辑[M]. 北京：高等教育出版社，2019.

[21]陶飞. 数字孪生及车间实践[M]. 北京：清华大学出版社，2021.

[22]吴遵高. 测量系统分析[M]. 北京：中国标准出版社，2004.

[23]薛留根. 现代统计模型[M]. 北京：科学出版社，2012.

[24]袁相碗. 数理逻辑和算法理论：计算机科学与人工智能的数学基础[M]. 北京：中国铁道出版社，2023.

[25]张宪超. 数据聚类[M]. 北京：科学出版社，2017.

[26]张志成. 逻辑学教程[M]. 北京：中国人民大学出版社，2006.

[27]朱扬勇，熊赟. 数据学[M]. 上海：复旦大学出版社，2009.

[28][美] Allison，Paul D. 缺失数据[M]. 林毓玲，译. 上海：上海人民出版社，2012.

[29][美] Ernest O. Doebelin. 测量系统应用与设计[M]. 王伯雄，译. 北京：电子工业出版

社，2007.

[30][英]Floris, Geerts. 数据质量管理基础[M]. 刘瑞虹，贾西贝，译. 北京：国防工业出版社，2016.

[31][美]Heam Donald. 计算机图形学[M]2版. 蔡士杰，吴春镕，等译. 北京：电子工业出版社，2002.

[32][美]Little Roderick, J. A，Rubin Donald B. 缺失数据统计分析[M]. 北京：中国统计出版社，2004.

[33][美]Ratner, Bruce. 统计挖掘与机器学习：大数据预测建模和分析技术[M]. 郑磊，刘子未，石仁达，等译. 北京：机械工业出版社，2021.

[34][英]Schonberger, Viktor Mayer. 大数据时代，生活、工作与思维的大变革[M]. 盛扬燕，周涛，译. 杭州：浙江人民出版社，2013.

[35]Chen, Kevin. Digital twin[M]. Montreal Quebec：Royal Collins Publishing Group Inc., 2021.

[36]Enders Craig K. Applied missing data analysis[M]. New York：Guiiford Press, 2010.

[37]Feffeman, Charles. Fitting smooth functions to data[M]. Providence, Rhode Island：American Mathematical Society, 2020.

[38]Goericke, Stephan. The future of software quality assurance[M]. Cham, Switzerland：SpringerOpen, 2020.

[39]Greenfield Tony. Design and analyse your experiment with Minitab[M]. London：Hodder Amold, 2007.

[40]King, Tim, Schwarzenbach, Julian. Managing data quality：a practical guide[M]. London, BCS：2020.

[41]Lee Yang W. Journey to data quality[M]. Cambridge, Massachuetts：Massachusetts Institute of Technology Press, 2006.

[42]Mitchell Tom M. Machine learning[M]. Boston：WCB/McGraw-Hill, 1997.

[43]Ohlhorst Frank. Big data analytics：turning big data into big money[M]. Hoboken, N J：Wiley, 2013.

[44]Thompson, Steven K., Sampling[M]. Hoboken. N J.：Wiley, 2012.

[45]Walle Alf H. Qualitative research in intelligence and marketing：the new strategic convergence[M]. Westport, Conn：Quorum Books, 2001.

[46]Wilfrid Hodges, Martin Hyland, Charles Steinhorn, John Truss. Logic：from foundations to applications, European logic colloquium[M]. Oxford：Oxford University Press, 1996.